AF839933

SUSTAINABLE AND FUNCTIONAL FOODS FROM PLANTS

Health Impact, Bioactive Compounds, and Production Technologies

Innovations in Agricultural and Biological Engineering

SUSTAINABLE AND FUNCTIONAL FOODS FROM PLANTS

Health Impact, Bioactive Compounds, and Production Technologies

Edited by

Megh R. Goyal, PhD, P.E.

Arijit Nath, PhD

Zoltán Kovács, PhD

APPLE ACADEMIC PRESS

First edition published 2024

Apple Academic Press Inc.
1265 Goldenrod Circle, NE,
Palm Bay, FL 32905 USA

760 Laurentian Drive, Unit 19,
Burlington, ON L7N 0A4, CANADA

CRC Press
6000 Broken Sound Parkway NW,
Suite 300, Boca Raton, FL 33487-2742 USA

4 Park Square, Milton Park,
Abingdon, Oxon, OX14 4RN UK

© 2024 by Apple Academic Press, Inc.

Apple Academic Press exclusively co-publishes with CRC Press, an imprint of Taylor & Francis Group, LLC

Reasonable efforts have been made to publish reliable data and information, but the authors, editors, and publisher cannot assume responsibility for the validity of all materials or the consequences of their use. The authors, editors, and publishers have attempted to trace the copyright holders of all material reproduced in this publication and apologize to copyright holders if permission to publish in this form has not been obtained. If any copyright material has not been acknowledged, please write and let us know so we may rectify in any future reprint.

Except as permitted under U.S. Copyright Law, no part of this book may be reprinted, reproduced, transmitted, or utilized in any form by any electronic, mechanical, or other means, now known or hereafter invented, including photocopying, microfilming, and recording, or in any information storage or retrieval system, without written permission from the publishers.

For permission to photocopy or use material electronically from this work, access www.copyright.com or contact the Copyright Clearance Center, Inc. (CCC), 222 Rosewood Drive, Danvers, MA 01923, 978-750-8400. For works that are not available on CCC please contact mpkbookspermissions@tandf.co.uk

Trademark notice: Product or corporate names may be trademarks or registered trademarks and are used only for identification and explanation without intent to infringe.

Library and Archives Canada Cataloguing in Publication

Title: Sustainable and functional foods from plants : health impact, bioactive compounds, and production technologies / edited by Megh R. Goyal, PhD, P.E., Arijit Nath, PhD, Zoltán Kovács, PhD.

Names: Goyal, Megh R., editor. | Nath, Arijit, editor. | Kovács, Zoltán (Lecturer in food science and technology), editor.

Series: Innovations in agricultural and biological engineering.

Description: First edition. | Series statement: Innovations in agricultural and biological engineering | Includes bibliographical references and index.

Identifiers: Canadiana (print) 20230593097 | Canadiana (ebook) 2023059316X | ISBN 9781774914540 (hardcover) | ISBN 9781774914557 (softcover) | ISBN 9781003415763 (ebook)

Subjects: LCSH: Functional foods. | LCSH: Plants, Edible. | LCSH: Nutrition.

Classification: LCC QP144.F85 S87 2024 | DDC 613.2—dc23

Library of Congress Cataloging-in-Publication Data

Names: Goyal, Megh R., editor. | Nath, Arijit, editor. | Kovács, Zoltán (Lecturer in food science and technology), editor.

Title: Sustainable and functional foods from plants : health impact, bioactive compounds, and production technologies / edited by Megh R. Goyal, PhD, P.E., Arijit Nath, PhD, Zoltán Kovács, PhD.

Description: First edition. | Palm Bay, FL : Apple Academic Press, 2024. | Series: Innovations in agricultural and biological engineering | Includes bibliographical references and index.

Identifiers: LCCN 2023055271 (print) | LCCN 2023055272 (ebook) | ISBN 9781774914540 (hardcover) | ISBN 9781774914557 (paperback) | ISBN 9781003415763 (ebook)

Subjects: LCSH: Functional foods. | Plants, Edible. | Sustainable agriculture. | Nutrition.

Classification: LCC QP144.F85 S875 2024 (print) | LCC QP144.F85 (ebook) | DDC 613.2--dc23/eng/20231229

LC record available at https://lccn.loc.gov/2023055271

LC ebook record available at https://lccn.loc.gov/2023055272

ISBN: 978-1-77491-454-0 (hbk)
ISBN: 978-1-77491-455-7 (pbk)
ISBN: 978-1-00341-576-3 (ebk)

ABOUT THE BOOK SERIES: INNOVATIONS IN AGRICULTURAL AND BIOLOGICAL ENGINEERING

Under this book series, Apple Academic Press Inc. is publishing book volumes over a span of 8–10 years in the specialty areas defined by the American Society of Agricultural and Biological Engineers (<asabe.org>). Apple Academic Press Inc. aims to be a principal source of books in agricultural and biological engineering. We welcome book proposals from readers in areas of their expertise.

The mission of this series is to provide knowledge and techniques for agricultural and biological engineers (ABEs). The book series offers high-quality reference and academic content on agricultural and biological engineering (ABE) that is accessible to academicians, researchers, scientists, university faculty and university-level students, and professionals around the world.

Agricultural and biological engineers ensure that the world has the necessities of life, including safe and plentiful food, clean air and water, renewable fuel and energy, safe working conditions, and a healthy environment by employing knowledge and expertise of the sciences, both pure and applied, and engineering principles. Biological engineering applies engineering practices to problems and opportunities presented by living things and the natural environment in agriculture.

ABE embraces a variety of the following specialty areas (<asabe.org>): aquaculture engineering, biological engineering, energy, farm machinery and power engineering, food and process engineering, forest engineering, information and electrical technologies, soil and water conservation engineering, natural resources engineering, nursery and greenhouse engineering, safety and health, and structures and environment.

For this book series, we welcome chapters on the following specialty areas (but not limited to):

1. Academia to industry to end-user loop in agricultural engineering
2. Agricultural mechanization
3. Aquaculture engineering
4. Biological engineering in agriculture

5. Biotechnology applications in agricultural engineering
6. Energy source engineering
7. Food and bioprocess engineering
8. Forest engineering
9. Hill land agriculture
10. Human factors in engineering
11. Information and electrical technologies
12. Irrigation and drainage engineering
13. Nanotechnology applications in agricultural engineering
14. Natural resources engineering
15. Nursery and greenhouse engineering
16. Potential of phytochemicals from agricultural and wild plants for human health
17. Power systems and machinery design
18. GPS and remote sensing potential in agricultural engineering
19. Robot engineering and drones in agriculture
20. Simulation and computer modeling
21. Smart engineering applications in agriculture
22. Soil and water engineering
23. Micro irrigation engineering
24. Structures and environment engineering
25. Waste management and recycling
26. Rural electrification.
27. Sanitary engineering
28. Farm to fork technologies in agriculture
29. Impact of global warming and climatic change on agriculture economy
30. Any other focus areas

For more information on this series, readers may contact:

Megh R. Goyal, PhD, PE
Book Series Senior Editor-in-Chief: Innovations in Agricultural and Biological Engineering
E-mail: goyalmegh@gmail.com

OTHER BOOKS ON AGRICULTURAL AND BIOLOGICAL ENGINEERING BY APPLE ACADEMIC PRESS, INC.

Management of Drip/Trickle or Micro Irrigation
Megh R. Goyal, PhD, PE, Senior Editor-in-Chief

Evapotranspiration: Principles and Applications for Water Management
Megh R. Goyal, PhD, PE and Eric W. Harmsen, PhD Editors

Book Series: Research Advances in Sustainable Micro Irrigation
Senior Editor-in-Chief: Megh R. Goyal, PhD, PE

- **Volume 1:** Sustainable Micro Irrigation: Principles and Practices
- **Volume 2:** Sustainable Practices in Surface and Subsurface Micro Irrigation
- **Volume 3:** Sustainable Micro Irrigation Management for Trees and Vines
- **Volume 4:** Management, Performance, and Applications of Micro Irrigation Systems
- **Volume 5:** Applications of Furrow and Micro Irrigation in Arid and Semi-Arid Regions
- **Volume 6:** Best Management Practices for Drip Irrigated Crops
- **Volume 7:** Closed Circuit Micro Irrigation Design: Theory and Applications
- **Volume 8:** Wastewater Management for Irrigation: Principles and Practices
- **Volume 9:** Water and Fertigation Management in Micro Irrigation
- **Volume 10:** Innovation in Micro Irrigation Technology

Book Series: Innovations and Challenges in Micro Irrigation
Senior Editor-in-Chief: Megh R. Goyal, PhD, PE

- Engineering Interventions in Sustainable Trickle Irrigation: Water Requirements, Uniformity, Fertigation, and Crop Performance

- Management Strategies for Water Use Efficiency and Micro Irrigated Crops: Principles, Practices, and Performance
- Micro Irrigation Engineering for Horticultural Crops: Policy Options, Scheduling, and Design
- Micro Irrigation Management: Technological Advances and Their Applications
- Micro Irrigation Scheduling and Practices
- Performance Evaluation of Micro Irrigation Management: Principles and Practices
- Potential of Solar Energy and Emerging Technologies in Sustainable Micro Irrigation
- Principles and Management of Clogging in Micro Irrigation
- Sustainable Micro Irrigation Design Systems for Agricultural Crops: Methods and Practices

Book Series: Innovations in Agricultural and Biological Engineering
Senior Editor-in-Chief: Megh R. Goyal, PhD, PE

- Advanced Research Methods in Food Processing Technologies
- Advances in Food Process Engineering: Novel Processing, Preservation and Decontamination of Foods
- Advances in Green and Sustainable Nanomaterials: Applications in Energy, Biomedicine, Agriculture, and Environmental Science
- Advances in Sustainable Food Packaging Technology
- Analytical Methods for Milk and Milk Products, 3-volume set:
    - o Volume 1: Sampling Methods, Chemical and Compositional Analysis
    - o Volume 2: Physicochemical Analysis of Concentrated, Coagulated and Fermented Products
    - o Volume 3: Microbiological Analysis, Isolation, and Characterization
- Applications of Supercritical Fluid Extraction in Food Processing
- Biological and Chemical Hazards in Food and Food Products: Prevention, Practices, and Management
- Bioremediation and Phytoremediation Technologies in Sustainable Soil Management, 4-volume set:
    - o Volume 1: Fundamental Aspects and Contaminated Sites
    - o Volume 2: Microbial Approaches and Recent Trends
    - o Volume 3: Inventive Techniques, Research Methods, and Case Studies

  - Volume 4: Degradation of Pesticides and Polychlorinated Biphenyls
- Dairy Engineering: Advanced Technologies and Their Applications
- Developing Technologies in Food Science: Status, Applications, and Challenges
- Emerging Technologies in Agricultural Engineering
- Engineering Interventions in Agricultural Processing
- Engineering Interventions in Foods and Plants
- Engineering Practices for Agricultural Production and Water Conservation: An Interdisciplinary Approach
- Engineering Practices for Management of Soil Salinity: Agricultural, Physiological, and Adaptive Approaches
- Engineering Practices for Milk Products: Dairyceuticals, Novel Technologies, and Quality
- Enzyme Inactivation in Food Processing: Technologies, Materials, and Applications
- Field Practices for Wastewater Use in Agriculture: Future Trends and Use of Biological Systems
- Flood Assessment: Modeling and Parameterization
- Food Engineering: Emerging Issues, Modeling, and Applications
- Food Process Engineering: Emerging Trends in Research and Their Applications
- Food Processing and Preservation Technology: Advances, Methods, and Applications
- Food Technology: Applied Research and Production Techniques
- Functional Dairy Ingredients and Nutraceuticals: Physicochemical, Technological, and Therapeutic Aspects
- Handbook of Research on Food Processing and Preservation Technologies, 5-volume set:
  - Volume 1: Nonthermal and Innovative Food Processing Methods
  - Volume 2: Nonthermal Food Preservation and Novel Processing Strategies
  - Volume 3: Computer-Aided Food Processing and Quality Evaluation Techniques
  - Volume 4: Design and Development of Specific Foods, Packaging Systems, and Food Safety
  - Volume 5: Emerging Techniques for Food Processing, Quality, and Safety Assurance

- Modeling Methods and Practices in Soil and Water Engineering
- Nanotechnology and Nanomaterial Applications in Food, Health, and Biomedical Sciences
- Nanotechnology Applications in Agricultural and Bioprocess Engineering: Farm to Table
- Nanotechnology Applications in Dairy Science: Packaging, Processing, and Preservation
- Nanotechnology Horizons in Food Process Engineering, 3-volume set:
    - Volume 1: Food Preservation, Food Packaging and Sustainable Agriculture
    - Volume 2: Scope, Biomaterials, and Human Health
    - Volume 3: Trends, Nanomaterials, and Food Delivery
- Nonthermal Plasma Technology for Food and Food Products
- Nonthermal Pulsed Electric Field Technology for the Food Industry
- Novel and Alternative Methods in Food Processing: Biotechnological, Physicochemical, and Mathematical Approaches
- Novel Dairy Processing Technologies: Techniques, Management, and Energy Conservation
- Novel Processing Methods for Plant-Based Health Foods: Extraction, Encapsulation and Health Benefits of Bioactive Compounds
- Novel Strategies to Improve Shelf-Life and Quality of Foods: Quality, Safety, and Health Aspects
- Phytochemicals and Medicinal Plants in Food Design: Strategies and Technologies for Improved Healthcare
- Processing of Fruits and Vegetables: From Farm to Fork
- Processing Technologies for Milk and Milk Products: Methods, Applications, and Energy Usage
- Quality Control in Fruit and Vegetable Processing: Methods and Strategies
- Scientific and Technical Terms in Bioengineering and Biological Engineering
- Soil and Water Engineering: Principles and Applications of Modeling
- Soil Salinity Management in Agriculture: Technological Advances and Applications
- State-of-the-Art Technologies in Food Science: Human Health, Emerging Issues and Specialty Topics
- Sustainable and Functional Foods from Plants: Health Impact, Bioactive Compounds, and Production Technologies

- Sustainable Green Nanomaterials: Synthesis, Characterization, and Engineering Applications
- Sustainable Biological Systems for Agriculture: Emerging Issues in Nanotechnology, Biofertilizers, Wastewater, and Farm Machines
- Sustainable Nanomaterials for Biomedical Engineering: Impacts, Challenges, and Future Prospects
- Sustainable Nanomaterials for Biosystems Engineering: Trends in Renewable Energy, Environment, and Agriculture
- Sustainable Use of Wastewater in Agriculture: Applications, Methods, and Technology
- Technological Interventions in Dairy Science: Innovative Approaches in Processing, Preservation, and Analysis of Milk Products
- Technological Interventions in Management of Irrigated Agriculture
- Technological Interventions in the Processing of Fruits and Vegetables
- Technological Processes for Marine Foods, From Water to Fork: Bioactive Compounds, Industrial Applications, and Genomics
- The Chemistry of Milk and Milk Products: Physicochemical Properties, Therapeutic Characteristics, and Processing Methods

ABOUT SENIOR-EDITOR-IN-CHIEF

Megh R. Goyal, PhD, PE

Retired Professor in Agricultural and Biomedical Engineering, University of Puerto Rico, Mayaguez Campus; Senior Acquisitions Editor, Biomedical Engineering and Agricultural Science, Apple Academic Press, Inc.

Megh R. Goyal, PhD, PE, is a Retired Professor in Agricultural and Biomedical Engineering from the General Engineering Department in the College of Engineering at the University of Puerto Rico–Mayaguez Campus. During his professional career, he has worked as a Soil Conservation Inspector; Research Assistant at Haryana Agricultural University and Ohio State University; Research Agricultural Engineer/Professor at the Department of Agricultural Engineering of UPRM; and Professor of Agricultural and Biomedical Engineering in the General Engineering Department of UPRM. He spent a one-year sabbatical leave in 2002–2003 at the Biomedical Engineering Department of Florida International University, Miami, USA.

Dr. Goyal was the first agricultural engineer to receive the professional license in agricultural engineering from the College of Engineers and Surveyors of Puerto Rico. In 2005, he was proclaimed the "Father of Irrigation Engineering in Puerto Rico for the Twentieth Century" by the American Society of Agricultural and Biological Engineers, Puerto Rico Section, for his pioneering work on micro irrigation, evapotranspiration, agroclimatology, and soil and water engineering.

During his professional career of 52 years, he has received many awards, including Scientist of the Year, Membership Grand Prize for the American Society of Agricultural Engineers Campaign, Felix Castro Rodriguez Academic Excellence Award, Man of Drip Irrigation by the Mayor of Municipalities of Mayaguez/Caguas/Ponce and Senate/Secretary of Agriculture of ELA, Puerto Rico, and many others. He has been recognized as one of the experts "who rendered meritorious service for the development of [the] irrigation sector in India" by the Water Technology Centre of Tamil Nadu Agricultural University in Coimbatore, India, and ASABE who bestowed on him the 2018 Netafim Microirrigation Award.

Dr. Goyal has authored more than 200 journal articles and edited more than 100 books. AAP has published many of his books, including *Management of Drip/Trickle or Micro Irrigation; Evapotranspiration: Principles and Applications for Water Management;* ten-volume set on *Research Advances in Sustainable Micro Irrigation.* He has also authored the textbooks *Elements of Agroclimatology* (Spanish) by UNISARC, Colombia, and two *Bibliographies on Drip Irrigation.* Dr. Goyal has also developed several book series with AAP, including Innovations in Agricultural & Biological Engineering (with over 60 titles in the series to date), Innovations and Challenges in Micro Irrigation; and Innovations in Plant Science for Better Health: From Soil to Fork.

Dr. Goyal received his BSc degree in Engineering from Punjab Agricultural University, Ludhiana, India, and his MSc and PhD degrees from the Ohio State University, Columbus, Ohio, USA. He also earned a Master of Divinity degree from the Puerto Rico Evangelical Seminary, Hato Rey, Puerto Rico, USA.

Readers may contact him at goyalmegh@gmail.com.

ABOUT THE EDITORS

Arijit Nath, PhD

Postdoctoral Researcher and Teaching Assistant, Department of Food Process Engineering, Institute of Food Science and Technology, Hungarian University of Agriculture and Life Sciences, Budapest, Hungary

Arijit Nath, PhD, is a Postdoctoral Researcher and Teaching Assistant in the Department of Food Process Engineering, Institute of Food Science and Technology, Hungarian University of Agriculture and Life Sciences, Budapest, Hungary. Dr. Nath has held a variety of positions, including postdoctoral research fellow in the Department of Food Engineering at Corvinus University of Budapest, Hungary; Department of Biochemical and Chemical Engineering at Technical University of Dortmund, Germany; Department of Refrigeration and Livestock Products Technology at Szent István University, Hungary; and chemical engineer in Soós Ernő Research and Development at University of Pannonia, Hungary. He has participated in several professional industrial training and workshops in the field of biotechnology. He has published 30 research and review articles in international journals and 12 book chapters. Dr. Nath has also presented the research outcomes in more than 70 international conferences and 15 conferences within India. He has received several best presenter awards at international conferences. He has also delivered lectures in several scientific communities with honor. Due to his strong proficiency in the field of biochemical and food engineering, he is serving as a reviewer in various scientific SCI journals. Dr. Nath is a member of the Asia-Pacific Chemical, Biological & Environmental Engineering Society; Indian Institute of Engineers; Indian Institute of Chemical Engineers; Biotech Research Society of India; European Membrane Society; Indian Membrane Society; and European Federation of Biotechnology. He has participated in teaching several theory subjects and is co-supervisor for undergraduate, postgraduate, and PhD students in the Department of Food Process Engineering, Hungarian University of Agriculture and Life Sciences, Hungary. His major research focus is on fermentation technology, enzymatic reaction, bioreactor design, probiotics and prebiotics, allergy-free food products, food waste valorization, membrane separation process, membrane-integrated bioreactors, environmental benign processes, and mathematical modeling of bioprocesses.

Dr. Nath was awarded his PhD degree in Chemical Engineering from Jadavpur University, India in 2014. During his PhD tenure, Dr. Nath was a Senior Research Fellow, supported by Council of Scientific and Industrial Research (CSIR), India; and Project Research Fellow, supported by University Grants Commission (UGC), India.

Zoltán Kovács, PhD

Department of Measurements and Process Control, Institute of Food Science and Technology, Hungarian University of Agriculture and Life Sciences, Budapest, Hungary

Zoltán Kovács, PhD, is a Professor in the Department of Measurements and Process Control, Institute of Food Science and Technology, Hungarian University of Agriculture and Life Sciences (MATE), Budapest, Hungary. He is also the International Affairs Advisor for the Institute. He received his MSc and PhD diplomas in Food Science at Corvinus University and completed his habilitation in the field of food science at Szent Istvan University, Hungary. He spent three years as a postdoctoral research fellow at Kobe University, Japan. Since his postdoc period, he has also been working with the near infrared spectroscopy (NIRS), using both the conventional and the novel aquaphotomics data evaluation techniques. Dr. Kovacs has won several research awards and fellowships, including pre- and postdoctoral awards and national excellence fellowships. Recently, he won the János Bolyai Research Scholarship of the Hungarian Academy of Sciences. He won the József Farkas Excellence Award of the Faculty of Food Science, Szent Istvan University, and the research group under his leadership won three Excellence Research Group grants for their research to develop measurement methods for quick determination of adulteration in various food products. Currently, Dr. Kovacs collaborates and leads different national and international research projects. He is the International Affairs Advisor of the Institute of Food Science and Technology, MATE. To date, Dr. Kovacs has published 60 impact factor articles, six book chapters in international journals and books and has contributed to over 100 international and Hungarian scientific conference presentations, for which he won best presentation awards. Dr. Kovacs is a member of Agro- and Biosystems Engineering Research Committee of Hungarian Academy of Science, and acts as the European Liaison of the International Aquaphotomics.

CONTENTS

Contributors .... *xix*
Abbreviations .... *xxiii*
Preface .... *xxvii*

PART I: Sustainable Foods and Health Impacts 1

1. Millets: Food Diversification and Processing Techniques 3
Damian Laryea, Sarah Theodora Esi Mills, Charles Cudjoe Dapuliga, and John-Lewis Zinia Zaukuu

2. New Horizons for Millets: Food and Nutrition Security 25
Damian Laryea, Sarah Theodora Esi Mills, Charles Cudjoe Dapuliga, and John-Lewis Zinia Zaukuu

3. Grapes and Wines as Functional Foods: Production, Quality, Consumption, and Bioactive Compounds 55
Krisztina Keller, Harshavardhan Reddy Kummitha, Nikoletta Tóth-Kaszás, and Zoltán Birkner

PART II: Sustainable Ethnic Foods Versus Gastronomy 101

4. Potential of Ackee in African Cuisine 103
Faustina Dufie Wireko-Manu, Yaw Gyau Akyereko, Jacob K. Agbenorhevi, Christian Arhinful, Baffoe Kyei-Asante, Joana Konadu Attafuah, Divina Allotey, and Ibok Oduro

5. Grape Wines in Hungarian Context: Production Technologies, Processing, and Health Benefits 129
Ildikó Galambos, Ildikó Ernszt, Zsuzsanna Marton, and Nóra Rodek

6. Health Impacts of Common Flavoring Agents in Indian Cuisine 163
Bedaprana Roy, Bidisha Chatterjee, Tamanna Sultana, and Arup Kumar Mitra

PART III: Potential of Bioactive Compounds in Medicinal Foods 193

7. Glycemic Index of Root and Tuber Crops: Increasing Relevance in Diet 195
Charles Cudjoe Dapuliga, John-Lewis Zinia Zaukuu, Damian Laryea, Yaw Gyau Akyereko, Faustina Dufie Wireko-Manu, and Ibok Oduro

8. Traditional Medicinal Herbs in North-Western Region of Himalayas: Case Study 223
Deepika Dimri and Ambika Kumar

9. Xanthohumol (*Humulus lupulus* L.): Potential of Bioactive Polyphenols in Beer 253
Anna Júlia Éliás, Zsanett Bodor, and Csilla Benedek

PART IV: Measurement Techniques for Quality of Natural Foods 291

10. Detection Methods of Adulteration in Natural Foods 293
John-Lewis Zinia Zaukuu, Zsanett Bodor, Flóra Vitális, and Zoltán Kovács

11. Analytical Methods for Nutritional Composition of Honey 323
Zsanett Bodor, John-Lewis Zinia Zaukuu, Balkis Aouadi, Zoltán Kovács, and Csilla Benedek

12. Potential of Aquaphotomics for Determination of Quality of Natural Foods 371
Jelena Muncan, Balkis Aouadi, and Roumiana Tsenkova

Index 421

CONTRIBUTORS

Jacob K. Agbenorhevi
Department of Food Science and Technology, College of Science, KNUST, Kumasi, Ghana

Yaw Gyau Akyereko
Department of Food Science and Technology, College of Science, KNUST, Kumasi, Ghana

Divina Allotey
Department of Food Science and Technology, College of Science, KNUST, Kumasi, Ghana

Balkis Aouadi
Department of Measurements and Process Control, Institute of Food Science and Technology, Hungarian University of Agriculture and Life Sciences, Budapest, Hungary

Christian Arhinful
Department of Food Science and Technology, College of Science, KNUST, Kumasi, Ghana

Baffoe Kyei-Asante
Department of Food Science and Technology, College of Science, KNUST, Kumasi, Ghana

Joana Konadu Attafuah
Department of Food Science and Technology, College of Science, KNUST, Kumasi, Ghana

Csilla Benedek
Department of Dietetics and Nutrition, Faculty of Health Sciences, Semmelweis University, Budapest, Hungary

Zoltán Birkner
University of Pannonia, Nagykanizsa Campus, University Center for Circular Economy, Nagykanizsa, Zrinyi Miklos, Hungary

Zsanett Bodor
Department of Measurements and Process Control, Institute of Food Science and Technology, Hungarian University of Agriculture and Life Sciences, Budapest, Hungary
Department of Dietetics and Nutrition, Faculty of Health Sciences, Semmelweis University, Budapest, Hungary

Bidisha Chatterjee
Department of Microbiology, St. Xavier's College (Autonomous), Kolkata, West Bengal, India

Charles Cudjoe Dapuliga
Department of Food Science and Technology, College of Science, Kwame Nkrumah University of Science and Technology KNUST, Kumasi, Ghana

Deepika Dimri
School of Environment and Natural Resources, Doon University, Dehradun, India

Anna Júlia Éliás
Doctoral School of Health Sciences, Semmelweis University, Budapest, Hungary

Ildikó Ernszt
Applied Management Institute, University Center for Circular Economy, University of Pannonia, Nagykanizsa, Hungary

Ildikó Galambos
Soós Ernő Research and Development Center, University Center for Circular Economy, University of Pannonia, Nagykanizsa, Hungary

Megh R. Goyal
University of Puerto Rico-Mayaguez, Mayaguez, Puerto Rico

Nikoletta Tóth-Kaszás
University of Pannonia - Nagykanizsa Campus, University Center for Circular Economy, Nagykanizsa, Hungary

Krisztina Keller
Department of Tourism, Marketing Institute, Corvinus University of Budapest, Budapest, Hungary

Zoltán Kovács
Department of Measurements and Process Control, Institute of Food Science and Technology. Hungarian University of Agriculture and Life Sciences, Budapest, Hungary

Ambika Kumar
Department of Chemistry, Bhagalpur National College Bhagalpur, Bihar, India

Harshavardhan Reddy Kummitha
Department of Tourism, Budapest Business School, Budapest, Hungary

Damian Laryea
Radiation Technology Center, Biotechnology and Nuclear Agriculture Research Institute, Ghana Atomic Energy Commission, Kwabenya, Ghana

Faustina Dufie Wireko-Manu
Department of Food Science and Technology, College of Science, KNUST, Kumasi, Ghana

Zsuzsanna Marton
Applied Management Institute, University Center for Circular Economy, University of Pannonia, Nagykanizsa, Hungary

Sarah Theodora Esi Mills
Council for Scientific and Industrial Research – Food Research Institute (CSIR-FRI), Accra, Ghana

Arup Kumar Mitra
Department of Microbiology, St. Xavier's College (Autonomous), , Kolkata, West Bengal, India

Jelena Muncan
Biomeasurement Technology Laboratory, Graduate School of Agricultural Science, Kobe University, Nada, Kobe, Japan

Arijit Nath
Department of Food Process Engineering, Institute of Food Science and Technology, Hungarian University of Agriculture and Life Sciences, Budapest, Hungary

Ibok Oduro
Department of Food Science and Technology, College of Science, KNUST, Kumasi, Ghana

Nóra Rodek
Applied Management Institute, University Center for Circular Economy, University of Pannonia, Nagykanizsa, Hungary

Bedaprana Roy
Department of Microbiology, St. Xavier's College (Autonomous), Kolkata, West Bengal, India

Tamanna Sultana
Department of Microbiology, St. Xavier's College (Autonomous), Kolkata, West Bengal, India

Roumiana Tsenkova
Biomeasurement Technology Laboratory, Kobe University, Nada, Kobe, Japan

John-Lewis Zinia Zaukuu
Department of Food Science and Technology, College of Science, Kwame Nkrumah University of Science and Technology, Kumasi, Ghana

Flóra Vitális
Department of Measurements and Process Control, Institute of Food Science and Technology, Hungarian University of Agriculture and Life Sciences, Budapest, Hungary

ABBREVIATIONS

AAS	atomic absorption spectrometry
AIF	apoptosis-inducing factor
AKT	protein kinase Akt
AMD	age-related macular degeneration
AOAC	Association of Official Agricultural Chemists
a_w	water activity
α-GST	α-glutathione S transferase
B(a)P	benzo(a)pyrene
Bax	Bcl-2-like protein
BC	before Christ
Bcl XL	B-cell lymphoma-extra large
Bcl-2	B-cell lymphoma 2
Bfgf	basic fibroblast growth factor
BSA	berry sensory analysis
BSJ	Bureau of Standards of Jamaica
CAC	Codex Alimentarius Commission
CAD	coronary artery disease
CCPs	critical control points
CHO	carbohydrate
CV	cardiovascular
CVD	cardiovascular disease
CYP2E1	cytochrome P450 protein
DA	descriptive analysis
DMOs	destination marketing organizations
DMSO	dimethyl sulfoxide
DNA	deoxyribonucleic acid
Dl	deciliter
EC	European Council
ECG	electrocardiogram
E-nose	electronic nose
ERK1/2	extracellular signal-regulated protein kinases 1 and 2
E-tongue	electronic tongue
EU	European Union
EUR	euro
FAAS	flame atomic absorption spectrophotometer
FAO	Food and Agriculture Organization of the United Nations
F&B	food and beverage

FDDF	freeze-dried defatted flour
F/G	fructose-glucose ratio
FTIR	Fourier transform near infrared spectroscopy
GC	gas chromatography
GC-FID	gas chromatography flame ionization detector
GI	glycemic index
GI	gastrointestinal
GL	glycemic load
GPC	gel permeation chromatography
GSPE	grape seed proanthocyanidin extract
G/W	glucose–water ratio
HACCP	hazard analysis and critical control points
HDL	high-density lipoprotein cholesterol
HIV	human immunodeficiency viruses
HMF	hydroxymethyl furfural
HPAEC-PAD	high-performance anion-exchange chromatography with pulsed amperometric detection
HPLC	high-performance liquid chromatography
HPLC-RP	high-performance liquid chromatography-reverse phase
ICP-MS	inductively coupled plasma mass spectrometry
ICP-OES	inductively coupled plasma-optical emission spectrometry
ICRISAT	International Crops Research Institute for the Semiarid Tropics
IEC	Electrotechnical Commission
IL	interleukin
IQ	2-amino-3-methylimidazo(4,5-f)quinoline
ISO	International Organization for Standardization
KNN	k-nearest neighbor
LAB	lactic acid bacteria
LC	liquid chromatography
LC-PAD	liquid chromatography-photodiode array detection
LDA	linear discriminant analysis
LDL	low-density lipoprotein cholesterol
LOC	level of concern
MCPA	methyl cyclopropyl acetic acid
MIP	molecularly imprinted polymers
MP-AES	microwave plasma-atomic emission spectroscopy
MS	mass spectrometry
NAD	nicotinamide adenine dinucleotide
NARS	National Agricultural Research Systems
NDMA	dimethyl nitrosamine
NIRS	near infrared spectroscopy
NF-Kb	nuclear factor kappa B
NGO	nongovernmental organization
Notch-1	ligand-activated transmembrane receptor

ODDF	oven-dried defatted flour
PARP	poly (ADP-ribose) polymerase
PBS	phosphate buffered saline
PCR	polymerase chain reaction
PLS	partial least squares
P21/WAF1	cyclin-dependent kinase inhibitor protein
P53	tumor suppressor protein 53
OIV	International Organization of Vine and Wine
QDA	Quantitative Descriptive Analysis
QDSA	Quantified Descriptive Sensory Analysis
QMS	quality management system
QuEChERS	quick easy cheap effective rugged safe
RASFF	Rapid Alert System for Food and Feed
RDA	Recommended Daily Allowance
RE	retinol equivalent
ROS	reactive oxygen species
RTC	ready to eat
SQFS	Safety Quality Food Standard
SVM	support vector machine
TLC	thin layer chromatography
TQM	total quality management
TSS	total soluble solids
Tween 20	polyoxyethylene(20)-sorbitan-laurate
UN	United Nations
UNDP	United Nations Development Program
UPLC-QToF	ultra-performance liquid chromatography-quadrupole time of flight
USAID	United States Agency for International Development
USDA	United States Department of Agriculture
USFDA	United States Food and Drugs Authority
UV/VIS	ultraviolet/visible
VEGF	vascular endothelial growth factor
WFA	Winemakers' Federation of Australia
WGCA	Winegrape Growers' Council of Australia
WHO	World Health Organization
XIAP	apoptosis-inhibiting protein

PREFACE

Meticulously selected plant-based products, be it food or medicine, continue to gain high prominence as they form a major component that could contribute to achieving the United Nations Sustainable Development Goal 2 of Zero Hunger. The nutritional value and the quality these products have are highly dependent on their origin (plant source) and processing mechanisms. Both quality assurance and quality control are integral to the obtainment of these high-caliber alternatives.

Structured into four main parts, this book navigates this intersection between food and functionality and provides insight into the nutritional composition of some key elements of plant-based diets (e.g., honey, millet, ackee, grapes, spices, beer, and herbs). This book also introduces the most abundant adulteration practices and points out the analytical methods of quality monitoring, their current trends, and their potential future applications.

In the first part, termed Plant-based Sustainable Health Foods, a primary focus is placed on millets, their nutritional and health benefits, as well as their potential as food security crops. As underutilized and often underestimated cereals, challenges pertaining to the prospective adoption of millets in food diversification are also presented, along with probable solutions. The chapters also shed some light on demographics of millet production and offer a glimpse into the impact of processing on the nutritional and organoleptic attributes of millet-based products. These grains are important because of their nutritional composition and due to the possibility of having them included in fortified diets or wholesome diets for nutritional target groups.

Next, advances made in terms of production and quality determination of two globally renowned beverages, wine and beer, have been discussed while mapping consumption trends and consumers' expectations and preferences. These products are produced from different raw materials but offer nutritional benefits and pleasure to consumers so that they are sometimes considered as recreational drinks. Owing to their pivotal role in guaranteeing customer satisfaction, instructions on how both HACCP and ISO can be implemented for prime wine quality are laid out. The chapters also compare emerging trends of consumer interest in these products, which are their most beneficial advantages.

The second part discusses some ethnic foodstuffs, ingredients, and condiments of distinguished importance for the gastronomy of African, European, and Far Eastern countries. A special section is devoted to urge readers to rediscover and galvanize culinary and industrial uses of indigenous crops such as ackee that grows in the wild in West Africa. Introducing the different compositional elements and sensorial attributes of ackee-derived products is also a focal point of our chapters. The contributors also report on the richness of ackee in terms of the phytochemical, fatty acids, and amino acids profiles and stipulate that the increased production and processing of this fruit may contribute to food and nutrition security in Africa. This undeniably comes with challenges, which were equally addressed in details to outline the full potentials of ackee.

In the same context, the main and by-products of wine production contain a wealth of beneficial ingredients and can therefore be considered as functional food. In addition to its historical background, vine cultivation, production, by-product utilization, and health impact related to wine and grape are also detailed with a special focus on Hungarian context. Wine is a beverage of great economic and nutritional importance.

A discussion of gastronomy would not be complete without India, famous for its exceptional colorful use of condiments. A selection of leaf and seed spices (*Ocimumtenuiflorum, Petroselinum crispum, Laurus nobilis, Coriandrum sativum, Cymbopogon citratus, Myristica fragrans, Brassica nigra, Cuminum cyminum, Trigonella foenumgraceum, and Foeniculum vulgare*) is discussed regarding their origin, chemical structure, and traditional usage. Their major health benefits and potential detrimental impacts, as recounted by recent research findings, were presented.

Titled "Plant-Based Bioactive Compounds and Medicinal Foods," the third part draws attention to root and tuber crops and their relevance as potential substitutes for other high-glycemic index sources. A selected few of these nutrient-rich crops (cassava, yam, cocoyam, sweet potato) and their corresponding nutritional value are laid out in an attempt to demystify consumers' assumptions about the actual glycemic index of root and tubers. Intrinsic and extrinsic factors impacting this primordial metric are also reported. Future prospects for the incorporation of these root and tuber crops in the dietary regimen have also been explored.

A subsequent chapter covers the topic of traditional medicinal herbs sourced in the North-Western region of the Himalayas and displays the curative features of some ethnomedicinal plants. An important aspect covered by this chapter consists in offering the locals' perspective on curing certain ailments with traditional remedies. Thus, findings from survey-based analysis within communities knowledgeable about the therapeutic effect of certain plants were presented. The findings discussed in this chapter could be elaborated through further research for their potential industrial and large-scale applications across the globe.

Next, a special emphasis is put on the benefits of beer phytochemicals, namely xanthohumol, a bioactive polyphenol that can prevent reactive oxidants from being formed. Phytochemicals are essential for eliminating free radicals and preventing damage to cells. Other equally important beer flavonoids and non-flavonoids, which have long been lauded as having functional properties, have been listed along with their origin and chemical structure. The contributing authors outline the different anticarcinogenic and pharmacological properties such constituents have as corroborated by in vitro and in vivo experiments. Generally, the chapter puts an emphasis on some of beneficial reasons why some consumers consider beer as a recreational drink for both its nutritional and pleasurable properties.

The fourth part of the book portrays some of the typically performed food adulteration practices, and focalizes on some of the emerging quality control methods. One of the chapters comprises a comprehensive list of fraud-prone plant-based products, the prime motives engendering such practices and the negative effects both on the quality of these products and on consumer's health. A description of the most important food legislations (such as the *Codex Alimentarius* and European directives) to ensure food security was also included. Mitigating food adulteration would not be as efficient if the applied methods can't keep pace with the constant refining of fraudulent practices. In this context, prominent techniques are laid out and the gaps to be addressed are identified. Sensor-based instruments (e-tongue, e-nose, and e-eye) were of particular interest to us, given their demonstrated efficacy in fraud detection. The advantages of these instruments were discussed from both an industrial and practical perspective such that possible avenues for their future improves could be explored for their increased applications.

As one of the prime targets of adulteration, a whole chapter is dedicated to the study of honey composition and the analytical techniques deployed for the determination of its quality-imparting components. Extensive research oriented toward compositional analysis of honey and cited in this book accentuates the pivotal role of both botanical and geographical origins of honey on the richness or lack thereof, of nutrients.

The last chapter unravels the basics of aquaphotomics, as a cutting edge near infrared-based technique and clarifies the analytical methodology dictating the outcome of aquaphotomics-based analysis. Aquaphotomics explores the potential of water to act as a molecular mirror. It often gives a compositional overview of the analyte in question. Lastly, we provide an overview of the applications with regards to the determination of the quality of plant-based food products such as, to name a few, the quality assessment of seeds, grains and starch, food preservation, the determination of moisture content and water activity as well as pinpointing defects and damage of food items.

The book is intended for undergraduate and postgraduate students, but it should also be valuable for plant science teaching staff, researchers, and food industry personnel working with physicochemical and compositional properties. This book should also help improve the quality and performance of plant-based medicines in a much wider context.

We hope that this book will be able to answer some of the concerns of the readers about these topics.

—Editors

PART I

SUSTAINABLE FOODS AND HEALTH IMPACTS

CHAPTER 1

MILLETS: FOOD DIVERSIFICATION AND PROCESSING TECHNIQUES

DAMIAN LARYEA, SARAH THEODORA ESI MILLS,
CHARLES CUDJOE DAPULIGA, and JOHN-LEWIS ZINIA ZAUKUU

ABSTRACT

With issues of food and nutrient insecurity across the globe, studies are being conducted on numerous crops to help solve this challenge. One of the underutilized crops is millet with nutritional and economic importance and benefits. It can be grown in harsh conditions in the era of global warming. Those of commercial importance have been pearl, barnyard, proso, finger, and foxtail millet, though there are still quite many varieties in existence with important benefits. Millets contain quite a substantial amount of protein to the extent that it is the source of protein for approximately 130 million individuals in Africa's sub-Saharan regions. Millets contain folic acid, vitamins A and B. This has attracted research into this crop. They have been used in the development of quite a number of food products including complementary or baby foods. Though the use of millets offers some challenges that have affected their promotion and consumption for many years, yet, there are possible solutions. Putting it at the focus or forefront of research will help solve these challenges and promote its consumption. This chapter aims to promote millet in food diversification; covering issues such as products millets could be used for including complementary foods, the effect of processing, challenges in the use of millet, and possible solutions. Hopefully, this will bring more attention to millet as a key crop in the fight against climate change and food and nutrient insecurity.

Sustainable and Functional Foods from Plants: Health Impact, Bioactive Compounds, and Production Technologies. Megh R. Goyal, Arijit Nath, and Zoltán Kovács (Eds.)
© 2024 Apple Academic Press, Inc. Co-published with CRC Press (Taylor & Francis)

1.1 INTRODUCTION

The 21st century has been faced with food security challenges that require innovative solutions. With sustainable development goal II, the United Nations Development Program (UNDP) aims to achieve zero-hunger globally. In this regard, much attention has shifted to Africa, where dietary and nutrient deficiencies have long been a standing issue.[57] Many approaches have been explored to combat food security but food diversification appears to be the best alternative option, especially in socioeconomic communities where access to affordable foods remains a frontline objective. Food diversification presents the possibility of food production such that it can be available to people of all classes, ages, and social settings. Although the potential of using food diversification for food security remains outstanding, major pitfalls such as using exotic raw materials instead of local ones are a primary drawback.

According to the Food and Agricultural Organization (FAO), >250,000 plant species have been identified globally, but only 7000 have been explored for food production. This sheds light on the potential benefits of underutilized crops. The terminology "underutilized crops," generally focuses on crop species, whose potential has not been fully utilized. This century has witnessed an unprecedented increase in cereal production that has resulted in a consequent drop in world hunger.[27] However, many cereal species in certain parts of the world are even unheard of and remain underutilized in spite of their economic and nutritional potential.

In Africa, a primary example is millet, which is a drought-resistant cereal crop often consumed by the poor. In order of important cereals, millet ranks sixth after wheat, rice, maize, sorghum, and barley.[65] There are a number of varieties of millets with at least 18 species that have been domesticated and cultivated. The barnyard (*Echinochloa frumentacea*), proso (*Panicum miliaceum*), finger (*Eleusine coracana*), pearl (*Pennisetum glaucum*), and foxtail (*Setaria italica*) millet varieties are of commercial importance.[38] Millets are the way forward for countries such as India and Africa in the semi-arid regions, where food and nutritional insecurity are among the major challenges faced, coupled with limited access to water resources.[32] Many opportunities such as income-generating activities that will boost economic development not just in the food and agriculture sector but in several other specific fields that cover pharmaceuticals, specialty chemistry, and therapeutics, are widely and readily available for exploitation given the genetic diversity of millets.[44]

Different varieties have varying quantities of nutrients; pearl millet, which is mostly used in processing contains more protein and energy than wheat and a very high amount of iron. It contains other micronutrients, such as folic acid, vitamins A, B1, B2, and B5. In general, all of the millets are reported to be rich in minerals, vitamins, carbohydrates, and amino acids, such as methionine, which is lacking in the diets of people who mostly consume starchy staples.[15] Some varieties of millets have been administered to cure issues of the spleen, stomach, stress-induced gastric mucosal lesions, and diabetes mellitus in traditional Chinese medicine.[28]

Millets have no gluten content, therefore, producing foods free of gluten for individuals with gluten-sensitive enteropathy and other allergies is considered another significant opportunity to capitalize upon. This will put significant pressure on food industries and all associated stakeholders to maximize potential by stepping up their game in meeting the needs of a growing and hungry global economy.[22,24]

This chapter aims to cover food products that have millet incorporated in them, complementary or baby foods with millet, their nutrient contents, and also how the different forms of processing affect the properties of the millet and in effect the properties of the products developed. It also covers some of the challenges associated with millet production, processing, and promotion, which when overcome could make it one of the most important cereals.

1.2 MILLET PRODUCTS: THEIR NUTRIENT CONTENT AND ACCEPTABILITY

There have been only a few products developed from millet since 2010, such as biscuits, cookies, malt mix, fruit bars, and some traditional foods, such as *Ogi* and *Kunu* from Nigeria and *Idli*, *Dhokla*, *Uthapam* from India. Extruded snacks have also been produced using millet as the main ingredient. Examples are the extruded snack produced using *Kodo* millet and chickpea flour blends[20] and instant *fura* produced using millet and soybean or Bambara groundnut.[17,19] All these food products have millet either playing a very important role or being part of the ingredients.

Ogi is a traditional Nigerian food. It is a gruel prepared from major ingredients, such as sorghum, maize, or millet. The name *Ogi* still applies if either of these ingredients is used. Another name for this gruel is *pap*. This gruel is consumed by both the young and old, although it is described by Omenna et al.[40] as a weaning food; mostly because it is generally fed to babies. The major ingredient is soaked, drained, wet-milled, allowed to ferment, and used in the production of this food. Both Sengev et al.[1] and Omenna et al.[40] have worked on *Ogi*.

Sengev et al.[1] improved the nutrient content of the product by adding crayfish to millet. Their study focused on *Ogi* because they observed lower breastfeeding rates among Nigerians in rural areas while infants were being fed more of *Ogi* which they had determined to lack enough protein. Although *Ogi* had a lot of energy, the product lacked the required amounts of proteins and other micronutrients. They asserted that this may result in kwashiorkor among younger children as they will not be obtaining a balanced diet. This prompted them to include crayfish in their *Ogi*.

Omenna et al.[40] decided to ferment cereal and soybean mix for the production of gruel by combining maize, millet, or sorghum with soybean in a ratio of 2:1, respectively. This was done to improve the nutrient availability and protein content of the product. Fermentation was found to increase the protein content slightly; but not as much as what was found in the study by Sengev et al.[1] However, it reduced the antinutrients, phytates, and tannins, therefore making the micronutrients in the product readily available. Millet could therefore be employed in the making of infant foods (such as *Ogi* or *pap*) but will need to be fortified with other ingredients to improve the protein content of the food.

Soybean may not be enough, as indicated in a study by Omenna et al.[40] to meet the 13% RDA.[1] Therefore, to obtain a more nutritious *Ogi*, there is a need to ferment the major ingredients to reduce antinutrients and then include crayfish to enhance the protein quality and other nutrients in the product.

To enhance the nutritional composition of some local food products (such as *Idli*, *Dhakla,* and *Uthapam*) in India, Gupta et al.[21] used pearl millet, finger millet, and *Moringa* leaves. They reported that these ingredients have various nutritional and health benefits because of the polyphenols and dietary fibers that are present. The food products produced had 20% each of pearl and finger millet and 10% of *Moringa* leaves. These proportions improved the calcium, vitamin C, and total phenolic contents of the local food products.

India has embarked on several research studies to improve existing food products and also to develop new products from millet. This has been necessary due to the decline in the consumption of millet, which is mostly due to the drudgery involved in millet processing. In doing this, all stakeholders and a number of technologies were involved.[13] This will ensure that products or technologies developed are disseminated quickly and effectively.

Finger millet flour was used to substitute white wheat flour in *Chapatti.* Chapatti is a steamed leavened bread, which is flat in shape. It is mostly made with white wheat flour, which often lacks several nutrients (micronutrients, etc.) due to the processing methods involved in obtaining them. This has resulted in the use of composite flours in products that mainly use white wheat flour. Composite flours could be prepared with roots and tuber flours or cereal flours, which have been known to be nutritious and will add to the total nutrition of the food product.

Panghal et al.[43] used finger millet to substitute white wheat flour in Chapatti up to 40%. This affected the sensory, pasting, and functional properties of the dough in making the Chapatti. Therefore, a substitution of up to 20% was recommended for a more acceptable bread with good dough handling properties. The inclusion of millet was to improve the nutrient composition of the bread, but no nutrients were determined in this study. It would be important to know what and how much of the nutrients of the Chapatti could be attributed to finger millet.

Akinjayeju et al.[3] prepared a dough meal from a combination of maize, soybean, millet, and cassava. Dough meal is prepared in Nigeria mostly with roots and tubers. The dough often lacks several micronutrients and is very low in protein. Dough meal serves different purposes in the meal of Nigerians. The intention of this study was to obtain a dough meal with high nutritional quality (protein, fiber, and micronutrients) to improve the health of consumers. The doughs obtained after the study were nutritious and had good pasting and functional properties. Dumplings produced from the dough meal were acceptable. They concluded that millet and soybean's inclusion in the dough meal enhanced the protein and amino acids, fiber, and other micronutrients (such as potassium, zinc, iron, and magnesium) in the meal.

Cookies are snacks and part of pastries making waves in Africa. The major ingredient is wheat flour and due to their sweetness and convenient nature, they are consumed a lot in Africa. Wheat is not grown in Africa, hence the need to import, which affects the economy. In addition, the white wheat flour used in the preparation of these cookies and other pastries lacks other very important nutrients (such as protein), micronutrients (such as vitamins), and minerals. To improve the nutrient composition of these pastry products indigenous, cereals are being used to substitute the white wheat flour to obtain composite flours. These composite flours have been researched for many years. Cereals or roots and tubers could be used to make composite flours, as long as the crops being used will improve upon the nutrient content of the composite flour and eventually the final product. This is being done to reduce the amount of wheat flour imported since it cannot be eliminated, due to the unique dough qualities wheat possesses when making pastry products.

In a study by Abioye et al.,[2] Finger millet flour produced from malted grains and roasted African yam beans was used to substitute wheat flour in the preparation of cookies. The malting technique was used to reduce antinutrients that exist in millet. The product with 10% malted Finger millet but no African yam bean was preferred. Similarly, the 100% white wheat flour cookies. Cookies with 20% malted Finger millet but no African yam bean was also highly rated in the preference test. The antinutrient composition of these products was within acceptable limits while nutrient contents were improved; crude protein, fiber, and mineral contents were increased slightly.

Manivel[37] also developed cookies using wheat, millet, corn meal, rye flour, and *Moringa* powder. One of the three combinations developed contained 30% millet flour in addition to 50% wheat flour and 20% *Moringa* powder. Although their focus was on *Moringa oleifera*, yet millet was employed in one of their formulations.

Sangma et al.[59] developed a malt mix from a combination of foxtail millet, roasted Bengal gram dal, milk powder, and honey. Although no nutritional contents of the product were determined; yet, they assessed the organoleptic qualities. After varying the contents of the ingredients, they found out that the most accepted product was the one with two parts of foxtail millet, 1 part of roasted Bengal gram dal with honey. They also observed that increasing the foxtail millet improved the organoleptic properties of the malt mix.

Dhokla is a traditional food in India, mostly for vegetarians, and is eaten as breakfast, a snack, or a main dish. It is made using fermented batter from rice or semolina and chickpea. The preparation of this product takes a while due to the fermentation process; therefore, Ransumithila and Saravanakumar[48] decided to develop an instant *Dhokla* mix to reduce the preparation time. They also had nutrition quality in mind and therefore they sought to improve the nutrient content of this meal by adding Foxtail millet. Foxtail millet was incorporated into the meal from 5 to 30%, at intervals of 5. After sensory evaluation, the product with 20% foxtail millet was the most acceptable. In addition, protein, fiber, and iron contents of the product were improved by the addition of millet.

An acceptable probiotic fruit bar was produced from a combination of fruits (guava and strawberry) pulps, skim milk powder, puffed cereals (*Bajra* and *Cholam*), millet, wheat flakes, jaggery, and liquid glucose. The probiotics added to this fruit bar were *Lactobacillus acidophilus*, *Lactobacillus delbrukii*, and *Streptococcus thermophilus*.[12] The product developed had a very high acceptance score and sufficient amounts of macronutrients (protein, fat, and carbohydrate) and micronutrients (beta-carotene, vitamin C, calcium, and iron).

RTC (Ready-to-Cook) flakes have also been developed by Patil et al.,[42] who found out that millet grains that were not dehulled contained high levels of dietary fiber compared with dehulled grains. The product developed had a substantial amount of protein and minerals, such as iron, and no trans-fat. The glycemic index (GI) of the product was found to be in the medium range (average of 52.11), while the glycemic load was 9.24, which is low.

Just as millet has been employed in the development of a number of products, it also has an important advantage in its use as a meat extender. Extended meat products are cheaper and are more nutritious forms of meat. They have gained importance due to the current socioeconomic conditions of most consumers in developing countries. The common ingredients used in these products are wheat, rye, or barley. In addition to these products being imported into most developing countries, they pose a risk to persons with gluten intolerance. Celiac disease has a direct correlation with gluten, which is found in these ingredients. Millet could be an alternative to the already-used cereals (wheat, rye, or barley) so that it has good functional properties, especially water-holding capacity, which will make the meat juicier.

With the presence of phenolic compounds, millet has the ability to prevent lipid oxidation by mopping-up free radicals, thereby extending the shelf-life of meat products. It also contain significant amounts of dietary fiber and resistant starches, which are good in reducing plasma LDL-cholesterol, thus reducing the risk of obesity, coronary, gastrointestinal disorders, and lowering the caloric density. In addition to all these benefits, millet has vitamins A and B that have been generally accepted in terms of sensory. In most products produced,[2,59] millet, therefore, has great potential for use as meat extenders in meat products.[62]

Millet also can reduce stunting and also can improve body mass index. This was observed in school-going children (between the ages of 10–14 years) fed with millet-based meals for 3 months. A significant improvement in stunting and body mass index in the children on the millet-based meals was observed compared with those that did not. The mid-day meals included millet–vegetable meal (made from pearl and little millet), a millet–lentil hot meal, millet *bi si belle bath* (made with pearl and little millet), a steam-cooked fermented savory cake and Finger millet *Idli*. Contrary to these meals, the children mostly consumed rice-based meals, which had been fortified. The millet-based meals were well-liked by the children. Products from millet could therefore improve the nutritional status of consumers.

1.3 MILLET AS AN INGREDIENT IN BABY FOODS

Due to the nutritional, functional, and health benefits of millet, it has also been employed in several complementary baby foods, for example, the traditional food *Ogi*, associated with the Nigerians. To enhance the nutrients in the *Ogi*, Sengev et al.[1] included crayfish, while Omenna et al.[40] fermented a millet and soybean mixture for the same meal. Millet is therefore not alien when it comes to the preparation of complementary foods that are developed from local cereals (such as millet) to offer a cheaper yet nutritious alternative to the commercially available complementary foods.

Millet has played a role in a nutritious complementary food developed by Rasane et al.[49] The product was made from a blend of Pearl millet, sorghum and oat, whole milk powder, whey protein concentrate, rice-corn cooked flour, and sugar. The product had enough protein content, vitamin, and minerals that could fulfill the nutrient requirement of children between the ages of 1–3 years. Complementary food has also been developed from a composite of millet, guinea corn, yellow corn, soybean, groundnut, and date palm. This blend was comparable to already existing or commonly known complementary foods in terms of both nutrition and acceptability.[8] An acceptable complementary food was made from 30% malted millet and 40% wheat in addition to skimmed milk powder, ghee, and whey protein concentrate.[36]

Manvesh et al.[58] also developed a pearl millet-based complementary food with the addition of vitamin A acetate and iron salt to improve its nutritional status. Another food was developed from a combination of barley, pearl millet, skimmed milk powder, sugar, flavoring agent, iron, and zinc salts.[7] The addition of vitamin A acetate, ascorbic acid, phytase, iron, and zinc salts could improve the absorption of essential micronutrients.[9]

Powder from *Moringa* leaf was added to pearl millet to produce a complementary food, of which samples with 2.5% *Moringa oleifera* leaf powder were most accepted. Millet has been found to be lacking lysine and threonine; hence, there is a need to include the *Moringa* to improve the protein quality.[39] Lysine and threonine are limited in millet, hence the need to identify food crops that could offer these amino acids to have a complementary food with high protein quality.

A nutritious complementary food from Finger millet, sorghum, amaranth, and pumpkin seeds has also been produced. The cereals were fermented, and this affected the pasting and functional properties. Peak viscosity was reduced but pasting temperature was increased. Protein solubility was increased due to the fermentation while high amounts of zinc and calcium were observed in the millet-amaranth-pumpkin combinations. Threonine and lysine were the limiting amino acids, hence alternative food sources with these amino acids could be added to the formulation. Fermentation is important in improving the nutrient content of millet-based complementary foods.[53]

Regarding complementary foods, protein quality, energy density, and digestibility of the product are very important factors. Complementary foods are an important meal for infants because at that critical stage, the inability to obtain important nutrients will result in malnutrition and its associated risks. This has prompted the use of processing technologies (such as extrusion, spray drying, or drum drying) in addition to the malting of millet to improve these qualities. Complementary food developed from a combination of Finger millet and soybean with 5% fish powder to further improve protein quality was developed. This product was able to meet the energy requirement of infants between the ages of 6 and 11.

The extrusion technology reduced the trypsin inhibitors, lectins, tannins, and phytates. Iron and zinc bioavailability was improved.[30] Complementary food has also been developed from a combination of Foxtail millet, Little millet, sorghum, and white ragi. Spray drying was used to make the instant product. This little millet-based complementary food was the best among the several combinations.[60]

Lactic acid bacteria (LAB) have been used on pearl millet–soybean slurry to develop a nutritious complementary food. The strains used had been selected previously for their ability to hydrolyze phytate, α-galactooligo saccharides, or starch. LAB could not reduce the phytate levels in the product, and as a result, more studies are needed on the use of LAB in hydrolyzing phytate and starch in millet–soybean-based complementary food.[56]

The notable points among these research studies were (1) in the development of complementary foods, malting of millet grains was part of the process and this reduced antinutrients, thereby enhancing the bioavailability of the nutrients in the millet grains, and thereby improving the nutrients in the complementary food; (2) millet still lacks certain amino acids, which are threonine and lysine that could be taken care of with the addition of an animal-based protein or choosing legumes with high protein quality.

1.4 EFFECTS OF PROCESSING METHODS ON NUTRIENTS AND OTHER PROPERTIES OF MILLET

Millets usually undergo some form of initial processing to become edible as well as to enhance their nutritional bioavailability, digestibility, and organoleptic, or sensorial properties (e.g., improved taste, flavor, texture, and palatability before consumption) and for preparation in other foods. Additionally, it is beneficial in inhibiting, removing, or destroying micro-organisms, retarding or preventing the occurrence of certain deleterious chemicals, physicochemical, and biochemical reactions, decreasing the content of antinutrients (such as phytates), and most importantly enhancing and preserving its nutritional value.[6] However, during food processing, there is a likelihood of undesirable changes occurring, thus affecting the nutritional and sensory characteristics. The type of processing method used and the duration of processing with respect to the particular foods will determine the degree of degradation or loss that occurred.[46]

Mechanical methods for processing millet have both positive and negative outcomes. In the dehulling or decorticating process, hydrothermal treatment easily facilitates the removal of the bran by hardening the texture of the endosperm. In this way, the grains can further be processed through milling to obtain flours of different particle sizes for use in other products including compositing it with other flours. This makes it possible to obtain a soft texture within a few minutes of processing, improving sensorial and consumption qualities.[29] On the contrary, these processes are associated with significant losses in the nutritional and chemical constituents of the grains particularly affecting its usage as a functional food. Notable changes observed include: the decrease in antioxidant capacity, total phytic acid, minerals, total phenols, polyphenols, crude, and dietary fiber that are majorly found in the grain periphery.[5,14,23]

The removal of phytates and tannins can improve iron bioavailability.[54] For pearl millets, the removal of antinutrients through dehulling is associated with an increased *in vivo* starch and protein as well as bioavailability of minerals, such as iron and zinc.[51] However, no negative effect has so far been reported on the fat and protein contents. The degree of milling and refining of millet flour is a factor impacting the decrease in the bioavailability of nutrients. It is therefore advised to use whole grains for flour to maintain the rich nutritional matrix stored in the bran.[61]

Traditional methods of soaking, germination, malting, and fermentation also have their corresponding effects on the millet grains primarily through the removal of antinutrients and improvement of nutritional constituents. Soaking for instance has been found to further decrease the presence of antinutrients, especially when combined with other processes of dehulling, milling, and cooking. It is the same for the malting or germination process, which in addition to removing antinutrients, also enhances the availability of minerals, vitamins (B and C), crude fat, and fiber. Soaking particularly increases the *in vitro* protein digestibility and bioaccessibility of iron and zinc, which are microminerals in Pearl and Finger millets, whereas malting results in increased beta-amylase activity and free alpha-amino nitrogen content in pearl millet when compared with sorghum.[25,47,55] The increase *in vitro* digestibility of protein for pearl millet is 26% from 14% and from 86 to 112% for starch digestibility and a doubling effect for iron solubility on account of germination. Other observations have been identified with the *in vitro* extraction and bioaccessibility of calcium, iron, and zinc in Pearl and Finger millets.

In Pearl millet, oxalate reductions occur from 0.502 to 0.068% and calcium solubility increases to 14.1 from 2.4 mg/100 g of grain,[61] due to elevated phytase activity, decreasing phytate contents in sprouts and the catabolism of polyphenols and saponins which otherwise would have negatively affected mineral availability. For proso millet, the process of malting is beneficial to increase the contents of lysine and tryptophan, total sugars, and non-protein nitrogen; and to decrease dry weight and starch contents.[55,61]

The presence and availability of calcium, iron, magnesium, and zinc are increased by the process of germination. The utilization of the nutrients in the grains by the growing sprouts and the reduction in the weight of nitrogen-containing complexes in the course of soaking and washing of the grains in addition to the breakdown of lipid and fatty acids at sprouting is said to reduce foxtail millet's amounts of fats and crude protein.[55] Both soaking and germination reduced the starch content of millet because of the action of amylase for starch hydrolysis. In general, germination reduces the overall tannin content of millets to 0.83% from 1.6%, softens the kernel structure, and increases its functional properties by improving the nutrient composition and concentration. The downside to soaking is the degradation and leaching in concentrations of iron, zinc, and phytates (affecting phytase activity), especially after the processes of dehulling and milling. Pearl millet grains when soaked normally lose 25% of iron and go through an appreciable decrease in calcium, phosphorus, and iron when soaked in acid.[61]

The fermentation process provides processors with the added advantage of manipulating the flavors and textures of millet. The fermentation of pearl millet while improving nutritional components (such as ash, fat, and fiber), moisture and protein, and enhancing flavonoids, significantly decreases the amounts of iron, zinc, sodium, and potassium after a period of 16 h.[21] For minerals, the in vitro accessibility from finger millet is rather insignificant contrasted with that of pearl millet. Fermentation of pearl millet for a period of 9 h decreases polyphenols present in the grains from 12 to 10% and phytic acid from 30 to 27%. Natural fermentation that occur by natural means also reduces phytic acid and the total polyphenol content in pearl millet.

Fermentation is widely associated with millets used in fermented food and beverages. The acid produced by microorganisms used in the fermentation process helps with the inhibition of microbial and pathogenic spoilage, balancing the amino acid content, and improving sensorial and nutritional qualities.[47] Amylase inhibitors, trypsin, phytic acid, and tannins are some antinutritional compounds that are reduced via fermentation. The effect of fermentation on starch digestibility is dependent on time, temperature, and the culture used. Collectively, these processes help to improve sensorial characteristics and starch digestibility and attain high nutrient quality with minimized antinutritional properties.[51]

Several heat application methods have been used in the processing of millet. Notable among them is the popping or puffing technique, which operates on a high-temperature short-time principle using sand as the medium for heat transfer. The result is the characterization of ready-to-eat millet meal with desirable colors, shapes, flavors, textures, aromas, and enhanced overall consumer acceptability due to the bursting open of the endosperm and gelatinization of starches present.[47,55]

Puffing is also beneficial in increasing dietary fiber, digestibility, and solubility of starch granules, inactivating destructive bacteria, and reducing the antinutrients of the final product.[64] Compared with the relatively high crude fat and fiber contents in raw foxtail millet, popped Foxtail millet has significantly lower levels. A recommendation pertaining to puffing is the use of novel technology with optimized conditions.

There is also roasting, which is much like puffing except that it results in higher volume expansion. Through roasting, certain antinutrients (such as alkaloids, goitrogenic agents, hemagglutinin, and glycosides), saponins, and trypsin inhibitors are removed thus improving the nutrient quality of millet grains and increasing the shelf-life.[23,64] Roasting is also beneficial in retaining the protein content in millet and improving the levels of zinc and iron.[51]

Boiling of millet grains in water until softened is perhaps the longest heating process for millet. The cooking process serves as a basic step from which several products including soups, gruels, and porridges can be obtained. Boiling is an important process associated with the increased amounts and availability of calcium, magnesium, zinc, and iron. Generally, cooking enhances palatability, increases organoleptic properties, and increases shelf-life by eliminating microbial load.[26]

Extrusion can be classified under heat processes associated with millet. It operates on the principle of high temperature for a short time and it results in the gelatinization and complete cooking of millet and millet-based products. The design of extruders is designed to allow easy room for biofortification processes with minerals and vitamins to make up for nutritional losses incurred during the transformation process.[47]

Mallasy et al.[35] indicated that fortified millet with whey protein is better in terms of elevated protein contents and sensorial preference as compared to millet only. Similar supplementation studies by Ali et al.[4] with soybean protein also revealed an enrichment of the amino acid profile of the composite flour particularly for lysine as much as a 1.5 to 2.4-fold increase and a decrease in the in vitro digestibility of protein with increasing substitutions of soybean protein. Millet-based composited and extruded snacks are ideal for low-cost production.

Preservation treatments employed in increasing the shelf-life of millets and millet-based products also have their associated pros and cons. These processes particularly work to inactivate the presence and functioning of endogenous enzymes and microorganisms and in effect prolong the shelf-life. Irradiation, refrigeration, and thermal processing are the three commonest techniques applied in the preservation of millet grains and flours; and are applied either alone or as a combined effect of more than one technique. It is very crucial that optimized conditions are strictly adhered to avoid the occurrences of undesirable changes in quality.

Thermal processing (such as hot-water blanching or dry-heating) of grains at temperatures of 98 and 100°C for 10 and 120 min, respectively enhances the shelf-life of pearl millet reducing the occurrence of undesirable lipid composition changes; and also the hydrothermal treatment of the grains of pearl millet inactivates the enzyme lipase thus enhancing shelf-life.[61] Pearl millet flour obtained from hydrothermal treatment of grains is characterized by better functional, pasting, and physical properties as compared with ordinary non-treated flour.[67] In finger millet, however, this process of hydrothermal treatment yields no notable changes in the overall nutritional composition, although some individual nutrients are somewhat slightly affected.[23]

Cooling processes (such as refrigeration) have also been found to maintain amino acids present in Pearl millet during storage irrespective of the processing methods in use.[34] In irradiation, food products turn out with better qualities, reduced antinutritional compounds, and improved nutritional quality as well as extended shelf-life. Its operation is the subject of packaged foods to alpha, beta, gamma, or X-rays at controlled conditions.[31] Through this process, the bioavailability of micronutrients and the quality of millets can be better enhanced and improved faster and more efficiently. Its effect on the amino acid profile of Pearl millet has been investigated and has been found to maintain the stability of all other amino acids present except for glutamic acid, leucine, and phenylalanine.[33] Hence, radiation influences millet so that it can decrease, to the point of elimination, the antinutrients present in millet grains, which will impact positively the bioavailability of nutrients, such as amino acids and minerals.[51]

The composite blending of millet flour with other flours combined with other treatments is another processing technique employed in the production of high-quality nutritious and healthy foods with desirable properties. For instance, biscuits have been formulated from Finger millet and wheat flour (60:40) with high dough characteristics and biscuit quality; porridges from press-dried cowpea and extruded millet with satisfactory characteristics (consistency, color, flavor, and smoothness) as weaning foods; breads from Barnyard millet, wheat and gluten (61:31:6) with wheat bread preferential equivalence; muffins from 60% Finger millet formulation and additives with better volume and quality characteristics of muffins; noodles from 30% finger millet formulation having significantly lower glycemic index that is good for diabetics.[50,54,61]

The combined processes of soaking, germination, and malting on the amount and activity of the antioxidants in millet grains cause positive effects as against the effects from dehulling and cooking. The antioxidant capacity of Finger millet is reported to increase twice as much after malting for 96 h and so does the germination of foxtail millet for 3 days leading to the enzyme activities breaking down complexes and producing higher antioxidant activity with better DPPH-scavenging activity.[11]

Antioxidant activities in finger and pearl millets may decrease when subjected to hydrothermal treatments, boiled, or roasted (effect of oxidation and degradation reactions taking place), and the dehulling of pearl millet also decreases its phenolic content and antioxidant potential because of the removal of the pericarp which is highly enriched in polyphenols and antioxidants.[10] The combination of germination, steaming, and roasting processes on little millet has increased flavonoids, tannins, and total phenolic contents.[45]

1.5 MILLETS: CHALLENGES AND PROBABLE SOLUTIONS

Despite the rich biodiversity of millets and their associated benefits that significantly contribute to many livelihoods and agricultural systems, especially in the West African regions, it remains underexploited among other major challenges. These challenges range from issues of social marginalization and poverty that tag small-scale farmers to governmental influences and policies that have come to favor the patronizing of other commercial crops, to individual consumer preferences and changes in dietary patterns, to agriculture-related issues of seed quality and improvement, postharvest processing, technological advancements, and value addition that undermine the full capacity of millet even as a food security cereal.

The economic decline in production and consumption quantities worldwide has rather created the stress of having to deploy effective strategies aimed at increasing consumption through many channels.[18,52] Of most importance, in terms of these challenges, is to change the mindset people have had for many years harbored about millet as a poor man's food and to enlighten farmers on better-improved techniques.

A leading cause in the decline of major consumption can be traced to "The Green Revolution" of the 1970s, which for some reason sought to give more prominence to other cash crops of rice, wheat, and maize "disregarding" the equal importance of millets. As a result of this and many other factors including rising incomes and growing urbanization, the position of millet formerly considered a staple in many countries, especially in Asia has dwindled. The consequential effect today is the numerous challenges encompassing millet and the brain-racking countering efforts by many institutions and bodies to foster the promotion of millet. By side-lining millet from top prioritized research and development programs for the longest time, policymakers are now at their best dedicating colossal amounts of resources, time and money inclusive, to create positive awareness for millet and if possible, blot out its tag as a "poor man's food." The aim is to simultaneously undertake multifaceted programs that will improve social, economic, financial, nutritional, and sustainability empowerment. This has taken on many forms including strengthening linkages between disconnected stakeholders for the purpose of bridging existing gaps among researchers and developers.

Further efforts have been observed in the area of actively and persuasively convincing decision makers on the significant role of the different millet varieties toward mainstream agricultural, nutritional, and food policies. Positive results are expected to show in the preservation and improvement of traditional methods associated with the production, processing, and usage of millets as well as the genetic diversity. They may be as simple as whatever and as complicated as advocating for the use of millet in school feeding programs or as a feeding support system for vulnerable groups in society.

The production, trade, and exports of millets are stipulated at figures less than 1%, although they are in the sixth position among the world's most important cereals. Pearl millet being the most produced of all millets has an average yield lower than that of other cereal crops.[16] In general, there has been a global decline in millet production for years now reflecting in the reduced land area delegated for its cultivation. Padulosi et al.[44] reported an overall decline in land area for millet farming at 0.97 million hectares from 5.29 to 1.35 from 2.3 million hectares for Finger millet over the period between 1951–1955 and 2006–2010.

An investigation on the worldwide usage of millet from the period 2009–2011 conducted by Sarker[52] revealed the usage of at least three-quarters of millet supplies as food for human consumption and the other quarter for the feeding of livestock. A portion of millet is used as livestock feed and in other processing-related activities in Africa; however, its consumption is still low in Sudan and Ethiopia, which are major producers of millet in Eastern and Southern Africa. Comparing these figures to that of the nineties, Sarkar reported a decline in food consumption quantities and an increase in use as livestock feed.

In India, consumption figures are reported to have fallen by 47 and 83% for Finger millets and other small millets, respectively.[16] Although some efforts have been put in place in attempts to bring rectification and boost consumption, yet, the story remains the same today not just for India but also for other countries.

Millet pricing is quite a controversial area seeing that prices at global levels are rather lower than at national levels for some countries, such as Ethiopia and Kenya. While prices for millet in Kenya were at $768 per ton in 2009, world prices in 2010 were at $400 per ton, a good indication of Kenya's low supply against high demands and one of the reasons why it is today regarded as the leading millet importer.[41]

There is also the challenge of capacity development of farmers, which has to do with educating farmers in certain critical areas pertaining to choosing seeds of good quality, enhanced agriculture practices, processing of grains, and the making of value-added products. The decline in the cultivation of millets, aside from the lack of interest, is affected by the quality of seed input. Over time, there has been a depletion of genetic diversification in millet species resulting in poor yields. This is because the seeds are not biologically sound. That is to say, the seeds are not particularly adjusted to suitably tolerate the changes in environmental parameters. Production efficiencies are in effect below capacity stemming from the use of poor-quality and low-yielding seeds as well as the practices of ineffective farming methods that are rudimentary and drudgery processes. Training of farmers is thus very vital to impart knowledge on the best practices to ensure better handling of seed at every farming stage.

It is important to bridge the gender-differentiated knowledge that men and women have about millet from their long exposure to their distinctive involvement in the production of millet. Knowledge in terms of seed variety/choice, crop yield, resistance to pests and insects, good market price, process simplicity and ease, and grain development when assessed and disclosed forms a rich source of information that is greatly important for the sustainability of agrobiodiversity over the long term.[63] Padulosi et al.[44] reported a significant initiative by MSSRF in establishing the notion of gene–seed–grain granary that tackles the inconsistencies of poor seed storage and food insecurity. Its concept is along the lines of establishing safe seed-storing environments by helping farmers properly store their seeds for the next cropping cycle rather than consume them entirely as food or use them as feed.

Time and effort involved in the preparation of millet meals at the household level is another constraint impeding the wider use of millets. This can be seen in the want of innovative and advanced processes and processing technologies to make available safe yet convenient or instant meals from millet grains at a commercial scale.[61] To avoid such tediousness, many homes and individuals have a preference for convenient or instant foods over millet.[66]

Successful innovations have been developed by a number of groups and organizations. This includes the building of a mill for the dehulling of millet grains with a seed recovery rate of 90%, and small-scale machinery that removes stones as part of improving postharvest processing. Other innovations on the lookout are the concept of a polisher to facilitate the consumption of millet.[44] In close association with this is the rationale of product development with millet. The many attributes of millet make it suitable for use in many products yet have not been exploited as it should. It is very important now more than ever especially in the face of threats of climate change and food insecurity to exploit this opportunity as a means of promoting consumption, increasing income, and empowering farmers both at national and household levels.[52]

1.6 SUMMARY AND FUTURE RESEARCH

As important as food diversification in ensuring food and nutrient security in the world, underutilized food ingredients, such as millet cannot be left out. Millets can be used in quite a number of food products and even compete with other cereals in terms of functional properties and nutrition. Research has also focused on it when it comes to complementary or baby foods due to the various nutrients and functionality they offer. In Africa, where wheat flour is imported, thereby affecting the economy, millet flour could be used to reduce the amount of wheat flour imported. Various forms of processing also have an effect on their nutrient content and functionality, and these should be considered when using millet as an ingredient in food product development. Though there have been challenges reported regarding the processing of millet, there have been some possible solutions. More research however will need to be done to establish various processing techniques to help promote millet for consumption. Breeding techniques to improve the millet varieties could also help with its processing and use at home. Focus on the millet varieties to develop a myriad number of food products, which is very common or popular with consumers, could really help with its promotion. This could contribute to achieving the sustainable development goal 2: zero-hunger, set by the United Nations Development Program. In this era of climate change, millet is indeed a very good option to consider given all the benefits it offers.

KEYWORDS

- **Africa**
- **food diversification**
- **millet**
- **sustainable development goal 2**
- **wheat**
- **zero hunger**

REFERENCES

1. Abraham, S. I.; Ifeanyi, N. Stephen S. Effect of Crayfish Inclusion on the Chemical and Sensory Properties of *ogi* Prepared from Maize, Millet and Sorghum. *Int. J. Nutr. Food Sci.* **2016,** *5* (6), 378–282.
2. Abioye, V. F.; Olatunde, S. J.; Elias, G. Quality Attributes of Cookies Produced from Composite Flours of Wheat, Germinated Finger Millet Flour and African Yam Bean. *Int. J. Res.—Granthaalayah* **2018,** *6* (11), 172–183.
3. Akinjayeju, O.; Fagbemi, T. N.; Ijarotimi, O. S.; Awolu, O. O. Optimization and Evaluation of Some Physicochemical and Nutritional Properties of Cereal-Based Soya-Fortified Flours for Dough Meal. *J. Adv. Food Sci. Technol.* **2019,** *6* (1), 40–59.

4. Ali, M. A.; El Tinay, A. H.; Mallasy, L. O.; Yagoub, A. E. Supplementation of Pearl Millet Flour with Soybean Protein: Effect of Cooking on In Vitro Protein Digestibility and Essential Amino Acids Composition. *Int. J. Food Sci. Technol.* **2010,** *45* (4), 740–744.
5. Bagdi, A.; Balázs, G.; Schmidt, J.; Szatmári, M.; Schoenlechner, R.; Berghofer, E.; Tömösközia, S. J. Protein Characterization and Nutrient Composition of *Hungarian proso* Millet Varieties and the Effect of Decortication. *Acta Alimentaria* **2011,** *40* (1), 128–141.
6. Boekel, M.; Fogliano, V.; Pellegrini, N.; Stanton, C.; Scholz, G.; Lalljie, S.; Somoza, V.; Knorr, D.; Jasti, P. R.; Eisenbrand, G. A Review on the Beneficial Aspects of Food Processing. *Mol. Nutr Food Res.* **2010,** *54* (9), 1215–1247.
7. Bonthala, V. S.; Muthamilarasan, M.; Roy, R.; Prasad, M. FmTFDb: Foxtail Millet Transcription Factors Database for Expediting Functional Genomics in Millets. *Mol. Biol. Rep.* **2014,** *41* (10), 6343–6348.
8. Ceyedika, N. C.; Calvin, N. N.; Joseph, A.; Abdullahi, R. M. Quality Assessment of Baby Food Produced from Cereals Enriched with Date Palm. *Sci. World J.* **2019,** *14* (1), 28–31.
9. Cercamondi, C. I.; Egli, I. M.; Mitchikpe, E.; Tossou, F.; Hessou, J.; Zeder, C.; Hounhouigan, J. D.; Hurrell, R. F. Iron Bioavailability from a Lipid-Based Complementary Food Fortificant Mixed with Millet Porridge Can Be Optimized by Adding Phytase and Ascorbic Acid But Not By Using a Mixture of Ferrous Sulfate and Sodium Iron EDTA. *J. Nutr.* **2013,** *143* (8), 1233–1239.
10. Chandrasekara, A.; Shahidi, F. Bioaccessibility and Antioxidant Potential of Millet Grain Phenolics as Affected by Simulated In Vitro Digestion and Microbial Fermentation. *J. Funct. Foods* **2012,** 4 (1), 226–237.
11. Coulibaly, A.; Chen, J. Evolution of Energetic Compounds, Antioxidant Capacity, Some Vitamins and Minerals, Phytase and Amylase Activity During the Germination of Foxtail Millet. *Am. J. Food Technol.* **2011,** *6* (1), 40–51.
12. Kavitha, B.; Vijayalakshmi, R.; Poorna, C. R.; Yalagala, I. M.; Sugasini, D. Nutritional Evaluation and Cell Viability of Formulated Probiotic Millet Fruit Bar. *J. Food Nutr. Disorder* **2018,** *7*, 2–10.
13. Dayakar, R. B.; Sangappa V. G. D.; Arlene C. V. A.; Tonapi, B., Eds. *Technologies of Millet Value-Added Products*. Center of Excellence on Sorghum, ICAR-Indian Institute of Millets Research: Hyderabad, India, 2016; p 48.
14. Dharmaraj, U.; Malleshi, N. G. Changes in Carbohydrates, Proteins and Lipids of Finger Millet After Hydrothermal Processing. *LWT-Food Sci. Technol.* **2011,** *44* (7), 1636–1642.
15. Dissanayake, B. D.; Jayawardena, H. S. Development of a Method for Manufacturing Noodles from Finger Millet. *Procedia Food Sci.* **2016,** *6*, 293–297.
16. Food and Agriculture Organization of the United Nations (FAO), International Fund for Agricultural Development (IFAD). *The State of Food Insecurity in the World: Strengthening the Enabling Environment for Food Security and Nutrition*; FAO: Rome, 2014; p 27.
17. Filli, K. B.; Nkama, I.; Jideani, V. A. The Effect of Extrusion Conditions on the Physical and Functional Properties of Millet–Groundnut Based *fura*. *Am. J. Food Sci. Technol.* **2013,** *1* (4), 87–101.

18. Fischer, H. W.; Reddy, N. N.; Rao, M. S. *The Role of Millets in Climate Risk Reduction: Evidence from Ananthapur—Andhra Pradesh*. Working Paper 3: Revitalizing Rainfed Agriculture Network (RRAN): India, 2016; p 53.
19. Filli, K. B.; Nkama, I.; Abubakar, U. M.; Jideani, V. A. Influence of Extrusion Variables on Some Functional Properties of Extruded Millet-Soybean for the Manufacture of *fura*: A Nigerian Traditional Food. *Afr. J. Food Sci.* **2010,** *4* (6), 342–352.
20. Geetha, R.; Mishra, H. N.; Srivastav, P. P. Twin Screw Extrusion of *kodo* Millet-Chickpea Blend: Process Parameter Optimization, Physicochemical and Functional Properties. *J. Food Sci. Technol.* **2014,** *51* (11), 3144–3153.
21. Gupta, A.; Singh, R.; Prasad, R.; Tripathi, J.; Verma, S. Nutritional Composition and Polyphenol Content of Food Products Enriched with Millets and Drumstick Leaves. *Int. J. Food Fermentation Technol.* **2017,** *7* (2), 337–342.
22. Harmayani, E.; Anal, A. K.; Wichienchot, S.; Bhat, R.; Gardjito, M.; Santoso, U.; Siripongvutikorn, S.; Puripaatanavong, J.; Payyappallimana, U. Healthy Food Traditions of Asia: Exploratory Case Studies from Indonesia, Thailand, Malaysia, and Nepal. *J. Ethnic Foods* **2019,** *6* (1), 1–8.
23. Hama, F.; Icard-Vernière, C.; Guyot, J. P.; Picq, C.; Diawara, B.; Mouquet-Rivier, C. Changes in Micro-and Macronutrient Composition of Pearl Millet and White Sorghum During in Field Versus Laboratory Decortication. *J. Cereal Sci.* **2011,** *54* (3), 425–433.
24. Olayanju, J. B. *Top Trends Driving Change in the Food Industry*, 2019. https://www.forbes.com/sites/juliabolayanju/2019/02/16/top-trends-driving-change-in-the-food-industry/ (accessed on Dec 19, 2021).
25. Krishnan, R.; Dharmaraj, U.; Malleshi, N. G. Influence of Decortication, Popping and Malting on Bioaccessibility of Calcium, Iron and Zinc in Finger Millet. *LWT-Food Sci. Technol.* **2012,** *48* (2), 169–174.
26. Khamgaonkar, S. G.; Singh, A.; Chand, K.; Shahi, N. C.; Lohani, U. C. Processing Technologies of Uttarakhand for Lesser-Known Crops: An Overview. *J. Acad. Ind. Res.* **2013,** *1* (8), 447–52.
27. Lenaerts, B.; Collard, B. C.; Demont, M. Improving Global Food Security Through Accelerated Plant Breeding. *Plant Sci.* **2019,** *287*, E-article ID: 110207.
28. Lin, H. C.; Sheu, S. Y.; Sheen, L. Y.; Sheu, P. W.; Chiang, W.; Kuo, T. F. The Gastroprotective Effect of the Foxtail Millet and Adlay Processing Product Against Stress-Induced Gastric Mucosal Lesions in Rats. *J. Trad. Complem. Med.* **2020,** *10* (4), 336–344.
29. Liu, J.; Tang, X.; Zhang, Y.; Zhao, W. Determination of the Volatile Composition in Brown Millet, Milled Millet and Millet Bran by Gas Chromatography/Mass Spectrometry. *Molecules* **2012,** *17* (3), 2271–2282.
30. Magala-Nyago, C. S. *Development of an Energy Dense Finger Millet Based Complementary Food Free of Antinutritional Factors*, 2010; p 2. http://dspace.mak.ac.ug/handle/10570/854 (accessed on Nov 30, 2021).
31. Manjula, K.; Bhagath, Y. B.; Nagalakshmi, K. Effect of Radiation Processing on Bioactive Components of Finger Millet Flour (*Eleusine coracana* L.). *Int. Food Res. J.* **2015,** *22* (2), 556.
32. Mkonda, M. Y.; He, X. Yields of the Major Food Crops: Implications to Food Security and Policy in Tanzania's Semi-Arid Agro-Ecological Zone. *Sustainability* **2017,** *9* (8), 1490–1498.
33. Mohamed, E. A.; Mohamed Ahmed, I. A.; Yagoub, A. E.; Babiker, E. E. Effects of Radiation Process on Total Protein and Amino Acids Composition of Raw and Processed Pearl Millet Flour During Storage. *Int. J. Food Sci. Technol.* **2010,** *45* (5), 906–912.

34. Mohamed, E. A.; Ahmed, I. A.; Babiker, E. E. Preservation of Millet Flour by Refrigeration: Changes in Total Protein and Amino Acids Composition During Storage. *Int. J.Nutr. Food Eng.* **2011,** *5* (4), 346–349.
35. Mallasy, L. O.; El Tinay, A. H.; Elkhalifa, A. E.; Ali, M. A.; Mohiedeen, I. E.; Yousif, N. E. Effect of Fermentation on Biochemical and Sensory Characteristics of Millet Flour Supplemented with Whey Protein. *Int. J. Food Sci. Technol.* **2010,** *45* (11), 2387–2393.
36. Murthy, B. S.; Hiremath, J. P.; Darshan, G. B. Malted Foxtail Millet (*Setaria italica*) and Wheat Flours in Formulation of Complementary Food. *Int. J. Sci. Environ.* **2016,** *5* (2), 708–716.
37. Manivel, K. Nutritional and Sensory Evaluation of *Moringa oleifera* Cookies. *Indian J. Public Health Res. Dev.* **2019,** *10*, 74–79.
38. O'Kennedy, M. M.; Crampton, B. G.; Ozias-Akins, P. Pearl Millet. *Compendium Transgenic Crop Plants* **2009,** *15*, 177–190.
39. Olaitan, N. I.; Eke, M. O.; Uja, E. M.; Oja, E. Quality Evaluation of Complementary Food Formulated from Moringa Oleifera Leaf Powder and Pearl Millet (Pennisetum Glaucum) Flour. *Int. J. Eng. Sci.* **2014,** *3* (11), 59–63.
40. Omenna, E.; Olanipekun, O.; Ogunwale, F. Nutritional and Sensory Properties of Co-Fermented Maize, Millet and Sorghum/ Soybean *pap-(ogi)*. *MOJ Food Process. Technol.* **2018,** *6* (1), 159–164.
41. Orr, A.; Mwema, C.; Gierend, A.; Nedumara, S. (Eds.) *Sorghum and Millets in Eastern and Southern Africa. Facts, Trends and Outlook*; ICRISAT Research Program, Markets, Institutions and Policies Working Paper 62; Patancheru, Hyderabad: International Crops Research Institute for the Semi-Arid Tropics (ICRISAT), 2016; p 12.
42. Patil, K. B.; Chimmad, B. V.; Itagi, S. Glycemic Index and Quality Evaluation of Little Millet (*Panicum miliare*) Flakes with Enhanced Shelf-Life. *J. Food Sci. Technol.* **2015,** *52* (9), 6078–6082.
43. Panghal, A.; Khatkar, B. S.; Yadav, D. N.; Chhikara, N. Effect of Finger Millet on Nutritional, Rheological, and Pasting Profile of Whole Wheat Flat Bread (*chapatti*). *Cereal Chem.* **2019,** *96* (1), 86–94.
44. Padulosi, S.; Mal, B.; King, O. I.; Gotor, E. Minor Millets as a Central Element for Sustainably Enhanced Incomes, Empowerment, and Nutrition in Rural India. *Sustainability* **2015,** *7* (7), 8904–8933.
45. Pradeep, S. R.; Guha, M. Effect of Processing Methods on the Nutraceutical and Antioxidant Properties of Little Millet (*Panicum sumatrense*) Extract. *Food Chem.* **2011,** *126* (4), 1643–1647.
46. Preedy, V. R., Ed. *Processing and Impact on Active Components in Food*; Elsevier: London, 2015; p 689.
47. Ramashia, S. E.; Anyasi, T. A.; Gwata, E. T.; Meddows-Taylor, S.; Jideani, A. I. Processing, Nutritional Composition and Health Benefits of Finger Millet in Sub-Saharan Africa. *Food Sci. Technol.* **2019,** *39*, 253–266.
48. Ransumithila, C.; Saravanakumar, R. Development of Value-Added Millet Based Nutritious Instant Dhokla Mix. *Int. J. Chem. Stud.* **2019,** *7* (3), 4878–4882.
49. Rasane P.; Jha, A.; Sabikhi, L.; Kumar, A.; Unnikrishnan, V. S. Nutritional Advantages of Oats and Opportunities for Its Processing As Value Added Foods—A Review. *J. Food Sci. Technol.* **2015,** *52* (2), 662–675.
50. Rajiv, J.; Soumya, C.; Indrani, D.; Venkateswara Rao, G. Effect of Replacement of Wheat Flour with Finger Millet Flour (*Eleusine coracana*) on the Batter Microscopy, Rheology and Quality Characteristics of Muffins. *J. Texture Stud.* **2011,** *42* (6), 478–489.

51. Rathore, S.; Singh, K.; Kumar, V. Millet Grain Processing, Utilization and Its Role in Health Promotion: A Review. *Int. J. Nutr. Food Sci.* **2016,** *5* (5), 318–329.
52. Sarker, A. *The Effect of Pre-processing on the Nutritive, Physical, and Sensory Properties of Proso Millet*; Doctoral dissertation, University of Guelph, Guelph, ON, Canada, 2015; p 122.
53. Simwaka, J. E.; Huiming, Z.; Masamba, K. G. Amino Acid Profile, Mineral, Pasting, Thermal and Protein Solubility Characteristics of Sorghum-Finger Millet Based Complementary Food as Affected by Fermentation. *J. Acad. Ind. Res. (JAIR)* **2015,** *3* (10), 504–510.
54. Singh, P.; Raghuvanshi, R. S. Finger Millet for Food and Nutritional Security. *Afr. J. Food Sci.* **2012,** *6* (4), 77–84.
55. Sarita, E. S.; Singh, E. Potential of Millets: Nutrients Composition and Health Benefits. *J. Sci. Innov. Res.* **2016,** 5 (2), 46–50.
56. Songré-Ouattara, L. T.; Mouquet-Rivier, C.; Humblot, C.; Rochette, I.; Diawara, B.; Guyot, J. P. Ability of Selected Lactic Acid Bacteria to Ferment a Pearl Millet–Soybean Slurry to Produce Gruels for Complementary Foods for Young Children. *J. Food Sci.* **2010,** *75* (5), M261–M269.
57. Shikuku, K. M.; Okello, J. J.; Wambugu, S.; Sindi, K.; Low, J. W.; McEwan, M. Nutrition and Food Security Impacts of Quality Seeds of Biofortified Orange-Fleshed Sweetpotato: Quasi-Experimental Evidence from Tanzania. *World Dev.* **2019,** *124*, 104–112.
58. Sihag, M. K.; Sharma, V.; Goyal, A.; Arora, S.; Singh, A. K.; Lal, D. Development of an Alternative Low-Cost Cereal-Based Weaning Food Fortified with Iron and Vitamin A (Retinol Acetate). *Indian J. Anim. Sci.* **2016,** *86* (4), 478–484.
59. Singma, J. J.; Suneetha, W. J.; Kumari, B. A.; Devi, K. S. Phytochemical Screening for Antioxidant Properties of Germinated *Foxtail millet*. *Curr. J. Appl. Sci. Technol.* **2019,** *38*, 1–6.
60. Suresha, K. B.; Jayaprakasha, H. M.; Ramya, K. G. Formulation of Ready-to-Cook (RTC) Malted Little Millet Based Complementary Health Food. *Food Sci. Res. J.* **2018,** *9* (2), 304–310.
61. Suleh, A. S.; Zhang, Q.; Chen, J.; Shen, Q. Millet Grains: Nutritional Quality, Processing, and Potential Health Benefits. *Compreh. Rev. Food Sci. Food Saf.* **2013,** *12* (3), 281–295.
62. Talukder, S.; Sharma, B. D. Scope of Millet Grains as an Extender in Meat Products. *Crit. Rev. Food Sci. Nutr.* **2015,** 55 (6), 735–739.
63. Team SOFA; Doss, C. The Role of Women in Agriculture: ESA-FAO Working Paper 11–02. Agricultural Development Economics Division; Rome: The Food and Agriculture Organization of the United Nations (FAO), 2011; p 29.
64. Thapliyal, V.; Singh, K. Finger Millet: Potential Millet for Food Security and Power House of Nutrients. *Int. Res. Agric. For.* **2015,** 2 (2), 45–54.
65. Ullah, A.; Ahmad, A.; Khaliq, T.; Akhtar, J. Recognizing Production Options for Pearl Millet in Pakistan Under Changing Climate Scenarios. *J. Integr. Agric.* **2017,** *16* (4), 762–773.
66. Vijayalakshmi, D.; Geetha, K.; Gowda, J.; Ravi, S. B.; Padulosi, S.; Mal, B. Empowerment of Women Farmers Through Value Addition on Minor Millets Genetic Resources: A Case Study in Karnataka. *Indian J. Plant Genet. Resour.* **2010,** *23* (1), 132–135.
67. Yadav, D. N.; Kaur, J.; Anand, T.; Singh, A. K. Storage Stability and Pasting Properties of Hydrothermally Treated Pearl Millet Flour. *Int. J. Food Sci. Technol.* **2012,** *47* (12), 2532–2537.

CHAPTER 2

NEW HORIZONS FOR MILLETS: FOOD AND NUTRITION SECURITY

DAMIAN LARYEA, SARAH THEODORA ESI MILLS,
CHARLES CUDJOE DAPULIGA, and JOHN-LEWIS ZINIA ZAUKUU

ABSTRACT

Food and nutrient insecurity are threats to human survival, and in the wake of climatic modifications, water scarceness, increasing world population, and rising food prices, there exists an urgent need to broaden the food basket by making use of underutilized crops. Millet is an underutilized cereal in the semi-arid tropics. It is mostly a staple crop with a major protein and energy source for approximately 130 million individuals in Africa's sub-Saharan regions. It also contains folic acid, Vitamins A, B1, B2, and B5. Pearl-based composite flour can be developed to provide an alternative to 100% wheat flour for bread. Despite these benefits, millet has not yet received much attention in research studies to improve it as a fortified crop and broaden its use in the human diet. The lack of attention has also resulted in dwindling production levels over the years. Millet is still generally regarded as "the poor man's food" but has a very great potential in food and nutrient security. This chapter focuses on the potential of millet as a food and nutrient security crop, crop origin and varieties, production, and nutritional and functional benefits of millet.

2.1 INTRODUCTION

Food and nutrient insecurity are a threat to human survival, and in the wake of climatic modifications, water scarcity, increasing world population, and

Sustainable and Functional Foods from Plants: Health Impact, Bioactive Compounds, and Production Technologies. Megh R. Goyal, Arijit Nath, and Zoltán Kovács (Eds.)
© 2024 Apple Academic Press, Inc. Co-published with CRC Press (Taylor & Francis)

rising food prices, there exists the need to broaden the food basket by making use of underutilized crops. Millet is an underutilized cereal that is grown in the semi-arid tropics. It refers to a number of annual growing grasses that are grown throughout the world. At least 18 species of millets are recorded to be domesticated and cultivated. Those of commercial importance are the barnyard (*Echinochloa frumentacea*), proso (*Panicum miliaceum*), finger (*Eleusine coracana*), pearl (*Pennisetum* glaucum), and foxtail (*Setaria italica*) millet varieties, with pearl being identified as the most extensively cultivated.[69] Africa has more native cereals than any other continent. The oldest West African cereal, fonio millet, has over the years been cultivated only in Africa specifically in Guinea and Nigeria, and Tef millet, which is confined to Ethiopia as one of the most important crops. A clear indication of the many other varieties not exported or grown worldwide.

Millets have received some attention for some time now mostly because they have been touted as a major food security crop for their affordability, availability, accessibility, and nutritional value. These are generally well adapted to an extended scope of ecologically different environments and can be cultivated amidst extremely harsh conditions of acidic and infertile soils, low and erratic precipitation, and high temperatures (millets are generally thermophilic and xerophilic). In addition to their ability to adapt to harsh conditions, they are very nutritious. They contain substantial amounts of protein and very high amounts of fiber to the extent they are referred to as fiber-rich crops.

Millets also contain substantial amounts of B-vitamins (riboflavin, niacin, and thiamine), vitamin A, iron, zinc, and other macrominerals (potassium, calcium, and phosphorus). Pearl millet which is the variety mostly consumed has a high amount of vitamin A (22–24 RE) compared with other cereals. Millets are gluten-free, contain phenolic acids, and have been known to have antimicrobial qualities and an ability to manage cardiovascular diseases. They also have good functional properties for use in food products such that pearl-based composite flour has been developed to provide an alternative to 100% wheat flour for bread, which will be good for gluten-intolerant bread consumers.[39] It also has the potential to be used as a meat extender. Due to the nutrient quality of millet and its functional properties, it has been used as an important ingredient in the development of complementary foods. Though there are antinutrients present in millet, there are processing methods that could help reduce them to bearable limits to make the nutrients bioavailable.

In spite of these potentials, millet production has dwindled over the years. In some parts of the world, millet cultivation areas have been used for other cereals, while consumer taste has shifted toward cereals other than millet. Millet has not received much research focus and funding as the other cereal crops despite its great potential. Review papers have looked at the potential of millet, such as the review by Saleh et al.,[95] which examined millet processing, nutrient quality, and health benefits of millet.

This chapter adds to the promotion of millet by reviewing recent research studies on how millet is the crop for today and the future, given the climate changes occurring. It focuses on types of millets, their potential as a food security crop, and the nutritional and functional properties associated with the crop.

2.2 MILLETS: DOMESTICATION AND AVAILABLE VARIETIES

The generic name for several annual growing grasses throughout the world is millet. It is one of the longest existing and domesticated crops worldwide mostly cultivated in dry and semi-dry regions of Africa and Asia. Its discovery dates back to 10,000 years ago and millets differ in a number of factors including taste, size, length of the growing period, climatic and soil requirements, and grain consistency.[57]

In ancient China, millet was regarded as one of the five sacred crops in addition to barley, sorghum, rice, and wheat. The family *Poaceae* (Gramineae), cultivated for its small edible grains has over 500 plus species including both wild and hybridized species; and the genus *Pennisetum* is said to have about 140 different varieties.[49] At least 18 species of millet are recorded to be domesticated and cultivated. Particularly of commercial importance are the Barnyard (*Echinochloa frumentacea*), proso (*Panicum miliaceum*), finger (*Eleusine coracana*), pearl (*Pennisetum americanum L.*), and foxtail (*Setaria italica*) millet varieties, with pearl being identified as the most extensively cultivated.[59,69,107] Africa has more native cereals than any other continent. The oldest West African cereal, fonio millet, has over the years been cultivated only in Africa specifically in Guinea and Nigeria; and Tef millet, confined to Ethiopia is one of the most important crops. A clear indication of the many other varieties not exported or grown worldwide.

As the most widely cultivated cereal, there are many controversies surrounding the origin of pearl millet. FAO[28] reports that while one school of thought believes that pearl millet originated from Africa and then migrated to India, others believe it is quite the contrary. Studies such as works done by Oumar et al.[67] and Manning et al.[58] rather support the hypothesis that the pearl millet indeed originated from Africa, specifically in the Northern–Central Sahel of West Africa. There are evidentiary facts dating its cultivation as far back as 2500–2000 BC to regions within the Lower Tilemsi Valley of North-Eastern Mali. Mauritania, Northern Niger, and Sudan (Western Butana plains) are some other areas identified with the early domestication of pearl millet.

Pearl millet spread to Sudan years prior to even becoming domesticated in Mali and it is from Sudan that the spread of pearl millet reached India and the other parts of the world.[106] In discovering the dates and types of millet first domesticated in India, two varieties (one small-grained, a pearl millet likely to have migrated from the Sahel regions variety, and the other, a large-grained variety native to India) were identified between 1900 and 1700 BC.[16,17,35,55,89] Rwanda and Ghana are named as part of the countries involved in the early domestication of pearl millet around 1600 BC and 1459 BC, respectively. The majority number of this variety found in many parts of Africa is further evidence that pearl millet is of African origin. Records for its cultivation in Brazil are documented to be in the 1960s and the 1850s for the United States.[59]

Today, it is widely spread across Africa and Asia with Arabia, Russia, Sudan, Egypt, Pakistan, Nigeria, China, and India, being the most important *Pennisetum glaucum* growing countries. Its prevalence and high cultivation rates rank it sixth among worldwide important cereals and third in India.[41,62,90,108] There are about 13 different cultivated species, and categorized into four different varieties—*typhoides*, *nigritarum*, *globosum*, and *leonis*—based on shape. The varieties *nigritarum*, *globosum*, and *leonis* are characterized by having candle-shaped inflorescences and *typhoides* with cylindrical-shaped inflorescences. The caryopses of *typhoides*, *globosum*, and *leonis* have terete cross-sections in addition to their unique shape while that of *nigritarum* is angular. All varieties except *typhoides* are grown within Africa. *Nigritarum* is typically grown across Northern Nigeria to Western Sudan and *globosum* in Benin, Ghana, Togo Niger, and Central Nigeria. Although peculiar to Sierra Leone, the *leonis* variety is found in other countries, such as Mauritania and Senegal. *Typhoides* are the predominant varieties grown in India and are widely distributed throughout Africa.[1,59]

The earliest naturalization of proso (*Panicum miliaceum*) and foxtail millets (*Setaria italica*) is associated with the semiarid regions of East Asia. In China, evidence including the discovery of biomolecular components, phytoliths, and tools for the cultivation of the millets as well as the usage of both millets for noodles was discovered in the Northern part of China at Cishan, about 10,300–8700 years ago.[52] Uncertainties about the first of the two species to be domesticated persist given limited taxonomic identification. For instance, although early discovery attributes the remains found at the Cishan site to foxtail millet based on its very small grain size (a characteristic of foxtail millet), the inability to microscopically observe the ash remains makes it difficult to conclusively declare the remains as foxtail millet or common millet or even both. This is partly also due to limited modern studies on the distinguishable features of foxtail and common millet.[53,94] It remains however that proso millet was East Asia's earliest dry farming cereal given its exceptional drought-resisting capacity.

Proso millet is divided into five varieties based on the shape and the morphology of its panicle. The varieties are *miliaceum*, *patentissimum*, *contractum*, *compactum*, and *ovatum*.[51,76] Further characterization by Vetriventhan et al.[103] revealed the existence of vast differences among *miliaceum* and *ovatum* varieties and a minimal difference between *compactum* and *ovatum*. While the *ovatum* variety is on average generally short plants with small panicles, the varieties *patentissimum* and *miliaceum* are averagely tall and quite difficult to distinguish from each other. The varieties *contractum*, *compactum*, and *ovatum* are characterized by the presence of compacted inflorescences. The difference is that while the species in *compactum* have erect cylindrical inflorescences, the species in *ovatum* have ovate and slightly curved inflorescences and the species of *contractum* have drooping inflorescences.

Among the millets, the foxtail variety happens to be the oldest domesticated and cultivated throughout Eurasia. It originated in Eastern Asia from China, and slowly spread westward toward Europe and then the U.S. The spread of foxtail millet from Northern China to East Siberia was around 3620–3370 BC, Korea around 3360 BC, Southern China around 2700–3000 BC, Taiwan around 2800 BC and the continuity to Europe and other parts of the world.[36,103] Although foxtail millet has over the years become a minor crop in comparison to the rapid growth and development of other modern crops including maize, it is still cultivated extensively as food and/or forage in many parts of the world serving as one of the most common and important crops even as a staple diet among the innate of South India.[22]

In Africa, foxtail and proso millets are not of much importance compared with the other millet species. Their cultivation is specifically limited to a few countries, such as Kenya and other upland areas in Eastern Africa. It is the third most cultivated millet in the world.

Three basic varieties and 10 sub-varieties are associated with foxtail millet based on inflorescence, plant height, and geographical distribution. The sub-varieties *Aristata, fusiformis*, and *glabra* belong to the variety *moharia* found in Afghanistan, Europe, Pakistan, and Southeast Russia; *erecta, glabra, nana*, and *profuse* in the variety *indica* are grown in certain South Asian areas, such as Sri Lanka and India; and c*ompacta, spongiosa*, and *assamense* in the variety *maxima* are well-known in Japan, Northern India, Eastern China, Nepal, Korea, and the USA. A fourth variety, *nana,* is available and refers to very short and slender poor-yielding wild green millet. The cultivars from India are typically more robust when compared with those found in Europe and the Far East.[21,101,105]

The domestication of finger millet is specific to the highland regions of Ethiopia and Uganda. This dates back to some 5000 years ago, and archeological remains found and examined show resemblance to the present-day finger millets which are still grown in Ethiopia.[98,101] Spreading toward Asia from Central and Eastern Africa, its cultivation reached India roughly 2000–3000 years ago, and further South-East Asia, China, Japan, and the Eastern and Southern African savanna. Today, it is widely cultivated across Africa in Eritrea, Kenya, Namibia, Niger, Northern Nigeria, Madagascar, Malawi, Mozambique, Senegal, and Zimbabwe as well as other countries, including Guam, Indonesia, Taiwan Sri Lanka, and the Himalayas. India holds the record for the second-largest diversification for finger millet.[33]

There are more than 25 countries in Africa and Asia associated with the cultivation of finger millet. They include Senegal, China, Nigeria, India, Kenya, Iran, Niger, Afghanistan, Mozambique, Sudan, Japan, Ethiopia, Zambia, Malaysia, Namibia, Burundi, Sri Lanka, Malawi, Madagascar, Nepal, Rwanda, Zimbabwe, and Eritrea. Four main varieties—*Compacta, Elongata*, *Plana,* and *Vulgaris*—are associated with finger millet based on the shape and compactness of its inflorescence. According to findings reported by Sood et al.,[88] all these varieties have evolved from the variety *coracana* through domestication under similar environmental conditions. There are three sub-varieties in the variety *Plana*, *confundere*, *grandigluma*, and *seriata* as there are in varieties *Elongata*, *laxa*, *reclusa,* and *sparsa*. In the variety *Vulgaris*, there are four sub-varieties, *digitate*, *incurvate*, *liliacea*, and *stellate* but no sub-variety has been found in *Compacta.*[33,88,102] Further classification can be obtained based on factors such as grain color (variation between dark brown and white), growth habit (either prostrate, erect, or decumbent), plant pigmentation, and time to flower.

The domestication of Barnyard millet is linked to Japan some 4000–5000 years ago with recent studies uncovering its origin of domestication to its wild growing grass, *Echinochloa crus-galli,* located in the tropics of Asia.[64] Despite its age-long period of domestication in Asia's and Africa's semi-dry regions, its cultivation is questionable as only 2 out of the 35 plus available species are majorly cultivated even as minor crops.[86,109]

The Japanese barnyard variety is cultivated in Japan, Korea and the north-eastern parts of China, and the Indian Barnyard millet is cultivated in India, Pakistan, and Nepal. Its cultivation is spread across Central Asia and to some extent in Africa, Europe, and the USA. Barnyard millet together with proso, little, and kodo millets are collectively called small or minor millets although they differ greatly in their detailed grain anatomy and other physical and chemical properties.[78,85] Little millets (*Panicum sumatrense*) and kodo millets (*Paspalum scrobiculatum*) are both believed to have been domesticated in South Asia about 5000 and 3700 years ago, respectively.[96]

Little millets have become native to India and hence its name Indian millet and continue to be produced elsewhere in China, East Asia, Malaysia, and the Caucasus.[46] The least known domesticated cereal in the world is the Guinea millet (*Brachiaria deflexa*) cultivated in and around the environs of Sierra Leone, and certain remote regions of Guinea as well as the highlands of Fouta Djallon. It is a semi-domesticated crop in the African savanna located in and around Cape Verde, Ethiopia, Eritrea, Senegal, Somalia, South Africa, western Asia, Pakistan, and India.[12]

2.3 MILLET AS A FOOD SECURITY CROP

Millets no longer remain just a subsistence product cultivated as a famine crop mainly to feed the poor and underprivileged in society. In several dry, semi-dry, and tropical zones, millets, having exceptional nutritional and nutraceutical properties, serve as an important staple food and fodder to animals. Millets, in recent times, have been touted as major food security crops for their affordability, availability, accessibility, and nutritional values. They are generally well adapted to an extended scope of ecologically different environments and can be cultivated amidst extremely harsh conditions of acidic and infertile soils, low and erratic precipitation, and high temperatures (Millets are generally thermophilic and xerophilic). Their strong rooting systems and ability to grow in skeletal soils of no more than 15 cm, as well as their short growing seasons, endows millets with the privilege of adapting and growing better in dry infertile soils with poor water-holding capacity than most other crops like rice which barely grows in such soils. Millets such as the pearl and finger varieties grow rather well in sandy and saline soils.[2]

The soils in West Africa are rather sandy and slightly acidic, yet the latter produces approximately 74% of the millets grown in Africa, even with very little rainfall of less than 50 cm. It demands very little water for production with a minimum water requirement as low as 30 cm compared with that of maize which requires rainfall around 50–60 cm and sorghum at 40 cm.[70] Its rainfall requirement is also 30% less than that of rice.[38] Specifically, pearl millets are well suited to efficiently utilize moisture compared with maize or sorghum and have a low annual rainfall of about 20–50 cm.[2] In general, the yields from rice are higher than other grains but poorer under rain-fed conditions where coarse grains such as pearl millet thrive better; and although their yields vary yearly, especially in droughts, the decline is minimal compared with larger declines recorded with rice.[80] It is also a proven fact that the water requirement of millet is 2.5 times lesser than that of rice.

With their short growth cycle and wide range of adaptability to survive in drought and dry land areas, they can be well-fitted in any of the multiple cropping systems even without the need for synthetic fertilizers. This helps to reduce the burden of the government to subsidize prices and rather invest in other forms of research aimed at promoting millet consumption and awareness. The fastest growing of all millets is Barnyard, maturing over 42–45 days, followed by proso maturing within 60–90 days, then pearl millet and finger millet taking about 80–100 and 100–130 days, respectively to mature. Foxtail has a broad maturity range between 70 and 120 days.[7] Regarded as eco-friendly crops, they barely require the use of pesticides as many of the cereals hardly attract pests (both during production and storage). It is of most importance to know that these characteristic qualities enable millets to survive, flourish, and withstand harsh conditions, place millets in the top upper category of climatic adaptation crops.[82,87]

Although climate change will have dramatic effects on the sustainability of our food systems affecting crop yields and productivity, these numerous challenges projected to accompany climate change, such as reduced rainfall, increased temperatures, and alterations in soil nutrients are less likely to affect the growth and production systems of millet as they presently have minimal requirements. It has also become very important to find means of reducing greenhouse gas emissions to control the drastic global temperature change as a means of helping to achieve food security. Even though the agricultural sector, as a whole, is a major contributor to the release of greenhouse gases, millets by their carbon sequestration can act as climate change mitigation tool against greenhouse gas emissions.[87] It is thus the ideal crop not just for situations with high incidences of famine but a food security and climate-smart crop and must be treated as such.

Aside from the issue of climate change that poses a threat to food security, the world is faced with the burden of having to nourish a rapidly growing populace estimated to hit 9.5 billion in about 30 years. Current estimates show that one in every nine people in the world suffers from hunger totaling a little over 820 million individuals where the majority of those affected are from developing countries.[31] This accentuates the ever-rising need for natively cultivated food and food items particularly in regions like Africa with important millet-producing lands.

In Africa, the issue is further compounded by other issues such as the poor level of mechanization which translates to a high cost of production, and the lack of financial support from internal and external donors that tend to decline the rate of numerous research. Statistical data showing the prevalence of under-nourishment in Africa in 2018 reveals a slight increase of 0.8% at 19.9% from that of 2010 at 19.1% while that of the world has seen a 1% decrease from 11.8% in 2010 to 10.8% in 2018.[31] This clearly indicates that although some progress is being made in the right direction worldwide, individual continents like Africa need a much more strategized action plan to help juxtapose issues of food insecurities. One of the possible solutions lies in giving priority to millet as a food security crop and thereby intensifying its cultivation and production.

Besides domesticated millets, certain varieties of wild millets have potential of helping to ensure food security. Although wild, some individuals and communities especially in the rural areas harvest and consume them as food, especially during periods of famine and drought crisis. This course of action by itself is a well-defined course of action in line with food security even at local levels. This presents an opportunity which when capitalized upon, will enhance the modification, adaptation, and variegation of millet through the use of genes from these undomesticated species instead of continuing to ignore and neglect them. Farmers in such communities are thus custodians of very crucial knowledge about wild millets, establishing a course that interchanges biological methods with native ideas which can bring tremendous contributions to agricultural development and livelihood enhancement. Wild millets, therefore, represent fundamental genetic resources that warrant some form of careful deliberation and incorporation in the conservation of millet biodiversity and rural development programs, for the reason that they have a role (the extent to which has not been discovered yet) to play in both food security and agricultural development.

The millet is not a strange crop to the African continent. Its usage as a primary food has found relevance in a wide range of simple to complex products by directly using the grains or milling into flour before its usage either alone or mixed with other flours to form the composite. They are not just used for the preparation of porridges, bread, steamed cooked products, and snacks but also in the beverage industry for making alcoholic and non-alcoholic drinks and in the livestock industry as feed.[28] In Kenya for instance, finger millet is in popular demand as processors turn to neighboring countries for supplies because of its discovery as an ideal food for infants and pregnant and nursing mothers owing to its high levels of iron, fiber, and calcium; and it is used as a therapeutic feed for diabetics and gluten intolerant people.[54]

There are many more advantages of millet to be maximized such as the use of pearl millet as a better composite feed for livestock compared with maize, given its higher protein content and even more to be discovered. The issues of rapid population growth and climate change, although threatening, undoubtingly provide the "perfect" opportunity to truly appreciate millets as the "miracle grains" they are so-called. Given the right policies, technologies, investments, research studies, and institutions, millet can be well capitalized upon as a food security crop associated with numerous benefits and opportunities.

Some pragmatic approaches as outlined by ICRISAT and NARS include reviewing relevant agricultural policies, raising awareness of new technologies used in millet production, improving seed production and delivery systems, sustaining and strengthening the development of new, high-yielding and well-adapted millet cultivars, and restoring degraded soils. So, upgrading, enhancing, and advancing agronomic practices is a must if millets are to gain their due significance in Africa.

Despite these distinctive characteristics of millets, most of the crop improvement strategies of agricultural programs, and policies are directed toward other staple crops for a variety of reasons, including governmental policy initiatives which favor the growth and production of other cereals like maize associated with greater risks of crop failure.[38] This has led to a drop in production in some regions suitable for millet cultivation, resulting in an overall worldwide decline in the consumption of millet over the years. For instance, dry or moist semi-arid lands of South Africa, which are ideal for millet production rather opt to plant maize as their preferred food choice and so does Malawi where finger millet was once widely grown as insurance against the failure of maize. As of 2010, statistics showed that 13% of Malawians had stopped growing millet owing to the drudgery involved, leaving only a small fraction of about 9% to grow finger millet.[70]

The story is the same in Zimbabwe and other countries such as the Southern part of India where declines in millet cultivation have been recorded and attributed to taste preferences for maize, regulatory requirements restricting the trade of principal foods to maize, the less manual work involved in milling maize, its high yield and the suitability of cultivating on heavier soils.[32,91] Notwithstanding, the promotion and management of millet in terms of its production, availability, and usage are critical to help meet the alimentary needs of millions of people worldwide but most especially in the semi-arid lands throughout sub-Saharan Africa. The minimal input requirements of millet must therefore serve as the basis upon which certain innovations and implementations relating to technological improvements and applications, awareness creation, nutrient management, value addition, integrated pest management, and customary genetic resource management are stressed upon with a focus on empowering small-scale farmers.[49] This will help boost and establish a form of a basis not only for enhancing food security through improved millet diversity but also for motivating farmers to shift to millet cultivation even on better lands and on a larger scale.

Per analysis, one-third of the world's total area is arid lands commonly found in Africa, Australia, and South Asia.[25,72] Forty-six out of the 54 countries in Africa are classified as semi-arid or arid. Some countries to an extent have about 99% of their lands being dry and sub-humid lands. The Republic of Moldova, Iraq, Turkmenistan, Burkina Faso, Kazakhstan, and Botswana are examples of such countries. Today, more than two billion people (one in three persons) rely on these dry and semi-dry lands with more than 1 billion people in some 100 countries threatened by desertification.[24] Water scarcity is highest in dry lands and increases with an increase in aridity. Currently, people living in dry lands have access to 1300 m^3 only, which is 700 cubic meters less than the basic well-being requirement.[81] This availability is projected to decrease in the upcoming years resulting in close to 50% of the entire human race residing in zones of escalated water crises and probably displacing between 24 million and 700 million people. Similarly, at least 24% of land—20% of which is cropland, and 20–25% of rangeland is globally degrading at a very fast pace on account of many factors, directly and/or indirectly, changing over time and varying by location.[77]

The annual loss in grains owing to land desertification and degradation is calculated at 20 million tons resulting in an overall annual loss of US$42 billion and greatly affecting the 1.5 billion estimated individuals depending on these degraded areas.[9,60] Apart from these environmental limitations, there is a lack of knowledge on the part of farmers in choosing foods and agricultural tactics adapted for the management, sustainability, and intensification of productivity in these areas. The obvious effect being crop failures which can be controlled through the usage of drought-resistant crops like millet which has suitable characteristics to withstand such seemingly unsuitable environmental circumstances. A robust argument can, therefore, be made in favor of increasing millet cultivation in these dry lands even for the sake of posterity considering its suitability for the advocacy of food and nutrition security.[26] This is of much importance considering Guinea and fonio millets, West African millets that have barely been given any form of substantial consideration or research even with their potential in fostering food security and the dynamics of agriculture.

2.4 MILLET PRODUCTION ACROSS THE WORLD AND AFRICA

The importance of millet ranks it sixth among worldwide cultivated cereals and third in Africa. However, in terms of global cereal production, trade, and exports, they are hardly recognized, with figures of less than 1%. Records as of 2016 revealed an annual production estimated at 30 million tons.[29] Productions from Africa account for 55% of the world's production and these are mainly produced by three West African countries, Nigeria, Niger, and Mali, which cultivate in 70% of the 59% of global land in Africa. Asian countries especially India are the next major millet producers with 42% global production under a global land area of 38%.[13,15,54,73] In total, developing countries cover 97% of the world's millet production and seed demand. Notwithstanding, the principal and single grower of millet is India although the overall production continues to decrease by the years following the shift to more remunerative crops. Although certain countries like Sudan cultivate large land areas of millet, their yields are relatively low, around 0.3 tons/ha, compared with Nigeria which produces six times more yields even on a smaller land area.[18]

By production in quantities, India ranks first globally followed by Niger, China, Mali, and Nigeria (Fig. 2.1). Pearl millet is the leading millet variety in terms of consumption and production. It is primarily cultivated in Africa and Asia on a total land estimation of at least 32 million hectares with India's contribution to total production estimated at 44% on the largest land area of approximately 8.8 million ha for pearl millet production.[48] Regardless, the typical yields from pearl millet are lesser compared with that of other cereals.

The total annual production of millet was around 28.5 million tons in 1996 and although these were mainly grown and harvested by farmers on very little land sizes primarily for individual purposes and in-house trade, it accounted for about 25–30% of global production.[70] The highest production record for millet was recorded in 2003 at a value of 34,820,407 tons and again in 2008 at roughly around the same value (34,398,213 tons). These values

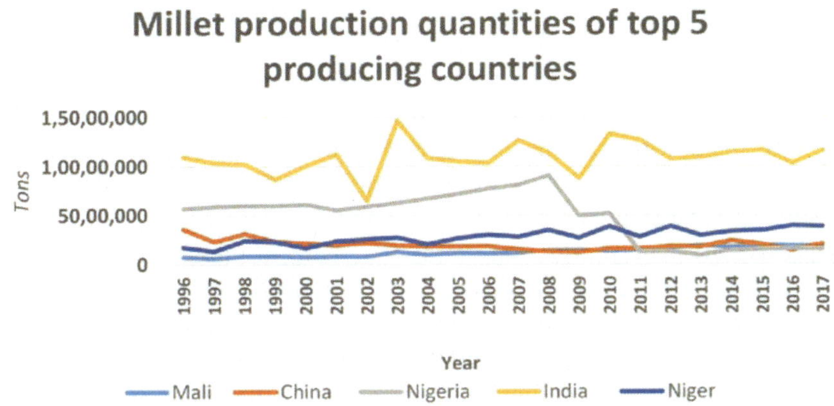

FIGURE 2.1 Millet production quantities of top 5 producing countries; Self-developed with concepts from FAOSTAT.[30]

have since seen fluctuations over the years, and today, millet production quantities as of 2017 remain at 28.4 million tons, as it were some 20 years back (Fig. 2.1). After showing substantial growth from the period of 1996 to 2008, Nigeria's millet production quantities, have drastically declined from 9 million tons to 1.5 million tons in 2017, removing it from the position as the second-largest producer to the fifth with a global share of 5.3%.

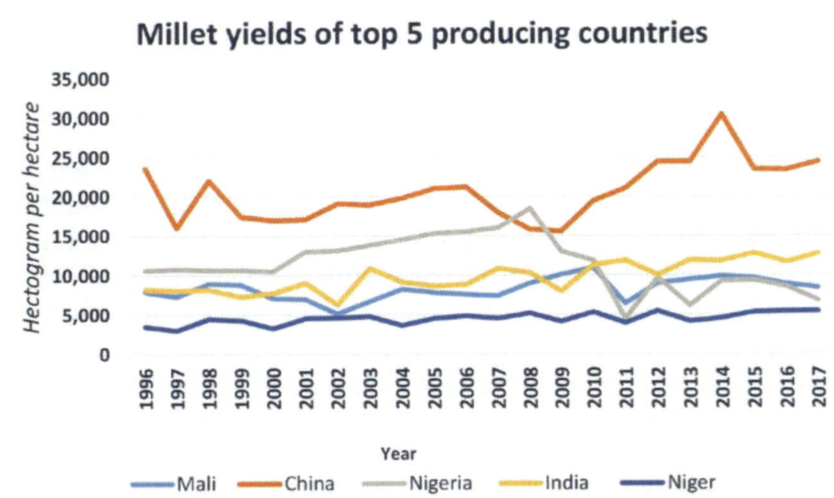

FIGURE 2.2 Millet yields of top five producing countries; Self-developed with concepts from FAOSTAT.[30]

Niger, on the other hand, has rather shown a positive increase in production quantities from 1.3 million tons in 1997 to at least twice that value, 3.7 million tons in 2017, ranking it as the second-largest producer of millet with a global share of 13.3%. China's global share of the millet economy is 7.0%, Mali at 6.3%, and India with the largest share of 40.6%.[30] Interestingly, the area of land used for the harvesting of millet in Nigeria (7.1%) and Mali (6.9%) is at least twice of what is cultivated in China (2.6%) but does not necessarily reflect in production yields. The current yields in China are not only much more than that of Nigeria and Mali but Niger, India, and the rest of the world (Fig. 2.2). This is owing to the use of improved technologies in addition to appropriate seed inputs exhibiting high-yielding and stress-resistant capacities.[54] The millet yield from India in 2013 was 30% more than the global yields resulting mainly from an upgrade in production owing to the widespread usage of improved varieties and techniques. Evidence is seen in India with a record of millet production of 1.2 t/ha and other countries recording yields of around 0.8 t/ha.[97]

Millets are very crucial to the agriculture and economic growth of developing countries where there is extensive usage of land for crop cultivation and those are thus often grown as catch crops in instances where delays in planting are observed and when the yield of other crops fail. Developing countries, such as Africa and India are reported to have the largest global millet production and seed demand of about 97% and a value of over US$ 13 million on the global millets market by 2022.[56] Projections are bound to be characterized by certain macroeconomic and microeconomic factors that will elevate the sales of millets, especially for their high nutritional values as a healthy and fiber-dense food product among customers.[49,74]

The overall growing preference of consumers for healthy yet uncompromised foods in terms of sensorial properties will become a major driver for millet demand for use in the making of innovative products. Several health benefits have been identified and associated with several millet varieties which cannot only serve as a reference point for the manufacturing and production of healthy and nutritious food products but also as functional foods.

The effect of millet has been studied in many capacities against certain diseases and conditions, such as cardiovascular disease, aging, cancers, diabetes, and celiac disease owing to its rich antioxidants and phenolic contents. Of specific importance are the tannins, phenols, and phytates with antioxidant activity sufficient to improve health and metabolic syndrome. In investigating the antimicrobial potential of millet, it was found that seed protein extracts from proso, foxtail, samia, Japanese barnyard, and particularly pearl millet were extremely functional in constraining the development and expressions of certain phytopathogenic fungi.[95]

Also, the phenolic acids in the seed coat of finger millet show antimicrobial activity against *Aspergillus flavus* and *Bacillus cereu*s. Therefore, it can be used in the production of antioxidants and food preservatives considering its positive impact of reducing the mycological contamination associated with grain sprouting and malt characteristics.[92,104] The ills of cardiovascular disease can also be controlled following the consumption of barnyard, finger, and proso millets as a function of their effect in reducing plasma triglycerides, blood glucose, and serum cholesterol as observed in hyperlipidemic rats.[50]

There is an urgent need to develop many of the millet marketing tackling issues related to but not limited to distance barriers between producing areas and their major urban centers as well as scattered and irregular supplies to expand the marginal scope of distribution in producing countries and to enhance international trade based on established quality standards.

2.5 NUTRIENT AND HEALTH BENEFITS OF MILLETS

Millets contain a substantial amount of protein, which is comparable to other cereals, such as rice, maize, and others. Millets are fiber-rich crops and contain a substantial amount of iron, zinc, and vitamins, such as vitamin A and B. Udeh et al.[68] reported that finger millets have about 22% dietary fiber. It also contains phenolic acids, such as hydroxybenzoic acids (e.g., gentisic, vanillic, gallic, syringic, and salicyclic) and hydroxycinnamic acids (e.g., chlorogenic, caffeic, sinapic, and p-coumaric). There have also been reports that millets contain flavonoids, such as quercetin, catechin, gallocatechin, epicatechin, myricetin, and isovitexin.

Finger millet (varieties: *Ravi*, *Rawana*, and *Oshadha*) contains approximately 13% of total dietary fiber with the majority of them (approximately 12%) being insoluble dietary fibers. They also contain β-glucans and fructans. Finger millet also contains approximately 8% of protein, 3% crude fiber, and 3% ash. Magnesium, potassium, calcium, iron, zinc, and phosphorus have also been reported in finger millet in a substantial amount. The protein and fat contents in finger millets are lower compared with pearl millet, sorghum, corn, wheat, and red rice. However, the ash and crude fiber contents are higher in finger millet than in the other cereal crops. Regarding the mineral contents, potassium, calcium, and phosphorus contents of the finger millets were found to be much more comparable to the other cereals (pearl millet, sorghum, corn, wheat, and red rice).[3]

The protein content of pearl millet in Benin ranges between 1.86 and 93.4 mg/g; calcium, between 31.77 and 728.71 mg/kg; magnesium between 340.27 and 4769.9 mg/kg; potassium between 3022 and 4401 mg/kg, iron between 13.85 and 2766.31 mg/kg and zinc between 43.34 and 809.38 mg/kg.[44]

Singh[83] compared the nutrient composition of four varieties of millets (such as pearl, finger, foxtail, and proso millets). It was observed that proso and foxtail millets had relatively higher protein content (12.5 and 12.3%, respectively) than pearl (10.6%) and finger millets (7.3%). Regarding fiber, foxtail millet had the highest content of 8% while pearl millet had the least (1.3%). With respect to minerals, pearl millet had the highest iron content of 16.9 mg/100 g with proso having the least (0.8 mg/100 g) and finger millet having the highest calcium content of 344 mg/100 g with proso millet having the least (14 mg/100 g).

Tables 2.1 and 2.2 indicate the nutrients present in the different types of millet, namely, barnyard, finger, foxtail, kodo, little, pearl, proso, Tef, and fonio millets. Pearl millets have the highest amounts of protein ranging from 7.30 to 16%, while proso, kodo, and little millets have ash contents ranging from 1.89 to 5.50% (Tables 2.1 and 2.2). Fat content is highest in barnyard millet with quantities ranging from 2.2 to 4.8%. Pearl millet is the most consumed food and has the most iron content of 8–180 mg/100 g which makes it a good ingredient for the development of complementary foods. Zinc could range from 1.4 to 10 mg/100 g in proso millets, from 2.04 to 11 mg/100 g in little millets, and from 2.14 to 9 mg/100 g in foxtail millets (Tables 2.1 and 2.2). The variations are due to the difference in cultivars used for the study and other climacteric conditions, such as soil and ecological zones. Micronutrients are required for the body's metabolism and particularly iron and zinc play very essential roles. Iron is required by the body for hemoglobin synthesis and required by infants for mental and physical well-being. Zinc is required in trace amounts for immune regulatory functions and helps with memory and wound healing.

The amount of nutrients in crops is not enough to describe their nutritional quality. Thus, the bioavailability of nutrients in the crop is a very important attribute. Cereals often have antinutrients, such as tannins and phytates which prevent the absorption of certain nutrients in crops, especially micronutrients. Phytates or phytic acid binds to iron and zinc to prevent their absorption while tannins bind to iron in crops preventing their absorption by the body. It is due to the presence of antinutrients that most of the scientists developing

TABLE 2.1 Nutrient Composition of Different Types of Millets: Self-developed with Data from Refs. [6, 8, 10, 14, 15, 19, 20, 23, 34, 37, 40, 42, 43, 45 ,48, 49, 63, 75, 93, 99, 100].

Type of millets	Proximate			Minerals		Fiber		Vitamins			
	Protein (%)	Ash (%)	Fat (%)	Iron (mg/100 g)	Zinc (mg/100 g)	Insoluble (g/100 g)	Soluble (g/100 g)	Vitamin A (RE*)	Thiamine (mg/g)	Riboflavin (mg/g)	Niacin (mg/g)
Barnyard	6.20–11.20	4.00–4.50	2.20–4.80	2.29–18.60	1.50–7.50	8.4	4.2	–	0.33	0.10–4.20	4.2
Finger millet	5.00–12.80	2.10–3.70	1.30–2.70	3.60–6.80	2.00–2.30	15.70–19.70	1.40–2.50	8.30–10.50	0.42	0.19–0.89	1.1
Foxtail	11.20–12.30	1.10–4.00	3.30–4.30	2.80–4.70	2.14–9.00	15.44–15.97	11.56–11.91	–	0.59	0.11–0.61	3.2
Kodo	8.30–11.00	2.60–5.00	1.29–4.20	0.50–3.60	0.70–1.50	4.29 ± 0.82	2.11 ± 0.34	–	0.15–0.29	0.09–0.20	1.29–2.00
Little	7.00–7.70	1.89–5.00	3.04–4.70	1.49–23.38	2.04–11.00	5.45 ± 0.48	2.27 ± 0.52	–	0.26–0.30	0.05–0.09	1.27–3.2
Pearl millet	7.30–16.00	1.40 –2.70	2.70–7.10	8–180	2.95–7.90	5	3	22–24	0.33–0.38	0.21–1.24	2.30–2.80

*RE, rectinol equivalent.

TABLE 2.2 Nutrient Composition of Different Types of Millets: Self-developed with data from Refs. [6, 8, 10, 14, 15, 19, 20, 23, 34, 37, 40, 42, 43, 45, 48, 49, 63, 75, 93, 99, 100].

Type of millets	Proximate	Minerals	Fiber	Vitamins	Type of millets	Proximate	Minerals	Fiber	Vitamins	Type of millets	Proximate
	Protein (%)	Ash (%)	Fat (%)	Iron (mg/100 g)	Zinc (mg/100 g)	Insoluble (g/100 g)	Soluble (g/100 g)	Vitamin A (RE*)	Thiamine (mg/g)	Riboflavin (mg/g)	Niacin (mg/g)
Fonio	7.00–9.00	1.00–4.00	0.50–2.00	3.60–13.36	3.00–4.23	–	–	–	0.30–0.48	0.05–0.10	3
Proso millet	12.5	2.00–5.50	1.10–3.10	0.80–2.90	1.40–10.0	–	–	–	0.20–0.41	0.18–4.54	2.30–4.50
Tef	10.0–11.0	2.80–3.10	2.00–4.00	1.14–150	0.96–6.80	–	–	8.00–9.00	0.30–0.39	0.2	2.5

*RE rectinol equivalent.

products from millet add a preprocessing step which is malting.[68] Other processing activities such as extrusion are also known to reduce the amounts of antinutrients in millets thus making their micronutrients bioavailable.

Millets are also regarded as fiber-rich crops with high fiber content. Mostly, all millet types have more insoluble fiber than soluble fiber. The difference could be very high when it comes to finger millet which has insoluble fiber amounts ranging from 15.7 to 19.7 g/100 g and soluble fiber ranging from 1.4 to 2.5 g/100 g (Tables 2.1 and 2.2). Millets with relatively higher fiber contents are finger and foxtail millets. Therefore, when developing a fiber-rich food, it will be good to include finger or foxtail millets for their high fiber content. Foxtail may be ideal since it has high and relatively balanced amounts of both insoluble and soluble fibers. While soluble fiber dissolves in water and is broken down to gel-like substances in the colon, insoluble fibers help food pass through the intestine more quickly. They both help with chronic and cardiovascular diseases. Pearl millet has 22 to 24 RE of vitamin A, the highest amongst all the types of millet. The B-vitamins, thiamine, riboflavin, and niacin are also present in all the millet types (Tables 2.1 and 2.2), making millet a very nutritious crop which could replace the other cereals, given its additional ability to withstand harsh climate conditions.

Consuming millet could help diabetic patients. This is because the major preventive measure or management of diabetics (both types 1 and 2) is ensured through dietary lifestyles. Consuming high-fiber foods (particularly the soluble fibers for type-2 diabetics) and products with more resistant starches or low glycemic index, which is the case for millet, is known to help manage the disease. One other important issue is that micronutrients play a very vital role in managing diabetes of which millet has many.[47] Protein from millet added to the diet has also been reported to increase insulin sensitivities and reduce blood glucose levels and triglyceride levels. This was reported by Nishizawa et al.[65] in their experiment on diabetic rats using different food components.

Seed protein extracts from proso, foxtail, samia, Japanese barnyard, and particularly pearl millet were extremely functional in constraining the development and expressions of certain phytopathogenic fungi,[95] which is indicative of the antimicrobial potential of millet. Phenolic acids isolated from finger millet's seed coat also inhibit the growth of *Aspergillus flavus* and *Bacillus cereus* and as such can be used in the production of antioxidants and food preservatives.[92,104] Barnyard, finger, and proso millets can also reduce the plasma triglycerides, blood glucose, and serum cholesterol as has been observed in rats,[50] and therefore, these may have the potential of managing cardiovascular diseases. The millets can be used to prepare gluten-free foods and food products for consumers with gluten intolerance or celiac diseases.

2.6 FUNCTIONAL AND PASTING PROPERTIES OF MILLETS AND MILLET-BASED FLOURS

Besides the nutrient and health benefits associated with millet, it also has some good functional and pasting properties that make it an important ingredient to use in food formulations or food product development. The qualities of a processed product, especially one with flour ingredients depend on or have a relationship with the individual functional properties of the ingredients.[71] Some existing functional properties include tapped or loose bulk densities, dispersibility, water and oil absorption capacities, water solubility index, etc. Pasting properties, which include but not limited to pasting temperature, breakdown, setback, final viscosities, and peak time also determine the cooking and baking properties of flour ingredients. They also collectively help to identify the cooking properties of flour composites formulated for products, such as bread cookies, complementary foods, and many others.[66]

A number of varieties of millet processed into flour have good functional and pasting properties, making it an important ingredient in food product development. Varieties of pearl millet from India were analyzed for functional properties and were found to have good water and oil absorption capacities and emulsion activity.[84] But the addition of millet to wheat flour has also been found to reduce viscosity and other dough properties of wheat-based composite flour. Therefore, wheat flour substitution can be done up to 30% millet flour, to produce bread of good dough and bread characteristics similar to that of 100% wheat flour.[5]

In a recent study, a pearl millet-based composite flour has been produced with good pasting and functional properties. A formulation containing pearl millet, kidney beans, Tiger-nut, in the ratio 85:10:5, and xanthan gum was found to be an equally better composite flour for bread production, compared with other wheat-based composite flours. This formulation had better functional properties than that with only wheat flour, although the latter had superior pasting properties.[39] Therefore, in the production of bread and other pastry products this formulation could be used to put an end to the use of wheat flour; especially in Africa where wheat flour is imported. It could also be a better alternative for consumers with gluten intolerance.

The use of millet, due to the antinutrients present involves the use of certain preprocessing and processing techniques to reduce antinutrients and enhance nutrient quality. However, this could influence the functional and pasting properties of millet. Therefore, pretreatments may be important in improving nutrient quality or reducing microbial activity, but care must be taken in order not to affect the pasting and functional properties of millet-produced flours. Gamma radiation has been reported to affect some functional properties (except for bulk density and water absorption capacity) and pasting properties of millet.[27] This is, however, variety-dependent and depends on the radiation dose. Radiation doses beyond 4 kGy could have these effects.

Fermentation and malting, which are other existing processing techniques, were found to reduce the bulk density of pearl millet flours but increase the swelling capacity, oil, and water absorption capacity. This was mainly due to changes observed in the microstructure of the millet; thus, a change from the irregular compact granular structure to a regular loose structure. This also affected the biscuit produced with the millet flours.[4] Indicating that knowing the functional properties of millet could help know the kind of product the flour could be used for or how the product will turn up in the end. In this study, the processing techniques used have enhanced the physicochemical properties of the millet flours used. Hence, the preprocessing steps, milling, and fermentation could be employed when making use of millet in food products.

Cooking millet in a 1% lime solution after soaking in a 1% lime solution for 24 h has also been found to improve oil, emulsion, and water absorption capacities, swelling power, and stability.[11] Dietary fiber extracted from foxtail millet bran has also been found to be a potential functional ingredient in food products. The bran of millet had good dietary fiber, and this fiber showed good adsorption capacities to lard, peanut oil, and cholesterol. This property of millet bran could be linked to its ability to reduce plasma LDL-cholesterol, reduce the risk of obesity, coronary, gastrointestinal disorders, and lower the caloric density of foods. It also had very good water-holding capacity and swelling power.[110] This shows the potential of bran from millet or the fiber from millet as a functional ingredient. Therefore, whole millet used in diet could serve as a functional ingredient.

Starch extracted from millet also has good functional and pasting properties compared with the other cereal crops.[61] Starch yield in pearl millet is about 56%, and they have been found to be a good ingredient in the production of cereal-based products because they have good functional properties. Nonetheless, these properties could be improved upon. In a study by Shaikh et al.,[79] the swelling power and solubility index were increased by a process known as succinylation. The peak viscosity was also increased. Therefore, just like any other cereal or root and tuber starches, starches from millet could have their functional and pasting properties enhanced through chemical modification.

2.7 SUMMARY AND FUTURE RESEARCH

There is no doubt that millet is the crop for this generation, especially regarding the degradation of the forests, conversion of forest zones to arid or semi-arid zones, limited water, and many more. Good nutritional and functional evidence exists of millet grains and their use in food products which have significantly impacted the nutritional status of consumers. The population continues to grow, and climate change is already affecting many across the world. Adopting millet will be very helpful as it is not just nutritious but also an environmentally friendly crop, requiring no use of pesticides or soil enrichment. Due to climate change, there is the anticipation of more harsh conditions and the production of greenhouse gases. Millets can help reduce greenhouse gases due to their carbon sequestration abilities. It is indeed a food and nutrient security crop.

Going forward, there are several wild millets which are localized and may need to be domesticated. This could lead to the discovery of cultivars or varieties of some types of millets with other important abilities. Research is ongoing to identify and differentiate amongst varieties of types of millet. To further make millet a "super-crop," the technology of biofortification could be adopted. This will enhance the nutrients in the millet. This technology has already been used in sweet potato breeding programs to obtain vitamin A-rich sweet potato known as the orange flesh sweet potato. Breeding programs targeting millet to come up with highly nutritious, tasty, and easy-to-process varieties will go a long way to bring millet to the forefront. Various organizations including governmental and non-governmental organizations as well as all stakeholders should come together to draft a policy to adopt millet as a food and nutrient security crop. For this process to be easy, more evidence will be needed. These include more research into the nutritional potential of millet such as feeding trials in human subjects, continued research into processing technologies for millet to maintain or improve the bioavailability of nutrients, and millets' potential ability to reduce greenhouse gases. Farmers could also be trained on the good cultural practices associated with the cultivation of millets, such as the selection of good seeds and management of land to ensure production efficiency. More advocacy and education will go a long way to promote millet as a food and nutrient security crop.

KEYWORDS

- **bioavailability**
- **cereal grain**
- **millet**
- **nutrition**
- **production efficiency**
- **soil enrichment**
- **vitamin**

REFERENCES

1. Abdoulkarim, L. S. *Effect of Isolation Distances on Varietal Seed Purity for open Pollinated Varieties of Pearl Millet (Pennisetum Glaucum (L.) R. Br.) at Farmers' Level in Northern Ghana*. M.Sc. Thesis for Northern Region, Ghana: School of Graduate Studies, Department of Horticulture, 2015; p 131.
2. Guigaz, M. *Memento de l'Agronome* (Moments of an Agronomist). Montpellier: CIRAD-GRET and *Ministère des Affaires Étrangères* (Ministry of External Affairs), 2002; p 34.
3. Jayawardana, S. A.; Samarasekera, J. K.; Hettiarachchi, G. H.; Gooneratne, J.; Mazumdar, S. D.; Banerjee, R. Dietary Fibers, Starch Fractions and Nutritional Composition of Finger Millet Varieties Cultivated in Sri Lanka. *J. Food Compos. Anal.* **2019,** *82*, 103249.
4. Adebiyi, J. A.; Obadina, A. O.; Mulaba-Bafubiandi, A. F.; Adebo, O. A.; Kayitesi, E. Effect of Fermentation and Malting on the Microstructure and Selected Physicochemical Properties of Pearl Millet (*Pennisetum glaucum*) Flour and Biscuit. *J. Cereal Sci.* **2016,** *70*, 132–139.
5. Aprodu, I.; Banu, I. Rheological, Thermo-Mechanical, and Baking Properties of Wheat-Millet Flour Blends. *Food Sci. Technol. Int.* **2015,** *21*, 342–353.
6. Arjun, D.; Kumar, R.; Singh, C. Finger Millet for Nutritional Security and Source of Food. *Adv. Food Sci. Technol.* **2014,** *2*, 184–190.
7. Baker. R. D. *Millet Production-Guide A-414*; New Mexico State University, College of Agriculture and Home Economics: Las Cruces, 2003; p 11.
8. Baye, K. *Teff: Nutrient Composition and Health Benefits*; Ethiopian Development Research Institute and International Food Policy Research Institute, 2014; p 11.

9. Bationo, A.; Hartemink, A. E.; Lungo, O.; Naimi, M.; Okoth, P.; Smaling, E. M., Thiombiano, L. *African Soils: Their Productivity and Profitability of Fertilizer Use*; Background Unpublished Paper for the African Fertilizer Summit 9–13 th June 2006; Abuja, Nigeria. IFDC, 2006; p 12.
10. Bangoura, M. L.; Ming, Z. H.; Atindana, J. N.; Xue, Z. K.; Tolno. M. B.; Wei, P. Extraction and Fractionation of Insoluble Fibers from Foxtail Millet (*Setaria italica* L.). *Am. J. Food Technol.* **2011,** *6*, 1034–1044.
11. Adéoti, K.; Kouhoundé. S. H.; Noumavo, P. A.; Baba-Moussa, F.; Toukourou, F. Nutritional Value and Physicochemical Composition of Pearl Millet (*Pennisetum Glaucum*) Produced in Benin. *J. Microbiol. Biotechnol. Food Sci.* **2021,** *6*, 92–96.
12. Awolu, O. O. Optimization of the Functional Characteristics, Pasting and Rheological Properties of Pearl Millet-Based Composite Flour. *Heliyon* **2017,** *3* (2), E-article 00240.
13. Ballogou, V. Y.; Soumanou, M. M.; Toukourou, F.; Hounhouigan, J. D. Structure and Nutritional Composition of Fonio (*Digitaria exilis*) Grains: A Review. *Int. Res. J. Biol. Sci.* **2013,** *2* (1), 73–79.
14. Bender, M. V. Millet Is Gone! Considering the Demise of Agriculture on Kilimanjaro. *Int. J. Afr. Historical Stud.* **2011,** *44* (2), 191–214.
15. Bhagavatula, S.; Rao, P. P.; Basavaraj, G.; Nagaraj, N. (Eds.) *Sorghum and Millet Economies in Asia-Facts, Trends and Outlook*; International Crops Research Institute for the Semi-Arid Tropics (ICRISAT): Andhra Pradesh, 2013; p 81.
16. Brink, M.; Belay, G. *Plant Resources of Tropical Africa 1, Cereals and Pulses*; PROTA Foundation, Wageningen, Netherlands and Backhuys Publishers, Leiden, Netherlands/ CTA, Wageningen, Netherlands, 2019; p 298.
17. Chauhan, M.; Sonawane, S. K.; Arya, S. S. Nutritional and Nutraceutical Properties of Millets: A Review. *Clin. J. Nutr. Dietetics* **2018,** *1* (1), 1–10.
18. Clotault, J.; Thuillet, A. C.; Buiron, M.; De Mita, S.; Couderc, M.; Haussmann, B. I.; Mariac, C.; Vigouroux, Y. Evolutionary History of Pearl Millet (*Pennisetum glaucum* L.) and Selection on Flowering Genes Since Its Domestication. *Mol. Biol. Evol.* **2012,** *29* (4), 1199–1212.
19. Coppens d'Eeckenbrugge, G.; Schiavo, M.; Caron, E.; Ongwen, D.; Ireri Kamau, J.; Rono, B.; Leclerc, C. Worldwide Interconnections of Africa Using Crops as Historical and Cultural Markers. *Les Cahiers d'Afrique de l'Est* (The East African Review) **2019,** *1* (52), 7–41.
20. Dame, Z. T. Analysis of Major and Trace Elements in Teff (*Eragrostis tef*). *J. King Saud Univ. Sci.* **2020,** *32* (1), 145–148.
21. Dayakar Rao, B.; Bhaskarachary, K.; Arlene Christina, G. D.; Sudha Devi, G.; Vilas, A. T.; Tonapi, A. (Eds.) *Nutritional and Health Benefits of Millets*; ICAR—Indian Institute of Millets Research (IIMR): Hyderabad, India; 2017; p 112.
22. Dayakar, R. B.; Sangappa, V. G. D.; Arlene, C. V. A.; Tonapi, B. (Eds.) *Technologies of Millet Value Added Products*; Center of Excellence on Sorghum, ICAR-Indian Institute of Millets Research: Hyderabad, India, 2016; p 48.
23. Deshpande, S. S.; Mohapatra, D.; Tripathi, M. K.; Sadvatha, R. H. Kodo Millet Nutritional Value and Utilization in Indian Foods. *J. Grain Process. Storage* **2015,** *2* (2), 16–23.
24. Diao, X., Jia, G. Origin and Domestication of Foxtail Millet. In: *Genetics and Genomics of Setaria. Plant Genetics and Genomics: Crops and Models*; Doust, A., Diao, X., Eds., Vo. 19; Springer Nature: Berlin, Germany, 2017; pp 61–72.

25. Diao, X., Jia, G. Foxtail Millet Breeding in China. In: *Genetics and Genomics of Setaria. Plant Genetics and Genomics: Crops and Models*; Doust, A., Diao, X., Eds., Vol. 19; Springer Nature: Berlin, Germany, 2017; pp 93–114. https://doi.org/10.1007/978–3-319–45105–3_4 Earth360. *Millets*, 2010. https://www.earth360.in/Millets.html (accessed on Oct 18, 2021).
26. Emmambux, M. N.; Taylor, J. R. Morphology, Physical, Chemical, and Functional Properties of Starches from Cereals, Legumes, and Tubers Cultivated in Africa: A Review. *Starch-Stärke* **2013,** *65* (9–10), 715–729.
27. Falade, K. O.; Kolawole, T. A. Effect of γ-Irradiation on Color, Functional and Physicochemical Properties of Pearl Millet (*Pennisetum glaucum* L.) Cultivars. *Food Bioprocess Technol.* **2013,** *6* (9), 2429–2438.
28. FAOSTAT. *Food and Agricultural Commodities Production Statistics*; Food and Agriculture Organization (FAO): Rome, 2005. http://www.fao.org/faostat/en/
29. FAOSTAT. *Data on Millet Producing Countries*, 2016. http://www.fao.org/faostat/en/#data/QC/visualize (accessed on Nov 20, 2021).
30. FAOSTAT. *Production-Yield Quantities of Millets in World + (Total) 1962–2018*, 2018. http://www.fao.org/faostat/en/#data/QC/visualize (accessed on Jan 10, 2022).
31. Finnis, E. Now It Is an Easy Life: Women's Accounts of Cassava, Millets, and Labor in South India. *Cult. Agric.* **2009,** *31* (2), 88–94.
32. Fuller, D. Q. Finger millet: Origins and Development. In: *The Encyclopedia of Global Archaeology*; Smith, C., Ed.; Springer: New York, 2013; pp 2783–2785.
33. Gebremariam, M. M.; Zarnkow, M.; Becker, T. Teff (*Eragrostis tef*) as a Raw Material for Malting, Brewing and Manufacturing of Gluten-Free Foods and Beverages: Review. *J. Food Sci. Technol.* **2014,** *51* (11), 2881–2895.
34. Giblin, J. D.; Fuller, D. Q. First and Second Millennium Agriculture in Rwanda: Archaeobotanical Finds and Radiocarbon Dates from Seven Sites. *Vegetation Hist. Archaeobot.* **2011,** *20* (4), 253–265.
35. Gomez, M. I.; Gupta, S. C. *Encyclopedia of Food Science, Food Technology and Nutrition*; FAO: Rome, 1993; pp 5–25.
36. Goron, T. L.; Raizada, M. N. Genetic Diversity and Genomic Resources Available for the Small Millet Crops to Accelerate a New Green Revolution. *Front. Plant Sci.* **2015,** *6*, 157–162.
37. Guedes, J. D.; Jiang, M.; He, K.; Wu, X.; Jiang, Z. Site of Baodun Yields Earliest Evidence for the Spread of Rice and Foxtail Millet Agriculture to South-West China. *Antiquity* **2013,** *87* (337), 758–771.
38. Gull, A.; Nayik, G. A.; Prasad. K.; Kumar, P. Nutritional, Technological, and Medical Approach of Finger Millet (*Eleusine coracana*). *Cogent Food Agric.* **2015,** *1* (1), E-article ID: 1090897.
39. Gupta, A.; Singh. R.; Prasad, R.; Tripathi, J.; Verma, S. Nutritional Composition and Polyphenol Content of Food Products Enriched with Millets and Drumstick Leaves. *Int. J. Food Ferment. Technol.* **2017,** *7* (2), 337–342.
40. Habiyaremye, C.; Matanguihan, J. B.; D'Alpoim Guedes, J.; Ganjyal, G. M.; Whiteman, M. R.; Kidwell, K. K.; Murphy, K. M. Proso Millet (*Panicum miliaceum* L.) and Its Potential for Cultivation in the Pacific Northwest, US: Review. *Front. Plant Sci.* **2017,** *7*, 1961–1970.
41. Izge, A. U.; Song, I. M. Pearl Millet Breeding and Production in Nigeria: Problems and Prospects. *J. Environ. Issues Agric. Dev. Countries* **2013,** *5* (2), 25–33.

42. Kajuna, S. T. A. R.; Mejia, D. (Eds.) *Millet: Post-harvest Operations*; Food and Agriculture Organization of the United Nations (FAO): Rome, 2001; p 50.
43. Kalaisekar, A.; Padmaja, P. G.; Bhagwat, V. R.; Patil, J. V., Eds. *Insect Pests of Millets. Systematics, Bionomics, and Management*; Academic Press: London, 2017; p 190.
44. Kalinová, J. Nutritionally Important Components of Proso Millet (*Panicum miliaceum* L.). *Food* **2007,** *1* (1), 91–100.
45. Kam, J.; Puranik, S.; Yadav, R.; Manwaring, H. R.; Pierre, S.; Srivastava, R. K., Yadav, R. S. Dietary Interventions for Type-2 Diabetes: How Millet Comes to Help. *Front. Plant Sci.* **2016,** *7*, 1454–1460.
46. Koutroulis, A. G. Dryland Changes Under Different Levels of Global Warming. *Sci. Total Environ.* **2019,** *655*, 482–511.
47. Krishnan, R.; Meera, M. S. Pearl Millet Minerals: Effects of Processing on Bioaccessibility. *J. Food Sci. Technol.* **2018,** *55* (9), 3362–3372.
48. Kumar, A.; Metwal, M.; Kaur, S.; Gupta, A. K.; Puranik, S.; Singh. S.; Singh, M.; Gupta, S.; Babu, B. K.; Sood, S.; Yadav, R. Nutraceutical Value of Finger Millet (*Eleusine coracana* L.), and Their Improvement Using Omics Approaches. *Front.Plant Sci.* **2016,** *7*, 934–941.
49. Kumar, A.; Tomer, V.; Kaur, A.; Kumar, V.; Gupta. K. Millets: A Solution to Agrarian and Nutritional Challenges. *Agric. Food Secur.* **2018,** *7* (1), 1–5.
50. Kundgol, N. G.; Kasturiba, B.; Math, K. K.; Kamatar, M. Y.; Usha, M. Impact of Decortication on Chemical Composition, Antioxidant Content and Antioxidant Activity of Little Millet. *Int. J. Adv. Res. Technol.* **2013,** *2*, 1705–1720.
51. Lee, S. H.; Chung, I. M.; Cha, Y. S.; Park, Y. Millet Consumption Decreased Serum Concentration of Triglyceride and C-Reactive Protein But Not Oxidative Status in Hyperlipidemic Rats. *Nutr. Res.* **2010,** *30* (4), 290–296.
52. Lü, H.; Li, Y.; Zhang, J.; Yang, X.; Ye, M.; Li, Q.; Wang, C.; Wu, N. Component and Simulation of the 4000-Year-Old Noodles Excavated from the Archaeological Site of Lajia in Qinghai, China. *Chin. Sci. Bull.* **2014,** *59* (35), 5136–5152.
53. Lu, H.; Zhang, J.; Liu, KB.; Wu, N.; Li, Y.; Zhou, K.; Ye, M.; Zhang, T.; Zhang, H.; Yang, X.; Shen, L. Earliest Domestication of Common Millet (*Panicum miliaceum*) in East Asia Extended to 10,000 Years Ago. *Proc. Natl. Acad. Sci.* **2009,** *106* (18), 7367–7372.
54. Macauley, H. *Cereal Crops: Rice, Maize, Millet, Sorghum, Wheat*, 2015. https://ir.ucc.edu.gh/xmlui/bitstream/handle/123456789/4510/Cereal_Crops-_Rice__Maize__Millet__Sorghum__Wheat.pdf?sequence=1&isAllowed=y (accessed on Jan 20, 2022).
55. Manimozhi-Selvi, V.; Nirmalakumari, A.; Senthil, N. Genetic Diversity for Zinc, Calcium and Iron Content of Selected Little Millet Genotypes. *J. Nutr. Food Sci.* **2015,** *5* (417), 10.
56. Manning, K.; Pelling, R.; Higham, T.; Schwenniger, J. L.; Fuller, D. Q. 4500-Year Old Domesticated Pearl Millet (*Pennisetum glaucum*) from the Tilemsi Valley—Mali: New Insights Into an Alternative Cereal Domestication Pathway. *J. Archaeol. Sci.* **2011,** *38* (2), 312–322.
57. Manning, K.; Timpson, A. The Demographic Response to Holocene Climate Change in the Sahara. *Quaternary Sci. Rev.* **2014,** *101*, 28–35.
58. MarketWatch. *Millet Market 2018 Global Industry Size, Demand, Growth Analysis Share, Revenue and Forecast 2023*, 2018. https://www.marketwatch.com/press-release/millet-market-share-2022-global-competitors-strategy-industry-trends-segments-regional-analysis-size-growth-review-key-players-profile-statistics-and-growth-to-2027-analysis-2022–01–25 (accessed on Dec 18, 2021).

59. Mathur, N. P.; Upadhyaya, D. H.; Guarino, L., Eds. *Global Strategy for the* Ex-Situ *Conservation of Pearl Millet and Its Wild Relatives*; Global Crop Diversity Trust: Rome, 2012; p 43.
60. Nagaraj, N.; Basavaraj, G.; Rao, P. P.; Bantilan, C.; Haldar, S. Sorghum and Pearl Millet Economy of India: Future Outlook and Options. *Econ. Polit. Wkly* **2013,** *28*, 74–81.
61. Nishizawa, N.; Togawa, T.; Park, K. O.; Sato, D.; Miyakoshi, Y.; Inagaki, K.; Ohmori, N.; Ito, Y.; Nagasawa, T. Dietary Japanese Millet Protein Ameliorates Plasma Levels of Adiponectin, Glucose, and Lipids in Type 2 Diabetic Mice. *Biosci. Biotechnol. Biochem.* **2009,** *23*, E-article ID: 0901071258.
62. Nkonya, E.; Anderson, W.; Kato, E.; Koo, J.; Mirzabaev, A.; Braun, J. V.; Meyer, S. Global Cost of Land Degradation. In: *Economics of Land Degradation and Improvement—A Global Assessment for Sustainable Development*; Springer Cham: Berlin, Germany, 2016; pp 117–165.
63. Nozawa, S.; Nakai, H.; Sato, Y. *Characterization of Microsatellite and ISSR Polymorphisms Among Echinochloa (L.) Species: Breeding Research (Japan)*; FAO: Rome, 2004; p 23.
64. O'Kennedy, M. M.; Crampton, B. G.; Ozias-Akins, P. Pearl Millet. *Compendium Transgenic Crop Plants* **2009,** *15*, 177–190.
65. Ocheme, O. B.; Oludamilola, O. O.; Gladys, M. E. Effect of Lime Soaking and Cooking (Nixtamalization) on the Proximate, Functional and Some Anti-Nutritional Properties of Millet Flour. *AU J. Technol.* **2010,** *14* (2), 131–138.
66. Okwudili, U. H.; Gyebi, D. K.; Obiefuna, J. A. Finger Millet Bioactive Compounds, Bioaccessibility, and Potential Health Effects: Review. *Czech J. Food Sci.* **2017,** *35* (1), 7–17.
67. Oladunmoye, O. O.; Aworh, O. C.; Maziya-Dixon, B.; Erukainure, O. L.; Elemo, G. N. Chemical and Functional Properties of Cassava Starch, Durum Wheat Semolina Flour, and Their Blends. *Food Sci. Nutr.* **2014,** *2* (2), 132–138.
68. Onyango, A. O. Finger Millet: Food Security Crop in the Arid and Semi-Arid Lands (ASALs) of Kenya. *World Environ.* **2016,** *6* (2), 62–70.
69. Orr, A.; Mwema, C.; Gierend, A.; Nedumaran, S. Sorghum and Millets in Eastern and Southern Africa. *Facts, Trends Outlook* **2016,** *76*, 14–20.
70. Oumar, I.; Mariac, C.; Pham, J. L.; Vigouroux, Y. Phylogeny and Origin of Pearl Millet (*Pennisetum glaucum* L.) as Revealed by Microsatellite Loci. *Theoret. Appl. Genet.* **2008,** *117* (4), 489–497.
71. Ponzio, N. R.; Puppo, M. C.; Ferrero, C. Mixtures of Two Argentinean Wheat Cultivars of Different Quality: A Study on Bread Making Performance. *Cereal Chem.* **2008,** *85* (5), 579–585.
72. Prăvălie, R.; Patriche, C.; Bandoc, G. Quantification of Land Degradation Sensitivity Areas in Southern and Central Southeastern Europe: New Results Based on Improving DISMED Methodology with New Climate Data. *Catena* **2017,** *158*, 309–320.
73. Pushparaj, F. S.; Urooj, A. Impact of Household Processing Methods on the Nutritional Characteristics of Pearl Millet (*Pennisetum typhoideum*): Review. *MOJ Food Process. Technol.* **2017,** *4* (1), 28–32.
74. Rao, P. P.; Basavaraj, G., Eds. *Status and Prospects of Millet Utilization in India and Global Scenario*; Global Consultation on Millets Promotion for Health and Nutritional Security; Directorate of Sorghum Research (DSR): Hyderabad, India, Dec 2013.

75. Reddy, V. G.; Upadhyaya, H. D.; Gowda, C. L. Morphological Characterization of World's Proso Millet Germplasm Collection. *J. SAT Agric. Res.* **2007,** *3*, 4–10.
76. Renganathan, V. G.; Vanniarajan, C. Exploring the Barnyard Millet (*Echinochloa frumentacea*) Segregating Population for Isolation of High Yielding, Iron and Zinc Content Genotype. *Int. J. Curr. Microbiol. Appl. Sci.* **2018,** *7* (4), 3611–3621.
77. Saleh, A. S.; Zhang, Q.; Chen, J.; Shen, Q. Millet Grains: Nutritional Quality, Processing, and Potential Health Benefits. *Compreh. Rev. Food Sci. Food Saf.* **2013,** *12* (3), 281–295.
78. Sarita, E. S.; Singh, E. Potential of Millets: Nutrients Composition and Health Benefits. *J. Sci. Innov. Res.* **2016,** *5* (2), 46–50.
79. Saxena, R.; Vanga, S. K.; Wang, J.; Orsat, V.; Raghavan, V. Millets for Food Security in the Context of Climate Change: Review. *Sustainability* **2018,** *10* (7), 2228–2235.
80. Serba, D. D.; Perumal, R.; Tesso, T. T.; Min, D. Status of Global Pearl Millet Breeding Programs and the Way Forward. *Crop Sci.* **2017,** *57* (6), 2891–2905.
81. Serna-Saldivar, S. O.; Espinosa-Ramírez, J. Grain Structure and Grain Chemical Composition. In: *Sorghum and Millets*; Taylor, J. R. N., Duodu, K. G., Eds., 2nd ed.; AACC International Press: London, 2019; pp 85–129.
82. Shaikh, M.; Ali, T. M.; Hasnain, A. Effects of Different Modification Reagents on Functional Properties of Pearl Millet Starches. *Starch-Stärke* **2017,** *69* (3–4), E-article 1600142.
83. Shanker, A. K.; Shanker, C., Eds. *Abiotic and Biotic Stress in Plants—Recent Advances and Future Perspectives*; InTech Open: Rijeka, Croatia, 2016; p 703.
84. Sharma, D. C. *Here Is How Millets Can Make India's Food Basket Climate-Resilient*, 2019. https://www.downtoearth.org.in/news/agriculture/here-is-how-millets-can-make-india-s-food-basket-climate-resilient-65372 (accessed on Nov 18, 2021).
85. Shetty, S. *Water, Food Security and Agricultural Policy in the Middle East and North Africa Region*; Middle East and North Africa Working Paper Series No. 47; World Bank: Washington, DC, 2006. https://openknowledge.worldbank.org/handle/10986/37139 (accessed on Oct 11, 2021).
86. Singh, M; Upadhyaya, H. D., Eds. *Genetic and Genomic Resources for Grain Cereals Improvement*; Academic Press: San Diego, 2016; p 339.
87. Siroha, A. K.; Sandhu, K. S.; Kaur, M. Physicochemical, Functional and Antioxidant Properties of Flour From Pearl Millet Varieties Grown in India. *J. Food Measurement Character.* **2016,** *10* (2), 311–318.
88. Siwela, M.; Taylor, J. R.; de Milliano, W. A.; Duodu, K. G. Influence of Phenolics in Finger Millet on Grain and Malt Fungal Load, and Malt Quality. *Food Chem.* **2010,** *121* (2), 443–449.
89. Sood, S.; Khulbe, R. K.; Gupta, A. K.; Agrawal, P. K.; Upadhyaya, H. D.; Bhatt, J. C. Barnyard millet: Potential Food and Feed Crop of Future. *Plant Breed.* **2015,** *134* (2), 135–147.
90. Sood, S.; Kumar, A.; Babu, B. K.; Gaur, V. S.; Pandey, D.; Kant, L.; Pattnayak, A. Gene Discovery and Advances in Finger Millet (*Eleusine coracana* L.) Genomics—An Important Nutria-Cereal of Future. *Front. Plant Sci.* **2016,** 7, 1634–1641.
91. Stevens, C. J.; Nixon, S.; Murray, M. A.; Fuller, D. Q. Early Millet Farmers in the Lower Tilemsi Valley, Northeastern Mali. In: *Archaeology of African Plant Use*; Stevens, C. J., Nixon, S., Murray, M. A., Fuller, D. Q., Eds.; Routledge: New York, 2016; pp 73–82.

92. Tadele, Z. Drought Adaptation in Millets. In: *Abiotic and Biotic Stress in Plants—Recent Advances and Future Perspectives*. Shanker, A. K.; Shanker, C. (Eds.); Rijeka, Croatia: InTech Open; **2016**; online; p 639–663.
93. Taylor, J.; Duodu, K. G., Eds. *Sorghum and Millets: Chemistry, Technology, and Nutritional Attributes*, 2nd ed.; Woodhead Publishing: Cambridge, 2018; p 440.
94. Taylor, J. R.; Emmambux, M. N. Gluten-Free Foods and Beverages from Millets. In: *Gluten-Free Cereal Products and Beverages*; Taylor, J. R., Emmambux, M. N., Eds.; Academic Press: San Diego, 2008; pp 119–148.
95. Ugare, R.; Chimmad, B.; Naik, R.; Bharati, P.; Itagi, S. Glycemic Index and Significance of Barnyard Millet (*Echinochloa frumentacae*) in Type-II Diabetics. *J. Food Sci. Technol.* **2014,** *51* (2), 392–395.
96. Upadhyaya, H. D.; Gowda, C. L.; Reddy, V. G. Morphological Diversity in Finger Millet Germplasm Introduced from Southern and Eastern Africa. *J. SAT Agric. Res.* **2007,** *3* (1), 1–3.
97. Upadhyaya, H. D.; Vetriventhan, M.; Deshpande, S. P.; Sivasubramani, S.; Wallace, J. G.; Buckler, E. S.; Hash, C. T.; Ramu, P. Population Genetics and Structure of a Global Foxtail Millet Germplasm Collection. *Plant Genome* **2015,** *8* (3), 7–11.
98. vander, S.; ten Brink, B.; Stehfest, E.; Bakkenes, M.; Sewell, A.; Bouwman, A.; Meijer, J.; Westhoek, H.; van den Berg, M.; van den Born, G. J.; Doelman, J. *Exploring Future Changes in Land Use and Land Condition and the Impacts on Food, Water, Climate Change and Biodiversity: Scenarios for the UNCCD Global Land Outlook*, 2017. https://www.pbl.nl/en/publications/exploring-future-changes-in-land-use (accessed on Dec, 6, 2021).
99. Van Schaik, L.; Dinnissen, R. *Terra Incognita: Land Degradation as Underestimated Threat Amplifier*; Netherlands Institute of International Relations: The Hague, Clingendael, 2014; p 36.
100. Vetriventhan, M.; Azevedo, V. C.; Upadhyaya, H. D.; Naresh, D. Variability in the Global Proso Millet (*Panicum miliaceum* L.) Germplasm Collection Conserved at the ICRISAT Gene Bank. *Agriculture* **2019,** *9* (5), 112–118.
101. Vetriventhan, M.; Upadhyaya, H. D.; Dwivedi, S. L.; Pattanashetti, S. K.; Singh, S. K. Finger and Foxtail Millets. In: *Genetic and Genomic Resources for Grain Cereals improvement*; Vetriventhan, M., Upadhyaya, H. D., Dwivedi, S. L., Pattanashetti, S. K., Singh, S. K., Eds.; Academic Press: London, 2016; pp 291–319.
102. Viswanath, V.; Urooj, A.; Malleshi, N. G. Evaluation of Antioxidant and Antimicrobial Properties of Finger Millet Polyphenols (*Eleusine coracana*). *Food Chem.* **2009,** *114* (1), 340–346.
103. Winchell, F.; Brass, M.; Manzo, A.; Beldados, A.; Perna, V.; Murphy, C.; Stevens, C.; Fuller, D. Q. On the Origins and Dissemination of Domesticated Sorghum and Pearl Millet Across Africa and into India: A View from the Butana Group of the Far Eastern Sahel. *Afr. Archaeol. Rev.* **2018,** *35* (4), 483–505.
104. World Health Organization (WHO). *The State of Food Security and Nutrition in the World—Safeguarding Against Economic Slowdowns and Downturns*; Food & Agriculture Organization (FAO): Rome, 2019; p 15.
105. Wrigley, C.; Corke, H.; Seetharaman, K.; Faubion, J. (Eds.) *Encyclopedia of Food Grains: The World of Food Grains*, 2nd ed.; Academic Press: Oxford, 2016; p 460.
106. Yadav, O. P.; Rai, K. N. Genetic Improvement of Pearl Millet in India. *Agric. Res.* **2013,** *2* (4), 275–292.

107. Yamaguchi, H.; Utano, A. Y.; Yasuda, K.; Yano, A.; Soejima, A. A Molecular Phylogeny of Wild and Cultivated *Echinochloa* in East Asia Inferred from Non-Coding Region Sequences of trnT-L-F. *Weed Biol. Manage.* **2005,** *5* (4), 210–218.
108. Zhao, Z. J. *Domestication of Millet—Paleoethno Botanic Data and Ecological Perspective: Archaeology in China and Sweden.* http://www.sdiarticle3.com/wp-content/uploads/2019/08/Ms_AJAAR_50520.pdf (accessed on Dec 1, 2021).
109. Zhu, Y.; Chu, J.; Lu, Z.; Lv, F.; Bie, X.; Zhang, C.; Zhao, H. Physicochemical and Functional Properties of Dietary Fiber from Foxtail Millet Bran. *J. Cereal Sci.* **2018,** *79*, 456–461.

CHAPTER 3

GRAPES AND WINES AS FUNCTIONAL FOODS: PRODUCTION, QUALITY, CONSUMPTION, AND BIOACTIVE COMPOUNDS

KRISZTINA KELLER, HARSHAVARDHAN REDDY KUMMITHA, NIKOLETTA TÓTH-KASZÁS, and ZOLTÁN BIRKNER

ABSTRACT

Grapes are the fifth largest fruit crop in the world. A total of 77.4 million ha of surface area were cultivated to produce a total of 77.8 million tons of grapes in 2018. Since wine consumption has increased worldwide in recent decades, a successful winery needs to know its consumers. This chapter, therefore, aims to introduce different grape types as well as the qualitative criteria which motivate wine consumption and incite consumers to spend on quality wines. We explore the motivations behind wine consumption and investigate different segmentation opportunities, particularly referring to the trips organized to consume wine. In addition, the quality of the wine was also tested, as there was a lot of criticism from consumers about food quality and food safety due to multiple examples of contaminated foods. Quality assurance systems have been introduced to build and maintain consumer confidence in food quality. This chapter's results discuss that the quality of the wineries' work depends on the following aspects: Physical environment, production practices, and the marketing environment.

Sustainable and Functional Foods from Plants: Health Impact, Bioactive Compounds, and Production Technologies. Megh R. Goyal, Arijit Nath, and Zoltán Kovács (Eds.)
© 2024 Apple Academic Press, Inc. Co-published with CRC Press (Taylor & Francis)

3.1 INTRODUCTION

With an estimated worldwide harvest of 7.78 million tons in 2018, grapes rank fifth behind crops of tomatoes, bananas, watermelon, and apples, respectively (Fig. 3.1). Grapes have already been cultivated in the Middle East 8000 years ago. The European grape species *Vitis vinifera* is the ancestor of many common grape species. The current evidence shows that grapes were introduced in Palestine and Egypt originating from the Caucasus region around 4000 BC. Gradually it spread all over the world by around 1000 BC, the grapes come in many colors and shapes. There are red, green, sultana, purple, and seedless varieties, as well as currants and sultanas from which raisins, grape jam, grape jelly, and grape juice are made, not to mention wine.[79] Although viticulture spread all over the world (Fig. 3.2), most of the world's grapes are still cultivated in the Mediterranean region. Nutrients in the grapes offer numerous health and nutritional benefits, such as prevention of cancer, heart conditions, high blood pressure, and constipation.[12]

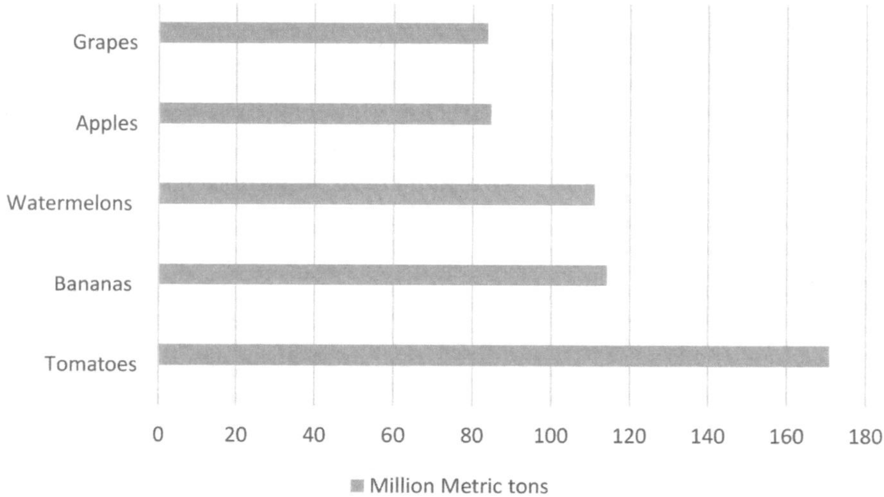

FIGURE 3.1 Worldwide harvest of top five crops in 2018. Source: Self-developed with concepts from Ref. [67].

The main environmental factors that must be taken into account in grape cultivation and viticulture efficiency, since they are essential for the plant's growth, are air pressure, the quality of soil and water, the availability of light, oxygen, heat, and carbon dioxide. Levels of wind force and of waste gasses in the given region are also of importance as their presence can be harmful to

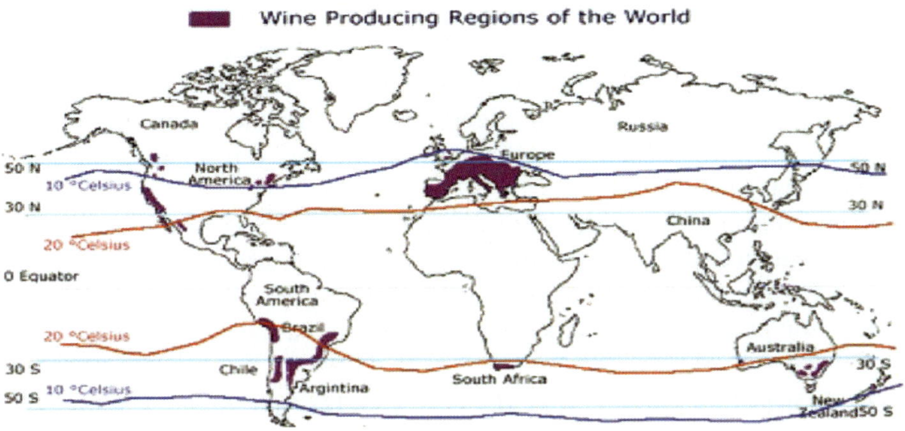

FIGURE 3.2 Wine-producing regions in the world. Source: Self-developed with concepts from Ref. [107].

the grapevine.[33] These elements not only affect the quantity of the producible wine grape but also the quality of prepared wines.

Suitable climatic conditions such as enough sunshine and available water during the plant growing period and adequate cold weather during the organism's life cycle of plant physical is critical stage for quality grape production. Over time, many grape-growing regions have existed specific climatic conditions situated between 30° and 50°C and between 30° and 40°C.[95] Although there is possibility to grow grapes outside of these climatic zones, grape cultivation takes a longer time. Under current climatic conditions, the ideal temperature for viticulture is 12–13°C (lower threshold) and 22–24°C (upper threshold), which allows proper ripening.[30] Water is a crucial factor in grape cultivation. Several factors determine the required water levels for grape cultivation such as climate, heat, soil, and grape species. Under certain climatic regions, there is sufficient rainfall for the plants to thrive, while other climate zones require irrigation.

Figure 3.2 shows that the periods from April to October in the Northern Hemisphere and from October to April in the Southern Hemisphere are the most suitable time periods for grape cultivation. The 10°C lower weather conditions would be suitable for grape harvesting northern hemisphere, whereas the 24°C or upper weather conditions would be suitable for some tropical areas for sustainable grape cultivation. The UN climate report in 1990 stated that more than 27 wine regions worldwide showed that average summer and winter temperatures have increased in the Northern Hemisphere by 1.5° and 1.4°C.[9] There is evidence that nighttime temperatures in many wine regions have risen more than daytime temperatures.[95] According to Robinson and Harding,[88] viticulture falls into three categories. A macroclimate region is the climate of a wider region, such as a region or country (most premium grapes are grown in Mediterranean, maritime, and continental climates). A microclimate is a climate within a vineyard that can affect it through management practices.

This chapter aims to introduce different types of grapes and qualitative criteria that motivate wine consumption and incites consumers to buy quality wines. Further, the quality of wineries based on the physical environment, production practices, and marketing environment has been discussed.

3.2 COMMON GRAPE SPECIES FOR CULTIVATION

The 90% of the world's cultivated grapes belong to the Vitaceae family. This species is grown in Mediterranean climatic conditions under a variety of soil and climatic conditions.[66] There are 60 interfertile species found in Asia, North America, and Europe.[93,104,110] Among these, *V. vinifera* is the most common species in the global grape production.[104] Including varieties that may be eaten as table grapes, dried to produce raisins, or crusted to make wine or juice, agronomical studies prove that *V. vinifera* had different breed morph type varieties even in the past centuries.[7,93]

The oldest grape species *V. vinifera* is the grape of antiquity often mentioned in Bible[109] The Caucasian region first time started cultivation of this species.[93,74,75] From that region, this species spread to the Euro–Mediterranean and Euro–Asiatic regions, from where wine culture spread all over the world.[36,45,93] The following are *V. vinifera* species found in the world[109]:

- **Asian species:** *Vitis embergeri*, golet Chain; *Vitis betulifolia*, Diels and Gilg
- **Caribbean:** *Vitis tiliifolia* Humb. and Bonpl (South America and Caribbean region).
- **Central and East China:** *Vitis aramata*, Diels and Gilg (China); *Viburnum davidii*, Romanet du caillaud (China); *Veleropilina r*eticulata, Pampanini Chain; *Vitis cognetiae*, Pulliat (South Korea, Japan); *Vitis* pentagona, Diels and Gilg (Laos, Vietnam, South China); *Vitis Amurensis*, Ruprecht (Sakhalin Island, Russia); *Vitis lanata*, Roxburgh (Northern India and China); *Vitis pedicellata*, Lawson. India
- **Middle Asia and Euro-Mediterranean:** *V. vinifera*, Linnaeu.
- **North Africa, Australia, and New Zeeland:** *V. vinifera* subsp.; *V. vinifera* silvestris.

- **North America**: *Vitis labrusca* (USA); *Vitis riparia* (Midwest USA); *Vitis rotundifolia* (Southeast USA); *Vitis aestivalis* (Midwest USA; *Vitis rupestris* (East USA and Canada); *Vitis mustangensis* (Texas, USA).

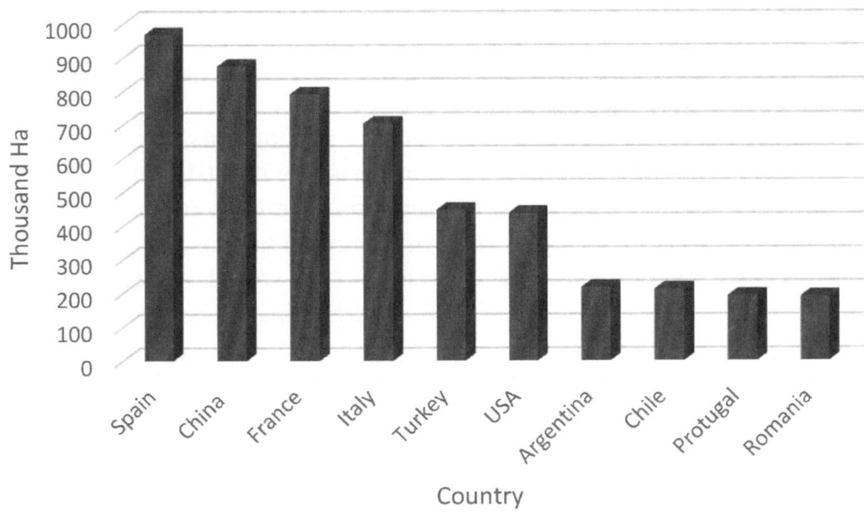

FIGURE 3.3 Area (in thousand ha) under grape cultivation among 10 leading countries in 2018. Source: Self-developed with concepts from Refs. [27, 51].

3.3 WORLDWIDE AREA UNDER GRAPE CULTIVATION

According to the statistics by the International Organization of Vine and Wine (OIV), 7.4 million ha of area is being used for the production of 77.8 million tons of grapes. Grapes can be consumed either fresh or as processed products such as wine, dried grapes, jam, and juices, etc. Although most of grapes are cultivated for making wine, another larger part is dried into raisins, and lastly table grapes as fresh grapes.[109]

Spain is the leading country for grape cultivation with 969,000 ha (Fig. 3.3). China, France, Italy, Turkey, USA are highlighted countries of grape cultivation among the top 10 countries that use their land for grape production.[27,51]

3.4 GRAPE PRODUCTION

Grape production in the majority of countries has slightly increased compared to that in 2017. According to International Organization of Vine and Wine

(OIV) data, 77.8 million tons of grapes were harvested from a total area of 7.4 million ha in 2018.

3.4.1 TRENDS AMONG THE GRAPE-PRODUCING COUNTRIES (TABLE 3.1)

According to the Mintel wine tourism report (2018), China has topped as the world's leading wine-producing country for many years (although Italy, the USA, and Spain briefly headed this list in 2014).

China is the top grape producer with 11.7 million tons, even though the production had slightly decreased in comparison with 2017 due to weather conditions. In contrast, Italy (8.6 million tons), the USA (6.9 million tons), France (6.2 million tons), Spain (6.9 million tons) had an increase in grape production since 2017. The production in Turkey was decreased since 2017 (to 3.9 million tons); while India (2.9%), Argentina (2.7%), Chile (2.5%), and Iran (2.3%) had a slight increase during the same time period. These top 10 countries produced 54.6% of the total grape production (77.8 million tons of grapes) in the world in 2017.

TABLE 3.1 Top 10 Grape-Producing Countries. Source: Self-developed with data from Ref. [51].

Country[a]	Production (million tons)				Grape production (%) in 2018		
	2015	2016	2017	2018	Tableb	Driedc	Wined
Argentina	2.5	1.9	2.1	2.7	0.9%	5.5%	93.7%
Chile	2.7	2.2	2.0	2.5	26.0%	3.9%	70.2%
China	13.2	12.6	13.1	11.7	84.1%	5.6%	10.3%
France	6.3	6.3	5.0	6.2	0.4%	0.0%	96.0%
India	2.6	2.6	2.9	2.9	92.6%	5.9%	1.5%
Iran	2.2	2.3	1.9	2.3	76.3%	23.7%	0.0%
Italy	8.2	8.4	6.9	8.6	13.5%	0.0%	86.5
Spain	6.0	6.3	5.0	6.9	4.0%	0.0%	96.0%
Turkey	3.7	4.0	4.2	3.9	56.1%	40.7%	3.2%
USA	6.9	7.0	6.7	6.9	16.3%	18.1%	65.6%
Rest of the world	45.7	46.4	50.2	45.4	36%	7%	57%

[a]World's top 10 grape-producing countries;

[b]Four kg of fresh grapes is converted into 1 kg of raisins;

[c]According to OIV conversation factors, 1.32 kg of fresh grapes is needed to obtain 1 L of wine; and

[d]Must and juices included.

Note: a + b + C + d = 100%

3.4.2 *WINE GRAPES*

The Romans were the first to differentiate grapes according to their color and type of seeds and categorized these as table, dried, or wine grapes.[104] The majority of grapes in the world are mostly grown for wine production. Wine is made in almost all regions are the world. According to Campbell and Guibert[17] regional wines are categorized as either old-world wines or new-world wines. Old-world wines are from regions with a rich history of wine production, typically from high-producing European regions of France, Italy, Germany, Spain Portugal, Austria, Greece, Lebanon, Israel, Croatia, Georgia, Romania, Hungary, and Switzerland.[4] New world wines mostly come from climatic conditions that are hotter most of the time and only sometimes cooler than the regions mentioned above. They are typically available in South Africa, North America, South America, Australia, and New Zealand.[4] New world wines use-different labeling for the adoption of both grapes and growing regions.[84]

TABLE 3.2 Ten Major Wine Producing Countries.

Country	2015	2016	2017	2018
	Million hectoliters (MhL)			
Argentina	13.4	9.4	11.8	14.5
Australia	11.9	13.1	13.7	12.9
Chile	12.9	10.1	9.5	12.9
China	13.3	13.2	11.6	9.1
France	47.0	45.3	36.3	48.6
Germany	8.8	9.0	7.5	10.3
Italy	50.0	50.9	42.5	54.8
South Africa	11.2	10.5	10.8	9.5
Spain	37.7	39.7	32.5	44.4
USA	21.7	23.7	23.3	23.9

Italy is the world's biggest wine producer with 54.8 MhL and due to changes in weather conditions, its production has increased since 2017 (Table 3.2). The wine production also increased in France (48.6 MhL), Spain (44.4 MhL), USA (23.9 MhL) Argentina (14.5 MhL), Chile (12.9 MhL), and Germany (10.3 MhL) compared to that in 2017. Australia (12.9 MhL), South Africa (9.5 MhL), and China (9.1) have slightly decreased their production.[51]

3.4.3 DRIED GRAPES (RAISINS)

The following methods are used to produce dried grapes:

- **Natural drying by the sun:** grapes are left on the vines until dried under natural conditions due to excessive sunshine in the given region;
- **Solar drying**: solar energy is used to get dry grapes all over the world;
- **Shade drying** is a natural way of drying grapes, mostly in China, India, and Australia[77];
- **Mechanical drying** is a rapid-drying agricultural technology.

Raisins are naturally rich in valuable nutrients and have a pleasant sweet taste. As part of the daily diet, these provide essential nutrients, soluble and insoluble fiber, and a unique composition of health-promoting phytochemicals.

Dried grapes are utilized in several ways depending on the quality and the type of seed of the grape. Some of the widely grown dried grapes species are Thompson seedless (Sultana grape), which is common in the Turkey region. Black Corinth variety of dried grapes is mostly produced in Zakynthos islands, while Muscat of Alexandria is common in South Africa and Australia.

Raisins are widely used in different parts of the world to make salads. There are different kinds of bakery product recipes that are available depending on the region, raisin breed, raisin cookies, cinnamon raisin rolls, etc.

Raisin juice is a concentrated raisin extract in the form of dark brown syrup. The process of producing raisin juice involves soaking raisins in water to leach a variety of compounds out of these and then evaporating the liquid, leaving a concentrated raisin juice.[92] Raisin juice contains high levels of thiamine, niacin, folic acid, vitamin B6, and vitamin B12, among others.

Raisin wine, also called straw wine, is a sweet wine made from different types of raisins (mostly Golden Raisins, Muscat Raisins, Black Flame Raisins, Red Raisins, and Green Raisins).[24] Historically, this beverage was traditionally made in France and North Italy. Later, this wine spread all over the world in the 20th century.

Raisins are widely used in Greece as a raw material to produce alcohol. Premium quality white raisins can be used for the production of raisin alcohol. Chopped and water-soaked raisins are fermented and, this way, pure ethanol can be prepared. The alcohol produced from these inferior types of raisins can be utilized as biofuel.[53,92]

TABLE 3.3 Main Countries for Production of Dried-Grapes.

Country	2015	2016	2017	2018
	Thousand tons (TT)			
Afghanistan	43.0	34.5	36.5	38.0
Argentina	17.4	30.7	26.7	37.8
Chile	63.2	56.2	33.0	30.0
China	190.0	165.0	180.0	190.0
Greece	27.6	28.0	28.0	27.0
Iran	100.7	170.0	110.0	150.0
South Africa	60.5	54.6	65.6	71.0
Turkey	333.5	384.0	400.8	381.0
USA	348.1	315.0	241.4	263.0
Uzbekistan	70.0	71.0	65.0	70.0

Around the world 1.3 million tons of dried grapes were produced in 2018 (Table 3.3). Among them, Turkey is the leading country to produce 381,000 tons of dried grapes that were slightly decreased (2%) when compared to production between 2015 and 2018,[51] followed by the USA, (263 TT), which had its national production decreased by 21%. China's (190 TT) decreased by (6%), Iran's (150 TT) increased by (60%), South Africa's (71 TT) increased by (54%), Uzbekistan's (70 TT) increased by (40%), Afghanistan's (38 TT) decreased by (-27%), Argentina's (37.8 TT) increased by (107%), Chile's (30 TT) decreased by (-55%), lastly Greece's (27 TT) increased by (13%) of their dried grape production.

3.4.4 *TABLE GRAPES*

Approximately 27.3 million tons of table grapes (also called fresh grapes) were produced in the world in 2018 (Table 3.4). Over 50 countries produced table grapes with China being the biggest producer with 9.5 million tons; China is followed by Turkey (1.9 Mt), India (1,9 MT), and Iran (1.7 Mt), Egypt (1.5 Mt).[51] Production of table grapes in 2018 has been affected due to different dynamics in leading countries such as China, Turkey, India, Egypt, and Brazil—in their cases, a decreasing trend was observable in the world market, while other countries such as Iran, Uzbekistan, Italy, the USA, and Chile have managed to slightly increase their production or either stagnated. Table grapes are packed with nutrients and contain natural sugars as well as small amounts of vitamins.

TABLE 3.4 Top Countries Producing Table Grapes.

Country	2015	2016	2017	2018
			Million tons	
Brazil	0.7	0.6	0.9	0.8
Chile	0.8	0.7	0.7	0.7
China	10.0	10.1	10.5	9.5
Egypt	1.6	1.6	1.6	1.5
India	1.8	2.0	2.0	1.9
Iran	1.8	1.6	1.4	1.7
Italy	1.1	1.1	1.0	1.1
Turkey	1.9	2.0	2.1	1.9
USA	0.9	0.9	0.9	0.9
Uzbekistan	1.2	1.2	1.2	1.2

3.5 DEMAND: MOTIVATION AND EXPECTATIONS OF THE CONSUMERS

In Chapter 5 of this book, the most relevant information on the supply of grapes has been presented. As is the case with all profit-oriented organizations, the purpose of grape production and wine preparation is to sell the end product. As observed previously in this chapter, the world's wine production capacity is quite high, which often means significant revenue to the producers and their countries. However, at the same time, if we are to make a profit from wine production, we need to know those who buy and consume the product. In this section, we are going to give an insight into the demand part. Within the frame of this analysis, the authors of this chapter have identified the motivations and habits of wine consumers, in accordance with their expectations for wine quality.

With worldwide wine consumption on the rise, a genuine understanding of consumer preferences and motivations is more crucial to wineries than ever before. To increase sales, they need to adopt new methods, among them organized travels to their vineyards.

Today's wine market is extremely complex, with varying demand and supply from all over the world. Consumers are increasingly demanding and satisfying their needs calls for quick, efficient, and differentiated responses from all actors in the supply chain. More and more aggressive offers are encountered in the wine market, and to ensure sustainable business operations, wine enterprises must offer added value.[76]

According to Pérez–Calderón et al.,[80] Schlüter and Norrild,[94] wine production and tourist activities fit together incredibly well. Wine tourism takes the form of visits to vineyards, wineries, and wine events.[47,48] As argued by Getz,[40] wineries are the main attractions, but they alone are not enough to attract tourists. As Carlsen[20] stated visitors are looking for not only wine but also for food and art in a winery's package deals. Wine tourism can be seen as a sensory experience that provides knowledge about wines and gives tourists enjoyment.[40]

Wine tourism is a multifaceted and complex activity that combines the production and consumption of agricultural products. The consumer not only consumes wine but also involves in the process of creating the experience. The picturesque countryside, the "rural idyll" and romance are important parts of the visit.[21]

Wine tourism is a relatively young industry whose important economic issue is how small and medium-sized wineries can become sustainable. "Winery visitation is a significant concern and it is defined as the visitor's intention to pay a first or return visit to a winery.".[16]

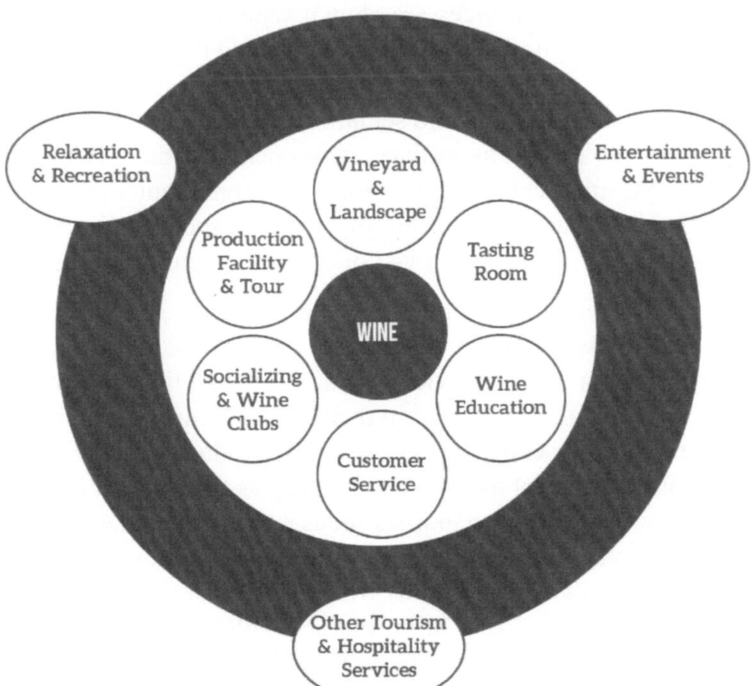

FIGURE 3.4 Basic, additional and complementary services in a wine region. Source: Self-developed with concepts from Ref. [16].

Wine tourists, who do not stay overnight, may come from the closest areas near the winery, while others arrive from remote locations. Demographically, they represent a relatively high socioeconomic level in terms of education, income, and profession.[14] Wine-related activities are of varying importance to each visitor since they are interested in different events and choose different products.[44]

Five factors can be considered the most essential for attracting tourists,[41] such as "the wineries are visitor friendly; there is a lot to see and do; attractive scenery; winery staff are knowledgeable about wine; and group tours of the wineries are offered." Other services such as wine education for winemakers, tasting room services, and educational opportunities for visitors to improve their wine knowledge and skills are emerging. Figure 3.4 shows an alternative vision of the benefits of a wine destination. In this figure, the reader can have the following observations:

- The central (core) product of the visit to the winery (the inner circle) is the wine itself, its quality,the possibility of its direct consumption, tasting, or purchase on the spot.
- Augmented wine services are shown within the middle circle. This concept goes beyond what Telfer[102] suggested regarding the vineyard–wine-making activity–tasting room threesome, including customer service, wine education, and socializing/wine club offers.
- Even though shown as tertiary components in the outermost layer, relaxation, and recreation facilities, entertainment and events, and other typical tourist and hospitality services such as accommodation, transportation services, and visitor centers appear to be significant.[16]

As we will observe in Chapter 9, the core, augmented, and ancillary services in a wine region are important not only to the tourists but also to every wine consumer. These services can highly raise the perceived quality of the wine. It is true, that the different methods of communication (labeling, branding, and media communication) and product packaging (linked services) are just as crucial in the wine industry as the product itself.

Today, the notion that wine tourists focus exclusively on wine consumption is no longer valid; instead, there is a valid perception that tourists are looking for an extensive and complex wine and tourist experience.[68]

Quality plays a crucial role in customer loyalty and satisfaction. Since visitors rely heavily on recommendations from friends and family, memorable experiences not only help return but also bring new customers to the wineries.[34] Wine tourists are looking for authentic and personalized experiences and they have high expectations in terms of variety and product quality.[56]

Several factors influence people's overall experiences during winery visits as well as their willingness to purchase available products. According to findings by Dodd and Gustafson,[31] the tendency to purchase wine is generally affected by the following: Service, wine characteristics, winery environment, and price.

Table 3.5 presents the concept of wine tourism experiences, indicating what geographical factors contribute to it. It is a combination of regional setting and activity site scales (spatial dimensions) with physical elements (natural and built environment) and human elements (social interaction environment).

TABLE 3.5 Perceptions and Dimensions of the Wine Tourism Experience.

		Geographical Elements	
		Physical and Built Environment	**Human and Social Environment**
Geographical Scale	**Personal Setting**	**1. Perceptions of the rural landscape, winescape and rural infrastructure**	**3. Perceptions of and social interactions with residents and other visitors in the region**
	Activity Site	**2. Perceptions of the servicescape and aesthetics at the winery**	**4. Perceptions of the sercie quality at the winery and social interactions with staff and other who are customers**

Source: Self-Developed with Concepts from Ref. [21].

To make the tourist experience memorable in the wine regions, service providers need to establish an emotional relationship with the visitors. In doing so, they must be authentic and credible in all aspects of wine making, from harvest to degustation.[96]

Fernandes and Cruz[34] indicated that despite the wide variety of wine tourism-related elements, tourists are most attracted by the physical characteristics and the attractive atmosphere of the wineries. The interest is generated by the interactions between the winery staff, tourists, and the environmental and physical elements of the wineries, creating a holistic tourist experience that goes far beyond wine tasting.

By providing complementary visitor-focused services, wine companies can[8] (a) build brand loyalty; (b) increase consumer awareness and familiarity; (c) foster a favorable image of the wine product; and (d) build strong customer engagement via on-the-spot experiences.

Consumer purchasing intention and satisfaction are linked, and their patterns are determined by both the product (wine) and the immediate and broader environment of its origin.[25] Macionis and Cambourne[63] found that although wine tourists' primary motivation is wine, yet many other motivations can be observed in wine tourism, such as: attending festivals, a sense of socialization, experiencing the countryside, seeing other attractions, learning about wine and wine making (education), eating at a winery, attending a winery tour, and personal meeting with the winemaker.

According to Gallenti and Galli,[39] such gastronomic itineraries should offer (i) a presentation of wines, other related products, and current production processes; (ii) demonstration of traditional cultivation techniques and crafts related to the past; and (iii) landscapes, cultural activities, and hospitality systems.

3.6 DEVELOPMENT OF WINE SELLING OPPORTUNITIES AT THE WINEYARDS

The main challenge for winemakers is to combine wine production with agricultural traditions, local values, and the beauty of the landscape. To facilitate this, they should strengthen their cooperation with those involved in tourism activities. In the meantime, however, they must not weaken their relationship with other winemakers but must work together on product development, innovation, quality improvement, and market development. However, wine companies are mostly individualistic and do not cooperate in areas such as marketing aimed at gaining foreign markets that would benefit all participants without weakening their identities. Managers of wineries need a strategic approach to analyze market potential, planned strategy, and level of investment. Export consortia or alliances with other local producers should be set up to make the most of the resources needed for promotional activities and sales organization.[73]

Getz[40] defined sustainable vineyard tourism as a combination of three perspectives: consumers (travel and holiday behavior); wine vineyard (a distribution or sales instrument); the location where tourism is linked "to the charm of wine" and sustainable wine production.

In general, vineyards are characterized by a high level of environmental sustainability as there is no high level of harmful emissions from wine making.[11] Nonetheless, local communities and the viticulture industry must be closely linked and this cooperation ensures that all the environmental concerns involved are taken into account.[3]

Wineries will have a positive impact by communicating their environmental sustainability with visitors, as they can contribute to increasing demand for sustainably produced wines while also increasing brand equity. Such positive results could spill over and encourage other wineries to adopt environmentally friendly production methods.[59] Pomarici et al.[83] examined the issue of the cost of switching to sustainable winery. It was found that the overall economic benefits of introducing a certification scheme for the California wineries involved in the investigation outweigh the additional costs.

Although there is an increased interest in sustainable wine practices, the high implementation costs are a major barrier to vineyards. The use of alternative energy to maintain vineyards remains limited. This can be enhanced through the usage of certification and labeling systems, which can show to the customers that the wines were produced using sustainable practices for saving water/energy, thus eliminating the use of synthetic pesticides or other additives.[3]

3.6.1 THE CELLAR ITSELF AS A PRODUCT

The production practices can influence the success of a vineyard business success According to Hall and Mitchell,[49] whose claim does not only refer to wine quality but also reflects on the possibilities to provide experiences to the visitors and customers. The process of wine making can be fascinating to those who do not know the methods or have not seen the equipment before.

Wine tourism is not only a visit to cellars and vineyards but also a series of unique experiences that include the natural landscape, atmosphere, local culture and traditions, wine, and local cuisine.[42]

Crescimanno and Galati[28] stated that preferences and customs for wine consumption have changed, as consumers are increasingly looking for quality wines and formulating more complex demands to blend experiential, symbolic and hedonistic elements. Gastronomy and wine are no longer just one of the complementary activities of travel, but these are the main reasons in many cases to visit a destination.

Wine tourists look for quality culinary offerings,[29] therefore gastronomic tourism needs to be supported as it strengthens local identity and culture and improves financial conditions in the region.

Nowadays, many wineries market landscapes, handicrafts, services and the characteristics of the destination in the form of a package of experiences.[16] Sales of wine for visitors depend on many aspects, such as characteristics of the grape, the terroir, the flavors of the vineyard, the reputation of the winemaker and the corporate brand, which often includes a family story as well.[62]

The experience of consuming wine is determined by the color, the bouquet, the taste, the tactile impression, the scent and the bottling of the wine.[19]

The selling process of a winery consists of the following three phases: Pre-purchase, postpurchase, and postpurchase stages. As Lemon and Verhoef[60] state, consumers gain experience not only when they consume wine, but also before buying the product and after tasting it.

This experience can also change tourists' behavior: visitors learn how to consume wine and enjoy the ritual.[38] Bruwer[15] sees the first encounter with a visitor at wineries as this may be the beginning of a relationship with a particular brand of wine. If the consumer is satisfied, he can become a loyal customer feeling satisfied about the decision to buy a particular wine brand.

Customer satisfaction plays an essential part in shaping the reputation of a wine brand and can also encourage wine buyers to share their experiences regarding a specific brand (Fig. 3.5).

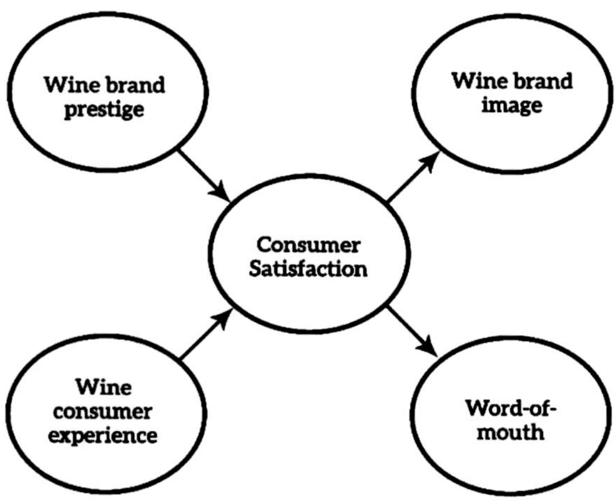

FIGURE 3.5 Consumer satisfaction. Source: Self-developed with concepts from Ref. [62].

Wine brand owners also need to develop interactive relationships to increase consumer satisfaction. This can be done, for example, through an online platform, via real-time communication, through follow-up with customers, and by motivating them to discover wine in new contexts. While pursuing these and communicating health benefits, brand owners should also pay attention to raising awareness regarding responsible consumption.[62]

Tourists who are more familiar with wine brands may be more satisfied with reputable and renowned wine brands than those who are not familiar with the characteristics of a particular wine or have no experience with such wines. Experienced and knowledgeable tourists in the field of wine brands can even anticipate what they will taste. In the case of a brand, they have not yet tried, they can try to predict what they are going to taste, due to their expertise with wine brands. If they already know the brand, they may develop the expectation of repeating what they have tasted and memorized from past experience.[62]

The Port Wine and Douro Wine brands, which are popular and respected in Portugal and abroad, are good examples of this. The selection, purchase and tasting of the original Port Wine is linked to the prestige associated with the status, uniqueness, and luxury of the brand.[62]

3.6.2 EVENTS FOR GRAPE AND WINE ACTIVITIES

Visitors can also enjoy several recreational activities offered in wine-making regions. Wine tourism growth is not solely incited by income prospects; it is largely due to the proliferation of local food and wine events, which have provided opportunities for local wineries to connect with tourists.[70] In today's wine tourism, due to today's advanced technology and increasingly sophisticated visitors, the emphasis is on experience-based entertainment rather than service-based entertainment, and the wine and the wine-making process are often embedded in events and attractions.[85] These events can be part of the marketing activities of the vineyard, which is an influencing factor in business success.[49]

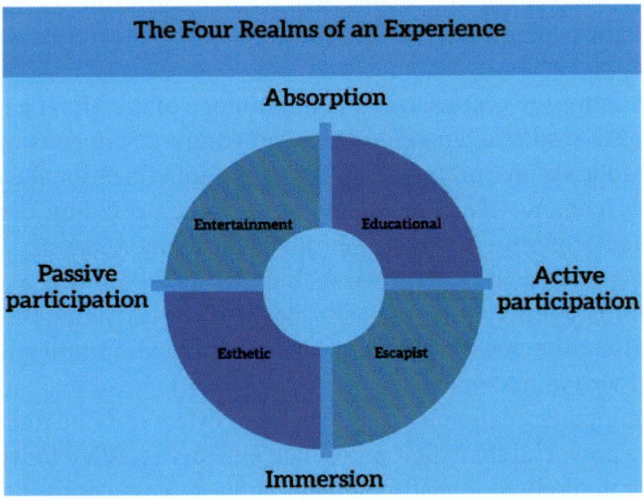

FIGURE 3.6 The four realms of a wine experience. Source: Self-developed with concepts from Ref. [82, p. 30].

Wine tourism includes visiting not only vineyards, wineries, and wine tasting but also wine festivals at wine tourism destinations.[48] According to Marzo-Navarro and Pedraja-Iglesias,[65] additional aspects can be added to this definition, such as getting to know the local attractions or the lifestyle of the local people, for example, by spending a night in the vineyard.

Wine tourists want to escape from their everyday lives during their holidays, and it is not enough for them to enjoy the aesthetic pleasure of the wine landscape and taste quality wines, it is also important for them to participate in leisure activities available in wine destinations.[86]

Wineries offer many additional activities for tourists, ranging from hot–air ballooning and cycling tours, vineyards tours by horse and carriage, grape picking, grape stomping, and harvesting grapes.[85]

The wine tourism supply chain involves accommodation, culinary experiences, and cultural and leisure activities, and each may include an educative element. The most widespread educative activity is wine tasting, while the second one is learning about appropriate food and wine pairings in the given area's restaurants. Some wineries also organize wine seminars and gastronomy courses with the help of their partners (chefs and farmers).[85]

In addition, festivals, the presentation of the history of wine in museums and galleries (whether sold or just exhibited) are not only entertaining but also educational.[69,81,85] One of the main attributes desired by wine tourists is personal development.[98]

The involvement of arts, music, and culture in the wine regions is well established.[22] According to Brown and Getz,[41] specifically wine-related shops and other businesses possess entertainment features within wine destinations.

Pine and Gilmore[82] emphasized the importance of the 4Es (i.e., the educational, escapist, aesthetic, and entertainment nature of the consumer experience). Wine tourists are motivated by the desire for educational experiences, whereby they enhance their knowledge and skills, providing evidence that the experience economy model is appropriate for wine tourism.

The 4Es can be seen in semipermeable quadrants divided by two intersecting continua. Participation intensity can be observed along the horizontal line, while along the one perpendicular to it, participant immersion (in both physical and virtual manner) is displayed (Fig. 3.6).

The combined recreational and educational offers of wine tourism operators are more powerful than their individual endeavors. Wine tourism service providers are unlikely to invest separately in these dimensions if they expect the tourism or wine organizations in the destination to hold educational and entertainment events. Individual businesses may do less to create entertaining experiences—or educational experiences—for patrons because they view this function as the domain of destination-wide organizations.

There is no one-size-fits-all approach for wine tourism service providers to apply the 4E approach. Businesses should enhance the visitor experience, depending on their capabilities.

The best approaches for entertainment are possibly wine blending events, with food-wine pairing, guided tours in and around the winery, and hiking opportunities in the vicinity of the vineyard being such for the educational, esthetics, and escapist aspects, respectively. Wine businesses can have an educational element not just regarding the wine, this may include farming, the socio-geological history of the region, or the flora and fauna as well. To enhance the entertainment and escapist experience they offer, businesses should consider not only professional art performances but also amateur events that can attract visitors and staff, such as cooking and wine blending competitions.

Regional organizations should engage all the supply-side stakeholders to create long-term events based on entertainment and escapist perceptions. When local businesses align their activities to create a package of services and events that may last several days and can include, among others, taking part in harvests and other work processes related to it, or even more playful activities like geocaching, give rise to truly meaningful local collaborations.[87]

The wine tourism research carried out by Quadri–Felitti and Fiore[86] has shown that the aesthetic dimension of the destination is of utmost importance to the guests. Tourism professionals can make better offers by knowing which wine tourism service provider is able to provide a high standard of service for which "E".

3.6.3 THEMATIC ROUTES BASED ON GRAPE AND WINE

Wine tourism has constantly been gaining ground since the mid-1990s, attracting tourists with relatively high purchasing power.[41] The wine routes are created to be dependent upon needs, either create, preserve, or improve reputation for quality in a given wine region, and are an example of how the wine producers come together to promote products and the destination.[106]

The Wine Route is an "advertised route," the result of collective efforts. By designating a delimited area, a wine trail aims to invite visitors to discover the wines and associated experiences of a specific area. This discovery takes place directly on wine farms and at other sites such as in tasting centers or wine museums.[6]

Vandecandelaere[106] stated from the point of view of consumers that the route can be considered as a tourist product that allows tourists to meet producers in a given wine region. In terms of supply, it can be considered a localized food system that brings together stakeholders to build the reputation of the destination and its products.

A wine route needs strategic organization and integrated development of wine producers. Successful initiatives are characterized by a well-thought-out organizational structure, modern business culture, and investments, which are rare in rural areas.

As destinations keep developing in a direction that makes them increasingly desirable from a purely touristic point of view, they necessarily limit their own business expansion capabilities within their area. Therefore, greater emphasis should be placed on developing partnerships between wineries and local entrepreneurs.

There is limited awareness among local residents of the benefits of developing wine tourism, and greater efforts are needed by managers of wineries and tourist facilities to communicate the positive impacts on local communities. Local residents will also visit the wineries more often if local winery leaders use traditional marketing (flyers) or social media (Facebook and Twitter) campaigns to raise awareness of the benefits their businesses bring to the local community.

3.7 IMPORTANCE AND NECESSITY OF QUALITY ASSURANCE IN WINE PRODUCTION

As seen above, quality plays a crucial role in customer loyalty and satisfaction, regardless of whether we are talking about tourists or regular buyers. This is true not only in the case of wine but also in the whole food industry.

Over the past decades, consumers have criticized the quality and safety of food products due to the presence of multiple contaminants. To create and retain consumer trust in food and food safety, the importance of quality management and food hygiene in the agri-food sector is paramount.[1] It is a proven fact that applying quality assurance during the grape-growing and wine-production processes can ensure greater certainty in terms of the level of final quality.[49] This resulted in a kind of quality turn that took place in viticulture in the last decades.

3.7.1 THE QUALITY TURN

The quality turn refers to a widespread shift from mass consumption to the increasing qualitative differentiation of food products.[43] This paradigm shift comprises a range of major transformations that will heavily affect the development of agri-food systems, notably the trend from a quantitative to a qualitative approach.[72]

The quality turn has brought on changes in the wine production processes and in the wine assessment system; as well. From that moment wine producers had to understand how to switch from a usual mode of production into a new, and in many cases, unknown, tradition-free production system, in which the quality of the produced wine is more important than its quantity.

Based on the above the evaluation of wine also had to change. This approach preferred the evaluation of quality criteria on reputation, prestige, and image. These were evaluated not only by considering qualitative attributes such as grape color or origin but also involving exterior characteristics, such as geographical indication, labeling, winery tradition, critics' recognition, and/or awards received. These criteria attribute a certain value to the wine and, indirectly, increase the willingness of the consumer to pay for it.[108]

3.7.2 THE PERCEPTION OF GRAPE AND WINE QUALITY

Speaking about grape quality, the evaluation is achieved by visual and taste evaluation of the fruit and assessment of total soluble solids (TSS) and acidity.[58]

According to Cozzolino et al.,[27] the abovementioned parameters are not sufficient as quality indicators, and quality cannot be judged only through tasting. Phenolic compounds can enhance the sensory properties of red wine, in particular its color and mouthfeel. Their levels in grapes are dependent on factors such as grape variety, ripening conditions, viticultural techniques, and wine-making practices.[46,89]

Physical and organoleptic characteristics should be taken into account in determining the quality of any kind of food. In the case of grapes, it is essentially true. The physical properties are important since they can influence the consumer's acceptability. These parameters are the weight, size, and color of the fruit. The intrinsic qualities such as sweetness and acidity make up the organoleptic (sensory) ingredients which are important both for the producer and the consumer alike.[91]

Based on the traditional approach, the evaluation of wine quality is measured by the yield, sugar content of the grapes, the pH value, and titratable acidity. The importance of these indicators has not changed for a long tim, even though other parameters such as color are predicted to become more important[61] and, as seen above, the quality turn can increase the value of other factors as well.

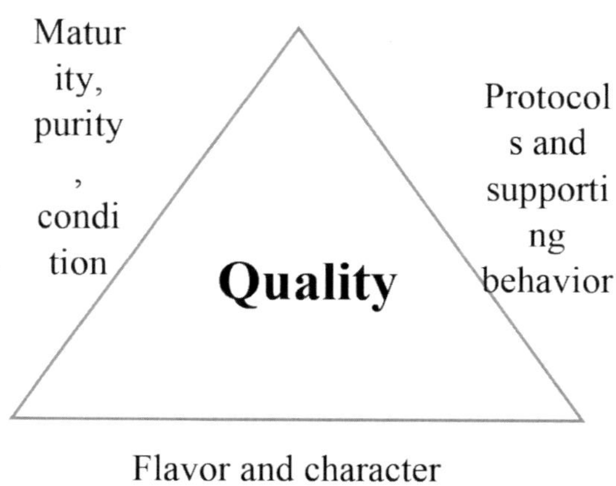

FIGURE 3.7 The quality triangle for Winegrape: Self-developed concept.

Although there are many approaches, it should also be recognized for the sake of clarity that certain characteristics of grapes and wine are difficult or impossible to quantify. Such a feature is the taste of grapes and wines, which makes the product really sought after. The characteristics of wine can be described by the following four impressions and judgments according to Spence[99]:

- **Hedonic**—how much a consumer likes wine.
- **Sensory**—physical characteristics of the wine, such as sweetness, acidity, alcohol content, and its impact during the tasting, like contraction, length, etc.
- **Analytic**—evaluation of the wine's age, complexity, balance, quality, and value.
- **Descriptive**—whether the wine is rather heavy or light, bitter or lascivious, masculine, or feminine.
- These four characteristics indicate that the quality of wine is commonly evaluated through the subjective feelings and impressions of the consumers; they are not much interested in the chemical composition of a wine and are neither in the workflows.

On the request of Winegrape Growers' Council of Australia (WGCA) and the Winemakers' Federation of Australia (WFA) Liaison Committee, Allan[2] created the Winegrape Quality Triangle (Fig. 3.7) and published with the title Winegrape Assessment in the Vineyard and at the Winery. According to Allan,[2] the maturity and purity is easily quantifiable by the following indicators:

- **Maturity**: TSS, pH, and titratable acidity, color (in case of red grape), cropping level;
- **Purity**: diseases, agrochemical residue, matter other than grapes, contamination, and varietal integrity;
- **Condition**: uneven ripening, temperature, spoilage, and damaged berries.

The protocols and supporting behaviors can be defined, but not easily measured. In the case of flavor and character, we can say that these are the most difficult to quantify since the taste can depend on consumer disposition, personality, age, sex, and perhaps also on the time of consumption.

3.7.3 SENSORY ANALYSIS

Sensory analyses are the most important and most widely used methods in evaluating the sensory properties of food. During these analyses the sensory tester assesses the properties of the food and begins to translate them to the customer's voice.[71]

Descriptive Analysis (DA) is a tool developed by sensory testing professionals in the late 1940s. Based on this, many DAs exist today, such as Flavor Profile, Texture Profile, Quantitative Descriptive Analysis (QDA), Time Intensity Descriptive Analysis, Free Choice Profiling, and Spectrum. These techniques use custom panels that measure the degree of organoleptic properties of not just the product itself but also between the products. Flavor Profile was the first basic DA technique. However, this also has its limitations as it is not easily quantifiable; and the test panel leader, as an active participant, can be considered a source of distortion. To eliminate these errors, QDA was conceived as a registered method in 1974, and has the great advantage of using linear rating scales that measure and quantify the product's sensory properties. The head of the board is only a helper and is not involved in the evaluation of the product. The new method, named Quantified Descriptive Sensory Analysis (QDSA) was based on the international standards ISO 1994 and ISO 11035, preparing the necessary description of the taste, aroma, and mouth sensors for the grape.[64,90]

Iland et al.[50] declared that sugars, acids, aromas, and phenolic compounds are important constituents of grapes. The quantity and distribution of these ingredients in the skin, pulp, and seeds of the berry greatly determine the style and quality of the wine. The ripening stage of the grapes plays an important role in the distribution and structure of the compounds as they can grow, decrease or modify during the development of the berries.[26] These changes have been demonstrated by chemical measurements of grape skin and seed tannins.[18,55]

We can see that, just as in the case of other ingredients of any other kinds of food, the quality of grape and wine is a crucial point for every winegrower. As described above, the consumers are mostly dealing with the taste, which is an individually different experience. That is why the professionals try to look for a receipt, which can best suit one of the most important customer aspects, taste. On the contrary, measuring and quantifying taste is a serious problem. Some analyses made an attempt to create a number-based measurement system, among which the Berry Sensory Analysis (BSA) is the most general for describing the maturity of grapes.

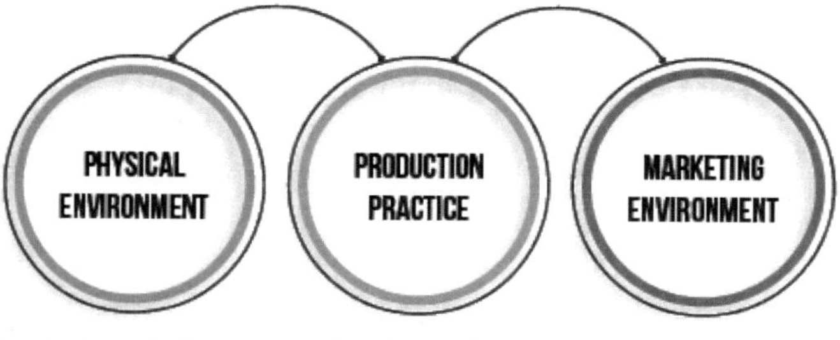

FIGURE 3.8 Business influences on the vineyard.

Source: Self-developed with concepts from Ref. [49].

BSA is a precise, rigid method of sensory evaluation that limits personal subjectivity and thus reduces the risk inherent in the evaluation process. It is typically used in conjunction with chemical analysis by winemakers to obtain accurate results on grape maturity. This technique helps winemakers make informed harvesting and wine-making decisions.[111]

Wen et al.[111] have described the typical method and all the steps that are needed for correct BSA analysis. The method requires a weekly sampling of the grape's ripening process and then increases it to 2 times a week as the fruit begins to mature. It is important that sampling is always done at the same time of the day and not immediately after rain. Each sampling represents the collection of 200 berries, following a representative pattern: From five-to-five berries, alternately from left to right, from the same points—right wing, left wing, central shadow, central sunshine, and mountain slope. The tasting maturity test deals with four parameters in the selected 200 samples. These include technological maturity, fruity aromatic ripeness, peel ripeness, and seed ripeness. From the 200 sampled berries, four to eight fruits are randomly selected for each tasting, and each aspect above should be scored on a scale of 1–4. This tasting should be repeated 3–5 times.[111] This sampling system and thinking can ensure that we can exactly follow the grape maturing in every cluster of our grape field, so we can define the proper date for the harvesting in each cluster.

The organoleptic analysis provides greater knowledge of wine (product expertise), while scoring allows a more accurate evaluation of wine (price valuation). The pairing of food and wine also makes it possible to determine the utility of the wine (communication–education) and the place of purchase and/or consumption (wine bar, wine shop, supermarket, restaurant, e-commerce, etc.).[35]

3.8 FACTORS INFLUENCING THE QUALITY OF WINE

In the previous paragraphs (Section 3.7), the importance of food and wine quality was described, and made a summary of the measurement possibilities of quality. This section analyzes what kind of factors have influenced the wine quality. According to Hall and Mitchell,[49] the aspects shown in Figure 3.8 influence the business success of a vineyard:

3.8.1 ROLE OF PHYSICAL FACTORS AND ENVIRONMENT ON GRAPES AND WINES

Physical environment refers to the factors that can influence the selection of a vineyard and vine variety. These factors will have an impact on the quality of wine a vineyard can produce. According to Hall and Mitchell,[49] physical environment refers to climatic factors (temperature, wind, rain amount, hail, and heat units in a growing season); soil type and quality; water supplies and availability; pests and diseases; and topography.

The physical factors primarily affect the growth and production of the grapes that are used as raw material for the production of wine. Kozma[57] stated that the living conditions of the vineyard and the quality conditions of the production site are 99% determined by the ecological factors, among which the climatic–physiographical factors are most important. According to Eperjesi et al.,[32] the following ecological factors should be taken into consideration:

- **Biotic factors**: Live or parasitic viruses, bacteria, fungi, and animal (mainly insect) pests directly affect the vine. Parasitic fungi (mildew, powdery mildew, and gray mold) are particularly dangerous.
- **Climatic factors**: light, heat, rain, microclimate. The best quality wine grape grows between the average isotherms of 10–16°C per year.
- **Physiographic factors**: geographical latitude, altitude, cardinal points, slope, proximity to large water bodies, and human facilities. The closer the incident angle of the sun's rays approaches 90°, the more light energy that the plant will receive.
- **Soil (edaphic) factors**: the grape is not very demanding on the soil, because it adapts well to different soils. There are cultivated vineyards all over the world on volcanic, sedimentary, and wind-bearing mother-of-medium-loam-heavy clay-loose loess, and sandy soils. Soils of different mechanical and chemical compositions with physical properties will imprint the wine's quality and character. Occasionally, the type of soil can make a significant contribution to the worldwide reputation of a wine specialty.
- **weather conditions of the given vintage** (in connection with climatic factors): the influence of climatic factors on the quantity and quality of the wine's chemical composition can be clearly traced. In favorable vintages, higher-than-average yields and higher quality were harvested in all varieties, while in unfavorable vintages the opposite was observed. It was also found that there is a definite rhythm in the performance of grapevines, which is due to climatic factors and internal biological causes. The effect of the vintage on the titratable acidity of the must is also significantly greater. In sunlight-rich, warmer, and moderately dry vintages, less acid was measured in all varieties and at all load levels than in light, cool, rainy years. The vintage also has a significant effect on the extract content of wine. In rainy, cool years, almost all varieties of wine had more sugar-free extracts than dry, warm years.

Katona[54] strongly agrees with these influencing factors. Based on his study, the climatic factors of viticulture constitute macroclimatic conditions and terrain effects (as amended) ensemble. The most characteristic environmental features (that modify climatic factors) are latitude, elevation and elevation effects, such as

- light, heat, humidity and air conditions;
- physiographic factor: altitude and exposure (slope direction and angle);
- edaphic factors (soil conditions);
- biotic factors: effects caused by the living environment;
- characteristics of a given breed; and
- production technology

It can be concluded that the physical influencing factors are mostly related to different characteristics of the production area, such as the location of vineyard,[32,57] and climatic, physiographic, edaphic, and biotic factors. Eperjesi et al.[32] added the vintage as a plus factor; while Kozma[57] mentioned the production technology and the characteristics of a given breed, which also can influence wine quality.

3.8.2 PRODUCTION PROCESS

Production methods can also influence wine quality since these will determine what kind of raw material goes into the wine and what level of purity is obtained at the end of the process. The production practices are related to (a) the management decisions about the development of a new vineyard or the redevelopment of an existing one; (b) cultural practices applied during the life cycle of the wine and vineyard; (c) the winery's oenological practices; and (d) financial considerations in the selection and management of vineyards and the wine-making practices.

According to Eperjesi et al.,[32] the characteristics for determining the cultivation and value of wine grape varieties are: length of growing season and ripening time, fertility, quality of the crop, environmental demand, resistance to diseases, and production safety. The quality of the fruit of the wine grape varieties is primarily determined by the sugar content of the berries, which is called maturity must be in terms of mass percentage.

The production process has the following steps:

a. So-called herbicide phyto–technical operations (pruning and green works);
b. Soil-acting, so-called agro–technical operations (tillage, fertilization, and irrigation);

c. Plant protection measures;
d. Harvesting (for the sake of wine quality, the leading grape-wine growers use the collection and transport of plastic boxes and take great care to ensure that the fruit goes into the vineyard intact); and
e. Plantation maintenance (capital replacement, maintenance of supporting equipment).

The quality of the vineyard can also be improved through green processes, such as: sorting the shoots, pruning the fruit shoots, trimming, cutting back the armpit shoots, and leafing around the curls.[32]

The maturity of the grape is crucial in the production process since it will determine all the activities that should be made in the vineyard. There are three stages in the maturity of the grapes: growth; full ripening; and overripe. These three stages cannot be sharply demarcated.

Growth is the initial stage of ripening. The blossoming berry loses its dark green color, becomes translucent, and begins to color the variety. Meanwhile, it becomes soft and elastic, and a waxy layer forms on its surface. Depending on the variety, the growth of the grapes usually begins 60–70 days after flowering. During the growing season, the migration of sugar-rich solutions into berries is usually very lively. The period of intense sugar growth is 30–40 days. During this period, the sugar content of the berries may increase by 0.3–0.5% by weight per day under favorable conditions, thus increasing the sugar content of the juice of the berries between 2° and 3° per week (with a potential alcohol content of 1.2–1.7% vol/vol). Contrary to the strong sugar influx, acidity decreases rapidly. At the end of the growth phase of the berries, the titratable acidity of the juice of the berries is 25–30 g/L. At the end of the growing season, the titratable acidity is reduced to 8–15 g/L, depending on the species and the weather, as a result of breathing. During respiration, malic acid is significantly reduced compared to tartaric acid.

During the full ripening period, the rate of sugar inflow gradually slows down, and acidification usually continues. The full ripeness of the berries is reached when sugar is no longer flowing from the leaves to the berries. By this time, the sugar and titratable acidity of the juice of the berries has reached the level typical of the variety and is due to the vintage and the previously mentioned viticultural factors. Vine yield per unit area and sugar yield (absolute sugar content) are highest at maturity. Sugar inflows are usually phased out by clogging the cluster and berry stalks, that is, clogging of the transport channels. However, the influx of sugar may suddenly cease due to the early autumn frost. After the foliage has been frozen, there is no sugar flow.

Overripe begins with the cessation of the influx of sugar. As a result of water loss and accumulation, the sugar content of the juice of the berries, as well as the concentration of acids, minerals, and nitrogen, is concentrated. The relative increase in sugar content at maturity depends on the degree of water loss. The reduction in yield may be offset by an improvement in quality upon ripening and may even increase in value. Technological ripeness in the vast majority of cases coincides with the total ripeness of the grapes. Unless there are compelling circumstances (such as strong rot), the harvest must be brought into line with the full maturity of the vine. Depending on the quality of the fruit, the type of wine suitable for the particular purpose may be made.[32]

3.8.3 MARKETING

The marketing environment refers to the winery's guest-centricity (the extent to which it seeks to meet customer expectations and needs).[49] This refers to (a) the intrinsic characteristics of the wine (red or white, variety, flavors, and fruit characteristics); and (b) the extrinsic characteristics of the wine (the place of production, the production practices, the price, market acceptance, which in turn is influenced by the media, consumer preferences, and fashion.

Although it is not common for vineyards to think about marketing, the market has a tremendous impact on vineyard and vineyard events, including not only the planting of vine varieties but also how they are handled and processed.[49]

Souza[97] declared that marketing strategies, that use several organic produced foods that are more nutritious and tastier than regular foods, have an impact on the consumers to buy wine. The addition of organic waste has a longer life and better organoleptic properties, because it contains more sugars and organic acids, is high in nutrients and solids, and therefore contains less water in foods, preventing the growth of microorganisms that can degrade the fruit.

This section will focus on the following marketing factors that can increase the profitability of viticulture and the companies dealing with wine production.

Branding, Place Branding: According to Hall and Mitchell,[49] "it is not the wine that the consumer needs to trust (although product quality is of course an important hygiene factor) nor is it the wine that is the focus of the relationship. The brand is important as a source of trust, but only when the brand is imbued with personal experiences with staff, can a true relationship be developed." This statement reveals the meaning of branding, including the branding of food products and wine. This approach allows wine brands to develop a personality and associate themselves with real people, places, and events in the minds of consumers; and this creates an added value.

The wine is linked to the place, where the grapes are grown and where the wine is produced. The character and taste of the wine are based on the interaction of wine-making and viticultural techniques, the grape variety, and the physical environment of the wine-growing region. The combination of these elements creates the "place" of the winery and strengthens the wine value chain. This sense of place goes beyond the physical characteristics of a place, much like terroir, which is more than a region's soil and climate.[49]

Labeling: Labels constitute an important part of the promotion of a winery. The appeal of labels at the point of sale can be crucial in attracting consumers to a given brand. Some stores may have thousands of different labels, and consumers often choose based on their first impression, which is the label for a wine.[49] In particular, front labels should evoke associations with the wine and the brand image. Wine label design is critical to the winery's marketing strategy, as it defines the wine's overall brand image and consumer perceptions. The image on the label must also influence other communication channels for the wine industry, including websites, advertising, and promotional material.[5] An adequate label should include the following:

- Company name;
- Color of the wine;
- Country name, region, or sub-region of the origin;
- Name of the vineyard, winemaker;
- Price;
- Varietal names or combinations; and
- Wine style (such as sweet, light, heavy, tannic, etc.).

The labeling of wines with a geographical indication indicates the region of origin, the grape variety, and the year of harvest. The wine may bear a description of the viticultural process, such as organic or biodynamic. However, consumers are given little information about the wine-making process and oenological procedures apart from the general terms used to describe the wine of the seeds. Labels usually only refer to the alcohol content. Other ingredients and additives are not listed or labeled (e.g., contains sulfite), some of which may be potentially unsafe for some consumer groups.[5]

Different Communicational Channels can help to make wineries even more popular. Nowadays, destination marketing organizations (DMOs) are becoming more and more prevalent in those geographical areas that appeal to tourists. Using this network and a region umbrella organization can be a good choice. Getting involved in these DMOs can result in more media presence, providing opportunities to feature more in local promotional materials. Usually, the wineries have to pay a lower membership fee to compensate for this, or they organize free tours and tastings for the media representatives.[49]

Television advertising for winemakers is quite rare for many reasons. One reason for this is that alcohol ads may be restricted or banned in certain media; on the contrary, television ads are one of the most expensive ways of communication. Radio has a more specific target audience since radio stations tend to address a narrow demographic segment and typically have a geographical demarcation as well. It is important to make sure that the audience of the radio corresponds to the target market of the brand according to demographic data. Billboards can reach a wide audience very efficiently, but they may not bring on a great deal of sales. Often, innovative concepts need to be invented and advertising may be enriched with humor, for instance.[49]

In many cases, PR can be more important for winemakers because it is less one way than promotion and can thus prove to be useful in establishing relationships with customers and other groups and people with an interest in the winery (e.g., with suppliers, equity holders, government, officials, workers, or the local public). PR is also beneficial for winemakers because they do not have to actually spend money to get the story across but can instead develop PR campaigns and marketing strategies in anticipation of the media picking up and sharing such stories. You may be communicating directly with your target audience and not relaying the message. There are many opportunities for sales promotions as well. In most cases, these are printed materials (because they are easy to use in a retail environment and relatively inexpensive to produce) or delivered personally by the sales staff or the owner or winemaker. In-person sales promotions are likely to have a greater impact by allowing consumers to have a direct relationship with the wine-growing company and thus greatly facilitate their relationship.[49]

Celebrity use in marketing communication is widespread in many product categories and is often used as reference groups. There are many ways to use reference groups, but there are two methods for wine. The first is when we introduce celebrities as "experts" and actually use them to validate wine quality. The second is the use of celebrities as models. In this case, consumers will be left with the impression of being-similar to the celebrity they see in the advertisement or that they shall also enjoy the celebrity's lifestyle by consuming the given wine. Sports stars, movie and television stars can also be included in wine recommendations.[49]

In the last decade, the role of social media has increased rapidly. This includes the different institutional blogs, websites, and professionals who have also prepared guidelines on the best practices for marketing.

Wine is an experiential product and many producers primarily use social media to reach consumers and communicate about the brand experience, product quality, and brand identity. Social media is particularly effective for wine consumers because it spreads by word of mouth among wine consumers.

The communication and socialization element of social media enables consumers to exchange information and inspire others to try various wines, making it a key channel for influencing wine sales.[101]

Table 3.6 presents the usage of various communication channels by vineries in four countries. From Table 3.6, it is concluded that personal selling is the most common in all countries, while sales via post and phone are common in Germany. Regarding social media, the US vineries are at the top; and within this category; Facebook is the most popular, while Twitter is the most common in New Zealand. Sharing blogs and videos is not very popular yet.

TABLE 3.6 Usage of Communication Channels in the Wine Business (Multiple Response).

Communication Channel	Germany	USA	New Zealand	Australia
Blogs	7.5	24.4	22.1	21.0
E-mail	85.9	97.9	90.6	83.2
FB	93.1	98.0	93.5	88.1
Instagram	1.1	35.2	28.6	21.9
Newsletter	33.7	61.0	54.7	41.5
Personally	97.9	87.7	89.6	82.0
Phone	82.5	80.5	65.1	49.4
Post	84.4	37.4	30.2	30.1
Social media	46.2	86.9	64.2	62.0
Twitter	13.8	67.3	76.6	55.7
Video sharing	15.5	39.5	45.5	30.1

Source: Self-developed with Data from Ref. [101].

Szolnoki et al.[101] compiled data from four countries on the use of communication channels in the wine business (Table 3.6). They have found more documented examples, which prove and demonstrate the positive impact of social media campaigns on business turnover.

3.9 QUALITY ASSURANCE STANDARDS

In Section 3.8 of this chapter, it was discussed why wine quality is so important, then Section 3.9 highlights the role of wine quality in convincing consumers to make a purchase. This section introduces important quality assurance standards that are relevant to winemakers, especially the advantages of having a kind of quality certificate.

3.9.1 BACKGROUND AND MOTIVATION

The most important motivational benefits of quality assurance systems for an enterprise are to: improve production processes, increase employee engagement and reduce costs. The external motivation factors come mainly in the form of customer expectations for certification and quality control systems, which can provide greater assurance to the purchasers to reduce the risk of objective liability and simplify control methods.[37]

Quality and safety are especially vital in the case of a product like wine; quality assurance throughout the entire procedure is essential to consumer acceptance, while safety is indispensable to protect human health. As detailed here, the diversity and quality of wines are the result of the variety of grapes, the specific quality of the soil, the location, the climate, and the processes designed to produce the wine. Wine safety must be assured already during the wine-making process by revealing and reducing the relative risks; that is, each step of the procedure, from the grape harvest to the in-between stages in the vineries, until the product is marketed, must be examined.[23]

According to Surak and Simpson[100] with the introduction of quality standards, the Total Quality Management (TQM) approach is embedded in corporate life, incorporating modern leadership and human resources principles, customer orientation, strategic planning, evidence-based decision making, and modern process control techniques. This allows companies to develop a high-quality quality strategy that improves competitiveness, enhances the value of products and services provided to customers, and improves business performance. These principles lay the foundation for customer-centricity that can stand its ground in the global marketplace.

The quality of a wine is generally related to its presentation, palatability, taste, aroma, color, composition, and other qualities.[23] However, the Spanish system of certification of origin (e.g., introduces a number of analytical standards for wines), which includes both objective (volatile acidity, alcohol, and sulfur content, etc.) and subjective (visual, olfactory, and tasting panels) standards, thus guaranteeing minimum quality requirements and territorial differentiation in a field that has in the past mainly focused on the local consumption and large-scale wine production.[78]

The importance of quality and the introduction of quality-management systems in wine production have become increasingly popular tools in recent decades. The most widely used systems in the wine sector are HACCP and ISO certification systems. Aggelogiannopoulos et al.[1] concluded that the implementation of a quality-management system compliant with ISO 9001:2000 and the adoption of a food safety system (such as HACCP) in the wine-making industry provides an appropriate framework for the organization to safeguard the qualitative properties and the safety of the final products. A quality-management system (QMS) should answer at least the following questions:

- What is required to be done?
- How will it be done?
- Is there a need for any special facility or equipment?
- Who has the authority?
- Which records are essential?

Speaking about a winery, the list of the above questions should be extended with the following aspects:

- Selection of winegrower;
- Collection of grapes;
- Cleaning and sanitation of plant facilities and equipment;
- Personal hygiene;
- Duties and responsibilities of seasonal personnel; and
- Distribution of final products.

These are referred to as work instruction manuals that help employees to describe each task in detail. As these operations have to be carried out in the exact same way every time, written instructions are necessary for proper implementation.[1]

Good Manufacturing Practice (GMP; developed in Australia) is a guideline of practices and processes for the safe manufacturing of any product. Additionally, GMP is a description of generally accepted procedures to be implemented throughout the Australian wine industry that, if left unchecked, could result in an unacceptable product. In fact, its adoption facilitates the development and implementation of advanced product quality-management and food safety programs, such as HACCP, Safety Quality Food Standard (SQF), and TQM.[103]

Having become familiar with the principles and importance of quality assurance standards, now the two most common quality assurance standards, the ISO and the HACCP, will be introduced.

3.9.2 ISO SYSTEM

The introduction of ISO:9000-based QMS for food companies is closely linked to compliance with quality-management procedures and regulatory requirements.[13] Winemaking is a time and labor-demanding process that requires full quality-management procedures so that the final wine product is enjoyable and acceptable. ISO has issued a number of wine-making standards that provide guidance on how to control and monitor the appropriate steps throughout the process (Fig. 3.9).

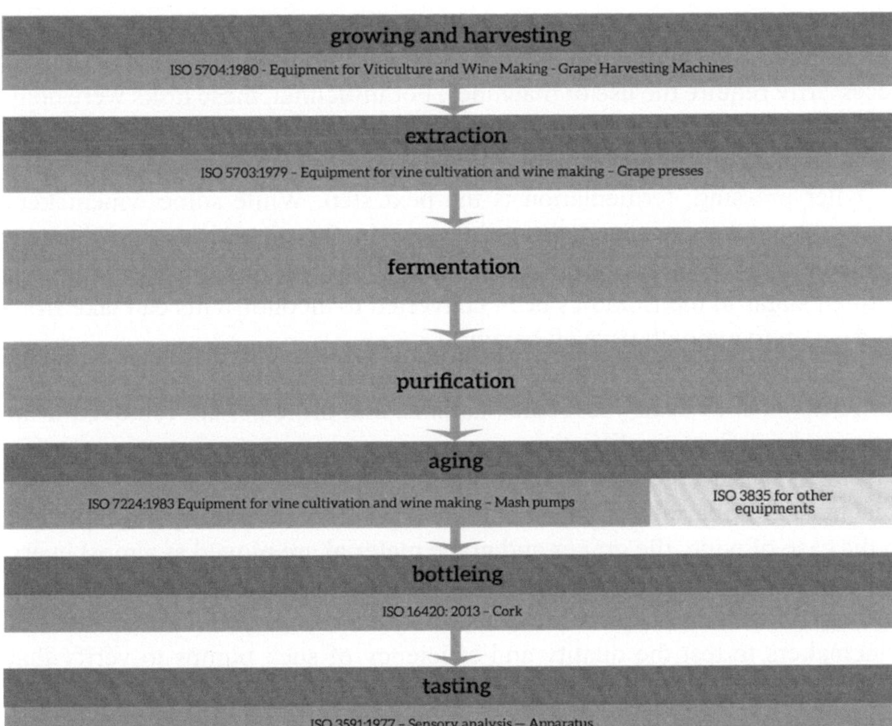

FIGURE 3.9 Steps in wine-making process and corresponding standards. Source: Self-developed with concepts from Ref. [52].

Wine making begins with growing and harvesting grapes, which eventually are used for beverages. The main concerns with grape harvesting are the efficiency of the process and its integrity throughout the process. To harvest large quantities of grapes at a time, winegrowers typically utilize a grape harvester. After the machine shakes the grapes, a tray collects the fallen grapes, which are then fed through a conveyer belt and multiple fans to rid them of leaves or other debris. ISO 5704:1980 (Grape and wine-making equipment; Grape harvesting machines; Test method) refers to test procedures for grape harvesters. The tests on grape harvesting machinery are intended to assess the efficiency in terms of grape loss and damage, to document their performance, to comprehend their total life cycle, and to observe their mechanical functioning and durability.

The next step in the process is extracting the fruit juices by pressing them. ISO 5703:1979 (Equipment for vine cultivation and wine making: Grape presses; Methods of test) provide test procedures for operating vine presses. It has been written to be applicable to both continuous and intermittent presses on grapes of various sizes and qualities to achieve higher net production of juice. During the initial steps, it is important to note that they do not necessarily require the use of machines. For millennia, these tasks were done by manual labor, dating back to the ancient Greeks and Romans. Vineyards and winemakers still occasionally use more workers than machinery.

After pressing, fermentation is the next step. While some winemakers allow this to happen from the wild yeast in the air, others expedite the fermentation by introducing cultured yeast into the juice. This continues until all sugar in the fluid has been converted to alcohol. This can take from 10 days up to a month or even beyond.

Fermentation is followed by purification, whereby all undesirable particles (such as dead yeast cells, tannins, and proteins) are removed from the wine by refining or filtration.

This is followed by the aging process, which can take place in stainless steel tanks, oak barrels, or actual bottles. The wine is moved by mash pumps. In the case of wine, the grapes and grape material are moved at almost every stage of the process. ISO 7224:1983 (Equipment for viticulture and wine making: Mash pumps) Test Methods for winegrowers provide methods for winemakers to test the quality and efficiency of such pumps to verify that they continue to be effective and do not impair the quality of the wine. The different types of equipment used in these stages are described in ISO 3835, which specifies the vocabulary of the process and includes concepts and technical drawings. It consists of a number of components:

- ISO 3835-1:1976 (Equipment for vine cultivation and wine making: Vocabulary, Part 1)
- ISO 3835-3:1980 (Equipment for vine cultivation and wine making: Vocabulary, Part 3)
- ISO 3835-4:1981 (Equipment for vine cultivation and wine making: Vocabulary, Part 4)

The wine is bottled after proper aging. It is essential that the wine is properly sealed in bottles to prevent the entry of chemicals into the air as it ages longer before being opened for drinking. For this, winemakers use corks because the bark of cork oak has qualities that leave it elastic despite the enormous pressure. The size and structure of the cork is governed by ISO 16420: 2013 (Cork: Cork stoppers for still wines (Mechanical and physical specifications); and those for sparkling wines (such as champagne) are covered under ISO 4710:2000 (Cork–Cylindrical stoppers for sparkling wines and gasified wines: Characteristics); and ISO Wine Making Standards Cork Cork-tree with harvested bark.

After making the wine and securing it with the right cork, winemakers can taste it. ISO 3591:1977 (Sensory analysis: Apparatus—Wine-tasting glass) indicates the exact dimensions and specifications of the wine glass used to test the taste of the wines so that it would release all its flavors and aromas.[52]

Numerous studies have shown that the main reasons for adopting the ISO 9000 (family of quality standards) are customer retention, marketing requirements, the necessity of process or organizational improvement, the desire for global adoption, and the absence of organization-wide focus. Implementing a quality-management system that meets the criteria of ISO 9001:2000 should be a key strategic decision for the organization. The structure and establishment of an organization's quality-management system are determined by the changing needs, specific objectives, products supplied, the process used, and the organization's size.[1]

3.9.3 HAZARD ANALYSIS AND CRITICAL CONTROL POINTS (HACCP)

The HACCP system is well-known in the food industry, and it can also be used in the wine sector. HACCP can be used from the grape harvest process to the wine production.[10]

HACCP is a very specific food security system that entails conducting analytical tests for microbiological, chemical, or physical hazards. Hazard analysis involves identifying the potential risks of a food processing operation and the identification of critical control points (CCPs) that need to be controlled to guarantee food safety.[105]

The HACCP standard includes many aspects, which have to be taken into consideration in viniculture. These aspects accompany the wine-making process and have instructions on how to avoid the different threats. Christaki and Tzia[23] collected the following hazards that are listed among the standards of wine-making steps:

- Grape growing: diseases;
- Harvesting: avoid premature, overmature harvesting, injury, and bad practices;
- Transportation: crush, oxidation, and foreign matters;
- Destemming: stem residues and inadequate cleaning;
- Crushing: pulp in the must and must oxidation;
- Juice separation: skin residues, extraction in the juice, lack of sugar, and acid;
- Skin pressing: deterioration of must quality and heavy sensory characteristics;
- Juice clarification: undesirable yeast in must and protein turbidity;
- Must storage: growth of mold and yeasts;
- Chemical adjustment of must: water concentration, sugar concentration, acidity adjustment, and must roughness;
- Filling of fermentation: must oxidation and foreign matters from pumps;
- SO_2 adjustment: smell of SO_2, not good distribution;
- Yeast inoculation: decrease in inoculated yeast activity, changes in wine quality by yeast product, yeast inoculated suffocation, and inactivation;
- Fermentation: growth of undesirable bacteria, stuck in fermentation, loss of aromatic components, production of H_2S, and oxidation due to income of oxygen in the fermenters;
- Red wine separation: brown turbidity of wine and increase in volatile acidity;
- Red pressing: sensory characteristics changes and low-quality wine;
- Racking: settling of tartaric potassium in wine and yeast revival;
- Blending: loss of sensory characteristics of blend wine and stabilization;

- Filtration: Fe_2 haze in the wine and filter taste development;
- Stabilization: brown turbidity and microbiological turbidity;
- Aging: changes in sensory characteristics, barrel flavor, and wine oxidation;
- Filling of bottles: loss of wine sensory characteristic due to oxidation, foreign matter, wine leakage from bottles, mold growth in wine bottles, and spoilage microorganisms in bottles;
- Corking: wine oxidation, contaminated corks, foreign materials in the wine from cork, and cork ruin during application; and
- Labelling: label default due to bottle's humidity, absence of information for legislation and consumers (% alcohol, composition, storage temperature, and instruction for use).

Simultaneous application of QMS and HACCP is widespread in the food and beverage (F&B) industry. The use of HACCP in the ISO-9001 quality-management system results in more effective food management than either ISO 9001 or HACCP alone; thus, increasing customer satisfaction and improving organizational efficiency. To facilitate the application of HACCP and ISO-9001 in food and beverage industry organizations, ISO published ISO 15161:2001, which provides guidance on the use of ISO 9001:2000 in the F&B industry.[1]

ISO 22000:2005 is a HACCP-based standard (based on ISO 9001:2000), which was specifically developed to guarantee food safety. TQM and ISO 9000 focus more on managerial aspects and are aimed at enhancing the quality overall. Furthermore, food producers are legally obliged to implement HACCP, while other schemes are applied on a voluntary basis in the food industry. The assurance of safe food production and transport appears to be a central objective of the F&B industry.[1]

Turning back to the initiation by Hall and Mitchell,[49] we can conclude that quality assurance systems mostly focus on the production processes and set recommendations or strict requirements in this regard. The physical environment can appeal only at the beginning point, in the grape-growing phase, mostly highlighting only the wine diseases. The third point, marketing aspects is also slightly touched upon, as the HACCP standard has regulations about the labeling.

As conclusion, we can say that the quality of wine has turned into a crucial aspect in the last decades, consumers have become very critical of food quality and food safety. Quality assurance systems are in place to develop and uphold customer confidence in the quality of the wine, both as a legal obligation and as a voluntary basis to improve quality.

3.10 SUMMARY

Grapes are among the most valuable and versatile kinds of fruit, ranked as the fifth most important in the world. The wine market is more complex today than it was in the past centuries as more and more actors become part of it from various parts of the world. Wines are inseparable from the place where they have been produced. Their character and taste are defined by the interrelationships of the grape cultivation and wine-making processes and types of grapes used and the environment. To make a winery or a wine region more visible to consumers, conscious brand image management activities are necessary, with an emphasis on increasing the overall sense of product quality.

Wine making is a painstaking and labor-intensive procedure, which requires careful quality management to assure that the end product is pleasurable. Several ISO wine-making standards have been issued to oversee all the necessary steps of the wine-making process. The well-known hazard analysis and critical control points (HACCP) system is applied in wine making. The joint application of the ISO and HACCP systems will increase consumer satisfaction and improve organizational effectiveness.

KEYWORDS

- **grape types**
- **grape production**
- **HACCP systems**
- **wine consumption**
- **wine-making standards**
- **wine quality assurance**

REFERENCES

1. Aggelogiannopoulos, D.; Drosinos, E. H.; Athanasopoulos, P. Implementation of a Quality Management System (QMS) According to the ISO 9000 Family in a Greek Small-Sized Winery: A Case Study. *Food Control* **2007,** *18* (97), 1077–1085.
2. Allan, W. Winegrape Assessment in the Vineyard at the Winery. *Aust. Viticult.* **2003,** *7* (6), 20–43.
3. Alant, K.; Bruwer, J. Wine Tourism Behavior in the Context of a Motivational Framework for Wine Regions and Cellar Doors. *J. Wine Res.* **2004,** *15* (1), 27–37.

4. Anderson, K.; Signe, N.; Vicente, P., Eds. *Global Wine Markets, 1860 to 2016: A Statistical Compendium*; University of Adelaide Press: Adelaide, 2017; pp 29–37.
5. Annunziata, A.; Pomarici, E.; Vecchio, R.; Mariani, A. Nutritional Information and Health Warnings on Wine Labels: Exploring Consumer Interest and Preferences. *Appetite* **2016,** *106*, 58–69.
6. Arroyo-García, R., Ruiz-Garcia, L.; Bolling, L.; Ocete, R.; López, M. A.; Arnold, C.; Ergul, A. Multiple Origins of Cultivated Grapevine (*Vitis vinifera* L.) Based on Chloroplast DNA Polymorphisms. *Mol. Ecol.* **2006,** *15* (2), 3707–3714.
7. Asero, V.; Patti, S. Wine Tourism Experience and Consumer Behavior: The Case of Sicily. *Tour. Anal.* **2011,** *16* (4), 431–442.
8. Ashenfelter, O.; Storchmann, K. *Wine and Climate Change*, 2014.
9. Balogh, J. M. Minőségirányítási rendszerek a borágazatban (Quality Management System in Wine Industry). *Taylor* **2016,** *8* (2), 79–86.
10. Banks, G.; Klinsrisuk, R.; Dilokwanich, S.; Stupples. Wines Without Latitude: Global and Local Forces and the Geography of the Thai Wine Industry. *EchoGéo* **2013,** *23*.
11. Bertelli, A. A. A.; Das, D. K. Grapes, Wines, Resveratrol, and Heart Health. *J. Cardiovasc. Pharmacol.* **2009,** *54* (6), 468–476.
12. Bolton, A. *Quality Management Systems for the Food Industry: A Guide to ISO 9001/2*; Springer Science & Business Media: Berlin, 1997; p 210.
13. Bruwer, J. Service Quality Perception and Satisfaction in a New Zealand Festival Scape: Buying Behavior Effects. *Tour. Anal.* **2013,** *18* (1), 61–77.
14. Bruwer, J; Karin A. The Hedonic Nature of Wine Tourism Consumption: An Experiential View. *Int. J. Wine Bus. Res.* **2009,** *21* (3), 235–257.
15. Byrd, E. T.; Bonnie C.; Yu-Chin J. H.; Keith D.; Sevil S. Wine Tourism: Motivating Visitors Through Core and Supplementary Services. *Tour. Manage.* **2016,** *52*, 19–29.
16. Campbell, G.; Guibert, N. Introduction: Old World Strategies Against New World Competition in a Globalizing Wine Industry. *Br. Food J.* **2006,** *108* (4), 233–242.
17. Caputi, L.; Carlin, S.; Ghiglieno, I.; Stefanini, M.; Valenti, L.; Vrhovsek, U.; Mattivi, F. Relationship of Changes in Rotund One Content During Grape Ripening and Winemaking to Manipulation of the Pepper's Character of Wine. *J. Agric. Food Chem.* **2011,** *59* (10), 5565–5571.
18. Cardebat, J.; Livat; F. Wine Experts' Rating: A Matter of Taste? *Int. J. Wine Busi. Res.* **2016,** *28* (1), 43–58.
19. Carlsen, P. J. A Review of Global Wine Tourism Research. *J. Wine Res.* **2004,** *15* (1), 5–13.
20. Carmichael, B. Understanding the Wine Tourism Experience for Winery Visitors in the Niagara Region, Ontario, Canada. *Tour. Geog.* **2005,** *7* (2), 185–204.
21. Charters, S. The Contemporary Wine Consumer. In: *Wine & Society: The Social and Cultural Context of a Drink*; Charters, S., Ed.; Elsevier Butterworth-Heinemann: Berlin, 2006; pp 205–226.
22. Christaki, T.; Tzia, C. Quality and Safety Assurance in Winemaking. *Food Control* **2002,** *13* (8), 503–517.
23. Christensen, L. P. *Raisin Grape Varieties: Raisin Production Manual*; Agricultural and Natural Resources Publication 3393; Oakland, CA: University of California, 2000; pp 38–47.

24. Conto, F.; Demetris V.; Mariantonietta F.; Alkis, T. Strengthening Regional Identities and Culture Through Wine Industry Cross Border Collaboration. *Br. Food J.* **2014,** *116* (11), 1788–1807.
25. Coombe, B. G.; Iland, P. G. Grape Berry Development and Wine Grape Quality. In: *Viticulture Volume I—Resources*; Dry. P. R., Coombe, B. G., Eds.; Wine Titles Pvt. Ltd.: Adelaide, 2004; pp 210–248.
26. Corigliano, M. A. The Route to Quality: Italian Gastronomy Networks in Operation. In: *Tourism and Gastronomy*; Hjalager, A. M., Richards, G., Eds.; Routledge: London, 2002; pp 156–165.
27. Cozzolino, D.; Dambergs, R. G.; Janik, L.; Cynkar, W. U.; Gishen, M. Analysis of Grapes and Wine by Near Infrared Spectroscopy. *J. Near Infrared Spectrosc.* **2006,** *14* (5), 279–289.
28. Crescimanno, M.; Galati, A. Competitiveness of Italian Wines in the International Market. *Bulgarian J. Agric. Sci.* **2014,** *20* (1), 12–22.
29. Dawson, H.; Holmes, M.; Jacobs, H.; Wade, R. I. Wine Tourism: Winery Visitation in the Wine Appellations of Ontario. *J. Vacation Market.* **2011,** *17* (3), 237–246.
30. de Oliveira, H. A. D. F. *Alternatives to Bunch Thinning in Yield Control and Its Effects on Quality of the Grapes and Wine Composition in Vitis vinifera L.*, 2018. https://sigarra.up.pt/fcup/pt/pub_geral.show_file?pi_doc_id=183169 (accessed on Mar 20, 2022).
31. Dodd, T.; Gustafson, A. W. Product, Environment and Service Attributes That Influence Customers Attitudes and Purchases at Wineries. *J. Food Products Market.* **1997,** *4* (3), 41–59.
32. Eperjesi, I.; Horváth, Cs.; Sidlovits, D.; Pásti, Gy.; Zilai, Z. *Borászati technológia (Wine Technology)*; Mezőgazda Kiadó (Farmer for Rent): Budapest, 2010; p 8.
33. Evans, R. G. *The Art of Protecting Grapevines from Low Temperature Injury*; Proceedings ASEV 50th Anniversary Annual Mtg.: Seattle, WA, 2000; pp 60–72.
34. Fernandes, T.; Cruz, M. Dimensions and Outcomes of Experience Quality in Tourism: The Case of Port Wine Cellars. *J. Retail. Consumer Serv.* **2016,** *31*, 371–379.
35. Festa, G.; Cuomo, M. T.; Metallo, G.; Festa, A. The Evolution of Wine Marketing Mix: From the 4Ps to the 4Es. *J. Bus. Res.* **2016,** *69* (5), 1550–1555.
36. Figueiral, I.; Bouby, L.; Buffat, L.; Petitot, H.; Terral, J. F. Archaeo-Botany, Vine Growing and Wine Producing in Roman Southern France: The Site of Gasquinoy (Béziers, Hérault). *J. Archaeol. Sci.* **2010,** *37*, 139–149.
37. Foti, V. T.; Pilato, M.; Timpanaro, G. Assessment of Result from Quality Control Systems in the Sicilian Winemaking Industry Through the Use of Multivariance Analysis. *New Medit.* **2011,** *10* (3), 39–48.
38. Fountain, J.; Fish, N.; Charters, S. Making a Connection: Tasting Rooms and Brand Loyalty. *Int. J. Wine Bus. Res.* **2008,** *20* (1), 8–21.
39. Gallenti, G.; Galli, M. The New Competitiveness in Rural Space: Some Methodological Remarks About a Case Study. *Agric. Medit.* **2002,** *132* (2), 148–165.
40. Getz, D. *Explore Wine Tourism: Management Development and Destinations*; Cognizant Communication Corporation: New York, 2000; pp 32–44.
41. Getz, D.; Brown, G. Critical Success Factors for Wine Tourism Regions: A Demand Analysis. *Tour. Manage.* **2006,** *27* (1), 146–158.
42. Giuseppe F.; Vrontis, D.; Thrassou, A.; Ciasullo, M. V. A Value Co-creation Model for Wine Tourism. *Int. J. Manage. Pract.* **2015,** *8* (3), 247–267.
43. Goodman, D. Rural Europe Redux? Reflections on Alternative Agro-Food Networks and Paradigm Change. *Sociologia Ruralis* (Rural Sociology) **2004,** *44* (1), 3–16.

44. Gu, Q.; Zhang, H. Q.; King B.; Huang S. Wine Tourism Involvement: A Segmentation of Chinese Tourists. *J. Travel Tour. Market.* **2018,** *35* (5), 633–648.
45. Guasch-Jané, M. R.; Andrés-Lacueva, C.; Jáuregui, O.; Lamuela-Raventós, R. M. First Evidence of White Wine in Ancient Egypt from Tutankhamun's Tomb. *J. Archaeol. Sci.* **2006,** *33* (8), 1075–1080.
46. Guidetti, R.; Beghi, R.; Bodria L. Evaluation of Grape Quality Parameters by a Simple vis/NIR System. *Trans. Am. Soc. Agric. Biol. Eng.* **2010,** *53* (2), 1–8.
47. Hall, C. M. Wine Tourism in New Zealand. In: *Proceedings of Tourism Down Under II: A Tourism Research Conference*; Dunedin, University of Otaga, 1996; pp 12–22.
48. Hall, C. M.; Macionis, N. Wine tourism in Australia and New Zealand. In: *Tourism and Recreation in Rural Areas;* Butler, R. W., Hall, C. M., Jenkins, J. M., Eds.; John Wiley & Sons: Chichester, 1998; pp 197–224.
49. Hall, C. M.; Mitchell, R. *Wine Marketing: A Practical Guide*; Elsevier Ltd.: Oxford, 2008; pp 43–50.
50. Iland, P. *Monitoring the Winemaking Process from Grapes to Wine Techniques and Concepts*, 2nd ed.; Patrick Iland Wine Promotions: Adelaide, 2004; pp 21–29.
51. International Organization of Vine and Wine (OIV). *2019 Wine Statistics*. http://www.oiv.int/en/oiv-life/2019-wine-productionnbspfirst-estimations (accessed on Mar 16, 2021).
52. *ISO Wine Making Standards*, **2016**. https://blog.ansi.org/?p=7224 (accessed on Mar 16, 2021).
53. Kana, K.; Kanellaki, M.; Kouinis, J.; Koutinas, A. A. Alcohol Production from Raisin Extracts: Volatile by Products. *J. Food Sci.* **1998,** *53* (6), 1723–1724.
54. Katona, Z. *Magyarország szőlőtermő területei és térszerkezet-változásainak térinformatikai minősítése.* Pécsi Tudományegyetem Természettudományi Kar Földtudományok Doktori Iskola, Doktori (PhD) értekezés (GIS classification of wine-growing areas and spatial structure changes in Hungary. Hungary: University of Pécs, Faculty of Science Doctoral School of Earth Sciences. Doctoral (PhD) dissertation), 2016; p 219.
55. Kennedy, J. A.; Hayasaka, Y.; Vidal, S.; Waters, E. J.; Jones, G. P. Composition of Grape Skin Proanthocyanins at Different Stages of Berry Development. *J. Agric. Food Chem.* **2001,** *49*, 5348–5355.
56. Kim, H.; Bonn, M. A. Authenticity: Do Tourist Perceptions of Winery Experiences Affect Behavioral Intentions? *Int. J. Contemp. Hosp. Manage.* **2016,** *28* (4), 839–859.
57. Kozma, P., Ed. *szőlő és termesztése, Part I: szőlőtermesztés történeti, biológiai és ökológiai alapjai* (Grapes and Their Cultivation, Part I: Historical, Biological and Ecological Foundations of Viticulture); Akadémiai Kiadó (Academic Publisher): Budapest, 2000; pp 318.
58. Krstic, M.; Moulds, G.; Panagiotopoulos, B.; West, S. *Growing Quality Grapes to Winery Specifications: Quality Measurement and Management Options for Grape Growers*; Wine Titles: Adelaide, 2003; pp 11–19.
59. Leddy, M. A.; Williams, P. *Investigating the Relationship Between Wine Tourism and Proactive Environmental Management at Wineries*, 2013 (May); online (pdf) unpublished report. https://www.semanticscholar.org/paper/Investigating-the-relationship-between-wine-tourism-Leddy/8475fd0782f80351bb5929f8dd3307e39dba194d (accessed on June 1, 2022).
60. Lemon, K. N.; Verhoef, P. C. Understanding Customer Experience Throughout the Customer Journey. *J. Market.* **2016,** *80* (6), 69–96.

61. Longbottom, M.; Simos, C.; Krstic, M.; Johnson, D. Grape Quality Assessments: A Survey of Current Practice. *Wine Viticult. J.* **2013,** *28* (3), 33–37.
62. Loureiro, S. M. C.; da Cunha, N. P. Wine Prestige and Experience in Enhancing Relationship Quality and Outcomes: Wine Tourism in Douro. *Int. J. Wine Bus. Res.* **2017,** *29* (4), 434–456.
63. Macionis N.; Cambourne, B. Wine Tourism, Just What Is It All About? *Wine Ind. J.* **1998,** *13* (1), 41–47.
64. Mantilla, O.; Collins, C.; Iland, P. G.; Johnson, T. E.; Bastian, S. E. P. Review: Berry Sensory Assessment: Concepts and Practices for Assessing Wine Grapes' Sensory Attributes. *Aust. J. Grape Wine Res.* **2012,** *18* (3), 245–255.
65. Marzo-Navarro, M.; Pedraja-Iglesias, M. Are There Different Profiles of Wine Tourists? An Initial Approach. *Int. J. Wine Bus. Res.* **2010,** *22* (4), 349–361.
66. McGovern, P. E. *Ancient Wine: The Search for the Origins of Viniculture*; Princeton University Press: Princeton, 2003; pp 31–39.
67. Miaschi, J. *The Most Popular Fruit in the World*, **2017.** http://worldatlas.com/articles/the-most-popular-fruit-in-the-world.html (accessed on Mar 16, 2021).
68. Mitchell, R.; Hall, C. M. The Post-Visit Consumer Behavior of New-Zealand Winery Visitors. *J. Wine Res.* **2004,** *15* (1), 39–49.
69. Mitchell, R.; Hall, C. M. Wine Tourism Research: The State of Play. *Tour. Rev. Int.* **2006,** *9* (4), 307–332.
70. Montella, M. M. *Wine Tourism and Sustainability: A Review*; Economics and Business Management: Rome, 2017; pp 21–29.
71. Moraes, L. C. Irradiação de alimentos (Irradiation of Foods). *Alimentos Tecnologia, São Paulo* (Foods & Technology) **2000,** *14* (87), 34–36.
72. Murdoch, J.; Marsden, T.; Banks, J. Quality, Nature, and Embeddedness: Some Theoretical Considerations in the Context of the Food Sector. *Econ. Geog.* **2000,** *76* (2), 107–125.
73. Musso, F.; Francionia, B. Agri-Food Clusters, Wine Tourism and Foreign Markets: The role of Local Networks for SME's Internationalization. *Procedia Econ. Finance* **2015,** *27*, 334–343.
74. Myles, S.; Boyko, A. R.; Owens, C. L.; Brown, P. J.; Grassi, F.; Aradhya, M. K.; Bustamante, C. D. Genetic Structure and Domestication History of the Grape. *Proc. Natl. Acad. Sci.* **2011,** *108* (9), 3530–3535.
75. Núñez, D. R.; Walker, M. J. Review of Palaeobotanical Findings of Early *Vitis* in the Mediterranean and of the Origins of Cultivated Grape-Vines, with Special Reference to New Pointers to Prehistoric Exploitation in the Western Mediterranean. *Rev. Palaeobot. Palynol.* **1989,** *61* (3–4), 205–237.
76. Nyaga, G. N.; Whipple, J. M. Relationship Quality and Performance Outcomes: Achieving a Sustainable Competitive Advantage. *J. Bus. Logist.* **2011,** *32* (4), 345–360.
77. Pangavhane D. R.; Sawhney R. L. Review of Research and Development Work on Solar Dryers for Grape Drying. *Energy Conv. Manage.* **2002,** *43* (1), 45–61.
78. Parga-Dans, E.; Alonso González, P. 'Marketing Quality' in the Food Sector: Towards a Critical Engagement with the Quality Turn in Wine. *Geoforum* **2017,** *85*, 5–8.
79. Patil, V. K.; Chakrawar, V. R.; Narwadkar, P. R.; Shinde, G, S. Grape. In: *Handbook of Fruit Science and Technology*; Salunkhe, D. K., Kadam. S. S., Eds.; CRC Press: Boca Raton, 1995; pp 23–54.

80. Pérez-Calderón, E.; Ortega-Rossell, F.; Milanés-Montero, P. The wine routes of Spain products club: The case of the Ribera of Guadiana Wine Route (Spain). In: *Wine and Tourism: A Strategic Segment for Sustainable Economic Development;* Peris-Ortiz M., Del Río Rama D. L. C. M., Rueda-Armengot C., Eds.; Springer International Publishing: Cham: 2016; pp 107–121.
81. Pikkemaat, B.; Peters, M.; Boksberger, P.; Secco, M. The Staging of Experiences in Wine Tourism. *J. Hosp. Market. Manage.* **2009,** *18* (2–3), 237–253.
82. Pine II, B. J.; Gilmore, J. H. *The Experience Economy: Work Is Theatre & Every Business*; 1st ed.; Harvard Business School Press: Harvard, 1999; p 272.
83. Pomarici, E.; Vecchio, R.; Mariani, A. Wineries' Perception of Sustainability Costs and Benefits: An Exploratory Study in California. *Sustainability* **2015,** *7* (12), 16164–16174.
84. Ponte, S. Governing Through Quality: Conventions and Supply Relations in the Value Chain for South African Wine. *Sociologia Ruralis* (Rural Sociology) **2009,** *49* (3), 236–257.
85. Quadri–Felitti, D. L.; Fiore, A. M. Destination loyalty: Effects of Wine Tourists' Experiences, Memories, and Satisfaction on Intentions. *Tour. Hosp. Res.* **2013,** *13* (1), 47–62.
86. Quadri–Felitti, D. L.; Fiore, A. M. Experience Economy Constructs as a Framework for Understanding Wine Tourism. *J. Vacat. Market.* **2012,** *18* (1), 3–15.
87. Quadri–Felitti, D. L.; Fiore, A. M. Wine Tourism Suppliers' and Visitors' Experiential Priorities. *Int. J. Contemp. Hosp. Manage.* **2016,** *28* (2), 397–417.
88. Robinson, J.; Harding, J., Eds. *The Oxford Companion to Wine*; American Chemical Society: Philadelphia, 2015; pp 37–46.
89. Rodrigriguez-Delgado, M. A.; Gonzalez-Hernandez, G.; Conde-Gonzales, J. E.; Perez-Trujullo, J. P. Principal Component Analysis of the Polyphenol Content in Young Red Wines. *Food Chem.* **2002,** *78* (4), 523–528.
90. Rousseau, J.; Delteil, D. Presentation d'une methode d'analyse sensorielle des baies de raisin: Principe, méthode, interpretation (Presentation of a Method Of Sensory Analysis of Grape Berries: Principle, Method, Interpretation. *Revue Française d'Oenologie* (French Review of Oenology) **2000,** *183*, 10–13.
91. Santillo, A. G.; Rodrigues, F. T.; Arthur, P. B.; Villavicencio, A. C. C. H. Sensory Analysis in Grapes Benitaka. *International Nuclear Atlantic Conference, INAC 2011*, Belo Horizonte, MG, Brazil, October 24–28, 2011; Unpublished paper; p 8.
92. Satisha J.; Sharma A. K.; Adsule P. G. Rootstock Influence on the Biochemical Composition and Polyphenol Oxidase Activity of 'Thompson Seedless' Grapes and Raisins. *Int. J. Fruit Sci.* **2014,** *14* (2), 133–146.
93. Savo, V.; Kumbaric, A.; Caneva, G. Grapevine (*Vitis vinifera* L.) Symbolism in the Ancient Euro-Mediterranean Cultures. *Econ. Bot.* **2016,** *70* (2), 190–197.
94. Schlüter, R.; Norrild, J. Ecotourism in Argentina: The Power of Wine to Promote a Region. In: *Tourism in Latin America: Cases of Success*; Netto, A. P.; Trigo, G. L. G., Eds.; Springer International Publishing: Cham, 2015; pp 71–84.
95. Schultz, H. R.; Jones, G. V. Climate Induced Historic and Future Changes in Viticulture. *J. Wine Res.* **2010,** *21* (2–3), 137–145.
96. Slåtten, T.; Mehmetoglu, M.; Svensson, G.; Sværi, S. Atmospheric Experiences That Emotionally Touch Customers: A Case Study from a Winter Park. *Manag. Serv. Qual. Int. J.* **2009,** *19* (6), 721–746.

97. Souza, L. H. L. M*anipulação inadequada dos alimentos: fator de contaminação* (Inadequate Handling of the Foods: Contamination Factors). *Higiene Alimentar* (Food Health) **2006,** *20* (146), 32–39.
98. Sparks, B. Planning a Wine Tourism Vacation? Factors That Help to Predict Tourist Behavioral Intentions. *Tour. Manage.* **2007,** *28* (5), 1180–1192.
99. Spence, C. Multisensory Experiential Wine Marketing. *Food Qual. Preference* **2019,** *71*, 106–116.
100. Surak, J. G.; Simpson, K. E. Using ISO 9000 Standards as a Quality Framework. *Food Technol.* **1994,** *48* (9), 63–65.
101. Szolnoki, G.; Dolan, R.; Forbes, S.; Thachd, L.; Goodman, S. Using Social Media for Consumer Interaction: An International Comparison of Winery Adoption and Activity. *Wine Econ. Policy* **2018,** *7*, 109–119.
102. Telfer, D. The Northeast Wine Route: Wine Tourism in Ontario, Canada and New York State. In: *Wine Tourism around the World: Development, Management and Markets;* Hall, C. M., Sharples, L., Cambourne B., Macionis, N., Eds.; Elsevier: Oxford, 2000; pp 253–271.
103. The Australian Wine Research Institute. *The Code of Good Manufacturing Practice for the Australian Grape and Wine Industry*, 2012.
104. This, P.; Lacombe, T.; Thomas, M. R. Historical Origins and Genetic Diversity of Wine Grapes. *Trends Genet.***2006,** *22* (9), 511–519.
105. Tzia, C.; Tsiapouris, A. *Application of the Hazard Analysis Critical Control Point (HACCP) System in the Food Industry*; Papasotiriou: Greece, Athens, 1996; pp 17–32.
106. Van Leeuwen, C.; Darriet, P. The Impact of Climate Change on Viticulture and Wine Quality. *J. Wine Econ.* **2016,** *11* (1), 150–167.
107. Vandecandelaere, E. *Les "routes des vins": des réseaux territoriaux pour valoriser levin au travers de son territoire–les cas du Languedoc Roussillon, de Mendoza et du Western Cape* (The "Wine Routes": Territorial Networks to Promote Wine Through Its Territory—The Cases of Languedoc Roussillon, Mendoza and the Western Cape). *Conference Systèmes Agroalimentaires Localisés: produits, entreprises et dynamiques locales* (Conference on Localized Agrifood Systems: Products, Companies and Local Dynamics), October 16–18, 2002; INRA, Montpellier; Unpublished Paper; p 8.
108. Vázquez, A. M.; González, P. A. Managing Collective Symbolic Capital Through Agro-Food Labelling: Strategies of Local Communities Facing Neoliberalism in Spain. *J. Rural Stud.* **2015,** *41*, 142–152.
109. Wang, H. Value as a Medical Tourism Driver. *Manag. Serv. Qual. Int. J.* **2012,** *22* (5), 465–491.
110. Weaver, R. J. *Grape Growing*; John Wiley & Sons: New York, 1976; pp 27–35.
111. Wen, J.; Harris, A. J.; Kalburgi, Y.; Zhang, N.; Xu, Y.; Zheng, W.; Zimmer, E. A. Chloroplast Phylogenomic of the New World Grape Species (*Vitis vinifera* L.). *J. System. Evol.* **2018,** *56* (4), 297–308.

PART II

SUSTAINABLE ETHNIC FOODS VERSUS GASTRONOMY

CHAPTER 4

POTENTIAL OF ACKEE IN AFRICAN CUISINE

FAUSTINA DUFIE WIREKO-MANU, YAW GYAU AKYEREKO, JACOB K. AGBENORHEVI, CHRISTIAN ARHINFUL, BAFFOE KYEI-ASANTE, JOANA KONADU ATTAFUAH, DIVINA ALLOTEY, and IBOK ODURO

ABSTRACT

Indigenous but underutilized food crops are currently receiving more research attention due to their potential to help feed the growing global population and achieve Sustainable Development Goal Two, which aims to "end hunger, achieve food security, improve nutrition, and promote sustainable agriculture by 2030." This objective includes domesticating, preserving, and making the best use of the currently available but underutilized agricultural commodities, as well as ensuring the doubling of agricultural commodity production. The ripe ackee (*Blighia sapida*) fruit aril is edible and has considerable nutritional value, but has been unexploited for food applications in the West African sub-region. Literature shows only a few people know and consume ackee fruit arils, as boiled or roasted. A couple of research has been done on ackee with results showing its potential as food in diverse ways. The potential of ackee as a food to improve diets and support food and nutrition security in Africa is highlighted in this chapter.

4.1 INTRODUCTION

At the onset of the new millennium, world leaders consented to a total of eight goals to serve as the developmental framework of the world for the next 15 years. Although in retrospect the program has been largely successful, food security, hunger, and poverty alleviation have been elusive.[21]

Sustainable and Functional Foods from Plants: Health Impact, Bioactive Compounds, and Production Technologies. Megh R. Goyal, Arijit Nath, and Zoltán Kovács (Eds.)
© 2024 Apple Academic Press, Inc. Co-published with CRC Press (Taylor & Francis)

Studies have revealed that only 12 commodities supply 75% of the world's food.[29] This implies that restoring investment and interest in the numerous local crops may provide a remedy to nutrition and food insecurity, and the increasing loss of biodiversity.[29] Some traditional or indigenous plant species or varieties have been relegated to the background in favor of exotic ones even though they may help to improve nutrition, health, and local economies, bring suppleness to climate change, improve agricultural biodiversity, and help preserve culture.[32]

The globe, especially Africa, can benefit from efforts to reduce hunger and enhance nutrition by preserving, domesticating, and using indigenous vegetable species and wild fruit. One such wild species is *Blighia sapida*, which has several local names including *Ackee*, *Guinep*, *Akye fufo*, *Ankye*, and many more in various local dialects.

Ackee (*B. sapida*) is one of the woody perennial multi-purpose fruit trees in the *Sapindaceae* family.[28] It is a native of the Guinean forest in West Africa but found in many other areas like Caribbean islands, Southern California, Florida, South and Central America, and India.[35] It produces fruits with arils and seeds, which have been reported to contain substantial amounts of nutrients.[5,35] The ripe ackee fruit aril is edible and has considerable nutritional value, but has been unexploited for food applications in the West African sub-regions.[18] Literature shows only a few people know and consume ackee fruit arils, as boiled or roasted.[24,20,62,63] A couple of research has been done on ackee with results showing the potential to contribute to nutrition and food security. Dossou et al.[18] reported that flour from ackee arils could be used as a functional ingredient in food systems.[18,19] Food security can be evenly achieved when indigenous wild fruits are exploited, domesticated, preserved, and utilized to the greatest extent.

This chapter explores the potential of ackee (a nutritious but underutilized fruit in Ghana) as food to enrich our daily diet, and it also focuses on food and nutrition security in Africa.

4.2 HISTORY OF ACKEE PLANT

Ackee (*B. sapida*) is a wild plant, which grows predominantly in African forests. Though the exact location of origin is unknown, some literature claims it is from the Guinean forests of West Africa.[50] It is believed that it had been transported on a slave ship to Jamaica from West Africa. In 1788, it was acknowledged that Thomas Clarke introduced ackee to the Eastern Parishes.[50] Captain William Blighia was the first to introduce the scientifically unnamed plant to the Kew Gardens in England in 1793. The tree was then named, as *B. sapida* by Koenig in honor of Captain William Blighia.[44,45,62] Research areas that have been explored include: identifying the natural toxic compounds in various parts of the plant and how they change depending on the processing techniques (such as cooking), identifying various phytochemicals and their medicinal and aesthetic potentials, developing propagation techniques and developing protocols for the production of ackee somatic embryos.[16,33,34,37,38,42,54,71,76]

Information regarding the origin of ackee in Ghana is lacking and the plant still remains underutilized. However, ongoing research into the potential use of the ackee plant is being done by the Department of Food Science and Technology, Kwame Nkrumah University of Science and Technology, College of Science, Kwame Nkrumah University of Science and Technology, Kumasi, Ghana in an attempt to include it in the food basket of the country.

4.3 MORPHOLOGY, DEVELOPMENT, AND DISTRIBUTION OF ACKEE

Ackee grows to 7–25 m high, 2.5 m in girth, with an intense green crown. The leaves have long stout stalks about 3.5–23 cm in length, having leaflets of about 5–15 cm long and 3.5–7.5 cm in diameter. The leaflets are obovate with the lowest part quite circular and near the base of the stalk.[43] The flowers are usually small, greenish-white in hue, and the flowering period is usually around October–March. Fruiting begins from March to September. The fruit in the shape of a capsule (i.e., obovoid) is green when young, turning yellow or bright red when mature, and ripened to about 3.5–6 cm × 3–5 cm in size, often splitting open on ripening on the tree (Fig. 4.1). The aril is found at the base of the seed and attaches it to the pod. The aril is the non-toxic part of

FIGURE 4.1 Ackee fruit: Self-developed with concepts from Keay.[34].

the fruit. The seed typically has a glossy testa (~2.5 cm × 2 cm in diameter), whereas the wrinkled arils may be pale yellow or cream in color about 2 cm long.[43] On average, an ackee pod contains three pegs, rarely four or five.[50]

The leaves have an alternate compound orientation having three to five pairs of glossy leaflets. The flowering part of the plant is about 7.61 to 17.77 cm with clusters of greenish-white fragrant flowers, each with five petals in axillary racemes (Figs. 4.2 and 4.3). Flowers are often insect pollinated. Ripe ackee fruit may be yellowish red in color with a wide-opened appearance displaying the seeds (Fig. 4.4). Unripe fruits are usually green to yellow in color and possess a closed appearance; making the seeds extremely difficult to be visualized.

Ackee's distribution is in the confines of the Guinea forests in West Africa (Ghana, Liberia, Senegal, Nigeria, Burkina Faso, Cameroon, etc.), the Caribbean islands (Haiti, Jamaica, Trinidad and Tobago), two states in the USA (Southern California and Florida), South and Central America, and India.[5,50] It requires annual rainfall ranging between 750 and 2500 mm, temperatures between 21 and 27°C and tolerates partially alkaline soils of pH range between 5.5 and 7.5, implying ackee's ability to grow well in the forest, semi-deciduous forest, savanna, and in the coastal belts of Ghana.[50]

FIGURE 4.2 Ackee tree bearing fruits: Self-developed with concepts from Mitchelle et al.[50]

FIGURE 4.3 Ackee leaves and Inflorescence (left), unripe ackee fruits (middle and right): Self-developed with concepts from Keay[43] and Mitchelle et al.[50]

FIGURE 4.4 Ripe ackee fruit with the different parts (seed: black; pod: light red; Aril: center whitish; and membrane): Self-developed with concepts from Mitchell et al.[50]

4.4 VARIETIES AND PROPAGATION OF ACKEE

Two varieties of ackee are typically known, such as cheese and butter varieties (Fig. 4.5). The cheese variety is characterized by hard, cream-colored aril, which maintains its shape during cooking; whereas the butter variety has soft, yellow arils which readily lose their shape during cooking.[50] For this reason, the cheese variety is often preferred for export than the butter variety.

FIGURE 4.5 Butter variety (left) and cheese variety (right): Self-developed with concepts from Mitchell et al.[50] and Keay.[43]

The plant can tolerate various types of soil but thrives in those well-drained and fertile. It can also grow in shallow clay soil, without an impenetrable layer or with lowered water table.[61] Propagation of the ackee plant is achieved either by direct seed planting unto the field, seeds raised on a nursery bed before transplanting unto the field, or by vegetative propagation (which employs stem cutting or air-layering).[50] From 3 to 6 years of seed cultivation, the ackee plant bears green immature fruits that turn bright red and splits open upon ripening. Despite having an all-year-round fruit-bearing period, the ackee plant has the following two major fruit-bearing seasons: Spring (from January to March) and summer (June to August).[53]

4.5 PHYSIOLOGY OF ACKEE

The fruit of the ackee (Fig. 4.4) is green when young, reaching 7.7–10-cm long, and becoming yellow to red at complete maturation.[50] The edible part of the fruit is the aril after removing the seed and membrane. The normal ackee fruit pod usually has two to five pegs in its pods.[11] There is usually a process of fruit splitting to show the inside seeds[10,25] as shown in Fig. 4.5.

Recent studies conducted on the physiological and biochemical characteristics or changes of ackee fruits show that sugars increase during the maturation, but reduce when fully ripe. Nevertheless, there is gradual sugar increment resulting from the conversion or transformation of complex sugars in the course of maturation and ripening.[26] Phenolic compounds and anthocyanins are reported to decrease drastically during ripening and maturation.[25] The decrease in total phenolic compounds and anthocyanins may be ascribed to the breakdown of these compounds as maturation and ripening occur.

4.6 DOMESTICATION OF ACKEE

Domestication is basically the act of exploiting wild plants for selection and cultivation taking into account their varieties, ecological adaptability, botanical characteristics, and usefulness.[28] According to the survey conducted by Ekue et al.[21] in Natemba and Yoruba, farmers in Benin are motivated for conserving ackee for the food, soap, shade and for medicinal purposes. Others also conserve ackee trees in their farms for the shade as well as the economic value. Study carried out in Jamaica revealed that about US $400 million is generated annually from the sale of the ackee fruit arils.[60]

With respect to botanical description and classification, farmers in Benin have identified variability in ackee trees in terms of plant height and diameter, bark color, fruits productivity, fruit size, phenology, and others based on the growing conditions of the cultivated area.[30] Ecological zone studies also revealed that plants in the Northern Sudan are taller and bear bigger fruits than those from the Southern region. The study further revealed that the local people in Sudan preferred the cultivated ackee fruits due to desirable qualities (such as: size, color, taste and oil content of the fruit arils) that are more predominant than the wild types.[20]

The research studies have been conducted on the issues regarding ackee's domestication with respect to botanical, ecological, geographical, cultivation and harvesting techniques and genetic factors.[20] This has really prepared the foundation for its domestication if considered as potential food security crop. Domestication of *B. sapida* in Africa can lead to income generation and job creation.

4.7 CHEMICAL COMPOSITION AND SAFETY RELATED ISSUES: B. SAPIDA

In their early stages of growth, the edible ackee arils contain higher concentrations of the poison hypoglycin. The fruit's toxic content eventually drops to a level that is safe for human eating when it ripens. Only eating the unripe ackee fruit results in toxic effect. To avoid any potential negative health impacts, it is essential to distinguish ripe ackee fruit from an unripe one.[20]

The metabolic syndrome known as "Jamaican vomiting sickness" can be brought on by eating unripe ackee fruit. This illness can manifest as excessive vomiting, impaired mental status, and hypoglycemia. The onset of hypoglycemia is slow at first but grows swiftly after a few hours.[9] Death has been reported during 12–48 h of infection. Seizures, hypothermia, comas, and even death have all been described in severe cases.

With intravenous fluids and dextrose, medical treatment serves as the primary form of supportive care. Although the disease's epidemiology has not been well documented, death and morbidity rates continue to be underreported.

After consuming an unripe ackee fruit, gastrointestinal (GI) symptoms often appear 6–48 h later, though in more severe cases, symptoms may appear sooner. While vomiting is common, it's crucial to remember that it does not necessarily rule out the diagnosis.

Ackee poisoning can typically be distinguished from other acute GI illnesses by the absence of diarrhea.[5] As shown in 85% of declared critical cases, seizures may be a predictor of disease severity.

Tremors, paresthesia, and altered mental status are a few examples of neurological symptoms. Quantifying the amount consumed is crucial because toxicity is dose dependent. If recovery takes place, it does so within a week of consumption. Ackee poisoning is treated with intravenous fluids and dextrose, if necessary.5

Blood glucose, serum electrolytes, liver function tests, renal function, lactate, ketones, and blood gasses are all included in the diagnostic examination. Laboratory tests, particularly those for glucose and blood electrolytes, must be closely monitored in symptomatic individuals. Before lab diagnostic fallouts, an early electrocardiogram (ECG) may show signs of electrolyte imbalance. The toxin hypoglycin and its metabolite, methylene cyclopropyl acetic acid (MCPA), may prevent fatty acids from being beta-oxidized, causing less glycogen to be stored in the body. Blood and urine tests can be used to check for hypoglycin and MCPA, but if they are quickly removed, they might not show positive results. Diagnostic tests that aid in determining the treatment or diagnosis of hypoglycin and MCPA toxicity in the body include high serum/urine carnitine (long-chain fatty acids transporter) concentrations and high urinary dicarboxylic acid concentrations.[9]

Unripe fruit's level of toxicity is not reduced by cooking. Ackee seeds are poisonous, and should not be eaten. The risk factors for ackee poisoning include eating an unripe ackee, buying tampered ackee, and reusing water that has previously been used to cook an unripe ackee.[14]

4.8 NUTRITIONAL COMPOSITION OF ACKEE'S ARILS

Ackee's arils nutritional composition is dependent on the variety, processing method, and/or origin of the plant.[18] Research has shown that the nutritional composition of ackee is comparable to that of many popular cereals, legumes, and oil seeds. Ackee arils contain a sufficient amount of protein, fat, fiber, micronutrients, and some beneficial phytochemicals essential for the growth, development, and prevention of diseases.

4.8.1 PROXIMATE COMPOSITION

Oyeleke et al.[59] indicated that the ackee's proximate composition is dependent on the origin, variety, and/or method of processing of the ackee arils. Studies conducted on the nutritional composition of ackee from Ghana and some other countries are presented in Table 4.1.

The moisture content of ackee arils flour ranges between 4.83 and 5.20%.[18] Similar studies also reported ackee with 3.95%,[59] 6.84%,[58] 6.96–8.39%[20], and 13.8%[4] of moisture content. The differences in the moisture content may be attributed to the analytical method used. The moisture content of the ackee arils flour is lower than or comparable with stable products such as *Phaselous vulgarus* (common bean) flour (12%) and cashew nut (7.2%) reported by Ouattara et al.[58] This implies that ackee arils flour has the potential of remaining on the shelf for a long time if processed into storable products. Also, the lower moisture content of the ackee arils flour offers advantages in terms of drying, milling, oil extraction, and many other processing operations. According to Codex Standard (1985), moisture content up to 15.5% is recommended for wheat flour.[15] Food with low moisture content is said to have a longer shelf-life since moisture, necessary for microbial growth, is limited. Therefore, the low moisture contents recorded for the various ackee flours are indicative of extended shelf-life.

TABLE 4.1 Proximate Composition of Ackee Aril Flour (w/w %): Self-developed with concepts from Akintayo et al.[4]; Falloon et al.[28]; Ouattara et al.[58]; and Oyeleke et al.[59]

Parameter/ Source	*Moisture*	*Protein*	*Fat*	*Fiber*	*Ash*	*Carbohydrate*
Akintayo et al.[4]	*13.80*	24.30	45.50	4.23	5.60	6.53
Dossou et al.[18]	4.83–5.20	10.94–11.67	51.6–56.66	3.83–4.08	8.56–9.05	14.41–20.62
Falloon et al.[28]	6.96–8.39	11.23	54.8	1.45	4.14	21.5
Oyeleke et al.[59]	3.95	15.27	51.58	16.14	6.20	6.86
Quattara et al.[58]	6.84	11.99	45.32	3.21	4.90	24.43

The quantity of fat in ackee arils ranges between 45.32 and 56.66%.[4,18,28,58,59] The fat content is very comparable to some high oil yielding crops, such as *Arachis hypogaea*—45%, *Moringa oleifera*—41.58%, sunflower—47.5%, peanut—46.0%, safflower—38.5%, cottonseed kernel—36.3% and flaxseed—34.0% according to McKevith[46] and Ouattara et al.[58] The comparatively high fat content of ackee arils makes it an alternative rich source for domestic and industrial applications. The application of this high fat content ackee arils in food will provide the medium for the solubilization of fat-soluble vitamins, thereby reducing vitamin deficiencies in human nutrition and contributing to food security. According to the United States Department of Agriculture (USDA) National Nutrient Database for Standard Reference,[75] the defatted ackee flour has a crude fat level that is similar to that of defatted peanut flour (21.90%). A study by Anderson-Foster et al.[5] also established that the characteristic properties of ackee oil position it as a potential pharmaceutical base and possible use as an alternative vegetable oil source to conventional vegetable oils. The defatted ackee aril flours have a reduced fat content which is implicative of increased shelf stability of the flour. The extraction of fat from ackee arils may generate income for farmers and create job opportunities for the actors along the value chain. If this is taken into account as a crop for food security in Africa, it will significantly enhance the indigenes' economic situation.

Ash levels in the flour samples made from ackee aril ranged from 4.14% to 9.05% a dry basis.[4,18,28,58,59] These values are very similar to those found for jack bean, pigeon pea, and cowpea flours, which have ash contents of 6.51, 4.58, and 4.73%, respectively,[7] suggesting that ackee arils flour could perform the same function as these flours in the food system if given the necessary consideration as an African food security crop. The high ash content of ackee aril flours presupposes that ackee is very rich in minerals and may therefore play a major responsibility in the prevention of micronutrient deficiencies in human nutrition. It also means that ackee aril flours have a significant destructive effect on spoilage microbial organisms that may infect the product during storage since the high ash content (high minerals) will create a high osmotic environment (through water absorption) that may prevent the thriving of these microbes.[68,69]

The protein content of ackee arils ranges from 10.94 to 24.30%,[4,18,28,58,59] depending on the varieties, geographical area, soil type, and sunlight destruction.[58] These plants (ackee), grown as ornamental plants in Africa, have protein contents, which are equivalent to protein–rich crops, such as: cowpea (18–35%) and pigeon pea (19.8%) reported by Stancheva et al.[70] and Singh et al.,[67] respectively. Admittedly, some varieties of ackee have low protein levels and may require protein complementation from other crops to meet the recommended daily dietary allowance. The Food and Agriculture Organization recommends minimum protein content of 12–15% in the diet.[77] The defatted ackee aril flours (22.04–23.00% protein content)[18] meet the minimum recommended protein requirement by WHO.

4.8.2 MINERAL COMPOSITION

According to Dossou et al.,[18] Potassium is a principal component.[4,28] Ackee also contains calcium, magnesium, sodium, zinc, and phosphorus,[4,18,36,59] which are comparable to predominant cereals crops like wheat, sorghum, millet, maize, and rice.[1] All these nutrients are essential for blood circulation, impulse transmission, osmotic balance, and bone development. Micronutrient deficiencies are major public health concerns in several countries in Africa,[69] and these minerals are readily available in this under-utilized fruit; ackee, which can help to manage such cases if given priority as a food security crop in Africa. Table 4.2 indicates the mineral composition of ackee fruit arils.

4.8.3 AMINO ACID PROFILE OF ACKEE ARILS

Ackee arils have been found to contain both essential and non-essential amino acids that are needed to promote growth and development. The amino acid constituents of unripe- and ripe ackee arils are shown in Table 4.3. Arginine is the most predominant essential amino acid in ripe arils, whereas unripe arils (poisonous for consumption at this stage) predominantly contain glutamic acid.[34] Glutamic acid is a precursor for the synthesis of other amino acids, NAD, glutathione, and glucose, and also functions as an antioxidant.[51] Arginine also plays a major role in the synthesis of protein, nitric oxide, polyamines, glutamate, proline, creatine, and agmatine, which are essential in metabolic, reparative, and immune response to conditions of trauma.[51] The glutamic acid (80.80 mg/100 g) and arginine (53.54 mg/100 g) content of ackee arils[34] are significantly higher than that of cowpea being 18.30 mg/100 g for glutamic acid and 7.30 mg/100 g for arginine.[39] This implies that the use of ackee arils in food systems may confer several health and physiological functions.

TABLE 4.2 Mineral Composition of Dried Ackee Arils: Self-developed with Concepts from Akintayo et al.[4]; Howélé et al.[36], and Oyeleke et al.[59]

Mineral	ODA Nigeria (mg/100 g)	SDA Cote d'Ivoire (mg/100 g)	SDA Nigeria (mg/100 g)
Ca (Calcium)	32.6	139.67	25.07
Cu (Copper)	8.1	NR	0.09
Fe (Iron)	17.5	17.33	1.95
K (Potassium)	951	1503.3	29.52
Mg (Magnesium)	271	215.3	21.12
Mn (Manganese)	NR	NR	0.7
Na (Sodium)	124	53.17	29.2
P (Phosphorous)	240	NR	0.49
Pb (Lead)	NR	NR	ND
Zn (Zinc)	6.5	4±0	0.2

ND, not detected; *NR*, not reported; *ODA*, oven dried arils; *SDA* sun dried arils.

TABLE 4.3 Amino Acid Profile of Unripe and Ripe Ackee Arils (Fresh Weight): Self-developed with Concepts from Golden et al.[34]

Amino Acid	Unripe Aril (mg/100 g)	Ripe Aril (mg/100 g)
Alanine	34.46 ± 2.84	14.42 ± 2.90
Arginine	53.54 ± 5.99	62.88 ± 5.87
Aspartic acid	4.37± 0.24	ND
Glutamic acid	80.80 ± 7.20	26.51 ± 0.35
Glycine	11.34 ± 0.48	7.13 ± 0.64
Histidine	6.38 ± 1.10	2.84 ± 0.29
Hydroxyproline	1.64 ± 0.15	1.40 ± 0.12
Hyp-A	124.4 ± 6.70	6.40 ± 1.10
Isoleucine	2.47 ± 0.55	0.18 ± 0.00
Leucine	5.92 ± 1.04	3.77 ± 0.38
Lysine	5.68 ± 0.75	10.56 ± 1.67
Methionine	2.32 ± 0.21	5.67 ± 0.45
Phenylalanine	6.03 ± 1.56	2.65 ± 0.27
Proline	1.45 ± 0.16	0.45 ± 0.01
Serine	9.58 ± 1.50	3.06 ± 0.28
Threonine	26.86 ± 4.09	5.73 ± 0.44
Tryptophan	1.46 ± 0.49	0.24 ± 0.00
Tyrosine	2.02 ± 0.40	0.58 ± 0.08
Valine	2.52 ± 0.15	0.80 ± 0.11

4.8.4 FATTY ACID PROFILE OF ACKEE ARILS

Profile of fatty acids of ackee aril by Anderson-Foster et al.[5] revealed that the arils are relatively high in stearic, palmitic, oleic, linoleic, and linolenic fatty acids compared with wheat, millet, sorghum, maize, buckwheat, barley and rice according to Becker[12] and Adeyeye and Ajewole.[3] The major fatty acids found in ackee arils are presented in Table 4.4. Ackee arils could complement or be used as an alternative in food systems where cereals are utilized.[24] These fatty acids are known to reduce the risk of coronary heart disease when consumed in a diet.

TABLE 4.4 Fatty Acid Profile of Cheese and Butter Ackee Arils.

Fatty Acid	Cheese Variety (mg/g)	Butter Variety (mg/g)
Palmitic acid (16:0)	80.2 ± 15.6^a	42.2 ± 9.0^a
Stearic acid (18:0)	47.0 ± 9.7^a	36.7 ± 9.4^a
Oleic acid (18:1)	143.5 ± 41.3^a	146.2 ± 4.3^a
Linoleic acid (18:2)	11.4 ± 2.8^a	7.2 ± 1.7^a
Linolenic acid (18:3)	1.7 ± 0.4^a	9.9 ± 2.3^a

Values in the same row with different letter superscripts are significantly different ($p < 0.05$).

4.8.5 SENSORY CHARACTERISTICS OF PRODUCTS DEVELOPED FROM ACKEE ARILS

Essuman et al.[27] studied the sensory attributes of different chips prepared from 100% wheat flour, 80%:20% wheat–ackee aril composite flour, 60%:40% wheat–ackee aril composite flour, 50%:50% wheat–ackee aril composite flour, and 40%:60% wheat–ackee aril composite flour. They revealed that the composite flours recorded comparable attributes or scores as the 100% wheat flour with respect to appearance, taste, texture, mouth feel, and aroma. It, however, indicated that 20, 40, and 60% ackee aril flour substitution had similar scores of 6.64, 6.72, and 6.68, respectively, for overall acceptability, and these were identified as the best substitution values. Although there was a marginal difference between the overall acceptability of the composite flour chips and that of the 100% wheat flour. The detailed results of the sensory quality of chips developed from wheat and ackee aril flours using a "9—point hedonic scale" are represented in Table 4.5.

TABLE 4.5 Sensory Characteristics of Ackee Aril Products

Quality Attributes	A	B	C	D	E
Appearance	6.88	6.46	6.40	6.30	6.60
Taste	6.80	6.12	6.08	5.38	6.02
Texture	6.86	6.65	6.74	6.44	6.66
Mouth feel	7.02	6.36	6.34	5.72	6.36
Aroma	6.58	6.30	6.52	6.52	6.70
Overall acceptability	7.28	6.64	6.72	6.18	6.68

Note: A = 100% wheat flour, B = 80%:20% wheat–ackee aril flour, C = 60%:40% wheat–ackee aril flour, D = 50%:50% wheat–ackee aril flour, and E = 40–60% wheat–ackee aril flour.

4.9 HEALTH BENEFITS OF ACKEE

Ackee contributes immensely to the health and nourishment of people in its growing communities or countries. The roots, the bark, the capsules, the leaves, and the seeds are potential medicinal sources for the treatment and management of about 22 diseases in Benin.[20] Among the 22 diseases, dental caries, malaria, internal hemorrhage, fever, malaria, yellow fever, burns, dysentery, inflammation of the eye, cutaneous infections, constipation, head lice, and whitlow are the most familiar ones, which have been treated or managed by ackee.[20]

The arils of ackee have been reported to be very potent in scavenging free radicals or oxidative species responsible for causing cardiovascular diseases, cerebral ischemia, cancer, rheumatic disorder, diabetic pregnancy, respiratory disorders, and aging.[41,73] Dossou et al.[18] affirm that the high phenolic content of the arils makes it a good crop with the potential to prevent cancer and cardiovascular diseases, reduce blood pressure, and absorb and neutralize free radicals, not to mention its functionality as an anticancer and antibacterial agent. These important health benefits position ackee as a potential plant for promoting human nutrition and health, and can contribute to food security in Africa and beyond in the accomplishment of the sustainable development goa2.

4.10 PHARMACOLOGICAL EFFECTS OF PHYTOCHEMICALS IN ACKEE

Ackee is extensively used by local traditional healers in Africa, particularly in West Africa.[22,26] The plant is used extensively for the treatment of several ailments including edema, intercostal pains, conjunctivitis, and ophthalmia.

The bark pulp is applied as a liniment for the treatment of edema, as well as intercostal pains. The bark is milled with capsicum and applied as a stimulant to relieve body pains. The leaf extracts are used in the treatment of ophthalmia and conjunctivitis.[68] In several West African settings, various parts of ackee are used either alone or in combination with other herbs in several herbal preparations for therapeutic effects. It has been suggested that such preparation has proved potent in the treatment of diseases like diarrhea, constipation, hernia, backache, stomach ache, psychosis, gonorrhea, and even cancer.[67]

Phytochemicals are secondary metabolites produced by plants to enhance their survival and often have therapeutic effects in humans; hence, to affirm the therapeutic effect of ackee in the treatment of the aforementioned diseases, Olusegun and Olutomi[56] decided to assay the various phytochemicals; chemical and antimicrobial component of the husk of *B. sapida* that conferred such medicinal potential of the plant. Their investigation affirmed the presence of alkaloids, balsam, saponins, saponin glycosides, and tannins as being present, whereas volatile oils, anthracene, and cardiac glycosides were found to be absent. Saponins and tannins were found to be high in the husk of ackee and their effects on disease control and management as anticancer.

Antioxidant agents have been reported by Olusegun and Olutomi[56] and Yildirim and Kutlu.[78] The anticancer and antioxidant activities of these compounds are attributed to the possession of hydrophilic glycosides, lipophilic and derived aglycone and sugar chains, and their ability to bind to biological substances such as protein, metallic ions, pigments, and basic and large molecular compounds. These structural features and chemical properties of saponins and tannins assist in preventing cellular DNA replication and the proliferation of cancerous cells in the human body.[78] The aqueous extract of the leaves rather than methanol extract was found to be effective in the inhibition of *Staphylococcus aureus* and *Escherichia coli*, two microbes that cause several infections in humans at concentrations of 25 mg/100 g.[78]

The effect of the aqueous extract of ackee root bark on the glycemic index was assessed using normoglycemic albino rats. The study proved that ackee root bark extract exhibited a substantial hypoglycemic effect on the albino rats studied. The data supported the use of the roots of the plant in the management of diabetes.[9,30]

Uses of ackee[70] theorizes the definition of food security along the following three parameters: Food availability, food accessibility, and food utilization. Ackee is available and accessible in Ghana, and it is grown along the streets as an ornamental plant. What is left of it, therefore, is its utilization. Studies have indicated that except for dried ackee arils used as food and ackee soap, no other commercial ackee products have been reported in the West African sub-regions.[18]

This fruit is, however, grown and eaten mostly in Jamaica and some West African countries, as it serves as a relatively cheap source of nutrients, that is, proteins, lipids, minerals, and vitamins.[65] As a result of the extensive research regarding the potential applications of ackee, it has been reported to have possible uses as food, medicine, wood for furniture; charcoal, construction, cosmetics, as well as fodder for livestock.[20] This emphasizes that the plant has massive potential for commercial ventures not only for farmers but also for the health and industrial sectors.

A. Current Uses

According to Ekue,[19] the arils of ackee are consumed in several ways in West Africa. The ackee arils could be eaten raw as a snack. They could also be included in soups as an alternative to sesame seeds, peanuts, or fish. They could be dried, ground/milled into powder, and added to soups both for their oil or thickening abilities. In Jamaica, the ackee fruit is usually parboiled prior to its use. The parboiled arils are cooked with rice or added to beef or salt pork in stews for consumption. They are also lightly fried in butter and eaten with codfish.[24,62]

In Nigeria, the extracts of the root bark have been found to exert hypoglycemic effects and hence corroborate the use of the ackee roots in the management of diabetes whereas ackee leaves and pulp are used to treat conjunctivitis.[9,30,64]

Ackee leaves have been reported to be suitable fodder for the West African dwarf breed of goats, most essential during the harmattan season.[2] The leaves serve as a cheap source of feed, usually consumed by these animals for their nourishment and growth, which drastically reduces the cost of rearing these animals.

B. Food Products Developed from Ackee (Figs 4.6 and 4.7)

Ackee flour can be used/applied in the food products, such as making composite flour, baking bread, biscuits, etc.[18] Dossou et al.[18] found that defatted flour had higher values for proximate composition and functional properties as compared to the full-fat flour. Also, freeze-dried defatted flour (FDDF) had the highest values for minerals and proximate composition while oven-dried defatted flour (ODDF) recorded the highest values for functional properties investigated with the exception of swelling power and foam stability. Sahe et al.[65] developed cassava-ackee-groundnut composite cookies from ackee aril flour with good sensory properties.

Ackee bread: ackee loaf (i.e., ackee–saltfish-filled bread-like pastry) is another product that has been developed as food.[62]

Ackee quiche: A pastry developed from ackee composite flour, baked with spiced vegetables and eggs.[62]

Again, fresh ackee aril is characterized by a nutty cheese flavor and has the potential to serve as a vegetable replacement for cheese in some food preparations.[2]

FIGURE 4.6 Ackee aril stew (left) and ackee aril soup (right): Self-developed with concepts from Rashford,[62] and Emanuel and Benkeblia.[24]

FIGURE 4.7 Ackee aril biscuit (left) and Ackee aril bread (Right): Self-developed with concepts from Rashford.[62]

C. Non-Food Uses of Ackee

Ackee is rich in saponins, bioactive compounds, which make soap lather in water, and as such it can be processed into soap. The immature ackee fruits are often crushed and used as soap (Fig. 4.8). Indigenously, the pods and seeds are burnt to potash and used to prepare crude local soap. This is a common practice in several West African countries. The soaps are sold on the local markets serving as a good source of revenue for local farmers and traders, especially women.[20]

Ash obtained from burnt ackee fruit pods or exocarp is also used as insect pest repellents for cowpea, corn, and other food crops cultivation. Wood obtained from the ackee tree is used for furniture, canoes, oars, and construction because of its ability to resist termite infestation. The seeds and fruit pods (exocarp) of ackee are used as a fish poison to anesthetize and impede their movement making them easy to catch.[20,62]

FIGURE 4.8 Ackee soap: Self-developed with concepts from Rashford.[62]

D. Other Potential Uses of Ackee

It has been reported by Dossou et al.[18] that the defatted flour had lower Hauser ratio (indicating good flow probability) and hence can be used in food industrial application. The study also indicated the use of ackee in the production of sausages and mayonnaise or thickeners for soup or stew.

The novel cassava–ackee–groundnut composite cookie made by Sahe et al.[65] could be adapted for mass production since the novel cookie has added nutrients (more proteins) than the raw cassava cookie.

Based on the nutritional and physicochemical quality characteristics of ackee flour, it would have good potential as a substitute for wheat in the pastry and bakery industry. It could be exclusively used for gluten-free pastry/baked products. Its oil could also be exploited for expanded use in the cosmetic industry and also as vegetable oil.

4.11 THE ACKEE INDUSTRY: CHALLENGES AND OPPORTUNITIES FOR COMMERCIALIZATION

In Africa, the major staple food crops are maize, sorghum, millet, rice, beans, cocoyam, yam, cassava, and plantain.[22] Ackee, a major staple and cash crop in Jamaica though available and used in Africa, is left out of this list due to its few challenges that have been addressed here.

The greatest challenge facing the ackee food value chain is the underutilization of this crop.[18,27,57] However, this may stem from several challenges along the value chain, such as Low ackee cultivation output by farmers; since ackee is largely not considered a staple or cash crop across Africa, the plant still grows in the wild or is only planted as an ornamental plant to provide shade and to serve as firewood.[18] Stakeholders in the ackee value chain need to be enlightened on the massive economic potential of ackee, this may go a long way to increase the crop cultivation and subsequent use along the entire food value chain in our bid to attain food security and eradicate poverty.

Lack of support from African governments for the crop in terms of provision of agriculture extension officers, farming machinery, and logistics may also serve as a deterring factor for farmers to go into ackee cultivation. The availability of trained agriculture extension officers, who help to educate farmers on the cultivation and supply of farming tools, equipment, and agrochemicals needed to increase yield; and produce the quality of staple crops (such as maize, rice, yam, etc.) has been a great motivation for farmers to go into the cultivation of such crops at the expense of other underutilized and wild crops like ackee.[13]

Low cultivation output implies low quantities of ackee produce, which does not entice aggregators to gather the produce to further the ackee food chain. Therefore, from an economic point of view, aggregation of ackee across Africa is simply an unprofitable business because the venture is not economically viable once the product is unavailable. Although once mature ackee produce is available all year long, ackee does not have any aggregators in the food value chain across Africa. Cocoa for example has several aggregators including Cocoa Marketing Company (CMC), which buy it in bulk for export.[46] At the processing level, the challenge is the lack of varied products. Most individuals have very limited knowledge of ackee and the various delicacies that can be prepared from its produce. Processing companies are limited not only by the lack of this knowledge but also by the low quantities available for industrial use.

The story is different in Jamaica, where ackee is canned and exported (Fig. 4.9), even though the industry is faced with numerous obstacles. Scientific and technological efforts have been made toward quality assurance but concerns for safety affect the expansion of ackee on the international market. The first export of processed ackee was initiated in 1954 when 100 cases were exported to Canada by Da Costa Brothers Ltd. Far ahead, the trade was extended to Britain that witnessed an even more profitable market.[17] The main consumers of this product are principally Jamaicans, who traveled, and established down in these countries.

In 1972, the United States Food and Drug Administration (USFDA) placed an alert on the importation of canned ackee, thereby prohibiting the direct exportation of ackee from Jamaica into the US markets. The restriction was made due to the toxicity of hypoglycin-A in ackee, whose upper limit was set at 100 mg/kg in canned ackee arils. An analytical technique for hypoglycin detection to a limit of 0.1 mg/100 g of fresh weight (fw) has been developed.[8] This analytical method or protocol provides the basis for effective checks on the conditional supply of ackee into the US market since 2000. Jamaica earned 4.3 million USD in 1999 and 8.5 USD in 2002 from the exportation of ackee. However, the exportation of ackee fruit arils was halted in 2005 for about a year due to technical reasons.

At a point, ackee could be exported by only certified processors, who have established food safety and quality control measures.[66] The Central Division of the Bureau of Standards of Jamaica (BSJ) stated that "the ackee industry is worth approximately J$ 400 million in the year 2005 [~4.5 million USD] with potential for further growth". However, the markets are actually variable since the industry is export-oriented and deeply reliant on the confidence of foreign consumers.[49]

Several efforts have been put in place to select some ackee trees and fruits suitable properties for the canning industry. This includes vegetative propagation at Research Stations[48] and the establishment of ackee orchards.[49]

FIGURE 4.9 Canned ackee arils: Self-developed with concepts from Rashford.[49]

Drying could also be a viable means of processing ackee at a commercial scale. The West African sub-regions, which have ackee growing wildly or consciously planted, employ drying methods for processing ackee arils. The dried ackee arils, however, have been found to have a short shelf-life of about 2 weeks on the open market,[6,20,60] although there is a need to develop a good packaging system for ackee died products.

Ackee arils hold enormous prospects for application in the food industry especially in West Africa because the derived ackee aril flour obtained showed properties, such as good foaming capacity, emulsion activity, and stability.[18] The flour also had a loose bulk density subsequently requiring less storage space implying a higher quantity can be packed for export and distribution.

Notwithstanding the challenges of ackee production and commercialization, it has great potential to be included in the stable food basket of the continent. This will eventually help to improve nutrition, health, and local economies, revitalize agricultural biodiversity, and help preserve the traditional knowledge and culture of growing countries.

4.12 SUMMARY

Ackee grows wild in most communities in Africa but has been used as a food and even non-food product for centuries by locals. The crop is underutilized even though it has good potential in Africa to contribute to food and nutrition security. The ease of propagation and domestication, coupled with existing indigenous and scientific knowledge for its varied uses, presents extensive possible applications of ackee in the food, cosmetics, and pharmaceutical industries among others. To maximize the full potential of ackee in Africa, governments should work together with scientists, the food industry, and all stakeholders to address challenges in production, and processing technologies including product development and food safety, packaging, and commercialization.

KEYWORDS

- **ackee fruit aril**
- **Africa**
- **food and nutrition security**
- **indigenous crops**

REFERENCES

1. Abdulrahman1, W. F.; Omoniyi, A. O. Proximate Analysis and Mineral Compositions of Different Cereals Available in Gwagwalada Market, F. C.T, Abuja, Nigeria. *Journal of Advances in Food Science & Technology* **2016,** *3* (2), 50–55.
2. Aderinola, O. A.; Farinu, G. O.; Akinlade, J. A.; Olayeni, T. B.; Ojebiyi, O. O.; Ogunniyi, P. O. Nutritional Potential of *Blighia sapida* Koenig (Ackee) Leaves as a Dry Season Feed Resource for West African Dwarf Goats in the Derived Savanna Zone of Nigeria. *Livestock Res. Rural Dev.* **2007,** *19* (6), 1–3.
3. Adeyeye, A.; Ajewole K. Chemical Composition and Fatty Acid Profiles of Cereals in Nigeria. *Food Chem.* **1992,** *1992*, 41–44.
4. Akintayo E. T.; Adebayo E. A.; Arogunde, L. A. Assessment of Dietary Exposure to the Natural Toxin Hypo-Glycine in Ackee (*Blighia sapida*) by Jamaicans. *Food Res. Int.* **2002,** *37*, 833–838.
5. Anderson-Foster, E. N.; Adebayo, A. S.; Justiz-Smith, N. Physicochemical Properties of Ackee Oil Extract and Its Potential Application as Emulsion Base. *Afr. J. Pharm. Pharmacol.* **2011,** *6* (3), 200–210.
6. Anonymous. Ackee Welcomed Back to US After Ban, 2006. https://www.accessdata.fda.gov/cms_ia/importalert_64.html (accessed on Dec 17, 2021).
7. Arawande, J. O; Borokini, F. B. Comparative Study on Chemical Composition and Functional Properties of Three Nigerian Legumes (Jack Beans, Pigeon Pea and Cowpea). *J. Emerg. Trends Eng. Appl. Sci. (JETEAS)* **2010,** *1* (1), 89–95.

8. Ashman, I. Develop Analytical and Processing Data to Support a Petition to the United States Food and Drug Administration for Reviewing the Import Alert on Processed Ackee Products, 1990; p 19. https://moam.info/ackee-bighia-sapida-and-jamaica-a-a-love-affair-for-life_5a80dbfd1723ddf8efe4e972.html (accessed on March 19, 2022).
9. Atolani, O.; Olatunji, G. A.; Fabiyi, O. A. *Blighia Sapida*: The Plant and Its Hypoglycins—An Overview. *J. Sci. Res.* **2009,** *39* (2), 15–25.
10. Barceloux, D. G. Akee Fruit and Jamaican Vomiting Sickness. In: *Medical Toxicology of Natural Substances: Foods, Fungi, Medicinal Herbs, Toxic Plants, and Venomous Animals*; Barceloux, D. G., Ed., Vol. 2; John Wiley & Sons: Hoboken, 2008; pp 34–38.
11. Barnett, W. L. Notes on Ackee. *J. Jamaica Agric. Soc.* **1939,** *43*, 82–85.
12. Becker, R. *Fatty Acids in Food Cereal Grains and Grain Products*; Taylor & Francis Group, LLC: Kainan, 2007; pp 303–316.
13. Bennett-Lartey, S. O., Oteng-Yeboah, A. A.; Bunso, G. Ghana Country Report on the State of Plant Genetic Resources for Food and Agriculture. *CSIR-Plant Genetic Resour. Res.* **2008,** *2008*, 1–33.
14. Bukya, A.; Vijayakumar, T. P. Properties of Industrial Fractions of Sesame Seed (*Sesamum indicum* L.). *Int. J. Agric. Food Sci.* **2013,** *3* (3), 86–89.
15. *CODEX Alimentarius* International Food Standards. Standard for Wheat Flour, 1985; p 4. https://www.fao.org/fao-who-codexalimentarius/sh-proxy/es/?lnk=1&url=https%253A%252F%252Fworkspace.fao.org%252Fsites%252Fcodex%252FStandards%252FCXS%2B152–1985%252FCXS_152e.pdf (accessed on April 2, 2022).
16. Chase, G. W.; Landen, W. O.; Soliman, A. G. Hypoglycin—A Content in the Aril, Seeds and Husks of the Ackee Fruit at Various Stages of Ripeness. *J. Assoc. Off. Anal. Chem.* **1990,** *73* (2), 318–319.
17. Davidson, M. Ackees and Avocados. *Jamaica J.* **1971,** *5*, 28–31.
18. Dossou, V. M.; Agbenorhevi, J. K.; Combey, S.; Afi-Koryoe, S. Ackee (*Blighia sapida*) Fruit Arils: Nutritional, Phytochemicals and Antioxidant Properties. *Int. J. Nutr. Food Sci.* **2014,** *3* (6), 534–537.
19. Ekué, M. R. M. *Blighia sapida*, Ackee: Conservation and Sustainable Use of Genetic Resources of Priority Food Tree Species in sub-Saharan Africa. *Bioversity Int.* **2011,** *2011*, 1–18.
20. Ekué, M .R. M.; Sinsin, B.; Eyog-Matig, O.; Finkeldey, R. Uses, Traditional Management, Perception of Variation and Preferences in Ackee (*Blighia sapida* K. D. Koenig) Fruit Traits in Benin: Implications for Domestication and Conservation. *J. Ethnobiol. Ethnomed.* **2010,** *6* (12), 1–14.
21. Ekué, M. R. M.; Gailing, O.; Vornam, B.; Finkeldey, R. Assessment of the Domestication State of Ackee (*Blighia sapida* K. D. Koenig) in Benin Based AFLP and Microsatellite Marker. *Conserv. Genet.* **2011,** *12*, 475–489.
22. Elbehri, A.; Kaminski, J.; Koroma, S.; Iafrate, M.; Benali, M. West Africa Staple Food Systems: An Overview of Trends and Indicators of Demand, Supply, and Trade. In: *Rebuilding West Africa's Food Potential*; Elbehri, A., Ed.; FAO/IFAD: Paris, France, 2013; p 42.
23. Emanuel, M. A.; Benkeblia, N. Adding Value to Tropical Fruits—The Case of the Jamaican Ackee Industry: Lessons for Policy and Practice. In: *Innovations in Tropical Food Processing*; CTA, Technical Centre for Agricultural and Rural Cooperation, 2012; pp 1–8.

24. Emanuel, M. A.; Gutierrez-Orozco, F.; Yahia, E. M.; Benkeblia, N. Assessment and Profiling of the Fatty Acids in Two Ackee Fruit (*Blighia sapida*) Varieties During Different Ripening Stages. *J. Sci. Food Agric.* **2 013,** *93* (4), 722–726.
25. Emanuel, M. A. Biochemical Variations in Ackee (*Blighia sapida*) During on Tree Ripening. In: *Fruit and Vegetable Phytochemicals: Chemistry and Human Health*; Yahia, E. M., Ed., Vol. 2; Wiley Blackwell Publishing: Hoboken, **2011**; pp 113–169.
26. Emanuel, M. A.; Benkeblia, N. Variation of Color, Total and Reducing Sugars in Cheese and Butter Ackee Cultivars During Five on Tree Ripening Stages. *Acta Horticult.* **2011,** *894*, 231–234.
27. Essuman, E. K.; Osei, J. A.; Gyimah, V. Proximate Composition and Sensory Qualities of Chips Produced from Ackee Aril Flour. *Am. J. Food Sci. Technol.* **2016,** *4* (2), *38–42.*
28. Falloon, O. C.; Baccus-Taylorb, G. S. H.; Minott, D. A. A Comparative Study of the Nutrient Composition of Tree-Ripened versus Rack-Ripened Ackees. *West Indian J. Eng.* **2013,** *36* (2), 69–75.
29. FAO (Food and Agriculture Organization of the United Nations). The State of the World's Biodiversity for Food and Agriculture; In: *FAO Commission on Genetic Resources for Food and Agriculture Assessments*; Bélanger, J., Pilling, D., Eds.; FAO: Rome, 2019; p 572.
30. FAO (Food and Agricultural Organization) Report. *Regional Overview of Food Security and Nutrition in the Near East and North Africa 2019—Rethinking Food Systems for Healthy Diets and Improved Nutrition*; FAO Publications: Cairo, Egypt, 2020; p 96.
31. Gbolade, A. A. Inventory of Antidiabetic Plants in Selected Districts of Lagos State, Nigeria. *J. Ethnopharmacol.* **2009,** *121*, 135–139.
32. Gil, J. D. B.; Reidsma, P.; Giller, K.; Todman, L.; Whitmore, A.; Van-Ittersum, M. Sustainable Development Goal 2: Improved Targets and Indicators for Agriculture and Food Security. *J. Human Environ.* **2019,** *48*, 685–698.
33. Golden, K. D.; Kean, E. A.; Terry, S. I. Jamaican Vomiting Sickness: A Study of Two Adult Cases. *Clin. Chim. Acta* **1984,** *142* (3), 293–298.
34. Golden, K. D.; Williams, O. J.; Bailey-Shaw, Y. High-Performance Liquid Chromatographic Analysis of Amino Acids in Ackee Fruit with Emphasis on the Toxic Amino Acid Hypoglycin-A. *J. Chromatogr. Sci.* **2002,** *40*, 441–446.
35. Goldson, A.; Bremmer, D.; Nelson, K.; Minott, D. A. Fat Profile of Jamaican Ackees, Oleic Acid Content and Possible Health Implications. *West Indian Med. J.* **2007,** *63* (1), 9–12.
36. Howélé, Q.; Bobelé, N.; Théodor, D.; Séraphi, K. C. Nutritional Composition Studies of Sun Dried *Blighia sapida* Arils from Côte d'Ivoire. *J. Appl. Biosci.* **2010,** *32*, 1989–1994.
37. Hassall, C. H.; Reyle, K. The Toxicity of the Ackee (*Blighia sapida*) and Its Relationship to the Vomiting Sickness of Jamaica: A Review. *West Indian Med. J.* **1955,** *4* (2), 83–90.
38. Hassall C. H.; Reyle, K. Hypoglycin-A and -B: Biologically Active Polypeptides from *Blighia sapida. Nature* **1954,** *173* (4399), 356–357.
39. Hussain, M. A.; Basahy, A. Y. Nutrient Composition and Ammo Acid Pattern of Cowpea (*Vignaun guiculata*) Grown in Saudi Arabia. *Int. J. Food Sci. Nutr.* **1998,** *49* (2), 117–124.
40. Hyatt K. Assessing the Competitiveness of Jamaican Ackee in Light of the Challenges Faced by Sugar and Bananas. In: *Food Safety and Value-Added Production and Marketing of Tropical Crops*; Badrie, N., Ed.; Proceeding of the 26th West Indies Agricultural Economics Conference, Caribbean Agro-Economics Society: Jamaica, 2007; p 25–35.

41. Jayasri, M. A.; Mathew, L.; Radha, A. Report on the Antioxidant Activities of Leaves and Rhizomes of *Costus pictus*. *Int. J. Integr. Biol.* **2009,** *5* (1), 20–26.
42. Kean, E. A.; Hare, E. R. γ-Gutamyl Transpeptidase of the Ackee Plant. *Phytochemistry* **1980,** *19* (2), 199–203.
43. Keay, R. J. Trees of Nigeria. *Nordic J. Bot.* **1989,** *11* (3), 322–322.
44. Koenig, K. D. *Blighia sapida*. *Ann. Bot.* **1806,** *2* (571), 16–17.
45. Lancashire, R. J. Jamaican Ackee, 2006. http://wwwchem.uwimona.edu.jm/lectures/ackee.html (accessed on Nov 20, 2021).
46. Monastyrnaya, E.; Joerin, J.; Dawoe, E.; Six, J. Assessing the Resilience of the Cocoa Value Chain in Ghana; ETHZ: Case Study Report, 2016; p 36. https://www.ethz.ch/con-tent/dam/ethz/special-interest/usys/ias/enhancing-resilience-dam/documents/Final%20Re-port.%20Cocoa%20in%20Ghana.pdf (accessed on March 25, 2022).
47. McKevith, B. Nutritional Aspects of Oilseeds. *Br. Nutr. Foundation Nutr. Bull.* **2005,** *30*, 13–26.
48. MAM (Ministry of Agriculture and Mining), Kingston, Jamaica, 1995. https://www.fao.org/3/i1500e/Jamaica.pdf (accessed on March 22, 2020).
49. MAL (Ministry of Agriculture and Land). *Viable Ackee Industry Must Be Protected*; Jamaica Information Service: Kingston, Jamaica, 2006; p 2.
50. Mitchell, S. A.; Seymour, W. A.; Ahmad, M. H. Ackee: Jamaica's Top Fruit. *Jamaican J. Sci. Technol.* **2008,** *33*, 84–89.
51. Morris, C. R.; Hamilton-Reeves, J.; Martindale, R. G.; Sarav, M.; Gautier, J. B. O. Acquired Amino Acid Deficiencies: A Focus on Arginine and Glutamine. *Nutr. Clin. Pract.* **2017,** *32* (1), 30S–47S.
52. Morton, J. F.; Dowling, C. F Akee, In: *Fruits of Warm Climates*; Morton, J. F., Winterville, N. C., Eds.; Echo Point Books & Media: Miami, 1987; pp 269–271.
53. Moya, J. Ackee (*Blighia sapida*) Poisoning in the Northern Province, Haiti. *Epidemiol. Bull.* **2001,** *22* (2), 8–9.
54. Odutuga, A. A.; Asemoto, H. N.; Musac, I.; Golden, K. D.; Kean, E. A. Fatty acid composition of ackee (*Blighia sapida* L.) fruit. *Jamaican J. Sci. Technol.* **1992,** *3*, 30–32.
55. Okogun J. I. The Chemistry of Nigerian Medicinal Plants. *Med. Plant Res. Nigeria* **1996,** *10* (5), 31–45.
56. Olusegun O. J. D.; Olutomi, O. P. Chemical, Phytochemical and Antimicrobial Screening of Extracts of *B. sapida* for Agricultural and Medicinal Relevance. *Nat. Sci.* **2013,** *11* (10), 12–17.
57. Osei, J. A.; Essuman, E. K.; Kyeremateng, D. O.; Agbenorhevi, J. K. Survey of Ackee Fruit Utilization in Ghana. *Jamaican J. Sci. Techno.* **2014,** *21* (24) 1–17.
58. Ouattara, H.; Niamke, B.; Dally, T.; Kati-Coulibaly, S. Nutritional Composition Studies of Sun Dried *Blighia sapida* aril from Cote d'Ivoire. *J. Appl. Biosci.* **2010,** *2010*, 1989–1993.
59. Oyeleke, G. O.; Oyetade, O. A.; Afolabi, F.; Adegoke, B. M. Nutrients, Antinutrients and Physicochemical Compositions of *Blighia sapida* Pulp and Pulp Oil. *J. Appl. Chem.* **2013,** *4* (1), 05–08.
60. Pen, M. Viable Ackee Industry Must Be Protected—BSJ Inspector, 2006. http://www.jis.gov.jm/agriculture/html/20060506T100000–0500_8777_JIS_VIABLE_ACKEE_INDUSTRY_MUST_BE_PROTECTED__BSJ_INSPECTOR.asp (accessed Nov 21, 2021).

61. RADA (Royal Academy of Dramatic Art); Protocol for Suppliers of Open Ackee Pods to Ackee-Canning Factories; 2006; Schedule 1, Issue 1, p 21.
62. Rashford, J. Those That Do Not Smile Will Kill Me: The Ethnobotany of the Ackee in Jamaica. *Econ. Bot.* **2001,** *55* (2), 190–211.
63. Royes, H. H.; Baccus, G. R. *Orchard Crops: The Jamaica Farmer's Guide*. Kingston-Jamaica: National Library of Jamaica; **1988**; pages 108.
64. Saidu, A. N.; Mann, A.; Onuegbu, C. D. Phytochemical Screening and Hypoglycemic Effect of Aqueous *Blighia sapida* Root Bark Extract on Normoglycemic Albino Rats. *Br J. Pharma. Res.* **2012,** *2* (2), 89–97.
65. Sahe, A. C.; Agbenorhevi, J. K.; Wireko-Manu, F. D. Quality Characteristics of Cassava-Ackee-Groundnut-Composite Cookies. *Am. J. Food Sci.Technol.* **2019,** 7 (2), 57–64.
66. Shakespeare-Blackmore, K. *Ackee Flies High Again*; The Gleaner: Kingston, 2007; p 263.
67. Singh, S. K.; Jadhav, P. V.; Nandanwar, R. S.; Patil, A. N.; Wandhare, M.; Naik, R. M.; Katkar, R. N. Assessment of Nutritional Quality Parameters in Selected Vegetable Type Pigeon Pea Genotypes. *J. Pharmacogn. Phytochem.* **2018,** *1*, 1446–1450.
68. Sochocka M.; Boratyriski, J. Osmoregulation—An Important Parameter of Bacterial Growth. *Natl. Inst. Heath* **2011,** *65*, 714–724.
69. Soetan, K. O.; Olaiya, C. O.; Oyewole, O. E. The Importance of Mineral Elements for Humans, Domestic Animals and Plants: A Review. *Afr. J. Food Sci.* **2010,** *4* (5), 200–222.
70. Stancheva, I.; Nedyalkova, K.; Geneva, M.; Donkova, R.; Hristozkova, M.; Perfanova, I.; Sichanova, M.; Petkova, G.; Djonova, E.; Valchovski, H. Nutritional Value of Cowpea (*Vigna Unguiculata*) Grain Grown Under Different Soil Moisture as Affected to the Dual Inoculation with Nitrogen Fixing Bacteria and Arbuscular Mycorrhizal Fungi. *Bulgarian J. Soil Sci.* **2016,** *1* (2), 112–121.
71. Stair, M. A.; Sidrak, G. Developing Techniques for Improving and Propagating Ackee (*Blighia sapida*). *Proceedings of the Second Annual Conference; Jamaica Society for Agricultural Sciences (JAGRIST)* **1997,** *4*, 10–13.
72. Statistical Institute of Jamaica: External Trade. *Statist. Inst. Jamaica* **2002,** *2*, 56–57.
73. Uddin, S. N.; Akond, M. A.; Mubassara, S.; Yesmin M. N. Antioxidant and Antibacterial Activities of *Trema cannabina*. *Middle-East J. Sci. Res.* **2008,** *3*, 105–108.
74. USAID (United States Agency for International Development), Washington, DC. http://www.usaid.gov/policy/ads/pps/foodsec/fs foodsec.html (accessed Nov 15, 2021).
75. USDA (United States Department of Agriculture), 2014; p 154. https://www.ars.usda.gov/ARSUserFiles/80400535/DATA/sr27/sr27_doc.pdf (accessed Oct 20, 2021).
76. Webster, S. A; Mitchell S. A.; Reid, W.; Ahmad, M. H. Somatic Embryogenesis from Leaf and Zygotic Embryo Explants of Cheese Ackee. *In Vitro Cell. Dev. Biol.—Plant* **2006,** *42* (5), 467–472.
77. WHO (World Health Organization). Protein and Amino Acid Requirements in Human Nutrition: Report of a Joint WHO/FAO/UNU Expert Consultation; Geneva, Switzerland. https://apps.who.int/iris/handle/10665/43411 (accessed March 22, 2022).
78. Yıldırım, I.; Kutlu, T. Anticancer Agents: Saponin and Tannin. *Int. J. Biol. Chem.* **2015,** *9* (6), 332–340.

CHAPTER 5

GRAPE WINES IN HUNGARIAN CONTEXT: PRODUCTION TECHNOLOGIES, PROCESSING, AND HEALTH BENEFITS

ILDIKÓ GALAMBOS, ILDIKÓ ERNSZT, ZSUZSANNA MARTON, and NÓRA RODEK

ABSTRACT

Hungary in the European Union plays a significant role for wine-making in the map. The climate in Hungary is a transitory one that features Continental, Oceanic and Mediterranean climate characteristics. Due to the diversified earth surface, micro climate, and soil conditions, a great variety of wines are produced. Regarding production, the ratio of white wines is 71.1%, while the ratio of red wines is 24.4%. With time progress demand and consumption of wine are increasing dramatically. With Hungary constitutes 1.9% of the wine production of the world. This ratio is 2.6 and 1.1% regarding the vineyard area and wine production of the EU. In 2015, wine production constituted approximately 1% of the world wine production. Hungary occupies the 16th–17th position among the 46 wine-producing countries of the International Organization of Vine and Wine. The process of grape treatment and wine production results several by-products. Different new products are produced from these by-products. They offer several health benefits and due to this reason, they can be considered as functional food. In recent years, the consumer's attention on functional food has been placed in Hungary. However, still it is in its infant situation and a relatively low proportion of the consumers is aware about beneficial effect of wine and food products-derived

Sustainable and Functional Foods from Plants: Health Impact, Bioactive Compounds, and Production Technologies. Megh R. Goyal, Arijit Nath, and Zoltán Kovács (Eds.)
© 2024 Apple Academic Press, Inc. Co-published with CRC Press (Taylor & Francis)

from grape by-products. It is believed that the present chapter will boost the concept about functional food from by-products of grape and wine production. This chapter will document: historical information of wines and grapes; their cultivation and production; process technology; by-product processing; and health impacts related with wine and grape.

5.1 INTRODUCTION

The consumption of grapes and wine is bounded with early stage of society. The sentence "Making great wine is an art, making good wine is a science, but the fermentation of grapes is a natural and unavoidable process" not only receive popularity among producer, it attracts consumers as well.[17] Wine may also be considered to be a medicine, a cult agent, and a noble beverage. Wine is associated with healing, beauty, religion, science, and arts. This beverage is attached with several socio-cultural issues, such as family-, social-, and business life.

Wine has a sacred role in mythology and in the Bible as well. In the Christian mass, the wine is transformed into the blood of Christ.[30] Proving its significance, wine have had a lot of divinities since the ancient times with different names: Bacchus in Rome, Dionysus in Greece, Bassareus in Lydia, and Dusares in Arabic region.[17]_The production of wine on Earth is widespread in all areas, where the climate is suitable for cultivation of grape. The first archeological and chemical traces of wine were discovered in Iran. Meanwhile, the first sign of winemaking of domesticated grapewine was explored on the territory of today's Armenia. In that region, the apparent wine press and fermentation vat were found at date back to 6100 years.[54]

From a long prior, wine as an everyday drink was considered. Because of alcohol content, ranging generally between 8.5 and 15%, it kills several harmful bacteria and virus. This is the fact that wine was used in armies during the scarcity of water. Besides, historians point out, that a clear connection can be assessed between poor vintages and the outbreaks of epidemics.[17] Until the 20th century, wine is used in pharmacy as medicine excipient, from then it was used exclusively as a solvent.[13] The oldest information of wine can be found in a Sumerian medical book, called pharmacopoeia, written at 2200 BC. According to this, sweet wine with honey was prescribed to heal coughs. In Babylon, Tabata curative drink also had wine content. In the ancient China, wine was also commonly used as a matrix of medicaments. Different plants, minerals, and animal parts were mixed with wine and considered a medical formulation. In Egypt, wine was used for sterilization purpose.[52] The Ebers

Medical Papyrus from Egypt, was written around 1550 BC, refers numerous wine-based elixirs, those cure loss of appetite and weaknesses. Meanwhile, wine was also used as a "masking" agent, which suppressed the bad taste of the different healing additives.[41] During 460–377 BC, Hippocrates suggested wine as a medicine, which has cooling, nourishing, purgative, and diuretic effect.[17,52]

Galen (129 A.D.–c. 200/c. 216), a physician in Marcus Aurelius, was convinced that moderate wine consumption can cure nearly all illnesses.[30] Wine is also operated as a disinfectant until the distillation was invented; the bonds on wounds were soaked in wine to avoid infection—as it is prescribed by famous Islamic doctors like Avicenna (980–1032 CE) or Albucasis (936–1013).[52] In the Middle Ages, it was believed that wine with the bones of a saint cures lunacy.[17] In the medical book, written by Henry de Mondeville during 13th–14th century, it is mentioned that consumption of wine is beneficial to produce blood, since they are similar to each other.[57]

In 1967, vinum album has ceased to be an official medicinal composition. Vinum Album was extremely easy to substitute with the mixture of water and ethanol. Its good effects on health were emphasized in the second part of the last century, when the Mediterranean diet was examined.[13] It was proven that French has 40% less heart attacks than the American people because of their special French diet along with red wine containing resveratrol.[31,60]

The above discussion portrays that wine has grabbed lot of attention around the world among all communities. Its beneficial effects are noteworthy. Lots of information about wine from the global point of view as well as specific country or parts of world are well documented. Unfortunately, information about wine from early stage of society to present context in Hungary is not documented.

The present chapter focuses on the historical information of wines and grapes, their cultivation and production, process technology, by-product processing, and health impacts of wines and grapes.

5.2 GRAPE AND WINE STATUS IN HUNGARY

5.2.1 HISTORICAL EVIDENCES

In Hungarian nation, grape has also a special significance, which appears even in the national anthem. In Hungary, the special importance of wine is proved by the ancient good wishes at New Year's Day. It is 'I wish you wine, wheat and peace!' Nowadays, Hungarians are convinced that wine is a part of their culture, as it is a national drink.[73]

The ratio of white wines is 71.1%, while the ratio of red wines is 24.4%. With time progress demand and consumption of wine are increasing dramatically. Hungary constitutes 1.9% of the wine production of the world. This ratio is 2.6 and 1.1% regarding the vineyard area and wine production of the EU. In 2015, wine production constituted approximately 1% of the world wine production. Hungary occupies the 16th–17th position among the 46 wine-producing countries of the International Organization of Vine and Wine.

From the 17th century, red wine appeared among medicines in the Hungarian medical books and herbarios.[44,45] In medical book written by Gergely Gellén at 1680, wine with honey was suggested for pregnant women to increase milk selection.[61]

Vitis Vinifera, the grapes which give the basis for wine, was developed from the previously grown *Vitis silvestris*. Vine and wine production have always been and is still an important economical source in Hungary. With its varied soil types and favorable climate conditions, Hungary has always belonged to Europe's traditional wine producing countries. In Hungary (and on the territory of the Carpathian Basin), the vineyards and wine-making memories date back to Roman times. The Roman Emperors Domitian and Probus also placed their attention to vineyard cultivation in their regulations. The Dionysos jug, which is a part of the 68.5 kg Roman treasure trove, found in Hungary at 1970s. At part of the Seuso treasure during A.D. 350–460, it also shows divine scenes of wine consumption. Furthermore, it was also suitable for filling wine.[77]

The conquering Hungarians have already found an advanced wine culture in the Carpathian Basin. One of the first wine production regions can be Tokaj region, which is still one of the most famous wine production regions in Hungary. The archeological findings also prove that Celtic grape production in Transdanubia, Hungary. Over the centuries, several historic events determined the Hungarian wine production scenario and had a negative impact on that, for example, the Turkish occupation, the Habsburgs' 1755 customs regulations on high tariffs for Hungarian wines, the massive destruction of phylloxera at the end of the 19th century, the Trianon peace treaty, the world wars—all of them have thrown back the blooming wine production of the country. After the Second World War, the large-scale grape production and wine making in Hungary targeted the Eastern Eutopean markets with a huge capacity. The change of regime (1989) has brought significant changes, that is, the quality approach replaced by the quantity one. Presently, around 150,000 people make their living from viticulture and wine-making. It signifies that together with the family members around 5% of the whole population of the country is involved in this sector.[5]

5.2.2 THE ECOLOGICAL FACTORS OF VITICULTURE

Mostly environmental factors determine the grape cultivation and productivity. Similar like others, different environmental factors, such as light, heat, water, oxygen, carbon dioxide, mineral, and waters are essential for the grape cultivation. The other elements of the environment, such as wind, combustion gas, and air pressure are not essential.[68]

For viticulture purposes, southern and southwestern slopes of high lands and foot hills, located in front of bigger mountain chains and basins, protected by mountains are the most desirable. For cultivation of grape plant, full shade is not desirable at all. Grape plant needs plenty of sunlight for its growth, but it utilizes diffused light in very efficient way. Based on the characteristics, parameters can be grouped into: (1) climate factors, (2) physiographic factors, (3) edaphic factors, and (4) biotical factors.[18] For photosynthesis purpose, only 1–3%, but not more than 5%, of the available light is used by grape plant. To produce more pigments and color, berries need more light. Moreover, sufficient light helps to build up thicker skin and more sugar (2.5–3.5% by weight) and less acid (2–3 g L^{-1}), which is more than berries, cultivated in shade. The average of the total annual sunshine duration of the vine-growing regions of Hungary is between 1800 and 2070 h and the sunshine hours during vegetation period are between 1250 and 1500 h.[51,58] In Table 5.1, the number of sun hours in grape cultivation or wine producing regions in Hungary[58] is represented.

TABLE 5.1 Number of Sun-Hours in Wine Regions of Hungary; Self-Developed Using Concepts Was from Rakonczás.[58]

Hungarian grape or wine production region	Sun hours (hours/year)
Southern Great Plain	above 2050
Pécs-Villány	2000–2050
Etyek Wine Region, Mór Wine Region	1900–2000
South of Lake Balaton, Balaton Uplands	1950–2000
Mátra-alja, Eger, Bükk-alja	1900–1950
Sopron Wine Region, Tokaj Wine Region	1850–1900

Heat is indispensable for life cycles in viniculture. The annual mean temperature, which is between 9 and 11°C in Hungary, is highly important for the cultivation of grape (Miskolc, Sopron: 9.5–9.9°C; Pécs, Szeged: 11.4°C). Successful vegetation periods and phases can only be achieved when adequate temperature conditions are available. The two third of the southern regions lies between the 10 and 11°C isotherms, whereas the northern regions are situated in areas with 9–10°C daily mean temperature. In Hungary, the mean temperature in July is between 20.1 and 22.7°C. The mean temperature of the grape growing season is between 14°C and 18°C. The best quality grape is produced in regions where the annual mean temperature is around 10–16°C. In such regions, where the mean temperature is 16–21°C, the yield is abundant; however, wines have less fragrance and aromas than those produced in cooler regions.[58]

In order to achieve the best possible wine grape quality, it is essential that the mean temperature in July is at least 18°C. Hungarian areas are suitable for very early, early and later maturing species. In order to produce 1 g of grape material, 250–300 g of water is required. In Hungary, 1 m^2 of leaf surface of the grape evaporates 0.5–1.5 L water on a normal day and 1.5–3.5 L on a dry and hot day. The unit leaf surface of the drought-tolerant vine varieties evaporates more water than those species not well tolerant heat. In case of excessive precipitation, the berries' meat becomes softer, its sugar content decreases, while its acid content increases, and the skin becomes thinner, greener, or lighter red or blue in color. Transportability and storability are also reduced. In case of water shortage, the berries small with thicker skin and their sugar, acid, lean, and color content are reduced. The annual precipitation in Hungary is 500–800 mm that is sufficient for grape cultivation; however, the distribution of the precipitation is not adequate. In Hungary, winter seasons often lack sufficient precipitation. Therefore, the soil's water content, having been filled up from winter precipitation, is less likely to last until May. Predominantly, air and wind affect grapes with their content (CO_2, O_2, moisture) and direction, respectively.[51]

5.2.3 VINICULTURE AND WINE REGIONS

Hungary is a traditional wine-making country in Europe. Practically, its entire territory is suitable for wine producing; however, there are great differences among certain wine regions. The climate in the Carpathian

Basin and Hungary is a transitory one that features Continental, Oceanic, and Mediterranean climate characteristics alike. Since the effects of these climatic attributes are uneven, the country's climate is varied. Approximately, two thirds of the country lies within a zone where the annual mean temperature is 10–11°C, which puts the country among the northern-situated wine-producing regions.

This may elude to the fact that mostly quality white wines are produced in Hungary, however, the southern wine regions within the country also offer great quality red wines. Due to the diversified earth surface, microclimate and soil conditions, a great variety of wines are produced in the particular wine regions. Almost 32.9% of the grape production area is located in the Southern Great Plain region, while 24.1% grape production area is located in the Northern Hungarian region. The distribution of grape production area is 13.4% in the Southern Transdanubian region, 11.6% in the Northern Transdanubian region, 7.8% in the Western Hungarian region, 7.3% in the Central Hungarian region, and 3.9% in the Northern Great Plain region.[19]

In Hungary, the wine-producing areas are divided into several categories. As per Act XVIII in 2004, there are 22 wine regions in Hungary for viticulture and wine industry. A wine region is a specific region, where the growing conditions are similar, for example, climate, soil composition, and terrain. According to these conditions, the produced wine will be specific and recognizable. Also, it is important that in a wine region, the wine-making traditions are typical. The size, location, and other characteristic features of wine producing regions are greatly differing. All wine-producing areas still grow the species that have been cultivated on their soil for centuries.[8]

The total area of the 22 wine production regions is 78,700 hectares that constitutes 86% of the total viticulture area. Only the wine regions of Eger, Mátra, Tokaj, and Kunság are bigger than that which reflects on one of the biggest problems of the Hungarian wine industry, namely, the fragmentation of lands. The average size of vineyards in wine regions is very small; the smallest is 0.17 hectares in the Balaton-felvidék Wine Region, whereas the largest is 1.67 hectares in the Eger Wine Region.[11] Table 5.2 represents the most significant red and white grape species, which are cultivated in Hungary. These are Kékfrankos (37.93%), Cabernet Sauvignon (14.11%), Merlot (10.35%), and Zweigelt (9.15%). Figure 5.1 shows the most significant white grape species of Hungary, which are Bianca (17%), Cserszegi spicy (16%), Italian Riesling (14%), and Furmint (14%).

TABLE 5.2 The Most Significant Red Grape Species in Hungary: Self-Developed with Concepts from Hegyir, 2016.[29]

Special	Area, ha	Distribution (%)
Blauburger	434	2.28
Cabernet Sauvignon	2690	14.11
Carbernet Franc	1370	7.19
Kadarka	371	1.95
Kékfrankos	7229	37.93
Kékoportó	1050	5.51
Merlot	1973	10.35
Pinot Noir	1094	5.74
Syrah	214	1.12
Zweigelt	1744	9.15
Total	18,169	95.32
All for red grapes	19,061	100.00

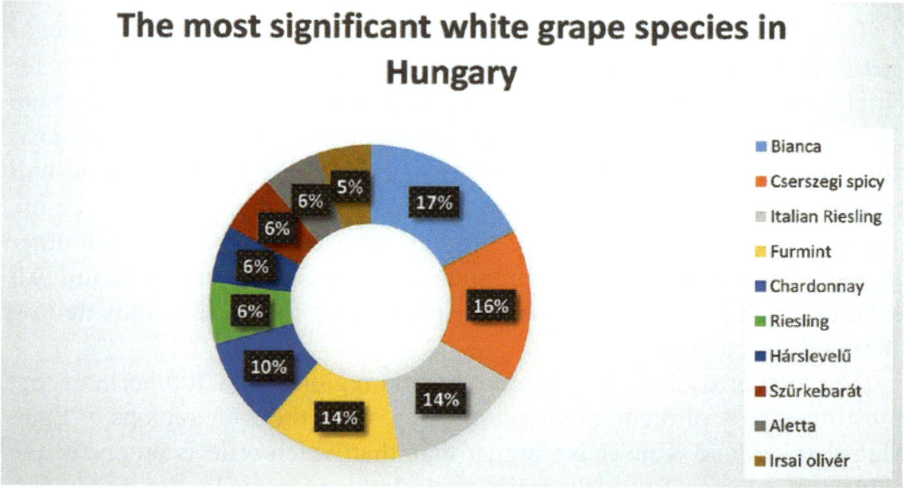

FIGURE 5.1 Significant white grape species in Hungary: self-developed with data from Hegyir, 2016.[29]

The Duna Wine Region was established in 2002 followed by the Pannon Wine region and the Balaton Wine Region in 2003. The former one includes the Csongrád, Hajós-Baja, and Kunság Wine Regions. The Balaton Wine Region is comprised of the Badacsony, Balatonfüred-Csopak, Balaton-felvidék, Nagy-Somló, and Balatonboglár Wine Regions. The Northern Hungarian Wine Region consists of the Ászár-Neszmély, Mór, Etyek-Budai,

Pannonhalma, and Sopron Wine Regions. The Pannon Wine Region incorporates the Szekszárd, Villány, Tolnai, and Pécs Wine Regions.[21]

5.2.3.1 THE DUNA WINE REGION

Three adjacent wine regions form this region are plain and covered mostly with sand with loess. This is the largest wine-growing zone of Hungary at present. Presently, all Hungarian types of grapes are found in the region.

5.2.3.2 THE EGER WINE REGION

The Mátraalja, Eger, and Bükkalja wine regions belong in this zone with famous wine types such as, Bull's Blood of Eger, Debrői Hárslevelű, and Welschriesling. Chardonnay, Muscat Ottonel, and Tramini, which are widely recognized international types, however, Pinot Noir is also gaining importance. The wine regions are as follows: the Mátra Wine Region, the Bükkalja Wine Region, and the Eger Wine Region.

5.2.3.3 THE BALATON WINE REGION

Balaton wine region includes the Badacsony Wine Region, the Balaton Uplands Wine Region, the Balatonfüred-Csopak Wine Region, the South of Lake Balaton Wine Region, the Zala Wine Region, and the Nagy-Somló Wine Region.[15]

Beside the wine regions on the northern and southern coasts of the lake, white wines are also produced in the wine regions in Zala and in Somló and the most popular one is the Welschriesling.

5.2.3.4 NORTHERN TRANSDANUBIAN WINE REGION

Wine regions around Buda and near the River Danube above Budapest belong to the Northern Transdanubian Wine Region. Exclusively, white wine is produced here. Internationally, popular wines (Chardonnay, Sauvignon Blanc, Rhine Riesling, Tramini, and Muscat Ottonel) play an important role, however, traditional types, such as Welschriesling, Leányka, and Ezerjó of Mór are also found here.

5.2.3.5 THE PANNON WINE REGION

This area is bordered by the River Danube, the River Dráva and the Lake Balaton, and includes the mountains of Villány and the southern slopes of the Mecsek. Both red and white wines are produced here but red wines are more important. Internationally known wines, such a Cabernet Sauvignon, Cabernet Franc, and Merlot became popular in the past decades and subsequently, Bordeaux-type cuvee wines began to be produced as well. Welschriesling, Hárslevelű, Rhine Riesling, and Chardonn 6ay are the most significant white wines that are produced in this region.

5.2.3.6 SOPRON WINE REGION

Only one Hungarian wine region belongs to this category. From geographical, viticulture and cultural perspective, Sopron and the wine region is strongly related to the Burgenland Wine Region of Austria. Sopron is considered to be the capital of the Kékfrankos grape type.

5.2.3.7 TOKAJ WINE REGION

In this region, sweet and heavy wines are produced. Aszú and szamorodni are made from Furmint, Hárslevelű, and Yellow Muscat, which are unique in Europe.[58]

5.2.4 VITICULTURE AND WINE PRODUCTION, AND TYPES OF WINES

In the Roman Age, there was a significant viticulture and wine production in the Carpathian Basin, so our ancestors found a well-developed viticulture in the region. The Hungarian wine production is based on three basic pillars: (a) the traditions of our conquering ancestors, (b) the viticulture launched by the Romans and preserved by survivors after the fall of the Roman Empire and (c) the knowledge and taste brought by Italian, French and German settlers, and monks.

In the wine industry, the share of-white wines is 71.1%, while that of red wines is 24.4%.[35] White wine grapes are cultivated in 68%, red wine grapes in 25%, and table grapes in 7% of the territory of the wine regions. The share of white wine grapes reaches 90% of the total amount produced in the Central Transdanubian Region, while the share ~~ratio~~ of red wine grapes exceeds 40% of the total grape production in the Southern Transdanubian Region.

According to the registry of vineyards compiled in 2001, there are 27 grape types with an aerial distribution of each covering at least 1% of the 88% of the total area of Hungarian vine lands (Table 5.3). The type structure sheds light on the heritage of the erstwhile industrial agriculture production. Welschriesling and Bluefrankisch are the most widespread types. Only four internationally accepted types exceed 1%; these are Merlot, Cabernet Sauvignon, Rhine Riesling, and Chardonnay. They only occupy approximately 10% of the Hungarian vineyards, thus these areas remain far from being overwhelmed by non-Hungarian types.[68]

The areal distribution of white wine grapes offering quality wines is as follows: Welschriesling, Müller-Turgau, Chardonnay, Muscat Ottonel, Hárslevelű, Furmint, Rhine Riesling, and Leányka. Regarding red wine grapes, the most widespread and cultivated is Bluefrankisch followed by Zweigelt, Portugieser, Merlot, Cabernet Sauvignon and Franc, and Pinot Noir.[22] Chardonnay, Sauvignon Blanc, Cabernet Sauvignon and Franc, and Merlot types were planted and used between 1996 and 2003. Concerning Hungaricums, the production of Furmint and Hárslevelű remained at the same level, while that of the Welschriesling had decreased. Among the new types, Cserszegi Fűszeres was very popular, whereas Bluefrankisch became fashionable from among the traditional ones. In the past years, Kadarka, Juhfark, and Kéknyelű types had also seen an increase in their reputation.[4] The volume of wine production is between 3.3 and 5.4 million hectoliters per annum. This is due to the fact that 94–95% of the grape is used for winemaking.

TABLE 5.3 Hungarian Wine Regions, Grape Types and Wines: Self-Developed with Concepts from Rakonczás, 2014.[58]

Wine region	Vineyards in 1988 (ha)	Vineyard in 2008 (ha)	Typical vine varieties and wines
Ászár-Neszmély Wine Region	2203	1494	Müller-Thurgau, Welschriesling, Leányka, Ezerjó, Rhine Riesling, and Chardonnay
Badacsony Wine Region	2048	1797	Welschriesling, Pinot Gris, Kéknyelű
Balatonboglár Wine Region	2765	2820	Királyleányka, Welschriesling, Chardonnay, Cabernet, Merlot
Balatonfelvidék Wine Region	1958	1510	Welschriesling, Pinot Gris, Chardonnay
Balatonfüred-Csopak Wine Region	2827	2270	Welschriesling, Müller-Thurgau, Chardonnay
Balatonmellék Wine Region	2186	1170	Welschriesling, Müller-Thurgau, Green Veltliner, Chardonnay, Zweigelt

TABLE 5.3 *(Continued)*

Wine region	Vineyards in 1988 (ha)	Vineyard in 2008 (ha)	Typical vine varieties and wines
Bükkalja Wine Region	2779	1590	Leányka, Welschriesling, Bluefrankisch
Csongrád Wine Region	5753	2360	Kövidinka, Rhine Riesling, Green Veltliner, Welschriesling, Bluefrankisch Zweigelt, Rhine Riesling of Ásotthalom, Chardonnay of Ásotthalom, Cabernet Franc of Csongrád, Kadarka of Csongrád, Welschriesling of Pusztamérges
Eger Wine Region	4011	5160	Bluefrankisch, Cabernet, Merlot, Portugieser, Leányka, Welschriesling
Etyek-Buda Wine Region	1481	1480	Chardonnay, Welschriesling, Rhine Riesling, Sauvignon Blanc
Hajós-Baja Wine Region	4082	2304	Rhine Riesling, Chardonnay,Kadarka, Bluefrankisch, Cserszegi Fűszeres, Cabernet Zweigelt
Kunság Wine Region	37,959	27,903	Ezerjó, Kövidinka, Aranysárfehér, Kadarka, Bluefrankisch
Mátra Wine Region (earlier Mátraalja Wine Region)	6642	7815 (7100)	Müller-Thurgau, Welschriesling, Muscat, Pinot Gris, Kékfrankos, Zweigelt, Cabernet Sauvignon
Mecsekalja Wine Region	1276	940	Zinfandel, Welschriesling, Chardonnay
Mór Wine Region	1263	890	Ezerjó, Királyleányka, Chardonnay
Pannonhalma-Sokoróalja Wine Region	1080	750	Welschriesling, Müller-Thurgau, Irsai Olivér, Királyleányka, Cserszegi Fűszeres
Somló Wine Region	527	690	Furmint, Welschriesling, Juhfark, Sárfehér
Sopron Wine Region	1839	1880	Bluefrankisch, Merlot, Zweigelt, Leányka, Chardonnay, Irsai Olivér, Green Veltliner
Szekszárd Wine Region	2793	2300	Kadarka, Bluefrankisch, Merlot, Cabernet Franc, Cabernet Sauvignon, Chardonnay, Welschriesling, Bull's Blood
Tokajhegyalja Wine Region	6642	5860	Furmint, Hárslevelű, Muscat Lunel
Tolna Wine Region	2133	3150	Chardonnay, Welschriesling, Green Veltliner, Müller-Thurgau, Rhine Riesling, Kadarka, Bluefrankisch, Zweigelt, Cabernet Sauvignon.
Villány Wine Region	2003	1890	Portugieser, Bluefrankisch, Cabernet Sauvignon, Merlot, Welschriesling, Hárslevelű
Total	**96,250**	**77,308**	**29 types**

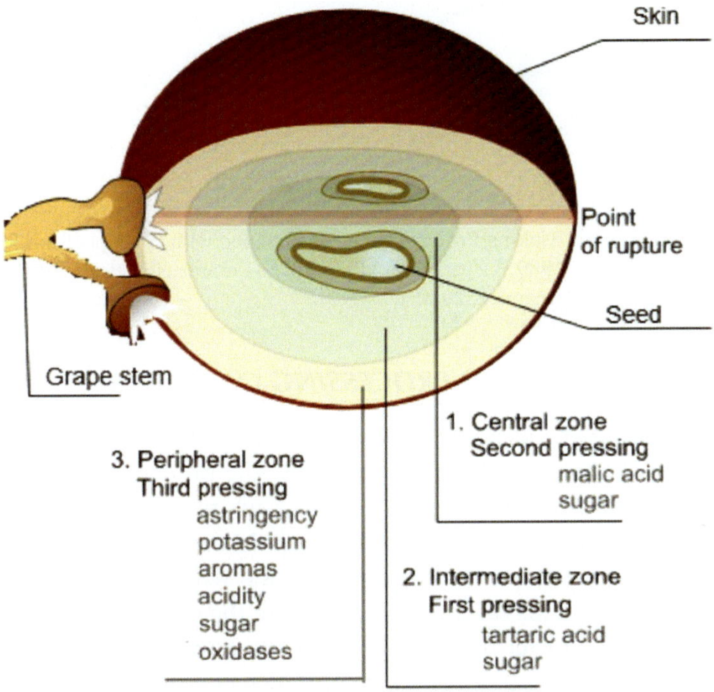

FIGURE 5.2 Structure of the wine grape berry: Modified from Ruiz Villareal.[59]

In present days, the farming is heavily influenced by Hungary's accession to EU. The European Union is in the first place in the world related with viticulture, wine production, and wine consumtion. Half of the EU budget is allocated to the agriculture; however, viticulture and oenology only receives 2–3% of such amount. The European Union provides subsidies to establishment, eliminate, and structurally reform vineyards.[71]

5.3 GRAPE PROCESSING TECHNOLOGIES: HUNGARIAN FEATURES

From botanically aspect grape is a berry and also woody perennial vine. The berries of Vitis vinifera (from the species of Vitis) is used as raw material for must and wine production. It has about 5000–10,000 species. A typical grape berry structure is presented in Figure 5.2. In Hungary, both craft wines and mass-produced (where wine is produced in huge volume) wines are available. Figure 5.3 shows steps in the production of wine from grapes.

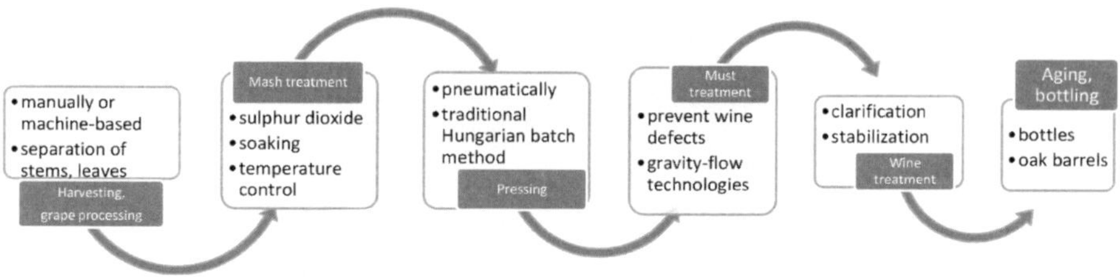

FIGURE 5.3 The Wine Making Processing: self-developed with concepts from Karasztiné, 2012,[28] Gruenhoff et al., 1989,[25] Eperjesi et al., 2010,[16] Jayaprakasha et al., 2003,[38] Kállay–Rácz, 2012,[39] Sanchez–Hernandez, 2010.[63]

5.3.1 HARVESTING AND PROCESSING OF GRAPES

The production begins with the harvesting, which can be carried out manually (picking) or using machine (Figure 5.3). The processing method varies depending on the type of the wine (such as white and red or rosé wine). The main difference is the wine coloring step. In case of red wine, the color components will dissolve from the skin into the wine. In first step of grape processing, stems and leaves are separated from the grapes to avoid the bitter taste, produced during fermentation.[28] After the removal of the stems, juice is separated by crashing and pressing of grapes and subsequently, skin and seed are separated.

5.3.2 MASH TREATMENT

The mash contains all the grape parts (skin and juice together) and it needs to be treated by sulfur dioxide, temperature and soaking. The time of soaking depends heavily on the grape variety. In the case of white wine, the mashing process is carried out in cold circumstances (at temperatures around 5–10°C) for a few hours without soaking. In case of light, rosé soaking is performed together with the mashing at the same conditions as white grape processing. During the production of the more *full-bodied, red rosé*, the step of soaking is necessary together with the skin, in this case, the mash extraction will be carried out for 4–24 h at 18–20°C. In case of red wine production, the mash is fermented and pressed only after the fermentation is completed.[16,39,63]

5.3.3 PRESSING

Separation of must from the mash can be done by pressing as follows: (a) in a static way (operation is performed in a cylinder lined with perforated plate, the must flows out through the perforations by its own weight); (b) in a dynamic way (continuous process, the mash is transported inside a perforated cylinder, the must leaves through the holes); and (c) pneumatically (pneumatic membrane pushes the mash through the cylinder, so the juice can be collected at the end of the process).

After the must separation, the remaining juice from the mash can only be obtained by external (pressure) force. Here, on the basis of the traditional Hungarian technology, the pressing will be carried out in a batch process in the grape press. In the above-mentioned methods, the following steps will be realized: filling, pressing, emptying, and washing. In the batch process, different fractions can be separated, such as: (a) vin de goutte: free run juice that can be released by the weight of the must and (b) vin de presse: separated during the pressing.

Typically, if the applied pressure during pressing process does not exceed two bars, then the resulting must will be suitable for wine production. Besides the batch operation, there is also a need for continuous technology. In this case, the mash flows through the cylinder and the marc is collected. In the press cylinder, a screw carries the material to be pressed in the perforated cylinder, and as a result of a gentle process, the pre-press (the first liquid flowing off the equipment) and the post-press is separated.

5.3.4 MUST TREATMENT

Following the must extraction in the must purifying step, several possibilities are available for the experts to prevent wine defects and causes arising later. During the must cleaning process, the residues from the grape fruit can be separated, for example, pieces from the berry skin, meat, seeds, or seed fragments. Their separation is basically carried out by gravity-flow technology. Its speed depends on the size of the floating particles, pieces, so the settling time can vary from a few hours to 24–36 h. As a complementary treatment, bentonite dosing can be applied. Bentonite is a clay mineral with a large surface, which can be used to accelerate sedimentation. During flotation, the particles in the suspension rise to the surface with the added inert gas (e.g., CO_2 or N_2) bubbles. The colloids in the must slow down the sedimentation. These colloids can be pekins, which can be degraded by enzymatic treatment.

Wine defects can be caused by inadequate, moldy, and rotten raw material (e.g., grey mold). Activated carbon is applied,[25] which helps to reduce the color and taste defects due to the high adsorption capacity based on its extremely large surface. The extremely high tannin content (typically as a consequence of the long soaking or over-pressure of the mash) can also be reduced by the adsorption. The applied materials can be gelatin or casein. Besides several other purifying technology steps, the temperature has a fundamental role in improving the must treatment and wine processing. Cooling can be carried out by using shell and tube heat exchanger, besides double-walled tanks can be applied for accurate temperature adjustment. The fermentation of the must can be initiated naturally or more frequently by inoculation with pure-bread yeast. During fermentation, the sugar content of the wine is converted into alcohol (ethanol, C_2H_6OH) and CO_2 is produced as a by-product.[38]

5.3.5 WINE TREATMENT

The aim of the wine treatment is the clarification and stabilization of the wine. In order to reach this goal, racking, filtration, or clay treatment can be applied. During sedimentation, the extracted materials will be removed together with the lees. With clarification colloidal, cloudy materials are removed allowing the wine to become mirror clean and stable.[16,39,63]

5.3.6 AGING OF WINE AND BOTTLING

During the aging of the wine, the tertiary aromas are developed. The most important factor in the wine aging is the presence of the air (oxygen). May wines have improved quality when stored in a barrel or bottle. They reach their peak values over time, but further maturation deteriorates the quality. During the aging process, acidity decreases the undesirable substances are eliminated by purification and stabilization. Wine is generally stored in oak barrels, which allows the flow of oxygen and the escape of water and alcohol. The essence of the wood can be recognized in the wine, too. Humidity affects the decrease of certain components. In a low humidity area, alcohol can be concentrated in the wine, while in high humidity area; the wine becomes weaker because of the decline of alcohol content.

For the proper quality, certain red table wines can be achieved by storing it in oak barrels for 2–3 years. For the best red wines, post-treatment can take about 2–20 years. The quality of rosé and dry red wines cannot be improved by aging in bottles or barrels, so they are matured only for a short period, followed by purification and bottling. More than 90% of table wines appear on the market before reaching the age of 2 years. In case of white wines, fresh aroma is desirable. For this reason, they are rarely or never stored in wooden barrels. The early bottling of white wines reduces the cost of storage and handling in barrels, and results in fresh and fruity flavor.[16,39,63]

5.4 CURRENT SITUATION OF THE HUNGARIAN VITICULTURE AND WINE PRODUCTION

Among 22 Hungarian wine regions, the number of wine-makers gradually comes up with outstanding wine by utilizing their professional skills, traditional recipe, and grape type depending on ecological conditions.[47] Hungary contributes 1.9% of the world wine production. This ratio is 2.6 and 1.1% regarding the vineyard area and wine production in the EU,[9] however, Hungary is a small country.[33] Between 2011 and 2014, Hungary achieved EUR 3808 per hectares when wine was made from Hungarian grape, whereas this figure reaches EUR 4215, when grape is purchased.[11] Therefore, development is required to achieve higher production values in Hungary as well. Hungary is not a leading wine-making country, constitutes 1.5% of the vineyards of the world and 1.5–2% of the wine production. It provides approximately 2.5% of the wine production of the European Union, however, Hungary is a significant wine exporter compared to the other countries of the region. Within a decade, this ratio was increased. In 2013, Hungary only contributed 1.59% of the EU wine production.[32] In 2015, wine production constituted approximately 1% of the world wine production and 1.65% of the EU wine production. Hungary occupied the 16th–17th position among the 46 wine-producing countries of the International Organization of Vine and Wine.[23]

FIGURE 5.4 Wine production in Hungary: self-developed with concepts from HNT, 2017.[34]

All in all, similar to other tradition wine-producing countries, Hungary's position in the world market is becoming weak. In 2012, the wine production sector contributed to the performance of the Hungarian agriculture by HUF 30 billion. Combined with tourism, catering and trading, this sector's potential is almost infinite.[1] The sector plays an important role in the 2014–2020 EU programs[45] that facilitate plantation and research in this field.[27]

5.5 WINE CONSUMPTION HABITS IN HUNGARY

The Hungarian wine market can be considered as a continuously developing and changing field. Not only the technological innovations, but also the consumption habits of the residents shape the market. The role of wine in the Hungarian history, culture, and gastronomy is dominant.[73] Nowadays, wine consumption is following the latest trends. It may say that it is a new golden age of wine consumption in Hungary. It is also spilled over into other sectors, such as tourism or healthcare. In 2017, the wine consumption per year per capita was 24.8 L, which is compared to the consumption in 2007 (28.5 L per year per capita), shows 4 L less.[36] Eventually, the total wine consumption has shown decrease since 2007. On the one hand, the world economic crisis had affected on discretional spending; on the other hand, the market of other alcoholic beverages has been expanding.

Currently, the structure of Hungarian wine market looks like an hourglass according to the experts. However, there is an oversupply of the premium and the low-quality wines; there is lower supply of mid-category products. The consumption habits rather tend toward premium quality; however, it requires the high degree of consciousness from the consumers.[72]

In 2016, representative research was conducted with 1500 respondents in Hungary by the Eszterházy Károly University. The research aimed to reveal the consumption habits and attitude of the Hungarian wine consumers. Regarding the frequency of wine consumption, the 22% of the Hungarians drinks wine every week or more times a week. These consumers purchase approximately 75% of the wine commercially sold (1.64 million hectolitre).[70] The drivers of wine consumption can be categorized in wide spectrum. There is a group either of consumers, who drinks wine only with friends, to relax and to mitigate stress, or due to health prevention issues. Besides the before mentioned motivations, other influencing factors can also determine wine purchase and consumption. The taste, origin, point of purchase, income level, gender, general knowledge about wines, and the way of consumption have an effect on the purchasing and consuming habits of wine.

In Hungary, men drink wine more than women; furthermore, the men also consume more often. By analyzing age, the youngest (18–29 years) and the oldest (over 65 years) people consume less wine than people aged between 30 and 65 years. Approximately 55% of the Hungarian men prefer drinking dry or semi-dry wine, while the third-quarter of the women loves to drink sweeter wines. Based on the color of wine, preferences of consumers can be identified. The Hungarians prefer red wine than white wine, but among heavy-consumers the white wine is far more popular.

The consumption of rosé wine is typical for moderate wine consumers. The most proportion of the consumers (ca. 80%) drinks wine at home or when visiting friends. Consuming wine at restaurants represents less than 10%; at wine-related events, this proportion is only 3%. The majority of the Hungarian consumers buy the wine in hiper- or supermarkets, while online purchasing of wine is not popular at all.[70] In spite of the fact that there is a growing interest toward wine and its consumption. The increasing interest and the demand for high quality wines have manifested in the popularity of wine in tourism sector. The most well-known national wine regions are Tokaj, Eger, Villány, and Badacsony.

5.6 GRAPE AND WINE AS FUNCTIONAL FOODS

The functional food is defined as a specific food supplying nutrients and energy-carriers, and having beneficial effects on health.[67] However, the expression of functional food has started to be in the spotlight in Hungary in the recent years only. The first reference to the idea goes back to the ancient times in India and China. The idea conceptually first appeared in Japan in the 1980s.[6] In 2010, representative research was conducted among the Hungarian consumers by Eszterházy Károly University, Hungary, with the aid of understanding the importance of functional food. Only 17% of the respondents have heard about the expression of functional food before. Those consumers who have been aware of the concept of functional food, 37% of them always take the health effects into consideration, while 50% of them do it occasionally. The one-third of the consumers has associated functional food with dairy products, 27% with fruits and vegetables, and 12% with meat products, and only 9% with bakery products. The research also studied the relationship between the demographics and attitude toward the functional foods. Educated consumers have more information about the effects of functional foods. However, the income has not affected the attitude of the respondents, the price of functional foods is very important for consumers.[67] Nevertheless, the latest representative research was carried out in Hungary in 2010, since then a positive shift could have been experienced. This change has happened in a broader context nowadays, due to conscious of health came to forefront. Consuming functional food is also part of the consciousness. According to international consumer behavior trends, just as in Hungary, the healthy lifestyle with healthy eating gets even more important role in everyday life. Consumers are empowered by the technology, such as social media, apps to choose most personalized solutions to preserve or improve their health condition.[49]

5.6.1 BIOACTIVE COMPOUNDS IN GRAPES AND WINES

The secret of the utilization of wine as a functional food lies in its compounds, which is determined by the ingredients of the grape primarily. The main components will be different in every year because of the various climate parameters, and will be affected by the treatment and processing of the grape. At the production of the wine, first must will be produced which is the freshly pressed juice of the grape fruit. It also contains solid particles, such as grape seeds and skin. Besides the basic compounds of the must (water, carbohydrate) polyphenols play the most important role as bioactive compounds; furthermore, minerals and vitamins have a positive effect on health.

5.6.1.1 POLYPHENOLS

Polyphenol compounds have favorable physiological effects. Polyphenols are major components of must and wine. They are transferred from grape must to the wine, where they retain their biological activity, maintaining their positive effect on health. Typically, they produce in much higher quantities in the must of blue grapes, also in red wines. Polyphenols are important because of their characteristics, such as sensitivity to oxidation, their aptitude to polymerization and reactions with other components. According to the chemical features, the following compounds are included under this category:

- **Non-flavonoid phenolic compounds** are hydroxybenzoic acid and its derivatives, hydroxy cinnamic acid and derivatives, other non-flavonoid-polyphenols (e.g., resveratrol). The characteristics of non-flavonoid phenols are the less astringent taste. Their presence can be detected mainly near the berry skin. Benzoic acid derivatives are the products of degradable anthocyanins. Cinnamic acid derivatives are involved in the compounds with the anthocyanins (acylated anthocyanins, can be found in the red wine grape). Outstanding health preserving effect has been noted for resveratrol which is effective against mammary and skin cancer cells.[42]

- **Flavonoid phenols** (e.g., flavonoid-catechins (3-flavons) leukoanthocyanins (flavan-3,4- diol), anthocyanins, flavons, and flavonols) are the most influential factors in the development of the bitter astringent taste and browning. Their typical representatives are catechins, leukoanthocyanins, and anthocyanin monomers. The procyanidins is produced by the polymerization of these compounds. Their reduction ability is excellent; furthermore, their antioxidant effect should be emphasized, which is resulted through their ability to create chelates with metal ions. Their occurrence is some kind of glucoside. Catechins are present in several forms in must and wine, and with increasing of their concentration, the activity of P vitamin increases.

5.6.1.2 TANNINS

Tannins are important flavor-forming compounds that greatly affect the stability of wine. They have an outstanding effect on inhibiting the enzymes of musts and wines. This antioxidize effect can be utilized in some wine treatment techniques. In case of the red grape, the degradation of the malic acid can occur; while in case of the white wine, it helps to reduce the need of the total sulfur. The hydrolysable tannins can be dissolved in the wine from the bleaching agents or oak barrels as they are not detectable in the grape, only in the wine. Non-hydrolysable tannins include monomer flavonoids of different degrees of polymerization. Due to the large number of hydroxyl groups, they are readily oxidized, thus affecting color tension. Procyanidins are oligomers of tannin-type polyphenol compounds (2–6 units). These polyphenol compounds can be the precursors of procianidins, which determine the color intensity, shade, and color depth of the white wines.

Tannins are important flavor-forming compounds that greatly affect the stability of wine. They have an outstanding effect on inhibiting the enzymes of musts and wines. In case of the red wine, the possibility is arising for the degradation of the malic acid, while at the white wine it helps to reduce the use of the total sulfur. Furthermore, minerals and vitamins can be found in grape and wine in very low quantities but they provide positive effects on health.

5.6.1.3 MINERALS

The mineral content of wine is contributed by inorganic ions. The most important anions are phosphate, chloride, and sulfate; while cations are potassium, sodium, calcium, and magnesium. In addition, iron, copper, aluminum, manganese, arsenic, zinc, and other metals can be found.

5.6.1.4 VITAMINS

Vitamins are useful for humans even in very small amounts. Due to their bioactivities, they are essential nutrients. Naturally, in the wine fat-soluble vitamins (A, D, E, and K) cannot be found. The most important water-soluble vitamin C is not present in significant amount in grape and wine. Many other water-soluble vitamins are present in the wine.[16,39,46] The quantity of some important vitamins present in Hungarian wine is represented in Table 5.4.

TABLE 5.4 Presence of Vitamins in Hungarian Wines: Self-Developed with Concepts from Kállay and Török, 1998.[40]

Vitamin	Red wines	White wines	Recommended daily intake (RDI)
B1 Thiamine	5–10 µg/L	10 µg/L	2 mg
B2 Riboflavin	177 µg/L	32 µg/L	3 mg
B6 Pyridoxine	0.35 mg/L	0.31 mg/L	5 mg
B12 Cobalamin	0.06 µg/L	0.07 µg/L	1 µg
H Biotin	2.1 µg/L	2 µg/L	10 µg
Pantothenic acid	0.98 mg/L	0.81 mg/L	10 µg

The beneficial effect of wine is attributed by many components, such as acids (e.g., tartaric acid, citric acid, and malic acid), aromas, vitamins, and polyphenol compounds (phenol acids, flavonoids, tannins, coumarins, and tannins). These compounds define the main characteristics of wine (such as taste, flavor, and the color). Comparing red wine to white wine, red wine contains six times more phenolic compounds. The antioxidant features of polyphenols help to protect cells from oxidative stress by blocking free radicals.[69,78] The direct health effects of the consumption of red wine have been in the focus of several studies and researches. The most frequently mentioned benefit of drinking wine is its cardioprotective effect. In Hungary, the cardiovascular diseases (CVD) is one of the leading cause of mortality, more than 50% of deaths are caused by CVD, and women are more concerned than men.[37] In addition to the cardiovascular effects, wine consumption can contribute to health prevention and has a positive impact on different diseases.

The Table 5.5 sums up the main health effects of modest wine consumption. It is proven that the consumption of wine, especially red wine has positive effect on health. The French paradoxon is well known, it describes the case that despite the relatively rich diet in saturated fats (which can be originated from the French cuisine, for example, consumption of cheese type Brie or Camembert, just to mention two from the approx. 350–450 types of French cheese); the probability of the coronary heart disease is relatively low. For this phenomenon, one reason could be the regular consumption of small volume (1–2 dL/day) of red wine.

TABLE 5.5 The Health Benefits of Red Wine: Self-Developed with Concepts from Guilford–Pezutto, 2011,[26]; Saleem–Basha, 2010.[62]

Effects on health and diseases	Description
Antioxidant effects	By increasing the antioxidant capacity of plasma helps to fight against oxidative stress being associated with diseases, such as heart failure, atherosclerosis, cancer, and neurologic degeneration.
	Makes the aging process slower.
Cancer	Resveratrol suppresses the proliferation of tumor and cancer cells.
Cardiovascular effects	The antioxidant, lipid, and anti-inflammatory effects of wine decreased cardiovascular diseases.
	Inhibits platelet aggregation
	Increases fibrinolysis
	Improves endothelial function
Diabetes	Polyphenols in wine help to prevent type 2 diabetes.
Gastrointestinal effect	Wine components have antibacterial effects, which reduce the risk of Helicobacter infection in intestine.
Inflammatory effects	Polyphenol and ethanol can modulate immune responses and induce the release of anti-inflammatory cytokine.
Lipid effects	Increases high-density lipoprotein (HDL) cholesterol, and decreases low-density lipoprotein (LDL) cholesterol
	Decreases lipid peroxidation.
Neurologic effects	The risk of development of neurological disorders, such as stroke, dementia may decrease by moderate wine consumption.
	Resveratrol causes oxidative stress decrease in neuronal cells and supports the efficiency of signaling pathways in learning process and memory.

5.6.2 THE BY-PRODUCTS OF WINE PROCESSING AND THEIR UTILIZATION

Grape processing and wine production have traditionally been a manual labor demanding sector, where large quantities of by-products are generated during each step. Although, today besides the production of handicraft wine, the proportion of large-scale industrial production is dominant. The utilization of by-products of wine processing is more and more widespread, thus helping to create a circular economy. The quantity of waste is reduced by utilizing certain by-products from wine industry. Many of these by-products can be considered as functional foods. Grape received lots of attention due to flavonoid and polyphenol contents, those have antioxidant potency.[14]

In Hungary's history and culture, a great importance has been attached to wine consumption. The wine was used for medical purpose in the ancient time. In 1967, the VI. Hungarian Pharmacopoeia did not contain "vinum album" as pharmaceutical ingredient. Later, positive effects of the wine were justified, nevertheless. Even the World Health Organization also recognized that the moderate alcohol consumption can have positive effects on the body, especially, related to coronary artery disease (CAD).[13]

5.6.2.1 GRAPE MARC

The grape marc is produced as a result of the pressing during the wine processing procedure, which is about the 15–20% of the weight of the wine.[10] In previous periods, it has simply served as animal feed or compost material; but presently, it is used as a feedstock to produce several bioactive compounds.

The fruit distillate can be produced by the evaporation from the non-separated grape marc. One of the examples is "pálinka," a Hungarian fruit-based brandy. The fruit distillate or "pálinka" can be produced using two-stage distillation or rectification. Products labeled as "craft brandy" ("kézműves pálinka") can only be generated by the first process, its basic requirement is having an own wort as a raw material. Usually, the volume of the cauldron is maximum 1000 L, the distillation unit has copper surface. It is important that the different vintages may not be mixed. After the two-stage procedure, the exact alcohol content can be adjusted by dilution.[2] In case of large scale production, usually rectification method is applied, which can be operated constantly. The result of the rectification process will be a solution with high alcohol content, from which—applying dilution—distillate or brandy for dietary purposes can be gained. The final alcohol content of the product (brandy) is between 37.5 and 55% both in case of traditional two-step distillation and rectification process, but the 40–52% alcohol content is typical.[48]

5.6.2.2 GRAPE SEEDS

The grape seeds[42] contain high concentration of chemopreventive agents, such as phenolic compounds (tannin, lignan, flavonoids, or resveratrol [in the red wine]). Therefore, it can help to fight against cancer.

The antibacterial and antioxidant activities of grape seed extract have been documented by Jayaprakasha et al.[38] while evaluating *Bacillus cereus*, *Bacillus coagulans*, *Bacillus subtilis*, *Staphylococcus aureus*, *Escherichia coli,* and *Pseudomonas aeruginosa*. They also found the phenomena that Gram-positive bacteria were completely inhibited at 850–1000 ppm with the use of grape-seed extract (GSE), while Gram-negative bacteria were inhibited with GSE at 1250–1500 ppm. Other studies[20,53] have proved its impact for the prevention of Cardiac Diseases and in retarding atherosclerosis. The effect of antioxidants was also studied in case of grape seed extract on coronary artery disease (CAD).

Nuttal et al.[53] used capsule formulation of an antioxidant polyphenolic extract in a single-blinded, tandomized, placebo-controlled cross-over study. After a washout period (min. 2 weeks), they repeated the treatment. The researchers continuously investigated the blood samples. Although the extract had no effect on serum vitamins (C and E), but increased the serum total antioxidant activity by 11%.

- **Grape seed powder**: Naturally, the grape seed is not tasty, hard, and has a long, a bit bitter after taste. Despite its organoleptic properties, grape seed powder is widely used based on its health effects. It can be consumed separately or mixed as an additive in different products, for example, dietary supplements, sport nutritions, etc.
- **Grape seed extract**: Using a standardized extraction method, grape seed proanthocyanidin extract (GSPE) can be produced. The free radical scavenging ability of GSPE was compared to the vitamin content (C, E, and beta-carotene). Applying in vitro and in vivo models, its cytotoxicity was demonstrated toward human breast, lung, and gastric adenocarcinoma cells.[3]

- **Grape seed oil** is one of the most common, but increasingly popular by-product in the wine making process. The grape seed oil is rich in valuable compounds with beneficial health effects.[24] Technology to prepare grape seed oil includes cold pressing method or multi-step extraction treatment. In the **cold pressing technology**, the grapes are pressed, and the seeds can be extracted from the remaining marc. After drying of the seeds, they will be pressed for gaining pure oil. The 50 kg of grape seeds is required for the extraction of 1 L of grape seed oil; otherwise, 500 kg of fresh grapes are needed. The oil content of the seed varies between 6 and 20%. Besides the extracted and refined grape seed oil, cold-pressed oil is gaining popularity and is taking over the leading role. This oil takes an active part in health preservation. It has been proved that the active ingredients (procyaidin) combination with vitamin E content play a role in improving the structure of the upper skin layer. Grape seed oil recovers wounds in skin, such as abrasions and minor burns. The color of the grape seed oil is green and even refined oils retain their characteristic flavor. The color of the cold pressed oil is deep green and the flavor is aromatic. Often the oil becomes cloudy, but this is not a sign of poor quality. Both virgin and refined grape seed oils can be used in baked and cold dishes.[50]

5.5.2.3 GRAPE SKIN

Mainly, the skin of the red and blue grape can be used as a by-product; the products made of them are rich in polyphenols and contain melatonin in high concentration, so they have significant antioxidant capacity. The available products are usually be powdered, and applied as dietary supplement, or utilized in the cosmetics industry. Extract of the fruit can be produced from the full grape with the help of extraction process. It has skin-rejuvenating outcome, although its efficiency is week. Nevertheless, several skin rejuvenating creams, soaps, face masks, face water, oil, shower gel, foam bath, lipstick, shampoo, slimming mud, peeling, and tonic, etc., contain this compound.[50]

5.6 VINOTHERAPY

Wine and the different parts of grape are utilized in the beauty industry.[60] Biomolecules in grape and wine have been used for therapeutically purpose in Egypt in 3200 BC. Meanwhile, vinotherapy became well-known only in the 20th century, when it was revealed in the famous wine region, Bordeaux in France. The first vinotherapy spa, Le Sources de Caudalie (The Sources of Caudalie) was established in France in 1998. Different elements of grapes were mixed with spring water and used for several therapeutic and beauty treatments, including baths, masks, massages, and scrubs for the skin.[66]

Massaging body with crushed grapes, using cosmetics made from grape, wine, and marc are unique in vinotherapy. This effect is based on the great amount of nutrients in the grape. The grape skin contains natural cell protection agents, flavonoids, antioxidant compounds, and polyphenols. They strengthen immune system, fight against free radicals and reduce the aging of human body.[76] The linoleic acid content of the dense grape seed oil is approx. 70%, which plays a significant role to regulate moisture level the skin. The minerals, like magnesium, calcium, potassium, and trace elements like (Se, Zn, Fe, and Mn) fill up the skin with energy.[76] In Hungary, vinotherapy is getting more and more popular. The palette of vinotherapeutic treatments is varied. Examples are bathing in Tokaj wine extract and essential oils or in red wine, skin nourishing masks and wrappings, massage with herbal pillows soaked in grape seed oil.[55]

5.7 SUMMARY

Grape and wine (our ancient functional foods) can offer several health benefits. Wine can even be considered as a symbol of fine life. It is a special drink that can take pride in having many deities in a unique way in history. It is not by accident, besides serving as a basic drink for millennia, it also became an essential accessory of religious ceremonies and social events. A transition could be seen: wine acted as a medicine, then it appeared as a "bottled poetry," today it is a "branded prestige consumer product." Almost all parts of grape can be used and have a positive effect on our health—either used externally or internally in some form. The moderate consumption of grape and wine stimulates brain function, strengthens bones, has a good effect on the heart, helps to fight against cancer and depression, is excellent for high blood pressure and stroke and even beautifies: it has a rejuvenating and renewing effect on the skin. It can be stated without exaggeration, that grape and wine are almost universal: they can be perfect example of functional foods. Wine and grapes do not only have economic importance, but their role is also constantly growing: the cosmetic industry has noticed their exceptional ingredients and offers vino-therapeutic treatments. Tourism builds on them within the framework of health tourism, but various wine tours, wine hotels, and wine tastings are also gaining renaissance; they determine the image of certain destinations basically. Referring to Hungary, nowadays grape- and wine production is blooming and is experiencing a new renaissance again with a new and high-level culture. Recently, there is a new and emerging trend, under which grapes and wine do not only play an important economic role, but also a cultural and civilization process is linked to them.

KEYWORDS

- **functional food**
- **grape**
- **viticulture**
- **wine**
- **wine's health effects**

REFERENCES

1. Ábel, I.; Hegedűsné, B. N. *Szőlőtermelő gazdaságok jövedelmezőségének vizsgálata* (Investigating the Profitability of Vine-Growing Farms). *Borászati füzetek (Wine Booklet)* **2017,** *3*, 29–33.
2. Act No. LXXIII of 2008 on *Pálinka, Grape Marc and Pálinka National Council*. https://net.jogtar.hu/jogszabaly?docid=A0800073.TV&celpara=&dbnum=1 (accessed June 22, 2021).
3. Bagchi, D.; Bagchi, M.; Stohs, S. J.; Das, K. K.; Sidhartha, D. Ray; Kuszynski, C. A.; Joshi, S. J.; Pruess, H. G. Free Radicals and Grape Seed Proanthocyanidin Extract: Importance in Human Health and Disease Prevention. *Toxicology* **2000,** *148* (2–3), 187–197.
4. Balinkó, E.; Tóth, I. *szőlőfajta használat alakulása Magyarországon 1973–2003 között* (Evolution of Grape Variety Use in Hungary Between 1973–2003). *Manuscript* 2004; pp 4–8.
5. Bényei, F.; Lőrincz, A.; Szendrődy, Győző; Sz. Nagy, L.; Zanathy, G. *Szőlőtermesztés* (Viticulture). Debrecen—Magyarország (Hungary): Mezőgazda Kiadó (Agriculture Publisher), 2011; p 188.
6. Biacs, P. Á. Production, Trade and Consumption of Functional Foods in Hungary. *Hungarian J. Nutr. Market.* **2017,** *4* (1–2), 3–6.

7. Bihari, T. *Borrégiók alakulnak* (Wine Regions Are Created). *Népszava* **2004,** *3*, 9–11.
8. Bodnár, L. *Az Egri Borvidék (The Eger Wine Region)*; Bodnár és Társa Geográfus Bt. (Bodnár and Associates Geographical Partnership): Vámospércs, 2001; p 12.
9. Boel, M. F. *Egy fenntartható európai borágazat irányába* (Towards a Sustainable European Wine Sector). Európai Bizottság (European Commission): Brüsszel (Brussels), 2006; pp 6–8.
10. *Bor és piac: Mire jó a törköly*? (Wine and the Market: What's the Use of Grape Marc?). https://borespiac.hu/2019/03/29/mire-jo-a-torkoly/ (accessed June 26, 2021).
11. Brazsil, D. *A magyar szőlő-bor ágazat 2016. évének értékelése* (Evaluation of the Hungarian Grape and Wine Sector in 2016). *Nak Konferencia: Élelmiszeripari Körkép—Alkoholos Italgyártás* (Food Industry Panorama Conference—Manufacture of Alcoholic Beverages): Budapest, 2016; pp 28–37.
12. Brazsil, D. *A magyarországi szőlő-bor ágazat stratégiája* (Strategy for the Hungarian Wine Sector). *Éves Duna Borrégió Konferencia* (Annual Danube Wine Region Conference), Kiskőrös, Hungary, 2017; p 6.
13. Csupor, D.; Szendrei, K. *A szőlő, a bor és az alkohol I* (Grapes, Wine and Alcohol: Part I). *Fitoterápia (Phytotherapy)* **2006,** *2*, 66–68.
14. Dey, G.; Negi, B.; Gandhi, A. Can fruit Wines be Considered as Functional Food. *Nat. Prod. Radiance* **2009,** *8* (4), 314–322.
15. Dupcsák, Zs.; Kerek, Z.; Marselek, S. *A szőlő- és bortermelés helyzete, lehetőségei* (The Situation and Potential of Grape and Wine Production). *Őstermelő. Gazdálkodók Lapja* (Primary Producer: Farmers' Journal) **2011,** *2*, 61–65.
16. Eperjesi, I.; Horváth, C.; Sidlovits, D.; Pásti, Gy.; Zilai, Z., Eds. *Borászati Technológia* (Wine-making technology). *Magyarország (Hungary):* Mezőgazda Kiadó (Agriculture Publisher), 2010; p 313.
17. Estreicher, S. K. *Wine from Neolithic Times to the 21st Century*; Algora Publishing: New York, 2006; p 3.
18. Feldman, Zs. *A szőlő- és borágazat helyzete* (The Situation in the Wine and Grape Sector). *XV. Lakiteleki Szőlészeti, Borászati és Marketing Konferencia* (XV. Lakitelek Viticulture, Winegrowing and Marketing Conference): Lakitelek, Hungary, 2014; p 240.
19. Fritz, J. *A. dél-dunántúli szőlő- és gyümölcstermesztés helye az országban* (The Place of Grape and Fruit Growing in South Transdanubia in the Country). Központi Statisztikai Hivatal Pécsi Igazgatósága (Directorate of the Central Statistical Office of Pécs), Pécs, Hungary, 2005; p 124.
20. Futterman, L. G., Lemberg, L. The Use of Antioxidants in Retarding Atherosclerosis: Fact or Fiction? *Am. J. Crit. Care* **1999,** *8*, 130–133.
21. FVM—*Földművelésügyi és Vidékfejlesztési Minisztérium* (Ministry of Agriculture and Rural Development); Agricultural Research Centers in Hungary: Budapest, Hungary, 2005; p 1–15.
22. Gaál, B.; Párdányi, M. *Bormarketing—A magyar borok marketingje* (Wine Marketing—Marketing of Hungarian Wines); Alfadat-Press Kft.: Tatabánya, Hungary, 2006; pp 23–28.
23. Gál, P. *A borstratégia premisszái* (The Premises of the Wine Strategy). *Éves Duna Borrégió Konferencia* (Annual Danube Wine Region Conference); Hegyközségek Nemzeti Tanácsa (National Council of Mountain Communities): Lakitelek, Hungary, 2016; pp 1–10.

24. Garavaglia, J; Markoski, M. M.; Oliveira, A; Marcadenti, A. Grape Seed Oil Compounds: Biological and Chemical Actions for Health. *Nutr. Metabol. Insights* **2016,** *9*, 59–64.
25. Gruenhoff, U.; Schneider, B. *Process for Preparing Wine*. Patent No.: 4,874,615; United States Patent Office; Oct 17, 1989. https://patents.justia.com/patent/4874615 (accessed June 22, 2021).
26. Guildford, J. M.; Pezzuto, J. M. Wine and Health: A Review. *Am. J. Enol. Cult.* **2011,** *2011*, 471–486.
27. Györe, D., Eds. *A közvetlen értékesítés szerepe az Egri borvidéken* (The Role of Direct Sales in the Eger Wine Region). *Disszertáció* (PhD Dissertation); Szent István University; Gödöllő, Hungary, 2014; p 151.
28. Harasztiné L. K. *Borászati technológiák eszközei* (Instruments for Winery Technologies). Eszterházy Károly College, 2012; p 1–102. https://www.tankonyvtar.hu/en/content/tamop425/2011_0001_521_Boraszati_technologia/ch06.html#id547165) (accessed June 22, 2021).
29. Heikefelt, C. *Chemical and Sensory Analysis of Juice, Cider from Different Apple Varieties*; MS Thesis for Swedish University of Agricultural Sciences: Alnarp, 2011; p 132.
30. Henderson, J. P.; Rex, D. *About Wine*, 2nd ed.; Delmar Cengage Learning: New York, 2012; p 688.
31. Higgins, L. M.; Llanos, E. A Healthy Indulgence: Wine Consumers and the Health Benefits of Wine. *Wine Econ. Policy* **2015,** *4* (3), 3–11.
32. HNT—*Hegyközségek Nemzeti Tanácsa* (National Council of Mountain Communities), 2014. http://hnt.hu/statisztikak/ (accessed June 20, 2021).
33. HNT—Hegyközségek Nemzeti Tanácsa. (National Council of Mountain Communities), 2016. http://hnt.hu/statisztikak/ (accessed June 20, 2021).
34. HNT—Hegyközségek Nemzeti Tanácsa.National (Council of Mountain Communities), 2017. http://hnt.hu/statisztikak/ (accessed June 20, 2021).
35. Horváth, Cs.; Mikulás, I. *A magyar borászat és az Európai Unió* (The Hungarian wine industry and the European Union); Magyar Kereskedelmi és Iparkamara (Hungarian Chamber of Commerce and Industry): Budapest, 2002; pp 96–101.
36. Hungarian Central Statistical Office (KSU) Database. https://www.ksh.hu/docs/hun/xstadat/xstadat_hosszu/elm11.html (accessed June 28, 2021).
37. Hungarian Central Statistical Office Database. https://www.ksh.hu/docs/hun/xstadat/xstadat_eves/i_wnh001.html (accessed June 28, 2021).
38. Jayaprakasha, G. K.; Selvi, T.; Sakariah, K. K. Antibacterial and Antioxidant Activities of Grape (*vitis vinifera*) Seed Extracts. *Food Res. Int.* **2003,** *36* (2), 117–122.
39. Kállay, M.; Rácz, L. *Bortechnológiai folyamatok és kémiai alapjaik* (Wine Technological Processes and Their Chemical Basis); Borkultúra Központ (Wine Culture Center): Hungary, Eger, 2012; p 119.
40. Kállay M., Török Z. *Rezveratrol izomerek meghatározása* (Determination of Resveratrol Isomers); Magyar Szőlő- és Borgazdaság (Hungarian Grape and Wine Industry): Hungary, Budapest, 1998; p 112.
41. Kate, E. *Historical Perspective on Health Benefits of Wine*, 2014. http://www.academicwino.com/2014/05/historical-perspective-health-benefits-of-wine.html/ (accessed June 10, 2021.)
42. Kaur, M; Agarwal, C; Agarwal, R. Anticancer and Cancer Chemopreventive Potential of Grape Seed Extract and Other Grape-Based Products. *J. Nutr.* **2009,** 139 (9), 1806–1812.

43. Keszler, B. *A régi Magyar orvosi nyelv forrásai és szótípusai* (Sources and Dictionaries of the Old Hungarian Medical Language). *Magyar Orvosi Nyelv* (Hungarian Medical Language) **2005,** *1*, 8–40.
44. Keszler, B.; Kaán L. M. Az igye-fogyott szegényeknek hasznokra (For the Yoke-Starved Poor). *Magyar Orvosi Nyelv* (Hungarian Medical Language) **2017,** *2* (1), 89–94.
45. Kis, M. Zs. Vidékfejlesztési Program 2015–2020 (Rural Development Programme 2015–2020). *Nak Országos Küldöttgyűlés* (National Assembly of Delegates): Gödöllő, Hungary, 2015; p 109.
46. Kiss, I. *Borászati termékek alternatív hasznosítása membrántechnika segítségével* (Alternative Recovery of Wine Products Using Membrane Technology); PhD Thesis; Corvinus University of Budapest: Budapest, Hungary, 2006; p 111.
47. Laposa, J.; Dékány, T. *Villány—borvidékek ékessége* (The Jewel of Wine Regions—Villány); Villány Város Önkormányzata (The Municipality of Villány): Hungary, 2001; p 6.
48. Magyar Élelmiszerkönyv (Hungarian Catalogue of Food Codex)—Alimentarius Hungaricus 2–109 számú irányelv Kézműves/kézmíves élelmiszerek általános jellemzői (General Characteristics of Artisanal/Handicraft Food). *Codex Alimentarius Hungaricus*, Directive 2–109), 2013. https://www.nak.hu/kiadvanyok/kiadvanyok/133-videkfejlesztesi-program-kezikonyv-1/file (accessed Aug 1, 2021).
49. Mellentin, J. Key Trends in Functional Foods & Beverages for 2019. https://www.nutraceuticalsworld.com/issues/2018–11/view_features/key-trends-in-functional-foods-beverages-for-2019 (accessed July 1, 2021).
50. Németh, R. *Szőlőmagolaj törkölyből—alternatív felhasználás* (Grape Seed Oil from Pomace—Alternative Uses), 2020; p 5. https://www.boraszportal.hu/hirszuret/szolomagolaj-torkolybol-alternativ-felhasznalas-2022 (accessed July 1, 2021).
51. Németi, S. *A magyar bor jövőképe 2. rész. Senki sem akar darázsfészekbe nyúlni* (The Vision of Hungarian Wine, Part 2: Nobody Wants to Stir Up a Hornets' Nest). *Borászati Füzetek* (Wine booklets) **2017,** *6*, 5–7.
52. Norrie, P. *The History of Wine as a Medicine. From Its Beginnings in China to the Present Day*; Cambridge Scholars Publishing: London, 2019; p 6–7.
53. Nuttall, S. L. An Evaluation of the Antioxidant Activity of a Standardized Grape Seed Extract, Leucoselect. In: *Grape Seed Extract for the Prevention of Cardiac Disease*; 1999; pp 91–94; *J. Clin. Pharm. Ther.* **1998,** *23*, 385–389.
54. Owen, J. Earliest Known Winery Found in Armenian Cave, 2011. https://news.nationalgeographic.com/news/2011/01/110111-oldest-wine-press-making-winery-armenia-science-ucla/ (accessed June 24, 2021).
55. Pécsek, B. *A régiós együttműködéstől a szőlőpépes pakolásig. Borturisztikai újdonságok* (From Regional Cooperation to Grape Pulp Wrapping: Wine Tourism Novelties). http://turizmusonline.hu/tt_itthon/cikk/borturisztikai_ujdonsagok (accessed June 26, 2021).
56. Piskóti, I.; Nagy, Sz. Innováció a funkcionális élelmiszerek területén (Innovation in Functional Foods). *VI—Nemzetközi Konferencia a közgazdász képzés megkezdésének 20. évfordulója alkalmából* (International Conference on the Occasion of the 20th Anniversary of the Start of Economics Education); University of Miskolc—Faculty of Economics: Hungary, 2007; pp 229–236.
57. Prioreschi, P. *A History of Medicine. Volume V. Medieval Medicine*; Horatius Press: Omaha, 2003; p 468. https://books.google.hu/books?id=wb_UMcH5C7EC&pg=PA149&dq=The+History+of+Wine++as+a+Medicine&hl=hu&sa=X&ved=0ahUKEwiIiKiN

r5bjAhWOIlsKHbMJCKYQ6AEIYDAI#v=onepage&q=wine&f=false (accessed June 26, 2021).

58. Rakonczás, N. *Szőlőtermesztés* (Viticulture). Debrecen: Debreceni Egyetem Mezőgazdaság-, Élelmiszertudományi és Környezetgazdálkodási Kar Kertészettudományi Intézet (University of Debrecen, Faculty of Agriculture, Food Science and Environmental Management, Institute of Horticulture): Debrecen University Press, 2014; pp 17–19.
59. Ruiz Villareal, M. *Wine_Grape_diagram_en.svg*, 2018. https://en.wikipedia.org/wiki/File:Wine_grape_diagram_en.svg (accessed 20 June, 2021).
60. Rutkowska-Podolowska, M.; Szczygiel, N.; Podplowski, G. Vinotherapy and Its Impact on Health Attitudes: The Innovation Perspective. *Inst. Econ. Res. Work. Papers* **2005,** *94*, 1–3.
61. S. Sárdi, M. Várandósság, szülés, születés. Szülés körüli tanácsok és eljárások a XVII.-XVIII. századi kéziratokban (Pregnancy, Childbirth, Birth. Child Births Advice and procedures In 17th and 18th Century Manuscripts). *Magyar Orvosi Nyelv* (Hungarian Medical Language) **2017,** *2*, 72–74. http://real-j.mtak.hu/11747/2/MONY_2017_02.pdf#page=19 (accessed June 27, 2021).
62. Saleem, T. S. M. et al.. Red Wine: A Drink to Your Heart. *J. Cardiovasc. Dis. Res.* **2010,** *1* (4), 171–176.
63. Sánches-Hernández, J. L.; Aparicio-Amador, J.; Alonso-Santos, J. L. The Shift Between Worlds of Production as an Innovative Process in the Wine Industry in Castile and Leon (Spain). *Geoforum* **2010,** *41* (3), 469–478.
64. Shapin, S. The Tastes of Wine: Towards a Cultural History. *Wineworld: New Essays Wine, Taste, Phil. Aesthetics* **2012,** *51*, 49–94.
65. Sivarooban, T.; Hettiarachchy, N. S.; Johnson, M. G. Physical and Antimicrobial Properties of Grape Seed Extract, Nisin, and Edta Incorporated Soy Protein Edible Films. *Food Res. Int.* **2009,** *41* (8), 781–785.
66. Smith, M.; Putzkó, L. *Health and Wellness Tourism*; Routledge: London, 2013; p 29.
67. Soós, G.; Biacs, P.; Kiss, A. Consumer Attitudes in Connection with the Consumption of Functional Foods. *Hungarian J. Nutr. Market.* **2013,** *1,* 13–21.
68. Sümegi, Zs. *Magyarország borászata* (Hungary's wine industry). Annual Danube Wine Region Conference; Lakitelek: *Hegyközségek Nemzeti Tanácsa* (National Council of Mountain Communities), 2014; p 18.
69. Szakály, Z.; Kiss, M.; Jasák, Helga: Functional Foods, Consumers Attitudes and Personalized Nutritions. *Hungarian J. Nutr. Market.* **2014,** *1* (1–2), 3–17.
70. Szolnoki, G.; Totth, G. *A Magyar borpiac és a borfogyasztói szokások elemzése. Szakmai prezentáció* (Analysis of the Hungarian Wine Market and Wine Consumption Habits. Professional Presentation.), 2017. http://hnt.hu/wp-content/uploads/2017/11/HNT_Szakmai_Nov14–1.pdf (accessed June 24, 2021).
71. Szűcs, I.; Farkasné, F. M. (Eds.) *Hatékonyság a mezőgazdaságban. Elmélet és gyakorlat* (Efficiency in Agriculture. Theory and Practice); Agroinform Kiadó (Publisher): Hungary, Budapest, 2008; pp 171–199.
72. The Hungarian Chamber of Agriculture: *Átalakult a hazai borpiac* (The Transformation of the Domestic Wine Market), 2016. https://www.nak.hu/agazati-hirek/elelmiszeripar/152-szesz-bor-soripar/92987-atalakult-a-hazai-borpiac (accessed June 27, 2021).
73. Tooth, G.; Szolnoki, G. *A magyarországi borfogyasztói szokások és a borpiac elemzése* (Analysis of Wine Consumption Patterns and the Wine Market in Hungary). *GAZDÁLKODÁS* **2019,** *63* (1), 22–39.

74. Törökné Kiss, K. Á.: Winemarketing, Wine Consumption and Wine Production in Hungary. *Hungarian J. Nutr. Market.* **2014,** *1* (1–2), 149–151.
75. Vidékfejlesztési Minisztérium—*Vidékfejlesztési Stratégia* (Ministry of Rural Development—RuralDevelopmentStrategy)*2012–2020*,2012;p83.http://videkstrategia.kormany.hu/download/4/37/30000/Nemzeti%20Vid%C3%A9kstrat%C3%A9gia.pdf004. évi XVIII. törvény a szőlőtermesztésről és a borgazdálkodásról (accessed June 26, 2021).
76. *Vinoterápia: Az egészség természetes* (Vinotherapy: Health Is Natural). http://www.naturstar.hu/hun/vinoterapia.html (accessed June 28, 2021).
77. Visy, Zs.; Mráv, Zs. *A Seuso-kincs és Pannonia* (The Sevso Treasure and Pannonia), 2nd ed.; Pécs, 2013; p 7.
78. Yoo, J. J.; Saliba, A. J.; Prenzler, P. D. Should Red Wine Be Considered a Functional Food? *Compreh. Rev. Food Sci. Food Saf.* **2010,** *9*, 530–551.

CHAPTER 6

HEALTH IMPACTS OF COMMON FLAVORING AGENTS IN INDIAN CUISINE

BEDAPRANA ROY, BIDISHA CHATTERJEE, TAMANNA SULTANA, and ARUP KUMAR MITRA

ABSTRACT

The Indian cuisine is enriched with different flavoring agents and seasonings with various active principles. Therefore, long-term consumption of these condiments might have both positive and negative impacts on our health. This review chapter has presented five types of leaf condiments (*Ocimum tenuiflorum, Petroselinum crispum, Laurus nobilis, Coriandrum sativum,* and *Cymbopogon citratus*) and five types of seed condiments (*Myristica fragrans, Brassica nigra, Cuminum cyminum, Trigonella foenumgraceum*, and *Foeniculum vulgare*); and has discussed their origins, usage, and their beneficial effects and detrimental effects (if any) when consumed for a long time. Multifarious health benefits have been discussed that are offered by these flavoring agents with respect to their anti-diabetic properties, anti-tumor properties, antioxidant properties, anti-inflammatory properties, etc. Recently, the anti-proliferative and anti-hypercholesterolemic properties of spices have gained overriding importance, as they attend to the key health concerns of mankind nowadays. Therefore, detailed research might be carried out in the future to purify the active ingredients in these flavoring agents and to understand closely their mode of action so that they can be used for medicinal purposes.

Sustainable and Functional Foods from Plants: Health Impact, Bioactive Compounds, and Production Technologies. Megh R. Goyal, Arijit Nath, and Zoltán Kovács (Eds.)
© 2024 Apple Academic Press, Inc. Co-published with CRC Press (Taylor & Francis)

6.1 INTRODUCTION

Spices and herbs, which have always been used since time immemorial in flavoring of several delicacies throughout the world, have bioactive substances to benefit human health. Before the discovery of allopathic medicine, there has been tremendous progress in Ayurveda for the use of herbal medications.[27] In recent years, World Health Organization (WHO) has also recognized traditional medication in different parts of the world and has set forth set of rules and guidelines so as to specify the practice of herbal medicines with safety. Researchers have been intrigued by the medicinal properties of these flavoring agents and have dived deep into determining and precisely pointing out the specific components present in these herbs responsible for curing various ailments.[47]

Treatment with chemical compounds having antimicrobial activity is expensive in developing countries for which people turn to herbal medicine. Many African tribes use Muthala (a traditional chewing stick),[50] which is scientifically known as *Diospyros lysioides* DESF[36] and/or *Euclea natalensis* A.D.C.[38] Several studies indicate the presence of antioxidative and antimicrobial, antibiofilm, antiadhesive, and antibacterial properties of herbal plants.[3,45,59]

Plants produce potential biologically active components. These secondary metabolites provide immunity to the host plant.[41] They are responsible for therapeutic activities, such as: lipid metabolism,[38,52] anti-diabetic,[78] anti-oxidant and anti-inflammatory,[76] antimicrobial,[37] anti-carcinogenic,[40] and anti-malarial.[57] Some common bioactive compounds present in plants are: alkaloids, anthraquinone, cardiac glycosides, flavonoid, phenol, phlobatannin, pyrrolizidine alkaloid, reducing sugar, saponin, steroid, tannin, terpenoid, and volatile oil.[70]

Although the traditional uses of these condiments are known, yet the bioactive compounds present in these herbs and spices have not been well described. This book chapter will give a better understanding of the essential phytochemistry, pharmacology, and toxicity associated with these condiments.

Authors of this chapter have discussed commonly used herbs and spices that are used in our daily diet. Emphasis was laid on the bioactive compounds reported in these condiments with reference to recent scientific research. This chapter also includes their potential effects in the health improvement while combating multidrug resistance and cancer.

6.2 OCIMUM TENUIFLORUM (TULSI)

Ocimum tenuiflorum is a culinary herb with restorative fragrance. It is indigenous to the Indian subcontinent and an important part of the Ayurvedic medicine. It is an erect plant of 30–60 cm in height. The inverse green leaves have a very specific taste and smell. It is a common plant in the Indian household that has an aesthetic value attached to it. It can retain carbon dioxide for a long time.[43] It has been used as a traditional medicinal plant from time immemorial and is greatly intertwined with holistic values.[67]

The oil in leaves and flowers of tulsi contains five different fatty acids (stearic, linolenic, palmitic, linoleic, and oleic acids). Tulsi also contains beta-carotene, calcium, and vitamin C. Along with other components, it includes volatile substances (estragol, eugenol, linalool, methyl chavicol), urolic acid, flavonoids, tannin, and camphor. The essential oil obtained from leaves contain eugenal, caryophylline, carvacrol, and limatrol along with eugenol and methylchavicol. Fatty acids and sitosterol are present in oil obtained from seeds. The oil obtained from leaves are also composed of n-triacontanol and ursolic acid. Apart from these, it also contains thymol, methyl chavicol, rosmarinic acid, citral, and linalool.[22] The chemical structure of some of these bioactive components in the essential oil obtained from tulsi has been represented in Figure 6.1.

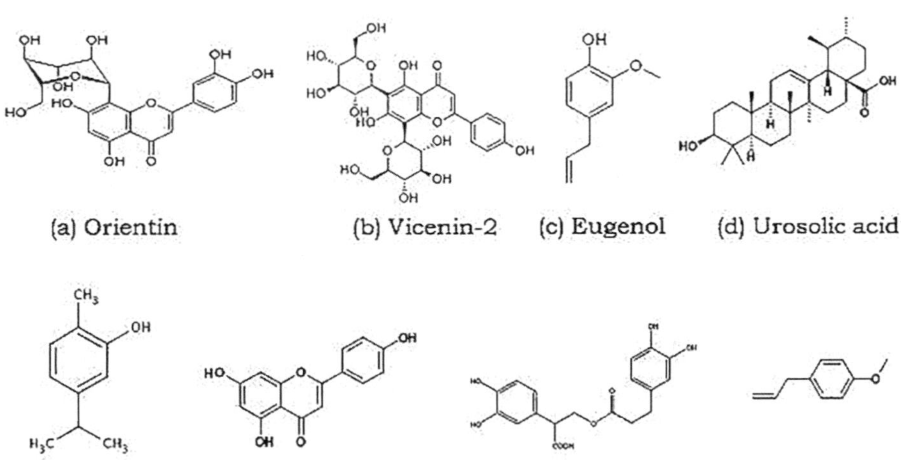

(a) Orientin (b) Vicenin-2 (c) Eugenol (d) Urosolic acid

(e) Carvacrol (f) Apigenin (g) Rosamarinic acid (h) Estragol

FIGURE 6.1 Chemical structure of few phytochemicals in tulsi, such as, (a) Orientin, (b) Vicenin (c) Eugenol, (d) Urosolic acid, (e) Carvacrol, (f) Apigenin, (g) Rosmarinic acid, and (h) Estragol Self-developed and compiled by authors.

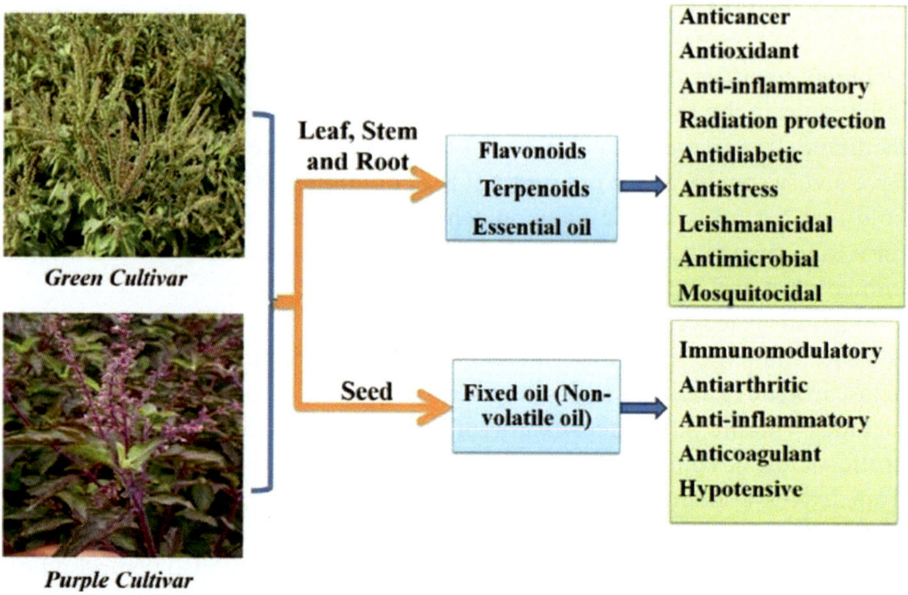

FIGURE 6.2 Health benefits of the essential biocompounds derived from *Ocimum tenuiflorum*: Self-developed by authors.

The leaves of tulsi plant are used to treat hepatic disorders and gastric problems. It is used as a mosquito repellant and for its larvicidal activities. Tulsi juice along with *triphala* is recommended for treatment of cataract, glaucoma, and chronic conjunctivitis. It can stimulate appetite. It can be used to treat common cold. It can remove excess cough from the respiratory tract. It can also lower level of uric acid and thus acts as an inflammatory agent. It acts as a nerve tonic to sharpen the memory.[22] Health benefits of essential compounds obtained from different plant parts of tulsi are shown in Figure 6.2.

Tulsi is also known for its antioxidant properties and provides protection against free radical induced damage. The reactive oxygen species (ROS) induce cell membrane damage by disrupting the membrane protein and lipids causing several diseases like cancer, diabetes, and liver cirrhosis. Antioxidants damage ROS and thus can decease lipid peroxidation and stimulates the activity of superoxide dismutase.[22]

The presence of antifungal and antibacterial activities are due to the essential oil. Significant activity against *Aspergillus sp*. have been reported. It also exhibits antibacterial activity against *Klebisella* sp., *Escherichia coli*, *Proteus* & *Staphylococcus aureus*, and *Vibrio cholerae*.[22] Ursolic acid showed anticancer property. The application of tulsi leaf extract visibly reduced the tumor cell size.[22] It has antilipidemic efficacy. It can reduce the LDL- cholesterol, triglycerides, and total cholesterol.[58] It has an antiestrogenic activity, which leads to antifertility.[22]

6.3 PETROSELINUM CRISPUM (PARSLEY)

Parsley belongs to the family Umbelliferae. It is mainly grown in Western Asia and Europe. It is an annual herb, which is widely used as culinary (Fig. 6.3). It is also known as heung choi, persil, petersil, ajmood, and salsa. The majority parts of the herb are used.[17]

FIGURE 6.3 *Petroselinum crispum* is a popular ingredient used as culinary. Self-developed and compiled by authors.

Parsley leaves and seeds contain essential oil.[79] Parsley seeds can produce 2–8% essential oil. The major constituent of this essential oil are α-pinene, myristicin, β-pinene, tetramethoxyallybenzene, and apiole.[39] Parsley seeds also contain petroselinic acid along with small quantity of linoleic acid, oleic acid, myristic acid, palmitic acid, myristoleic acid, stearic acid, and 7-octadecenoic acid, which together constitute 13–22% of the fixed oil.[17,42] Parsley leaves can produce 0.04–0.4% essential oil. The major bioconstituents of essential oil are α – pinene, p-cymene, β-pinene, α-terpenoline, myrcene, menthatriene, limonene, α-terpineol, α-phellandrene, apiole, and myristicin. Parsley leaf contains apiin (apigenin-7-O-(6-O-apiosyl) glucoside) as the main component of flavonoids.[17] The leaf also contains xanthotoxin, proteins, isopimpinellin, furocoumarins, fats, and sugar.[42] It is also rich in different types of minerals, for example, calcium, potassium, thiamine, iron, riboflavin, and vitamins A and C, and niacin.[49] Some of these have been depicted in Figure 6.4.

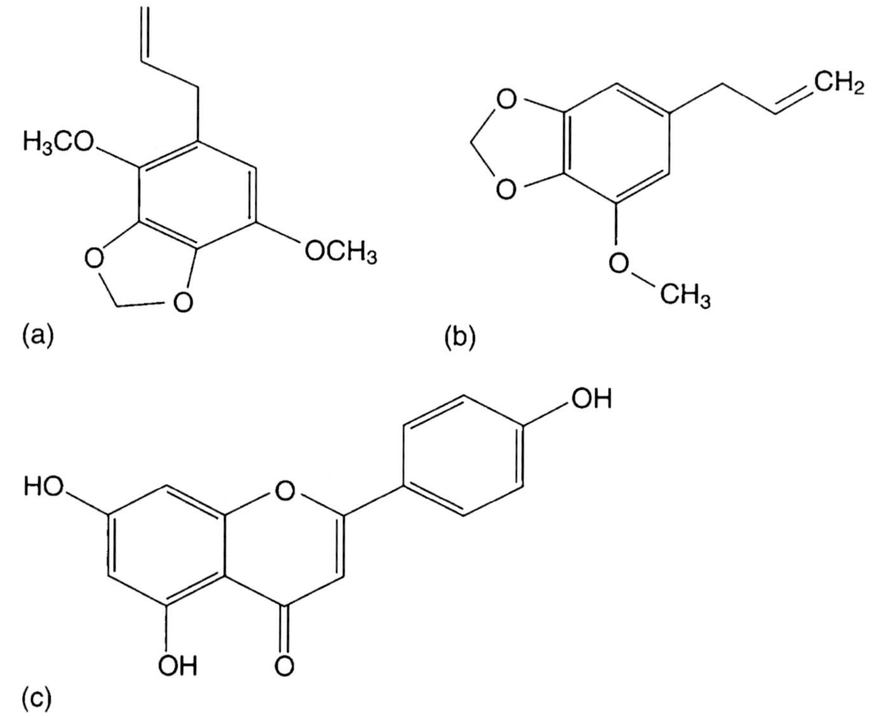

FIGURE 6.4 Phytochemical compounds isolated from *Petroselinium crispum*, such as: (a) apiole, (b) myristicin, and (c) apigenin.

The presence of phenolic compounds and these essential oils enhance the medical importance of parsley. It is considered as an anti-inflammatory, diuretic, antimicrobial, and hypoglycemic properties.[77] In India, it is also used as an expectorant. Parsley juice inhibits histamine production and is thus used as an antiallergic. It also acts as a liver tonic that helps to break kidney stone. The roots can be used as purgative and for elimination of bloating. It can also cause weight reduction, eliminate flatulence, and colic. Apart from these, it can also be used as a breath freshener, sooth eyes, heal bruises, relieve earache, and toothache. It can act as a mosquito repellent and reduce pain from insect bites. It is also used to treat hair growth.[17]

Parsley has anticancer, carminative, antioxidant, antiallergic, diuretic, and antipyretic properties.[17] It can induce glutathione S. transferase (GST:

a detoxifying enzyme) and also inhibit B(a)P induced tumorigenesis.[81,82] It also indirectly inhibits various procarcinogens, like aflatoxin B1.[64]

Regular food plants of Apiaceae Family (e.g., parsley, carrot, parsnip, celeriac) contain aliphatic C17-polyacetylenes (falcarinol type), which possess antibacterial, antifungal and antimycobacterial, anti-inflammatory, serotonergic, antiplatelet-aggregatory, and neurogenic properties. In vivo study suggests that these poyacetylenes show cytotoxicity against human cancer cells, their bioavailability, and have potential anticancer effects. It was thus concluded that they may act as effective nutraceuticals.[14]

The constituents of parsley have strong antioxidant properties as mentioned earlier. Flavonoid-rich samples show positive correlation between chemical property and antioxidant activity. Essential oil has significant importance in scavenging effect.[29] It has been reported that the presence of parsley induces erythrocyte glutathione reductase and superoxide dismutase activities.[53] It can also inhibit malonaldehyde formation due to UV-irradiation.[79] Simmer, stew, and soup-making techniques induce antioxidant property; whereas frying or grilling the food reduces it.[12] Various extracts of parsley for scavenging activity showed good results for DPPH and OH^- radicals.[65]

There may be some adverse side effects related to the consumption of parsley. Although there is no report on cross-reaction between allopathic medicine and parsley, yet sensitive people may develop allergy. As we know parsley contain furocoumarins, therefore fair-skinned humans may develop photosensitivity on immediate exposure to sun rays after they were in immediate contact with freshly obtained herb. Overconsumption of essential oils from parsley may cause toxicity. Patients suffering from kidney disease should consult a doctor before parsley consumption. Pregnant women should avoid parsley due to its abortifacient property.[17]

6.4 LAURUS NOBILIS (BAY LEAF)

Bay leaf belongs to the family Lauraceae. They are short perennial shrubs. Apart from being an important ingredient in Indian cooking (Fig. 6.5), bay leaf has been utilized as medicine.[62] This is why apart from being produced in Udaipur (India), bay leaves are also exported from other countries, like Nepal, to India.[13] These are native plants of Southern Mediterranean region, the subtropics and tropics of Eastern Asia, South and North America, the Balkans, and Asia Minor.[8] Bay leaf contains carbohydrate, steroid, flavonoids, alkaloids, citric acid, tannins, triterpenoids, eugenol, and essential oil.

The essential oil secreted from leaves and other parts of the plant imparts a distinct aroma to the bay leaf. It is the main source of vitamin A, vitamin C, vitamin E, iron, calcium, potassium, and fat in trace amounts.[8]

FIGURE 6.5 Bay leaves as herbs and spice. Self-developed and compiled by authors.

Bay leaf in combination with various other herbs have been extensively used in making sherbet and cooking with vegetables, fish, and meat. The leaf of the bay leaf contains a volatile oil, which can be applied as an irritant, as an insecticide, as germicide, or as a coolant. The essential oil possesses antimicrobial and antioxidant properties, which can enhance shelf-life of food materials.[24]

Bay leaf has many health benefits. It has been used to treat gastrointestinal problems like flatulence, bloating, and incomplete digestion. It is widely used as a diuretic agent.[25] They have been used to treat rheumatism, skin rash, earache, nausea, anemia, vomiting, fever, colic, body odor, and diarrhea because of its sharp, aromatic, restorative, and flatulence-relieving qualities.[8]

Bay leaf can improve our immune system by anti-inflammatory, antidiarrheal, and anti-diabetic characteristics. It has antioxidants like vitamin C, vitamin E, and carotenoids that can reduce blood cholesterol and control uric acid level. It also contains sesquiterpene lactones that inhibit NO production and thus improves the liver glutathione S-transferase activity.[28] Anticonvulsant (like pinene, eugenol, and methyl-eugenol) and sedation and motor impairment (like eugenol, cineole, and methyl-eugenol) are present in its essential oil components.[73]

Bay leaf can control diabetes and migraine.[28] Fresh leaves are used to treat kidney congestion. According to traditional medicine, it can also treat arthritis, cold, headache, cataract, fungal disease, colic ulcer, anorexia, neuralgia, diarrhea, and appetizer.[62] Therefore, bay leaf is used to cure various microbial infections. It can also inhibit carcinogenic growth. Bay leaf contains vitamin A, therefore it can also cure night blindness.[8]

It is also known to act as an antimutagenic due to the presence of an antimutagen 3-kaempferyl p-coumarate, which is obtained from the ethyl acetate extract of bay leaf after experimental identification and chromatographical purification.[71] Antiviral[10] and anticholinergic[31] activities have been identified in various components of its essential oil.

Ethanol extract of bay leaf has strong antioxidant activity. This activity has been evaluated by scavenging of free radical, reducing power, scavenging of hydrogen peroxide, metal chelating, and scavenging of superoxide anion radical assays. It was concluded that the phenolic compounds present in ethanolic extract possessed the antioxidant activity. Essential oil present in bay leaves showed analgesic and anti-inflammatory effects.[74]

Bay leaf powder showed significant immunostimulant activity. This activity was observed in non-specific immune responses including phagocytosis in blood leukocytes, lysozymes and extracellular and intracellular respiratory burst activities, along with protein levels.[10] Although in vitro antibacterial activity was detected in seed oil, essential oil, and methanolic extract of seed oil, yet the latter was more potent.[56] The antibacterial activity of bay leaf was reported against *S. aureus, Bacillus subtilis*, and *Staphylococcus intermedius*, and it may be due to 1,8-ceneol (major constituent of bay leaf).[20] The antifungal activity of bay leaf was determined in vitro against seven plant pathogens at various concentrations. At 250 μg/ml, maximum antifungal activity was observed against *Botrytis cinerea.*[63] Insect repellant activity of essential oil has been reported.[26] Bay leaf also showed acaricidal activity against *Psoroptes cuniculi* at 10% concentration having 73% mortality.[46]

Bay leaf and its essential oil is generally considered as safe when consumed as a condiment in food. The leaf as a whole may choke a person but is safe when taken in powdered form in food. The intact dried leaf cannot be digested. Diabetic people should better avoid bay leaf as it may intervene with control of blood sugar. Bay leaf in presence of anesthesia and other medicines used during and post-surgery might affect the central nervous system. The discontinued use of bay leaf has been suggested at least 2 weeks prior to surgery.[8]

6.5 CORIANDRUM SATIVUM (CORIANDER)

Coriander or *Coriandrum sativum* of the family Apiaceae (Fig. 6.6) is an annual herb used extensively in India.[69] Its origin is the Mediterranean and Middle East. It is widely used as a condiment in Latin American, Chinese, and Indian cuisine. It is largely cultivated in India which is its biggest market and also the largest exporter.[69] In India, the coriander cultivating states are Rajasthan, Uttar Pradesh, Andhra Pradesh, Karnataka, Madhya Pradesh, and Tamil Nadu. Coriander is also grown in several parts of the world including Europe and America.[44]

Coriander mainly contains moisture, carbohydrate, protein and fats, vitamin A, vitamin C, niacin, thiamine, riboflavin, sodium, potassium, calcium, iron, and phosphorous.[16] Apart from these, it contains various types of glycolipids, such as, acylated steryl glucoside, glucocerebroside, and steryl glucoside.[66] It has the highest selenium content of 23.53 ppm.[55] Some other minerals have also been detected including magnesium, sulfur, aluminum, tin, silicon, zinc, manganese, and copper by ray fluorescent analysis.[4]

Coriander contains essential oil consisting of oxygenated monoterpenes and monoterpene hydrocarbons.[7] Coriandrol or d-linalool present in essential oil is the primary oxygenated compound. The other major components of the essential oil are α-pinene, camphor, γ-terpinene, geraniol, and geranyl acetate.[21] The minor components of essential oil are α- terpinene, acetic acid esters, β-pinene, borneol, camphene, dipentene, limonene, myrcene, n-decylaldehyde, and p-cymol.[44] Some of these components have been shown in Figure 6.7.

FIGURE 6.6 Coriander is used widely in the Indian cuisine for its aromatic flavor and multiple health benefits. Compiled by authors.

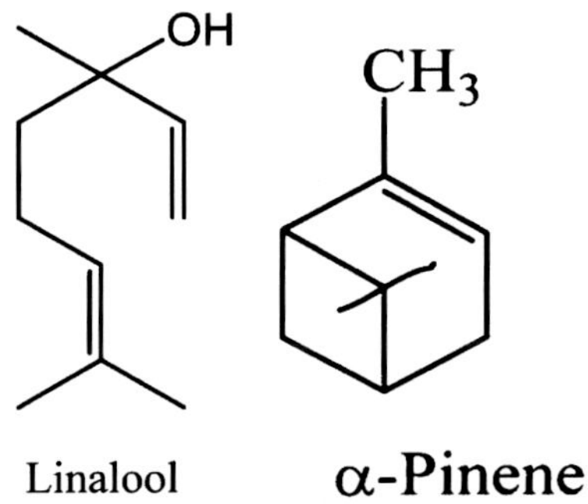

FIGURE 6.7 Bioactive compounds reported in *C. sativum*. Self-developed by authors.

Different parts of coriander like seeds, oil, and leaves have different uses in traditional medicine. Coriander helps in digestion and is used for the treatment of digestion disorders like flatulence, vomiting, dyspepsia, and loss of appetite.[35] It also helps in the treatment of dysentery, ulcerative colitis, typhoid fever, and hepatitis. Dried coriander can inhibit acidity. It is also used as a diuretic.[2] It is beneficial in lowering blood cholesterol. Coriander is used as an anti-inflammatory, antimicrobial, and analgesic. It is also used to treat rheumatism, conjunctivitis, and facial problems (including blackheads, pimple, and dry skin).[44]

A significant antibacterial activity of coriander has been reported against *S. aureus*, *E. coli*, and *Salmonella typhi*.[5] Another report stated that Gram-positive bacteria (*Listeria monocytogenes* and *S. aureus*) were significantly inhibited by coriander oil but not much effect was seen in case of Gram-negative bacteria (*Pseudomonas fragi*, *E. coli*, *S. typhi*).[19] The antifungal activity of coriander oil is still under debate. Significant anti-helminthic activity was observed against nematode eggs but extract with water and alcohol exhibited better results against than the crude aqueous extract.[44]

Anti-diabetic activity has been reported in coriander. It increases the glycogen synthase activity along with an increase in glycogen phosphorylase and gluconeogenic enzymes which decreases gluconeogenesis and glycogenolysis. It also increases the enzyme activity of glycolysis and pentose phosphate pathway thus glucose is metabolized at faster rate. Coriander also exhibits anti-cholesterol activity. It enhances lipid breakdown and reduce lipid uptake. Thus, it can be used as a preventive medication against hypolipidemia.[44]

Various components of coriander can modulate the synthesis of IFN-γ and stimulate the human peripheral blood mononuclear cells.[11] Aqueous extract of coriander exhibits muscle relaxation and also acts as a sedative. It also showed a dose-dependance antifertility activity. Along with this, it could also be used to treat irritable bowel disorders.[44] There have been no reports stating the toxicity of coriander. However, long exposure to the vapors or extracts may irritate the skin and eye. Till date, the use of coriander in food has not been stated harmful.[44]

6.6 CYMBOPOGON CITRATUS (LEMONGRASS)

Cymbopogon citratus (Fig. 6.8) belongs to the Poaceae family. This herbaceous plant is a perennial grass that is extensively grown in the tropical and savannah regions in Southeast Asia. It is popularly consumed as an aromatic drink and used in traditional cuisine for its tangy lemon flavor. Lemongrass leaves are used to prepare tea, which can relieve gastrointestinal and gut problems. In many countries, it is advised to consume lemongrass to treat feverish conditions and as a relaxant and sleeping-aid. It helps to overcome the problem of mood swings, insomnia, and has an anti-depressant agent.

FIGURE 6.8 Lemongrass is widely used in cuisines all over the world for its fragrant leaves and stalks, which are used as a flavoring component. Self-developed and compiled by authors.

Lemongrass is consumed throughout the world, because of its popularity on the basis of its nutritional, medicinal, and cosmetic values. Studies on its components have reported the presence of tannins, saponins, flavonoids, phenols, anthraquinones, alkaloids, deoxy-sugars, and various essential oils that contribute to its nutritional value. The presence of secondary metabolites has been indicated in the various pharmacological effects of this plant.

Cymbopogon citratus contains myrcene (an antibacterial and pain reliever), citronellal, citronellol, and geraniol.[23,32] Several researchers have shown the sedative, fungistatic activities, central nervous system depressor, analgesic, and antimicrobial activity of leaves of *C. citratus*. The neral, gereniol, geranial, myrcene, limonene, and citronellal have been specifically recognized as the main precursors behind the essential oils of *C. citratus* (Fig. 6.9).

Oral application of lemongrass showed its toxicity level to be extremely low thereby confirming it to be safe for consumption.[18,23,32] It was described in a recent investigation that *C. citratus* has a gastroprotective function. Essential oils extracted from citratus leaves may prevent gastric damage, which was induced by substances like ethanol or aspirin. Although lower doses of these essential oils do not show gastroprotective function against the injury caused by ethanol, there has been evidences that document the occurrence of oxidative stress and impairment of prostaglandin synthesis, which in turn contributes to mucosal damage in the gastrointestinal regions because of ethanol. In antioxidant studies by the use of DPPH, it was found that lemongrass showed its action as a cryoprotectant that prevented gastric mucosal damage. The mechanism of action of ethanolic damage of the gastric mucosal cavity is carried out by the release of vasoactive compounds, which were the causes of vascular damage necrosis of the gastric cells leading to the formation of ulcer.

A number of studies have documented the cryoprotective role of these essential oils, thus proving to be a promising solution for the combat of gastropathy by nonsteroidal anti-inflammatory drugs.[30]

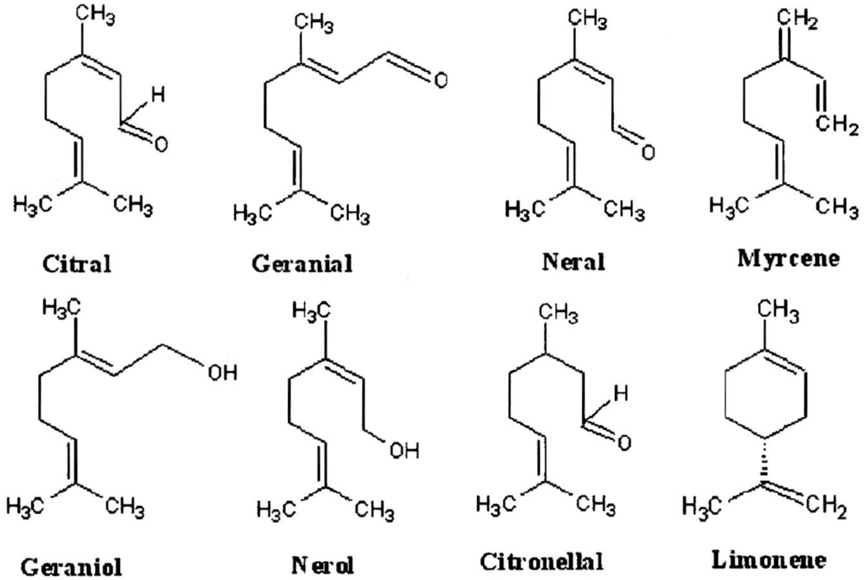

FIGURE 6.9 Main bioactive compounds in essential oil in *C. citratus*. Self-developed by authors.

C. citratus is useful in the treatment of fungal infections with promising antioxidant effects. Studies on animal subjects indicate that the essential oils can also have positive effects on anxiety and can act as sedative and anticonvulsive agents. Even inhalation of the lemongrass oil might also have a calming effect on the nervous system. It can have positive-effects on the brain cognitive functions, such as, power and continuity of attention memory qualities. *Cymbopogon citratus* is consumed as a popular decoction owing to its aromatic flavor and hypotension property. Its aqueous extract has been reported to be a radioprotector and its antimutagenic properties have been experimentally demonstrated.[28]

6.7 FOENICULUM VULGARE (FENNEL)

Fennel seeds are widely used as culinary spice in India. They are indigenous to the Mediterranean shores but are also found in various parts of the world, where dry soils are present, preferably near the banks of a river. Fennel has pale white bulb along with a green stalk (Fig. 6.10). It is mainly used because of its peculiar flavor and various health benefits. All parts of this plant are edible and add flavor to several food recipes. Traditionally, fennel was used for the treatment of cough and cold, nausea, flatulence, diarrhea, constipation, and also to increase breast milk production, which was mainly due to the fact that it contains a chemical compound called anethole which not only has an antimicrobial activity but also mimics estrogen and increases prolactin. According to the United States Department of Agriculture (USDA) National Nutritional Database, fennel contains vitamin C, vitamin B-6, and vitamin A. It is rich in potassium and contains considerable amounts of sodium, calcium, and magnesium. Trace amounts of iron, phosphorus, copper, zinc, manganese, niacin, selenium, pantothenic acid, choline, folate, beta-carotene, vitamin E, vitamin K, and lutein, along with high levels of dietary nitrates are present in it.

Along with above-mentioned components, fennel also contains flavonoids, like quercetin, which is antioxidant, antibacterial, and anti-inflammatory in action; phenolic compounds, like Rosmarinus acid and chlorogenic acids, which are again antioxidant and anti-inflammatory in action; terpenes, like limonene (improves wound healing); and essential amino acids, like leucine, isoleucine, phenylalanine, and tryptophan.[60] The fennel plant has medicinal properties, and it belongs to the Umbelliferae (Apiaceae) family. Consumption often has been an ancient practice because of its flavor.

FIGURE 6.10 Fennel seeds for cooking in Indian cuisine. Compiled by authors.

Number of studies have shown that *F. vulgare* is effective in controlling several infectious disorders of different origins, such as, bacteria, fungi, virus mycobacteria, protozoa, etc.[3,6,9,15,54] It was found to be effective in reducing damage of liver, reduction in blood sugar levels, and even reducing tumor growth. Also, *F. vulgare* has exceptional memory-enhancing effect and has been known to decrease stress. Experiments on animal subjects suggest that extensive consumption of *F. vulgare* is not injurious to health.

Injection of fennel can be done in numerous forms, such as, raw grilled or boiled. Sometimes even baked forms are also used in certain dishes. A diet consisting of a desired quantity of fennel could have several health advantages specifically due to the presence of essential fatty acids. Methanol extracts of fennel seeds exhibited anti-inflammatory activity, which was assessed using three different screening protocols, such as: (i) carrageenan-induced paw edema; (ii) arachidonic acid-induced ear edema; and (iii) formaldehyde-induced arthritis. These are extensively used for testing non-steroidal anti-inflammatory drugs. For acute inflammation, methanolic extract (200 mg/kg) exhibited remarkable inhibition of paw edema (69%) induced by carrageenan injection as compared to the control group of animals. As fennel has estrogen like activity, therefore it may reduce menstrual pain (dysmenorrhea), alleviate symptoms of menopause, and also increases libido.

Anethole, a component of fennel, is a competitor of Dopamine, which in turn is a blocker of the hormone prolactin that stimulates production of breast milk. Thus, it helps in increasing breast milk production. Fennel has been found to inhibit the growth of *E. coli* and *Salmonella* sp. Being an abundant source of antioxidants (such as, caffeoylquinic acid and quercetin), fennel has been found to increase the activity of superoxide dismutase and catalase in rats. The extract of fennel seed was seen to inhibit the growth of breast and liver cancer cells. The principal active component being anethole that has been found to reduce tumor weight and tumor volume in mice and also inhibit the growth of prostate cancer cells. Anethole blocks tumor and metastasis mainly by increasing the body's immune machinery that responds to cancer cells.

Fennel has sometimes been found to bring about premature sexual development in young children. As fennel increases the levels of prolactin (the hormone that stimulates production of breast milk), it may also cause hyperprolactinemia. Atopic individuals allergic to plants of the same families can also be allergic to fennel. It is known to cause allergic rhino-conjunctivitis and asthma to people, who are allergic to it. Fennel seeds contain some bioactives, which protect skin against ultraviolet rays from the sun. This can lead to rashes and redness on skin, referred to as photodermatitis.[61] Fennel also contains estragole that has been shown to be a potential carcinogen. Although its activity is almost neutralized by the other components of fennel itself.

6.8 BRASSICA NIGRA (MUSTARD)

The seeds of the mustard plant (Fig. 6.11) are used as a spice in different regional Indian foods. Cultivated in various parts of the world, mustard grows well in temperate regions like India, Pakistan, Nepal, Canada, Hungary, etc. Mustard oil is also extracted from mustard seeds that is extensively used in Indian cuisine. The seeds are used in roasted form as a spice and planted to obtain the plants, which are cooked to obtain *Sarson ka saag*, a major delicacy of Punjab.

Mustard contains carotenoids, phenolic compounds, and glucosinolates. These phytochemicals act as detoxifiers due to their antioxidant property and help reduce oxidative stress. Due to this reason, it is considered to be good

FIGURE 6.11 Mustard seeds with innumerable health benefits. Compiled by authors.

for human health.[8] Mustard has high content of carbohydrates, sugar, dietary fibers, fat, and protein. It also has different kinds of vitamins (A, B complex, C, E, K), minerals (Ca, Fe, Mg, P, Na, Zn, etc.), and omega-3-fatty acids in considerable amounts.

Glucosinolates in mustard seeds are broken into isothiocyanates myrosinase enzymes. These isothiocyanates can reduce the risk of cancers by preventing the growth of cancer cells and inhibiting the formation of cancer cells.[18]

Sinigrin (allyl-glucosinolate or 2-propenyl-glucosinolate) in Brassicaceae family is a natural glucosinolate that has number of medicinal properties. Studies conducted on pharmacological activities of Sinigrin show that it has anticancer, antibacterial, antifungal, anti-inflammatory antioxidant, biofumigation, and also wound-healing properties. Sinigrin is also the precursor of allyl-isothiocyanate (AITC), which may inhibit proliferation of food-borne pathogens, reduce risks of liver cancer, and inhibit cell structure, etc.

Mustard is a very rich source of essential oils and plant sterols (such as, cempesterol, sitosterol, avenasterol, brassicasterol, and stigmasterol). Mustard seeds may induce increased activity of superoxide dismutase (due to the presence of Manganese that acts as a cofactor of the enzyme), glutathione peroxidase, and catalase in turn helping in healing of psoriasis-induced inflammations and lesions. Both copper and iron in mustard help in red blood corpuscles formation and cellular metabolism. Mustard can also help in healing contact dermatitis, in easing bronchitis and bronchial pneumonia, and in promoting blood circulation in patients.

Despite of having various health benefits, mustard can result in tachypnea that is rapid and shallow breathing, sweating, dizziness, agitation, and dropping of blood pressure.[51] Mustard can lead to increased thyroid issues due to the disruption of the thyroid hormone and its function, both by mustard seeds and leaves. Mustard seed contains high levels of oxalate that can interfere with the calcium absorption channels in the body leading to serious health complications due to build-up of body fluid and crystallization.

6.9 MYRISTICA FRAGRANS (NUTMEG)

The seed of Nutmeg (Fig. 6.12) is dried and then grounded to form a fragrant spice that is used in number of baked foods and confectioneries. The seed covering (crimson-colored aril) of nutmeg is dried also to be used as a spice called Mace in number of delicacies. These are good sources of copper, potassium, calcium, manganese, iron, and vitamins (A, B complex, and C). Nutmeg also contains myristicin (Fig. 6.13), eugenol-elemicin, safrole, and volatile oils (such as, pinene, terpinol, cineole, etc.) that add to its flavor.

Cumin seeds contain important flavonoids like apigenin and luteolin that have potent antioxidant activity; cumin-aldehyde (Fig. 6.15) is also known to scavenge superoxide ions. Studies indicate that cumin seeds are effective in inhibiting the induction of gastric squamous cell carcinomas. In rat-model study, cumin has shown to suppress colon carcinogenesis. Lipid peroxidation was seen to be inhibited by cumin, which suggests its chemo-preventive property.[34]

Cumin-aldehyde and cumin oil have shown antimicrobial activity against *Klebsiella pneumonae,* and also exhibit strong larvicidal activity.

Cumin does not have much negative impacts on human health, and is considered to be highly safe for consumption. In larger doses, there are few risks associated in rare cases, like slowing down blood clotting and thus can make bleeding disorders worse. Also, due to its anti-diabetic action, it may reduce the blood sugar levels of non-diabetic person. It suppresses testosterone secretion levels that could make men less fertile. It may also trigger miscarriage in pregnant women.

Intake of cumin as a dietary constituent had a pronounced impact on the output of bile acid (quantity secreted per unit time); and secretion of bile acid was found to be as high as 70% with respect to the control. Thus, we can conclude that cumin helps in digestion by stimulating the biliary secretion of bile acids. In vitro studies showed inhibition of free radicals and the formation of AGE by cumin. It also possesses anti-diabetic activity and is extremely effective in managing the oxidative stress and AGE formation. Cumin-aldehyde from cumin seeds exhibited inhibitory activity, which was tested against aldose reductase and α-glucosidase. Research is being conducted toward comparing its impact with quercetin (inhibitor of aldose reductase) and carbose (inhibitor of α-glucosidase). *Cuminum cyminum* has been conventionally used to treat indigestion and hypertension. The aqueous extract of cumin seed has anti-hypertensive properties. Its role has been evaluated in the expression of nitric oxide synthase and in treating inflammation and oxidative stress.

6.11 TRIGONELLA FOENUMGRAECUM (FENUGREEK)

Fenugreek (Fig. 6.16) is a common Indian spice and an aromatic herb. Both its leaves (dried or fresh) and seeds (dried) are used as spice in several dishes and delicacies of the Indian subcontinent, and South and Central Asia. Fenugreek is indigenous to Southern Europe, Western Asia, and the Mediterranean region. Fenugreek has a diverse range of beneficial nutrients, such as, iron, manganese, and copper, and several phytochemicals (Fig. 6.17), such as, choline, trigonelline, yamogenin, gitogenin, tigogenin, neotigogenins, diosgenin, and luteolin. The health benefits conferred by Fenugreek are due to the presence of saponins and dietary fibers in it. It also contains amino acids like histidine, 4-hydroxyisoleucine, lysine, arginine, sterols (e.g., cholesterol and sitosterol), vitamins (A, B, C), and nicotinic acid.

Fenugreek seeds are composed of alkaloids, such as: trigonelline, carpaine, and gentianine. The fenugreek seeds are rich source of fiber and protein. The fiber is further subdivided into gum (gel fiber) and neutral detergent fiber. The fenugreek seeds are also composed of fiber-galcatomannan, 4-hydroxyisoleucine amino acid, and fenugreekine (component exhibiting probable hypoglycemic activity).

Fenugreek seeds contain 4-hydroxyleucine, which produces a significant decrease in blood sugar level by increasing the insulin production, thus behaving as a secretagogue. Fenugreek seeds combined with metformin used for the treatment of Type 2 *Mellitus* has been recently reported, showing significant change in the glycemic profile. Therefore, the economical and availability of Fenugreek seed powder in India has led to the conclusion that it might be used as a favorable form of treatment.

FIGURE 6.16 Fenugreek leaves and seeds that are popular in Indian cuisine for their characteristic aroma. Compiled by authors.

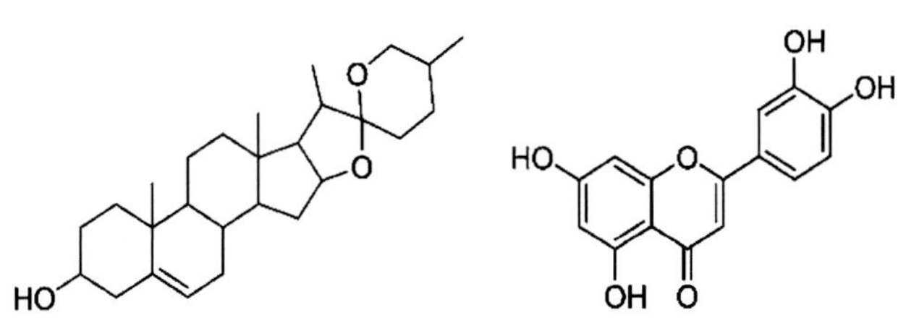

FIGURE 6.17 Phytochemicals obtained from fenugreek. Compiled by authors.

Scientists conducting research independently have inferred that Fenugreek has been shown to have anti-atherogenic, anti-anorexic, anti-hyperlipidemic, anti-inflammatory, and several other beneficial effects.[75,80] Evidence suggests that fenugreek seed extracts are effective in preventing increase of blood sugar levels, reduced lipid profile, and it also exhibited antioxidant properties on the tissues of liver and kidney. Fenugreek has a strong antioxidant potential due to the presence of flavonoids and polyphenols in it.[72] It has been seen that fenugreek can protect erythrocytes, from hemolysis, and from hydrogen peroxide-induced oxidation.[1] Cryogrinding techniques help in improving the antioxidant activity of the extract. It also appears to facilitate the lesion-healing process due to its antioxidant properties. It shows a potent anti-plasmodia effect.[57]

Number of studies with the various solvents indicate that ethanolic extract of fenugreek has been shown to have anti-neoplastic effect on breast cancer cells, where they reduce viability, induce early apoptosis, and degrade cellular DNA.[68] Presence of galactomannan in fenugreek fiber promotes satiety in obese and overweight patients by delaying the time of gastric emptying and also by decreasing energy intake (by causing marked drop in hunger, consumption of food).

Purified methanolic extract of Fenugreek has been proved to be hypertensive in rat model system. It also shows anti-cataract property in rats with chemical-induced cataract. This was done by restoration of glutathione levels and oxidative stress-relieving enzymes (SOD, CAT) in the eye lens and impeding lipid peroxidation.

In comparison to other extracts, fenugreek seed extract (in polyphenol) prevents ethanol-induced toxicity in human liver cells. Administration of fenugreek extract to 60 days' ethanol fed cells was able to increase the levels of serum lipid hydroperoxides, and it prevented the leakage of enzymes and the rise in lipid peroxidation levels to a remarkable extent.[68]

Just as any other biochemical compound, fenugreek and its phytochemicals may also have some toxicological side effects. In vivo experiments on rats, mice rabbits, and human trials have shown fenugreek to cause cases of congenital malformations, testicular toxicity, and antifertility effects in males, anti-implantation, and abortifacient activity in females due to the presence of saponins, which is further indicative of the fact for the limited use of fenugreek during pregnancy. Few diagnostic reports also suggest that fenugreek can have neuro-developmental, neurobehavioral, and also neuropathological side-effects. Therefore, taking into account the toxic dosage, consent from dietician, and careful use of this spice in daily life, it is expected to deliver health-related benefits.

6.12 SUMMARY

In this chapter, authors have elaborately discussed the traditional use of these condiments, their scientific classification system, the phytochemicals extracted from these condiments, and most importantly, they have focused on the medicinal importance of these flavoring agents based on the scientific evidences put forward in latest scientific reports and reviews. In this review, health benefits of *Ocimumtenui florum, Petroselinum crispum, Coriandrum sativum, Laurus nobilis, Cymbopogon citratus, Myristica fragrans, Brassica nigra, Cuminum cyminum, Trigonella foenum-graceum*, and *Foeniculum vulgare* have been discussed. However, number of other condiments are still unexplored though they have been traditionally used in treating various ailments.

Herbs and spices commonly used in the Indian cuisine as flavoring agents confer innumerable health benefits. However, the key ingredients or the principle bioactive compounds responsible for these properties remain unexplored. Scientists have only recently turned their attention toward these condiments for gaining scientific insight and providing strong scientific conclusions regarding the health benefits associated with consuming these traditional herbs and spices. As a result, there needs to be enough human trials to substantiate and support the hypothesized beneficial properties of the flavoring agents to mankind. This chapter provides encouragement and motivation for further scientific inquiry and calls for active research in this field to be undertaken for developing new chemotherapeutic agents and treatment strategies.

To procure scientific evidence regarding the importance of bioavailability of the active principals, investigations on the effect of the food array and interactions between its respective components, the enzymatic, chemical, and bacterial environment within the gut were monitored with respect to application of a single dose. Furthermore, the analysis used to determine if antagonism or synergy has occurred among their various bioactive compounds is another factor that needs to be considered and studied.

Efforts to demarcate the appropriate extraction and purification process of the specific phytochemicals from these spices and herbs should be undertaken. Thin-layer chromatography, high-performance liquid chromatography, colum chromatography, x-ray crystallography, and other such sophisticated techniques shall be utilized to extract and purify the phytochemical compounds from these flavoring agents. Studies on specifically implying the effective solvent system for the condiments and subsequent preparation of their extracts is also required for exhibiting optimal activity for conferring health benefits.

In future, it is expected that more research will be taken up related to the health benefits of condiments based on the bioactive components present in it resulting in an improved immunity. A wide variety of future prospects are awaiting in this field that will motivate the young scientists toward exploring the health benefits in a flavorful way!

ACKNOWLEDGMENT:

Authors acknowledge Rev. Dr. Dominic Savio, S. J., Principal at St. Xavier's College (Autonomous), Kolkata, for his constant encouragement and support.

KEYWORDS

- **anti-diabetic**
- **anti-inflammatory**
- **antioxidant**
- **flavoring agents**
- **health benefits**
- **medicinal properties**
- **spices**

REFERENCES

1. Abdullah, M. A.; Al-Bayati, F. H.; Ali, N. A. W.; Baharuddin, N. A. Wound Healing Potential of *Ageratum conyzoides, Trigonella foenumgraceum* and *Ginkgo biloba* Paste in Rats. *Dentika Dental J.* **2007,** *12*, 22–25.
2. Aissaoui, A. E. I.; Hilaly, J.; Israili, Z.; Lyoussi, B. Acute Diuretic Effect of Continuous Intravenous Infusion of an Aqueous Extract of *Coriandrum sativum* in Anesthetized Rats. *J. Ethnopharmacol.* **2008,** *115*, 89–95.
3. Al-Bakri, A. G.; Othman, G.; Afifi, F. U. Determination of the Antibiofilm, Antiadhesive, and Anti-MRSA Activities of Seven *Salvia* Species. *Pharmacogn. Mag.* **2010,** *6* (24), 264–270.

4. Al-Bataina, B. A.; Masiat, A. O.; Ai-Kofahi, M. M. Element Analysis and Biological Studies on Ten Oriental Spices Using XRF and Ames Test. *J. Trace Elements Med. Biol.* **2003,** *17* (2), 85–90.
5. Al-Jedah, J. H.; Ali, M. Z.; Robinson, R. K. The Inhibitory Action of Spices Against Pathogens That Might be Capable of Growth in a Fish Sauce (*Mehiawah*) from the Middle East. *Int. J. Food Microbiol.* **2000,** *57*, 129–33.
6. Balasubramanian, N.; Avinash, S. D. Antibacterial Principles from *Myristica fragrans* Seeds. *J. Med. Food* **2006,** *9* (3), 395–399.
7. Bandoni, A. L.; Mizrahi, I.; Juarez, M. A. Composition and Quality of Essential Oil of Coriander (*Coriandrum sativum* L.) from Argentina. *J. Essent. Oil Res.* **1998,** *10*, 581–584.
8. Batool, S.; Khera, R. A.; Hanif, M. A.; Ayub, M. A. Bay Leaf. *Med. Plants of South Asia* **2020,** *2020*, 63–74.
9. Baytop, T. *Therapy with Medicinal Plants in Turkey: Past and Present*; Nobel Tıp -Press: Istanbul Pakistan, 2000; pp 13–31.
10. Bhat, S.; Kaushal, P.; Kaur, M.; Sharma, H. K. Coriander (*Coriandrum sativum* L.): Processing, Nutritional and Functional Aspects. *Afr. J. Plant Sci.* **2014,** *8* (1), 25–33.
11. Bilen, S.; Bulut, M. Effects of Laurel (*Laurus nobilis*) on the Non-Specific Immune Responses of Rainbow Trout (*Oncorhynchus mykiss*, Walbaum). *J. Anim. Vet. Adv.* **2010,** *9*, 1275–1279.
12. Charles, D. J. P. *Handbook of Herbs and Spices*, 2nd ed., Volume 1; Woodhead Publishing Series in Food Science, Technology and Nutrition: London, 2012; pp 430–451.
13. Cherng, J. M.; Chiang, W.; Clliang, L. E. Immunomodulatory Activities of Common Vegetables and Spices of Umbelliferae and Its Related Coumarins and Flavonoids. *Food Chem.* **2008,** *106*, 944–950.
14. Chohan, M.; Forster, G.; Opara, E. I. Determination of the Antioxidant Capacity of Culinary Herbs Subjected to Various Cooking and Storage Processes Using the ABTS (*+) Radical Cation Assay. *Plant Foods Human Nutr.* **2008,** *63*, 47–52.
15. Choudhary, D.; Kala, S.; Todaria, N.; Dasgupta, S.; Kollmair, M. Effects of Harvesting on Productivity of Bay Leaf Tree (*Cinnamomum tamala*): Case from Udayapur District of Nepal. *J. For. Res.* **2014,** *25*, 163–170.
16. Christensen, L. P. Aliphatic C(17)-Polyacetylenes of the Falcarinol Type as Potential Health Promoting Compounds in Food Plants of the Apiaceae Family. *Recent Patents Food Nutr. Agric.* **2011,** *3*, 64–67.
17. Conti, B.; Canale, A.; Bertorli, A.; Gozzini, F.; Piselli, L. Essential Oil Composition and Larvicidal Activity of Six Mediterranean Aromatic Plants Against the Mosquito *Aedes albopictus*. *Parasitol. Res.* **2010,** *107*, 1455–1461.
18. Coskuner, Y.; Karababa, E. Physical Properties of Coriander Seeds (*Coriandrum sativum* L). *J. Food Eng.* **2006,** *80*, 408–416.
19. Dashwood, R. H. Frontiers in Polyphenols and Cancer Prevention. *J. Nutr.* **2007,** *137*, 267S–269S.
20. Deepika, B. A.; Ramamurthy, J. Comparative Clinical Data for Gingivitis Treatment Using Gels from *Ocimum sanctum* (Tulsi). *Bioinformation* **2021,** *17* (12), 1091–1098.
21. Delaquis, P. J.; Stanich, K.; Girard. B.; Mazza. G. Antimicrobial Activity of Individual and Mixed Fractions of Dill, Coriander and Eucalyptus Essential Oils. *Int. J. Food Microbiol.* **2002,** *74*, 101–109.

22. Derwich, E.; Benziane, Z.; Boukir, A.; Mohamed, S.; Abdellah, B. Chemical Composition and Antibacterial Activity of Leaves Essential Oil of Laurus *nobilis* from Morocco. *Aust. J. Basic Appl. Sci.* **2009,** *3*, 3818–3824.
23. Diederichsen, A. *Promoting the Conservation and Use of Underutilized and Neglected Crops, Part 3: Coriander* (*Coriandrum sativum*); International Plant Genetic Resources Institute: Rome, 1996; p 231.
24. Dryden, M. S. Complicated Skin and Soft Tissue Infection. *J. Antimicrob. Chemother.* **2010,** *65* (3), 35–44.
25. EL, S. N.; Karagozlu, N.; Karakaya, S.; Sahın, S. Antioxidant and Antimicrobial Activities of Essential Oils Extracted from *Laurus nobilis* L. Leaves by Using Solvent-Free Microwave and Hydro-Distillation. *Food Nutr. Sci.* **2014,** *5* (2), 97–106.
26. Elmastaş, M.; Gülçin, I.; Işildak, Ö.; Küfrevioğlu, Ö.; İbaoğlu, K.; Aboul-Enein, H. Radical Scavenging Activity and Antioxidant Capacity of Bay Leaf Extract. *J. Iran. Chem. Soc.* **2006,** *3*, 258–266.
27. Erler, F.; Ulug, I.; Yalcinkaya, B. Repellent Activity of Five Essential Oils Against *Culex pipiens*. *Fitoterapia* **2006,** *77*, 491–494.
28. Eron, L.; Lipsky, B.; Low, D.; Nathwani, D.; Tice, A.; Volturo, G. Biological Effects of Essential Oils: A Review. *Food Chem. Toxicol.* **2008,** *46*, 446–75.
29. Fang, F.; Sang, S.; Chen, K. Y.; Gosslau, A.; Ho, C.-T.; Rosen, R. T. Isolation and Identification of Cytotoxic Compounds from *Laurus nobilis*. *Food Chem.* **2005,** *93*, 497–501.
30. Fejes, S.; Kéry, A.; Blázovics, A.; Lugasi, A.; Lemberkovics, E.; Petri, G.; Szöke, E. Investigation of the In Vitro Antioxidant Effect of *Petroselinum crispum* (Mill.). *Acta Pharmaceutica Hungarica* **1998,** *68*, 150–156.
31. Fernandes, CN.; De, Souza, H. F.; De, Oliveria, G.; Costa, J.; Kerntopf, M. R.; Campos, A. R. Investigation of the Mechanisms Underlying the Gastroprotective Effect of *Cymbopogon citratus* Essential Oil. *J. Young Pharma* **2012,** *4*, 28–32.
32. Ferreira, A.; Proença, C.; Serralheiro, M.; Araujo, M. The In Vitro Screening for Acetylcholinesterase Inhibition and Antioxidant Activity of Medicinal Plants from Portugal. *J. Ethnopharmacol.* **2006,** *108*, 31–37.
33. Gardoia, P.; Ichikawa, H.; Malani, N.; Sethu, G.; Aggarwal, B. B. From Ancient Medicine to Modern Medicine: Ayurvedic Concepts of Health and Their Role in Inflammation and Cancer. *J. Soc. Integr. Oncol.* **2007,** *5*, 25–37.
34. Jabeen, Q.; Bashir, S.; Lyoussi, B.; Gilaani, A. Coriander Fruit Exhibit Gut Modulatory Blood Pressure Lowering and Diuretic Activities. *J. Ethnopharmacol.* **2009,** *122,* 123–30.
35. Khan, M. R.; Mutasa, S. L.; Ndaalio, G.; Wevers, H. Antibiotic Action of Constituents of the Roots of *Eucleanatalensis*. *Pak. J. Sci. Ind. Res.* **1978,** *21*, 197–199.
36. Lai, P. K.; Roy, J. Antimicrobial and Chemopreventive Properties of Herbs and Spices. *Curr. Med. Chem.* **2004,** *11*, 1451–1460.
37. Lall, N.; Meyer, J. J. M. Antibacterial Activity of Water and Acetone Extracts of the Roots of *Eucleanatalensis*. *J. Ethnopharmacol.* **2000,** *72*, 313–316.
38. Lamarti, A.; Badoc, A.; Bouriquet, R. Chemotaxonomic Evaluation of *Petroselinum crispum* (Parsley) Marketed in France. *J. Essential Oil Res.* **1991,** *3*, 425–433.
39. Lampe, J. W. Spicing Up a Vegetarian Diet: Chemopreventive Effects of Phytochemicals. *Am. J. Clin. Nutr.* **2003,** *78*, 579S–583S.

40. Leung, A. F. *Encyclopedia of Common Natural Ingredients. Uses in Food, Drugs and Cosmetics*; John Wiley & Sons: New York, 1980; pp 7–9.
41. Lopamudra, S.; Preetha, B. Review Paper on Tulsi Plant (*Ocimum sanctum* L.). *Indian J. Nat. Sci.* **2020,** *10* (60), 122–132.
42. Maa, Q. G.; Xu, K.; Sang, Z. P.; Wei, R. R.; Liu, W. M.; Su, Y. L.; Yang, J. B.; Wang, A. G.; Ji, T. F.; Li, L. J. Alkenes with Anti-Oxidative Activities from *Murrayakoenigii* (L.) Spreng. *Bioorg. Med. Chem. Lett.* **2016,** *26*, 799– 803.
43. Macchioni, F.; Perrucci, S.; Cioni, P.; Morelli, I.; Castilho, P.; Cecchi, F. Composition and Acaricidal Activity of *Laurus novocanariensis* and *Laurus nobilis* Essential Oils Against *Psoroptes cuniculi*. *J. Essent. Oil Res.* **2006,** *18*, 111–114.
44. Mahan, S.; M, H, Sarwar, M. H.; Niaz, A. Q.; Safia, M. The Role of Oil Seeds Nutrition in Human Health: A Critical Review. *J. Cereals Oilseeds* **2013,** *4* (8), 97–100.
45. Malwal, M.; Sarin, R. Antimicrobial Efficacy of *Murrayakoenigii* (Linn.) Spreng. Root Extracts. *Indian J. Nat. Prod. Resour.* **2011,** *2,* 48–51.
46. Manjunatha, H.; Srinivasan, K. Hypolipidemic and Antioxidant Effects of Dietary Curcumin and Capsaicin in Induced Hypercholesterolemic Rats. *Lipids* **2007,** *42,* 1133–1142.
47. Marczal, G.; Balogh, M.; Verzr-Petri, G. Phenol-Ether Components of Diuretic Effect in Parsley, Part I. *Acta Agronomica Academia Scientiarum Hungaricae* **1977,** *26*, 7–13.
48. Modi, S.; Sharma, R. K. Coriander. In: *Handbook of Herbs and Spices*, 2nd ed., Vol. 1, 2012; pp 216–249; online. https://dokumen.pub/handbook-of-herbs-and-spices-vol-1-2nd-ed-9780857090393–0857090399.html (accessed March 31, 2022).
49. More, G.; Tshikalange, T. E.; Lall, N.; Botha, F.; Meyer, J. J. Antimicrobial Activity of Medicinal Plants Against Oral Microorganisms. *J. Ethnopharmacol.* **2008,** *119*, 473–477.
50. Naidu, K. A.; Thippeswamy, N. B. Inhibition of Human Low Density Lipoprotein Oxidation by Active Principles from Spices. *Mol. Cell. Biochem.* **2002,** *229*, 19–23.
51. Nielsen, S. E.; Young, J. F.; Daneshvar, B.; Lauridsen, S. T.; Knuthsen, P.; Sandström, B.; Dragsted, L. O. Effect of Parsley (*Petroselinum crispum*) Intake on Urinary Apigenin Excretion, Blood Antioxidant Enzymes and Biomarkers for Oxidative Stress in Human Subjects. *Br. J. Nutr.* **1999,** *81*, 447–55.
52. Olajide, O. A.; Ajayi, F. F.; Ekhelar, A. I.; Awe, S. O.; Makinde, J. M.; Alada, A. R. A. Biological Effects of *Myristica fragrans* (Nutmeg) Extract. *Phytother. Res.* **1999,** *13* (4), 344–345.
53. Ozcan, B.; Esen, M.; Sangun, M. K.; Coleri, A.; Caliskan, M. Effective Antibacterial and Antioxidant Properties of Methanolic Extract of *Laurus Nobilis* Seed Oil. *J. Environ. Biol.* **2010,** *31* (5), 637–641.
54. Ozcan, M. M.; Unver, A.; Ucar, T.; Arslan, D. Mineral Content of Some Herbs and Herbal Teas by Infusion and Decoction. *Food Chem.* **2008,** *106*, 1120–1127.
55. Palaniswamy, M.; Pradeep, B. V.; Sathya, R.; Angayarkanni, J. In Vitro Anti-Plasmodial Activity of *Trigonella foenum*-graecum. *Evid. Based Complem. Alternat. Med.* **2008,** *7*, 441–445.
56. Pandey, B. P., Ed. *Economic Botany*; Chand and Company Ltd.: New Delhi, 1990; p 294.
57. Panghal, M.; Kaushal, V.; Yadav, J. P. In Vitro Antimicrobial Activity of Ten Medicinal Plants Against Clinical Isolates of Oral Cancer Cases. *Ann. Clin. Microbiol. Antimicrob.* **2011,** *10*, 21–25.

Seed yams account for up to 63% of the overall variable costs of production and are heavy for transport. The tuber division ratio is very small, less than 1:10 relative to some 1:300 cereals. Lower quality plantation material leads to problems with viruses, nematodes, and fungi leading to low tuber yields, accompanied by poor shelf life.[37] In West Africa, the seed yam processing is primarily informal and completely market-driven. Yams are cultivated primarily for direct human consumption and are sold in all growing regions as fresh produce. Figure 7.2 shows the life cycle of yam. Yam is one of the leading tubers in Africa in terms of production (about 20 million Metric tons) and consumption, and there are more than 200 varieties of yams in the world.[4]

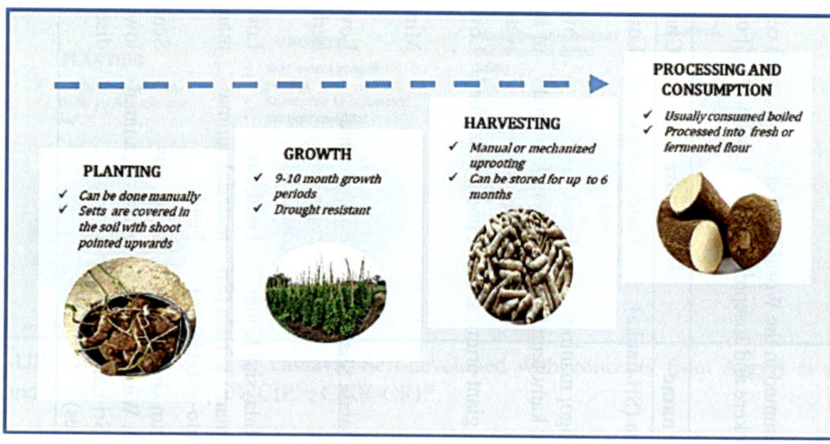

FIGURE 7.2 Life cycle of cam: Self-developed with concepts from Ashwar et al.[15]; Brand-Miller and Buyken[24]; Ravi et al.[84].

FIGURE 7.3 Life cycle of Cocoyam: Self-developed with concepts from Adefegha[4]; Chandrasekara and Joseph[29]; Ojinnaka et al.[76]; Ravi et al.[84].

Cocoyam is a perennial herbaceous plant with an underground starchy corm and long robust petioles that support a whorl of heart-shaped leaves growing up to a height of 1.5–2 m. They have been touted as a food source for over 400 million people all over the world especially in the tropical and subtropical parts.[77] Cocoyam is commonly grown in the tropics and in more than 65 countries around the world.[2] As of 2011, world production of cocoyam stood at 10 million metric tons.[44] Cultivated on approximately 1.3 million hectares, Nigeria is the world's leading producer with 34%, whilst Ghana ranks fourth with 14%.[44] Cocoyam is mostly grown in the Eastern, Western, Ashanti, and Brong Ahafo regions in Ghana. Over 90% of both cocoyam's production and sales are dominated by the females. Generally, growth in the first 2 months is slow, starting with the development of shoots and ending with the formation of corms.[4] Figure 7.3 shows the life cycle of cocoyam. Production, however, is on the decline. Reasons for the reduction include labor-intensive production systems.

Majority of the farmers practice subsistence farming making use of traditional farm equipment like hoes, and cutlasses.[79] Land preparation, for example, could take weeks. Youth migration from rural areas has also made labor scarce.[80] Another challenge is the dearth of quality planting materials, and the lack of improved varieties, lower cocoyam yields resulting from decreasing soil fertility.

FIGURE 7.4 Life cycle of sweet potato: Self-developed with concepts from Adeola and Ohizua[5]; Ellis et al.[39]; Foster-Powell[50]; Stathers et al.[90].

Sweet potato is ranked as the sixth most important food crop and along with cassava, fifth, and potato, third, they are the only root and tuber crops in the top five list. Ninety-five percent of sweet potatoes are produced by developing countries.[83] Global production for sweet potatoes stands at about 105 million metric tons as of 2013.[32] It is estimated that Asia alone produces about 90 million metric tons.[83] The position of sweet potato in terms of food for humans and cheap source of feed for animals is growing exponentially especially in some African countries as well as Asia. Sweet potato just like cocoyam is making tremendous contributions toward the fight against food insecurity in Sub-Saharan Africa. Sweet potato also notably survives harsh conditions and requires little agrochemical inputs.[22,32] Ghana produces about 396,000 metric tons of sweet potato annually, only the Western region does not grow sweet potatoes. The vines of sweet potatoes and the roots are used as planting materials. They can be grown all year round and they are notable for their short growth cycle of 80–120 days. Figure 7.4 shows the life cycle of sweet potatoes. In a bid to improve the value of sweet potatoes, the International Potato Centre (CIP) has developed some appreciable number of varieties of sweet potatoes.[32]

Notable are the *Apumuden and Bohye* varieties which are excellent sources of beta-carotene, a precursor for Vitamin A. Sweet potato can also be used as an ornamental plant due to the distinct shape and colors of the leaves of some of its varieties.[3] Sweet potatoes like other roots and tubers previously discussed, share similar production challenges. Some of these challenges are highly intensive labor input, lack of access to good quality planting materials, and lack of adequate knowledge of pest and disease control to mention a few.[3] Efforts are being made however to bridge this knowledge gap. The CIP in collaboration with institutions across Africa is actively involved in training programs yearly to bridge this gap.[33]

Several varieties of the four major roots and tubers have been produced. Most of these varieties have been integrated into farming in rural communities, but their GI is unknown including the already-existing cultivars consumed in Africa and the rest of the world. Though there have been several studies on GI for roots and tubers, most do not specify varieties used. This makes it difficult to identify which varieties of the roots and tubers have a relatively lower GI. Roots and tubers with relatively lower GI could be used as parents to breed low GI roots and tubers. It is assumed that people in rural communities have higher life expectancy due to their consumption of these staple foods; however, there are no substantial reports to support these claims. Before westernization, cardiovascular diseases such as diabetes were not common in Africa due to the consumption of traditional foods, often made from roots and tubers. Production of roots and tubers needs to be increased with more research into the mechanization of their farming, especially cassava and cocoyam.

7.3.3 FACTORS THAT AFFECT THE GI OF ROOT AND TUBER CROPS

The glycemic index of root and tuber crops can be broadly classified into intrinsic and extrinsic factors. Intrinsic factors refer to the inherent attributes of the food that will impact the GI when consumed. Examples of these intrinsic values include starch structure, amylose to amylopectin ratio, dietary fiber content, moisture content, protein, fats, starch granule size, starch digestibility, and organic acids.[56] Root and tubers generally contain more amylose than amylopectin. During retrogradation, amylose can form more resistant starch (R3) which contributes to the dietary fiber of the food.[24] Dietary fiber and organic acids reduce the glycemic index of root and tuber.[26]

Extrinsic factors are mainly processing methods that affect the GI of the food. Preliminary processing operations such as fermentation, particle size reduction, and drying as well as various cooking methods have been reported to impact glycemic index. Fermentation has been reported to reduce glycemic index due to the formation of organic acids.[40] However, further thermal processes such as steaming, and roasting break down the hydrogen bonds of the organic acids, hereby reducing the glycemic lowering potential of the food.[8]

Studies have shown that particle size reduction and drying have no significant effect on the GI of cassava. Boiling, roasting, and baking have touted to increase glycemic index of root and tubers. However, the dry heat cooking methods increase GI significantly more than the moist heat methods.[56] Frying reduces GI of root and tubers.[34] There is a dearth of information on the glycemic indices of other processing methods on root and tubers such as extrusion, microwave, irradiation, drum drying, and spray drying. The effects of these advanced methods of processing on GI need to be documented so as to further improve the root and tuber value chains.

7.4 GLYCEMIC INDEX OF SELECTED ROOT AND TUBER CROPS

7.4.1 CASSAVA

Cassava is a primary carbohydrate staple source for many African families and other parts of the world like Brazil and India. Nigeria contributes the highest in the production of cassava worldwide.[74] The cassava root is processed into various food products as well as feed for animals. Beer has gained popularity as one of the innovative products that can be made out of the roots. The raw cassava root has a GI of 46, which is a low index score. However, processing into various forms increases the GI.[41]

The research study was carried out to determine the GI level of fermented white cassava granule (gari). The fermented cassava granule glycemic index was 74 after a period of the 72-h fermentation process. The fermented cassava dough "fufu" is the most commonly consumed form of cassava in Nigeria and Kenya. The fufu has a nutritional advantage in that there are an additional protein and reduced cyanide content in the fermented meal (36). Boiled cassava has a GI of 47. It contains a substantial amount of vitamin C, niacin, and iron; this is the only preparation method of Cassava that has managed to retain more content of vitamin C of the root. The cassava flour has a GI score of 62 and can be fortified with minerals and proteins. The fermented form of cassava which is commonly prepared in Ghana (konkote) and Nigeria (lafun) has a GI score ranging from 47 to 59. This range is due to the numbers of hours in which the konkote/ lafun has been fermented. The more fermented the cassava is the more likely the GI increases.[74] Roasted cassava contains a high amount of resistant starch with a glycemic index of 43 which is classified as low. This lowers obesity risks and even diabetes. On the other hand, cassava has the capability to improve blood sugar in addition to its appetite-reducing capacity.[74]

7.4.2 YAM

Yam provides food security and sustaining livelihood systems for millions of West Africans. The yams are mostly consumed after processing, which includes boiling, roasting, pounding, and mashing. The glycemic index score for raw yam is low (36–51).[60] However, its preparation method determines the final GI scores. When the yam is boiled the GI is 51. Mashing, however, increases the GI scores. Studies have reported that mashed yams have a GI score of 72.[11] When the yam is mashed or in fermented flour form, it is easier to be digested compared to when it is boiled. Studies have also found that cooking time has a significant effect on the GI of yams.[60] In a study by Kaushal,[60] yams were placed on the oven for more than 12 h, and the other yams were boiled for about 4 h. The digestion rate of these two yams was significantly different as the one heated for 12 h had a higher GI score than the ones boiled for 4 h.

Yams have a lot of nutritional advantages for the human body as the tuber is rich in vitamins, fiber, and minerals. Yams are rich in manganese and potassium, which are essential minerals for bone growth, heart functions, and metabolism. The tuber contains vitamin C, a water-soluble antioxidant which is essential for the repair of damaged body tissues. Yam is used in food preparations as raw yam, cooked soup flour, or powder. The yam tuber has a variety of bioactive elements, namely, mucin, dioscorin, dioscin, allantois, choline, diosgenine, and polyphenol. Yam tuber mucilage contains mostly soluble fiber and glycoprotein. Several researchers have found that yam extracts are hypoglycemic, nonoxidant and have antimicrobial activities.[67] Yams can improve the proliferation of gastrointestinal cells and rapidly increase the production of enzymes in the small intestine. Yams are somewhat similar to potatoes in terms of nutritional content. These are around 70% water and 27% carbohydrate.[62] They supply large amounts of potassium and vitamin C. Dietary fiber in yam is higher than in potato.

In West Africa, a thick paste of yam is made by cooking and pounding in a large mortar and pestle. Pounded yam is a favorite food in the West African yam belt that is often prepared on special occasions or to honor guests.[21] Most cultures celebrate the new crop with yam festivals. The fresh yam tubers are peeled, dried, chopped, and milled in flour to make the amala or telibowo bread. Like other processing roots, the yam tubers are high in starch but also contain almost 1–3% of the protein in fresh samples. Processing of the raw yam tuber, to the more instant and stable form, could be easier in the preparation of tedium, enhancing ojojo's popularity and commercial potential, that is a fried yam snack.[52] In addition to their significance as a source of food, the contributions that yams make to the development of cultural heritage in certain parts Western Africa cannot be underrated. Notable of mention is the famous Yam Festival, a tradition that has also spread abroad where there is a significant population of tribal people who observe it.[55]

Yams can be stored comparatively longer than other fresh tropical items, so stored yam is sold to farmers or marketers year-round. Boiling, frying, and baking are common preparation methods. Boiled or baked Yam can be eaten with sauce or palm oil with vegetables. Boiled yam can be mashed into a mortar and consumed as "fufu" or "utar." Modern food processing equipment is now available for mashing and boiling yam into fufu by pressing a button. Yam cultivars containing toxic chemicals including dioscorene are cut and soaked for several hours in saltwater before consumption. Yam tubers are often processed into a variety of food products, such as yam flour, enjoyed in several parts of the tropical regions. Industrial yam processing and utilization include starch, livestock, and poultry feed, as well as yam flour production.[21]

7.4.3 COCOYAM

Cocoyam is recognized and used worldwide as food, especially in the tropics by over 400 million people in communities in Oceania, Africa, and Asia.[47] There are two edible types of the tuber: taro (*C. essculanta*) and tannia (*Xanthosoma sagittifolium*). Taro corms and tannia cormels can be solely pounded and cooked or blended with yam and plantain and eaten with soup as "fufu." Cormels of tannia can be used in different forms of food. They can be fried, baked, roasted, boiled, pounded in a paste, prepared as flour chips, porridge, and soup. The grain is small in size, with an easy digestibility of starch estimated at 98.8%.

The glycemic index of raw unprocessed cocoyam is 41. When cooked, the cocoyam GI increases up to 88, depending on the cooking process and time.[92] The unfermented cocoyam flour has a medium glycemic index score of 52, roasted cocoyam commonly consumed in the tropics have a GI between 60 and 88.[11] Fried cocoyam chips also have a GI ranging from 58 to 73. Table 7.3 summarizes the glycemic indices of the root and tubers discussed in this study. The tuber is known to possess more health value as compared to other tubers. It is a high source of proteins, vitamins, and minerals. The tuber leaves are used for medicinal purposes. It has more calories compared to cassava and potatoes. The calories come from complex amylose and amylopectin carbohydrate and are rich in fiber with lower fat content.[92]

Taro corm contains lower amounts of fats and proteins, but taro protein is a little higher than that of sweet potato, Yam, and Cassava.[64] The protein of taro is rich in certain essential amino acids, nutritionally, but low in methionine, isoleucine, and tryptophan. It has been reported that a daily meal of taro corm provides adequate amounts of iron and calcium. Tannia cormels can be boiled, baked, and roasted in different forms, across the world's producing and consuming regions, leaves, petioles, inflorescence, and stolons are an essential delicacy and in some countries are an important source of vitamins, especially folic acid.[76]

Figure 7.5 shows images of various processing methods for cocoyam. In commercial food formulations, the nutrient quality of taro corms has been utilized, for instance, the vitamins of taro (A and B), Ca, and phosphorus have been used as a basis for commercial food preparations for infants. Taro flour has also been used for the production of polyvinyl plastics, polyethylene, and polystyrene more than maize, cassava, and potato starch.[49] Similarly, in 1978, Onwueme reported that, due to its pentosan content, taro flour has been used effectively in bakery and sweets. It is used as a nourishing ingredient for ice creams and in the production of syrups and alcoholic beverages, and as a filler for taro and plastics gums. In the pharmaceutical, paper, and plastic industries, taro mucilage has been reported to be beneficial.[84] Taro is nutritionally better than tannia but not less than cassava or yam. Primarily taro starch granule of cocoyam is smaller and quite easy to digest, making it perfect to be a source of carbohydrate for people with digestive issues, especially the elderly people.

Compared with many other tuber and roots crops, cocoyam has incredible health and medicinal advantages. Taro contains a sufficient level of vitamin C, zinc, potassium, folate, and thiamin compared to yam and cassava. It is a source of fiber that may help to control blood sugar in people with diabetes and may reduce blood lipids that constitute a cardiac disease risk factor. Taro is a major energy and starch source.[57] Since taro starch is very small in granules, it increases digestibility and is used to make baby foods and special diets for diabetic people. Several researchers have studied and profiled cocoyams' therapeutic properties.[65] Other applications of taro in therapeutic practice, particularly the leaves, are used as a poultice and styptic as well as in the treatment of wasp stings with its stem sap. The corm can also be mashed and cooked into a paste known as *poi* that is hypothesized to have the ability to act as a probiotic.

7.4.4 SWEET POTATO

In the last several decades, several research groups have been looking for plant-based medicines. Although a great deal of study is focused solely on herbal medicine, the fact that many vegetables and fruits also have medicinal properties should not be ignored.[39] *Ipomoea batatas*, widely known as sweet potatoes from the *Convolvulaceae* family, is a large, starchy, and sweet root vegetable. The nutritious tuberous root with smooth skin is broad and tapered. Sweet potatoes are recognized for their short period of 90–120 days of growth, good nutritional content, and sweetness. As energy and phytochemical origin in human food and animal feeding, sweet potato played a significant role. The plant has medicinal value and different parts of the plant are used as herbal medicine. The sweet potato leaves have found use in the Akan tribe of Ghana in the treatment of type 2 diabetes and in the treatment of inflammatory and infectious-oral diseases in Brazil.[63]

FIGURE 7.5 Some root and tuber processing methods: Self-developed with concepts from Dixon et al., 2010[35]; Edoh et al., 2016[37]; Ojinnaka et al., 2009[76]; Onyeka, 2014[78]; Stathers et al., 2013[90].

Sweet potatoes like yams are prepared by boiling, baking, eaten raw by making a smoothie or salad and microwaving. The GI for raw sweet potatoes is between 46 and 60; this is subjected to change as different methods of cooking are used. In the case of baked sweet potatoes, the GI rises to 64 while that of microwaved ones increases to 66. Roasted sweet potatoes have the lowest GI score of 44.[9] Figure 7.5 shows various processing methods for root and tuber crops.

The main reason, why the GI for the sweet potato changes just like other root and tuber, is due to the changes in starch structure during heating.[41] When heat is applied, it breaks the starch granule of the potato, making it easier to be digested, thereby contributing to a rise in blood glucose. As sweet potato is one of the low glycemic index foods, it also has a variety of health benefits, and it makes a significant contribution when included in one's diet. According to research, a cup of mashed sweet potato has 757 mg of potassium, 248 calories, 10 g of fiber, iron, magnesium, phosphorus, calcium, and potassium. Allen et al. (2012)[9] explained the nutritional advantages of sweet potatoes as it contains anthocyanin, carotenoids, and phenolic acids; these are valuable antioxidants that scavenge free radicals in the body. Including sweet potatoes in diet helps to maintain healthy skin.

Sweet potato flour contains almost 0% fat and has a low glycemic index of about 46. The flour is mostly helpful to people who are on weight-loss diets.[36] The leaves also contain significant amounts of iron, zinc, vitamins A and C, and proteins. However, they are highly perishable and are usually extracted by the fermentation process. The lactic acid fermentation, which enhances nutrient availability, helps in releasing the vitamins and minerals from the sweet potato leaves as they are fermented.[9]

Different types of sweet potatoes have been eaten raw to treat diabetes, blood deficiency, and lastly hypertension, in regions of Kagawa, Japan. Sweet potato stems have been used to treat prostatitis. Arunachal Pradesh, India, ethnic groups of Monpa are using sweet potato tubers as a basic food and sweet potato leaves as fish feed. Sweet potato is often used as a primary food, a starchy vegetable along with its fleshy roots, petioles, and tender leaves, animal feed, snack food, a basis for the extraction and the fermentation of industrial starch and other processed items. Apart from protein and niacin, sweet potato is rich in other nutrients. It provides more than 90% of the nutrients per calorie needed for most populations.[49]

Sweet potato is also an essential source of calories in different parts of East, West, and Central Africa and is eaten by people of different ages. In Nigeria, the modern utilization of sweet potatoes includes: (1) boiled, cooked, and eaten with stew, (2) boiled or pounded with either fermented cassava commonly known as "foofoo" or pounded yam, (3) milled and dried to sweeten the porridge, and (4) chip-sliced, dried, and browned with vegetable oil or boiled with beans.[68]

While sweet potato can contribute in essential nutrients, it is typically consumed due to its sensory characteristics, as a replacement or supplement to maize, wheat, or rice, as the primary ingredient of conventional, but rarely consumed dishes in several developing nations. Potential sweet potato foods include bread pudding, cake, casserole, pie, muffins, sweet potatoes, and refrigerated potatoes pieces.

Sweet potatoes are high in complex carbohydrates, irons, and vitamins including beta-carotene, vitamin B2, vitamin C, vitamin E, and dietary fiber.[39,88] Sweet potato comes in different flesh and skin colors. The flesh colors range from yellow, white, orange, and purple. The orange flesh varieties are touted for their high beta carotene content, which are also notably provitamin A. Nonetheless, the yellow and purple varieties also have some amounts of beta-carotene.

TABLE 7.3 Glycemic Index of Selected Root and Tuber Crops in Different Processed State: Self-Developed with Concepts from Refs. [4,11,30,40,45,48,50,56,58,62,73,92,96]

Crop type	Raw	Boiled	Roasted	Flour (unfermented)	Flour (fermented)	Fried
Cassava	46–57	46–65	43–58	32–67	47–59	65–78
Cocoyam	41–65	44–77	60–88	52–59	79–86	58–73
Sweet potato	46–60	41–76	89–91	37–46	93–98	61–83
Yam	36–51	36–59	62–67	26–46	66–78	48–88

Production and consumption of the orange flesh variety, due to its high beta carotenes, and given the vitamin A deficiency issues in sub-Saharan Africa, have been promoted by various research institutes. It has been praised as the staple crop that can aid in the eradication of vitamin A deficiency in sub-Saharan Africa and other developing countries. Due to this, several research studies have been developing a myriad number of food products from sweet potato.[51] Some of the products from sweet potato include pastry products, noodles, and complementary foods.[68]

Different researchers have developed pasta from a combination of sweet potato flour and soy protein.[64,68] The researcher also developed consumer acceptable pastry products (chips and doughnuts) and juice from the sweet potato leaves as well as cookies from composite flour of wheat and sweet potato flours.[86] Wine and soft drinks have also been produced from the purple flesh sweet potato.[67] Due to vitamin A deficiency issues, some research studies (such as Argyri et al.[13]; Mahmoud and Anany[70]; and Laryea et al.[63]) have focused on developing complementary food from the orange flesh sweet potato. Argyri et al. in 2016[63] developed his complementary food from a combination of sweet potato, soybean, soybean oil, and fishmeal employing the use of oven-toasting.

Mahmoud and Anany[70] developed their product from a combination of rice, fava beans, sweet potato flour, and peanut oil, while Laryea et al.[63] developed theirs from a combination of sweet potato flour, millet, and soybean employing the drum dryer as the processing method to make the product instant. These complementary foods seek to complement the nutrients (particularly vitamin A) of children under the age of 5, since these are the most affected, including pregnant mothers. Based on their nutritional profile, these products when promoted could indeed help solve the vitamin A deficiency issues in sub-Saharan Africa.[39]

7.5 USE OF ROOT AND TUBERS FOR COMPOSITE FLOURS IN THE BAKING INDUSTRY

Two concepts of composite flour have been developed over the years. A composite flour is a mixture or blend of wheat flour and flour from any other food commodity for baking purposes.[7] Researchers identified composite flour as the term to refer to wheat flour mixed with non-wheat flours such as maize, cassava, and soybean flours.[75] Now, a composite flour is known as a mixture of flours excluding wheat flour for baking purposes. In accordance with both definitions,[91] a composite meal is a blend of tuber meals rich in starch (yam, potatoes, cassava) and/or flours rich in protein (groundnut, soya bean) and/or cereals (millet, rice, maize, sorghum) that may or may not include wheat meal itself. To minimize wheat imports in third-world countries and increase the nutritional content of bakery foods, experiments were carried out using these flour mixes consisting of wheat flour and legume, cereal, oilseed, and tuber flours.

The production of foodstuffs using a mix of flours is on the rise and is attracting investigators in the manufacture of baking industry. Ultimately, composite flour is an excellent new approach to using uncommon foodstuffs in the production of wheat-based products with different properties.[86] The composite meal has been found to be useful in developing nations because it lowers the import of wheat flour and promotes the processing of local crops into flour. The FAO stated that the incorporation of these flour composites into different products could be of economic benefit by decreasing or even replacing imports of wheat and satisfying demands for pastry or bread products by using domestic products rather than wheat. Substituting of wheat flour with cassava flour has been reported to reduce GI of the composite as substitution increased to about 20%.[14]

The texture characteristics of products baked from these flour mixes differed from those made from wheat flour. However, their nutrient compositions and appearances were similar to those of wheat flour.[1] Although wheat is regarded as a significant provider of good calories and other nutrients, yet it is considered nutritionally deficient because cereal proteins lack essential amino acids such as threonine and lysine. The result of compositing wheat flour with other food flours is an increase in nutrient content of the resulting flour or food product. For instance, the protein content of both cassava groundnut and cassava soya bread is greater than regular wheat bread.[6]

In many underdeveloped countries, bread consumption has steadily increased due to modifying eating habits, constantly growing community, and a large portion of the total increased budget can now be spent on food. There have been several attempts targeted at encouraging the replacement of wheat flour with local food flours to reduce the import of wheat flour. Cassava, barley, and corn flour are the most widely studied for composite flour bread production.[85] In addition to cereals and legumes, cassava, cocoyam, taro, and other tuber crops have been identified as alternative sources of essential bread raw materials.

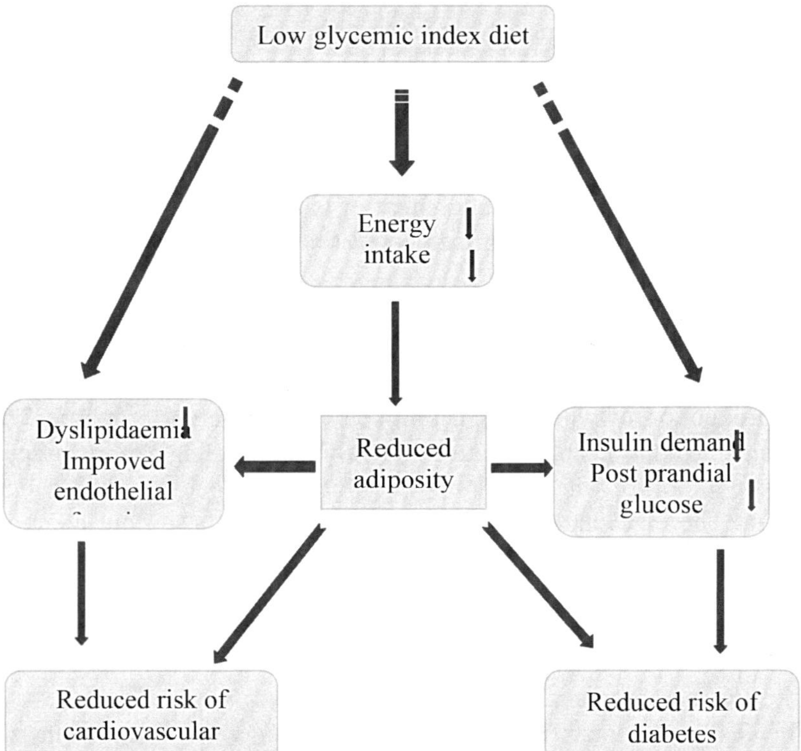

FIGURE 7.6 Pathways through which a diet with low GI may decrease the risk of cardiovascular disease, obesity, and diabetes.

7.5.1 ADVANTAGES OF CONSUMING LOW GLYCEMIC INDEX FOODS

In the treatment of hyperglycemia, low glycemic index (GI) diets are helpful. Cardiovascular diseases are the leading cause of death in diabetes, so considering the effect of GI on blood lipids is significant.[23] The importance of low GI foods to the management and prevention of diabetes treatment has gained significant attention, with low-GI foods promoting less secretion of insulin and decreased incretin levels relative to high-GI carbohydrates.[69,38] Figure 7.6 summarizes the pathways in which a low GI diet may reduce CVDs and risk of diabetes. Usually, traditional dietary treatment for obesity includes consumption of diets low in fat. Although diets low in fat can lead to short-term weight loss, it does not prove to be effective in the long term, perhaps because such diets invoke physiological changes that prevent weight recovery.

In comparison, low GI weight reduction diets can have better effects on hunger and metabolism, which will increase the chance of long-term success.[64] Low GI diets have also been reported to be useful in the management of reduced glucose tolerance.[24] Studies have shown about 9% reduction in low density lipids for subjects who consume low GI foods; this reduction is proportional to improved cholesterol levels which lower the risk of heart disease.[25]

7.6 FUTURE PROSPECTS FOR ROOT AND TUBER CROPS

The food industry is constantly changing and adopting an inventive and business-oriented approach, focused on fermented root and tuber crops. Such foods provide and retain large amounts of nutritionally healthy foods that enrich the human diet in a wide variety of texture and flavors. Appropriate microbial protection and packaging innovations can make their delivery easier throughout the world. In contrast, genomic methods in the fermented food industry provide plenty of possibilities. These include proteomics, transcriptomics, and post-genomic methods and are significant new tools for the optimization and the better understanding of mechanisms. A broad scope is available to develop industries to produce new fermented food from roots and tubers worldwide with a perspective of root and tuber production and nutritional quality.

- Identification of essential physio-chemical factors that modulate the product acceptability of low glycemic indices foods.
- Low cost, suitable technologies for processing root and tuber crops having low glycemic indices and for exporting these foods to developing nations should be evolved.
- One of the major issues about the functional use of the GI is what practical benefits can be anticipated from a lower postprandial glycemic excursion. Studies that have found long-term implications of eating low glycemic indices were missing and there was an absence of distinctions between mixed foods (produced from multiple root

and tuber crops). Regarding the absence of distinctions among mixed foods and the lack of consistency of a mixed meal's glycemic impact, the main issue is the use of full instead of discrete zones to test post-prandial responses. Some research that compared incremental areas showed a good GI predictability. Clinical tests can be carried out over a long period where it can be proven that there is a better adherence in patient populations when diets with low GI were prescribed. Multi-disciplinary, coordinated research, and development strategies to improve production, post-harvest, storage, and processing efficiency, as well as the performance of the root and tuber crops and their future value-added products, are critical questions that should be discussed globally. In addition, extensive, effective agricultural and consumer awareness should be conducted in respect of leveraging the nutritional, economic, and health benefits of root and tuber crops, and how it can be used in eradicating vitamin deficiency and food insecurity. Other key issues that urgently need to be discussed include:

- Research which will produce foods from roots and tuber crops that are nutritious and consumer acceptable value-added foods.
- The provision of appropriate recipes and samples of quality added foods to consumers and farmers are also important in order to increase root and tuber production and consumption.
- Urgent efforts to classify and develop the properties of flour and starch crops by sensory scientists, horticulturists, plant breeders, and nutritionists.

7.7 SUMMARY

Root and tuber crops such as taro, sweet potatoes, yam, and cassava are some major sources of nutrients and energy supply in the dynamic food chain in Sub-Saharan Africa. The macro and micronutrient content of the crops make them invaluable to the health and wellbeing of their consumers. Root and tuber crops can provide about half of the daily energy requirement and some important vitamins such as vitamin C. Their leaves have been lauded for the presence of phenolic compounds and antioxidants that are noted to have anticancer activities through scavenging radical activity. The crops have also been reported to be hypocholesterolemia, hypoglycemic, immunomodulatory and possess antimicrobial activities, marking their potential as health foods. In their raw state, root and tubers, especially yam, cocoyam, and sweet potato have low glycemic indexes, but their processing methods have been reported to increase this even further. This is particularly so for the fermented flour of sweet potato, cassava, and yam. Additionally, root and tubers can be processed using diverse methods that present the advantage of variation in their final products. This is major advantage that can be explored as consumers hardly patronize root and tuber crop products for reasons such as physical appearances and other ideological fallacies, irrespective of their nutritional benefits.

ACKNOWLEDGMENTS

The authors appreciate the efforts of McGill University (Canada) and International Fund for Agricultural Development (IFAD) for their contribution to the glycemic index research across Sub-Saharan Africa. We also wish to appreciate Kwame Nkrumah University of Science and Technology Research Fund (KREF) for their contribution to studying the glycemic index of the cassava flour and wheat flour Composite breads in Ghana.

KEYWORDS

- **bread**
- **carbohydrate**
- **diabetes**
- **glycemic index**
- **nutrition**
- **roots**
- **tuber crops**

REFERENCES

1. Abdelgadir, M.; Elbagir, M.; Jarvi, M.; Abbas, M. Glycemic and Insulin Responses of Six Traditional Sudanese Carbohydrate-Rich Meals in Subjects with Type-2 *Diabetes Mellitus*. *Diabetic Med.* **2005,** *22* (2), 213–217.
2. Abdulrashid, M. and Agwunobi, L. N. Taro Cocoyam (*Colocasia esculenta*) Meal as Feed Ingredient in Poultry. *Pak. J. Nutr.* (ANSInet: Asian Network for Scientific Information) **2009,** *8* (5), 668–673.
3. Adeleke, R. O.; Odedeji, J. O. Functional Properties of Wheat and Sweet Potato Flour Blends. *Pak. J. Nutr.* **2010,** *9* (6), 535–538.
4. Adefegha, S. A. Impact of Pasting on Starch Composition, Estimated Glycemic Index, Phenolic Constituents, Antioxidant Activities and Antidiabetic Properties of Flour Produced from Cocoyam (*Colocasia esculenta*) Corm. *J. Food Biochem.* **2018,** *42* (4), 1–11.

5. Adeola, A. A.; Ohizua, E. R. Physical, Chemical, and Sensory Properties of Biscuits Prepared from Flour Blends of Unripe Cooking Banana, Pigeon Pea, and Sweet Potato. *Food Sci. Nutr.* **2018,** *6* (3), 532–540.
6. Akinoso, R.; Olatoye, K. K. Energy Utilization and Conservation in Instant-Pounded Yam Flour Production. *Int. Food Res. J.* (Universiti Putra Malaysia, Faculty of Food Science & Technology) **2013,** *20* (2), 575.
7. Akonor, P. T.; Tortoe, C.; Hagan, L. Proximate Composition and sensory evaluation of Root and Tuber Composite Flour Noodles. *Cogent Food Agric.* **2017,** *3* (1), 1–6.
8. Akyereko, Y. G.; Wireko-Manu, F. D.; Oduro, I. Influence of Processing Methods on Food Components and Glycemic Index of Cassava-Based Traditional Foods. *J. Food Nutr. Sci.* **2020,** *8* (1) 6–14.
9. Allen, J. C.; Corbitt, A. D.; Maloney K. P.; Butt, M. S.; Truong, V. D., Glycemic Index of Sweet Potato as Affected by Cooking Methods. *Open Nutr. J.* **2012,** *6* (1), 1–11
10. Amagloh, F. K.; Hardacre, A.; Weber, J.; Brough, L.; Coad, J. Sweet Potato-Based Complementary Food for Infants in Low-Income Countries. *Food Nutr. Bull.* **2012,** *33* (1), 3–10.
11. Amua, Q.; Amokaha, R. A.; Abeni, O. P., Analysis of Glycemic Indices and Loads of Food Prepared from Two Widely Consumed Nigerian Yam Tubers: *Dioscorea rotundata* and *Dioscorea alata*. *Chem. Res. J.* **2019,** *4* (4), 58–69.
12. Araya, H.; Contreras, P.; Alvina, M.; Vera, G.; Pak, N. Comparison Between an In Vitro Method to Determine Carbohydrate Digestion Rate and the Glycemic Response in Young Men. *Eur. J. Clin. Nutr.* **2002,** *56* (8), 735–741.
13. Argyri, K.; Athanasatou, A.; Bouga, M.; Kapsokefalou, M. The Potential of In Vitro Digestion Method for Predicting Glycemic Response of Foods and Meals. *Nutrients* **2016,** *8* (4), 209–214.
14. Aryeetey, E.; Asjom-Shoae, J.; Keshteli, A. H.; Sadeghi, O.; Hamid, A.; Esmaillzadeh, A.; Adibi, P. Association Between Dietary Insulin Index and Load with Obesity in Adults. *Eur. J. Nutr.* **2020,** *59* (4), 1563–1575.
15. Ashwar, B.; Gani, A.; Shah, A.; Wani, I.; Masoodi, F. A. Preparation, Health Benefits and Applications of Resistant Starch: A Review. *Starch-Stärke* **2016,** *68* (3–4), 287–301.
16. Asiedu, R.; Sartie, A. Crops That Feed the World, Part 1: Yams. *Food Secur.* **2010,** *2* (4), 305–315.
17. Balagopalan, C. Cassava Utilization in Food, Feed and Industry. In: *Cassava: Biology, Production and Utilization*; CABI Publishing: Oxon, 2002, *26* (2), 301–318.
18. Barine, K. K. D.; Yorte, G. S. In Vitro Starch Hydrolysis and Prediction of Glycemic Index (PGI) in Amala and Plantain-Based Baked Products. *J. Food Res.* **2002,** *5* (2), 73–80.
19. Behera, S. S.; Ray, R. C. Nutritional and Potential Health Benefits of Konjac Glucomannan, a Promising Polysaccharide of Elephant Foot Yam, *Amorphophallus konjac*: A Review. *Food Rev. Int.* **2017,** *33* (1), 22–43.
20. Björck, I.; Liljeberg, H.; Östman, E. Low Glycemic-Index Foods. *Br. J. Nutr.* **2000,** *83* (S1), S149–S155.
21. Bovell-Benjamin, A. C. Sweet Potato: A Review of Its Past, Present, and Future Role in Human Nutrition. *Adv. Food Nutr. Res.* **2007,** *52* (4), 1–59.
22. Bradshaw. J. E., Ed. Root and Tuber Crops. In: *Handbook of Plant Breeding*; Springer Science & Business Media: Lebot, 2010; pp 87–97.

23. Brand-Miller, J. The Glycemic Index as a Measure of Health and Nutritional Quality: An Australian Perspective. *Cereal Foods World* **2007,** *52* (2), 41–44.
24. Brand-Miller, J.; Buyken, A. E. The Glycemic Index Issue: Current Opinion in Lipidology. *LWW* **2012,** *23* (1), 62–67.
25. Brand-Miller, J. C.; Holt, S. H.; Pawlak, D. B.; McMillan, J. Glycemic Index and Obesity. *Am. J. Clin. Nutr.* **2002,** *76* (1), 281S–285S.
26. Brennan, C. S. Dietary Fiber, Glycemic Response, and Diabetes. *Mol. Nutr. Food Res.* **2005,** *49* (6), 560–570.
27. Brouns, F.; Bjorck, I.; Frayn, K. N.; Gibbs, A. L.; Lang, V.; Slama, G. Glycemic Index Methodology. *Nutr. Res. Rev.* **2005,** *18* (1), 145–171.
28. Buttar, H. S.; Li, T.; Ravi, N. Prevention of Cardiovascular Diseases: Role of Exercise, Dietary Interventions, Obesity and Smoking Cessation. *Exp. Clin. Cardiol.* **2005,** *10* (4), 229–249.
29. Chandrasekara, A.; Josheph, T. Roots and Tuber Crops as Functional Foods: A Review on Phytochemical Constituents and Their Potential Health Benefits. *Int. J. Food Sci.* **2016,** *3* (1), 1–8.
30. Chen, Y. J.; Sun, F.; Wong, S. H. Glycemic Index and Glycemic Load of Selected Chinese Traditional Foods. *World J. Gastroenterol.* **2010,** *16* (12), 1512–1518.
31. Chung, H.-J.; Liu, Q.; Pauls, K.; Fan, M.; Yada, R. In Vitro Starch Digestibility, Expected Glycemic Index and Some Physicochemical Properties of Starch and Flour from Common Bean (*Phaseolus vulgaris* L.) Varieties Grown in Canada. *Food Res. Int.* **2008,** *41* (9), 869–875.
32. CIP. *Facts and Figures About Sweet Potato*; International Potato Centre, 2013. https://cipotato.org/sweetpotato/sweetpotato-facts-and-figures/Page-1 (accessed Dec 20, 2021).
33. CSIR-CRI. *Technical Report on Sweet Potato Genotypes Proposed for Release*; Council for Scientific and Industrial Research-Crops Research Institute, 2005; p 12.
34. Diamante, L. M.; Shi, S.; Hellmann, A.; Busch, J. Vacuum Frying Foods: Products, Process and Optimization. *Int. Food Res. J.* **2015,** *22* (1), 1–8.
35. Dixon, A.; Okechukwu, R. U.; Akoroda, M. O.; Ilona, P.; Ogbe, F.; Egesi, C. N.; Kulakow, P. A.; Ssemakula, G.; Iluebbey, P. Improved Cassava Variety Handbook. *Int. Inst. Trop. Agric. Rep.* **2010,** *1,* 1–12.
36. Dutta, S. Sweet Potatoes for *Diabetes mellitus*: A Systematic Review. *Pharmacophore* **2015,** *6* (1), 60–72.
37. Edoh-Ognakossan, K.; Kouakou, J.; Mazalo Pali, A.; Nanga-Nanga, S.; Plagne-Ismail, C. *Cassava: Production and Processing*. Yaounde, Cameroon: CTA/ ISF Cameroon, 2016. https://hdl.handle.net/10568/100793 (acessed March 31, 2022).
38. Eleazu, C. O. The Concept of Low Glycemic Index and Glycemic Load Foods as Panacea for Type 2 *diabetes mellitus*: Prospects, Challenges and Solutions. *Afr. Health Sci.* **2016,** *16* (2), 468–479.
39. Ellis, W. O., Oduro, I.; Owusu, D. Development of Crackers from Cassava and Sweet Potato Flours Using *Moringa oleifera* and *Ipomoea batatas* Leaves as Fortificant. *Am. J. Food Nutr.* **2011,** *1* (3), 114–122.
40. Eli-Cophie, D.; Agbenorhevi, J. K.; Annan, R. A. Glycemic Index of Some Local Staples in Ghana. *Food Sci. Nutr.* **2017,** *5* (1), 131–138.
41. Ellong, E. N.; Billard, C.; Adenet, S. Comparison of Physicochemical, Organoleptic and Nutritional Abilities of Eight Sweet Potato (*Ipomoea batatas*) Varieties. *Food Nutr. Sci.* **2014,** *5* (2), 190–196.

42. Falade, K. O.; Akingbala, J. O. Improved Nutrition and National Development Through the Utilization of Cassava in Baked Foods. *Using Food Sci. Technol. Improve Nutr. Promote Natl. Dev. Int. Union Food Sci. Technol.* **2008,** *34* (4), 1–12.
43. Falade, K. O.; Akingbala, J. O. Utilization of Cassava for Food. *Food Rev. Int.* **2010,** *27* (1), 51–83.
44. FAOSTAT, 2013. faostat.fao.org/site/291/default.aspx (accessed 20 Jan, 2022).
45. Fasanmade, A. A.; Anyakudo, M. M. C. Glycemic Indices of Selected Nigerian Flour Meal Products in Male Type 2 Diabetic Subjects. *Diabetologia Croatica* **2007,** *36* (2), 33–38.
46. Ferreira, A. M.; Penalva, L. C.; Iglesia, I.; Urtasun, L. Correlation of Phenotypic Tests with the Presence of the blaZ Gene for Detection of Beta-Lactamase. *Braz. J. Microbiol.* **2016,** *174*, 1–8.
47. Ferrer-Mairal, A.; Penalva, L. C.; Iglesia, I.; Urtasun, L.; Remon, S.; Cortés, E.; Moreno, L A. In Vitro and In Vivo Assessment of the Glycemic Index of Bakery Products: Influence of the Reformulation of Ingredients. *Eur. J. Nutr.* **2012,** *51* (8), 947–954.
48. Feskens, E. J. M.; Du, H. Dietary Glycemic Index from an Epidemiological Point of View. *Int. J. Obes.* **2006,** *30* (S3), S66–S71.
49. Flores, H. E.; Walker, T. S.; Guimarães, R. L.; Bais, H. P.; Vivanco, J. M. Andean Root and Tuber Crops: Underground Rainbows. *HortScience* **2003,** *38* (2), 161–167.
50. Foster-Powell, K.; Holt, S. H. A.; Brand-Miller, J. C. International Table of Glycemic Index and Glycemic Load Values: 2002. *Am. J. Clin. Nutr.* **2002,** *76* (1), 5–56.
51. Glato, K.; Aidam, A.; Kane, N. A.; Bassirou, D.; Couderc, M.; Barnaud, A.; Vigouroux, Y. Structure of Sweet Potato (*Ipomoea batatas*) Diversity in West Africa Covaries with a Climatic Gradient. *PloS One* **2017,** *12* (5), 1–7.
52. Go, H. K.; Rahman, M.; Kim, G.; Song, C.; Kim, J.; Kim, S.; Kang, H. Antidiabetic Effects of Yam (*Dioscorea batatas*) and Its Active Constituent, Allantoin, in a Rat Model of Streptozotocin-Induced Diabetes. *Nutrients* **2015,** *7* (10), 8532–8544.
53. Goñi, I.; Garcia-Alonso, A.; Saura-Calixto, F. Starch Hydrolysis Procedure to Estimate Glycemic Index. *Nutr. Res.* **1997,** *17* (3), 427–437.
54. Hoover, R. Composition, Molecular Structure, and Physicochemical Properties of Tuber and Root Starches: A Review. *Carbohyd. Polym.* **2001,** *45* (3), 253–267.
55. Hsu, C.; Hurang, S.; Chen, W.; Weng, Y.; Tseng, C. Qualities and Antioxidant Properties of Bread as Affected by the Incorporation of Yam Flour in the Formulation. *Int. J. Food Sci. Technol.* **2004,** *39* (2), 231–238.
56. Itam, E. H.; Itam, A. H.; Odey, M. O.; Ejemot, N. R.; Asenye, M. F.; Ezike, N. N. Effect of Processing Method on the Glycemic Index of Some Carbohydrate Staples (*Manihot esculanta*, *Ipomoea batata* and *Dioscorea rotundata*) in Both Normal and Diabetic Subjects. *Ann. Biol. Res.* **2012,** *3* (12), 5507–5510.
57. Iyagba, A. G. Review on Root and Tuber Crop Production and Their Weed Management Among Small Scale Farmers in Nigeria. *ARPN J. Agric. Biol. Sci.* **2010,** *5* (4), 52–57.
58. Jimoh, A. K.; Adediran, O. S.; Adebisi, S. A.; Biliaminu, S. A.; Okesina, A. B., Effect of Food Processing on Glycemic Response to White Yam (*Dioscorea rotunda*) Meals. *Diabetologia Croatica* **2008,** *37* (3), 67–72.
59. Karikari, S. K. Cocoyam Cultivation in Ghana. *World Crops* **1971,** *23* (3), 118–122.
60. Kaushal, P.; Kumar, V.; Sharma, H. K., Utilization of Taro (*Colocasia esculenta*): A Review. *J. Food Sci. Technol.* **2015,** *52* (1), 27–40.
61. Kimura, S. Glycemic Carbohydrate and Health: Background and Synopsis of the Symposium. *Nutr. Rev.* **2003,** *61* (5), S1–S4.

62. Kouassi, N. K.; Tiahou, G. G.; Abodo, J. R. F.; Camara, C.; Amani, G. N. Influence of the Variety and Cooking Method on Glycemic Index of Yam. *Pak. J. Nutr.* **2009,** *8* (7), 993–999.
63. Laryea, D.; Wireko-Manu, F. D.; Oduro, I. Formulation and Characterization of Sweet Potato-Based Complementary Food. *Cogent Food Agric.* **2018,** *4* (1), 1–15.
64. Laurie, S. M.; Van Heerden, S. M. Consumer Acceptability of Four Products Made From Beta-Carotene-Rich Sweet Potato. *Afr. J. Food Sci.* **2012,** *6* (4), 96–103.
65. Lauzon, R. D.; Shiraishi, K.; Yamazaki, M.; Sawayama, S.; Sugiyama, N.; Kawabata, A. Physicochemical Properties of Cocoyam Starch. *Food Hydrocoll.* **1995,** *9* (2), 77–81.
66. Lebot, V. Tropical Root and Tuber Crops: Cassava, Sweet Potato, Yams and Aroids. *CABI* **2009,** *17*, 1–7.
67. Lee, H.-S.; Kwon, K.; Kim, B.; Kim, J. Quality Characteristics of Instant Nuroong-gi to Which *Dioscorea japonica* Powder Was Added. *Korean J. Food Preserv.* **2009,** *16* (5), 680–685.
68. Limroongreungrat, K.; Huang, Y.-W. Pasta Products Made from Sweet Potato Fortified with Soy Protein. *LWT-Food Sci. Technol.* **2007,** *40* (2), 200–206
69. Ludwig, D. S. Clinical Update: The Low Glycemic Index Diet. *Lancet* **2007,** *369* (9565), 890–892.
70. Mahmoud, A. H.; Anany, A. M. E. Nutritional and Sensory Evaluation of a Complementary Food Formulated from Rice, *faba* Beans, Sweet Potato Flour, and Peanut Oil. *Food Nutr. Bull.* **2014,** *35* (4), 403–413.
71. Mettler, S.; Lamprecht, R.; Kurt, N.; Wenk, C. Colombani, P. C. The Influence of the Subjects Training State on the Glycemic Index. *Eur. J. Clin. Nutr.* **2007,** *61* (1), 19–24.
72. Miller, C. J.; Dunn, E. V.; Hashim, I. B. The Glycemic Index of Dates and Date/Yoghurt Mixed Meals: Are Dates the Candy That Grows on Trees? *Eur. J. Clin. Nutr.* **2003,** *57* (3), 427–431.
73. Nwosu, J. N.; Owuamanam, C.; Omeire, G. C.; Eke, C. C. Quality Parameters of Bread Produced from Substitution of Wheat Flour with Cassava Flour Using Soybean. *Am. J. Res. Commun.* **2014,** *2* (3), 99–118.
74. Ogbuji, C. A.; David-Chukwu, N. P., Glycemic Indices of Different Cassava Food Products. *Eur. J. Basic Appl. Sci.* **2016,** *3* (3), 1–11.
75. Ohimain, E. I. The Prospects and Challenges of Composite Flour for Bread Production in Nigeria. *Glob. J. Human Soc. Sci.* **2014,** *14* (3), 42–52.
76. Ojinnaka, M. C.; Akobundu, E. N. T.; Iwe, M. O. Cocoyam Starch Modification Effects on Functional, Sensory and Cookies Qualities. *Pak. J. Nutr.* (ANSInet) **2009,** *8* (5), 558–567.
77. Onokpise, O. U.; Wutoh, J.; Ndzana, X.; Tambong, J.; Wilson, J. G. *Evaluation of Macabo Cocoyam Germplasm in Cameroon: Perspectives on New Crops and New Uses*; ASHS Press: Alexandria, **1999**; *3* (1), 394–396.
78. Onyeka, J. Status of cocoyam (*Colocasia esculenta* and *Xanthosoma* spp) in West and Central Africa: Production, Household Importance and the Threat from Leaf Blight, CGIAR Research Program on Roots, Tubers and Bananas (RTB), 2014. https://cgspace.cgiar.org/handle/10568/83409# (accessed Jan 20, 2022)
79. Opare, L. U. Edible aroids—Post operation; FAO, Rome, 2000. https://www.scirp.org/journal/paperinformation.aspx?paperid=43369 (accessed Dec 21, 2021).
80. Owusu-Darko, P. G.; Paterson, A.; Omenyo, E. L. Cocoyam (corms): An Underexploited Food and Feed Resource. *J. Agric. Chem. Environ.* **2014,** *3* (01), 22–27.

81. Pawlak, D. B.; Ebbeling, C. B.; Ludwig, D. S. Should Obese Patients Be Counselled to Follow a Low Glycemic Index Diet? *Obes. Rev.* **2002,** *3* (4), 235–243.
82. Pereira, B. L. B.; Leonel, M. Resistant Starch in Cassava Products. *Food Sci. Technol.* **2014,** *34* (2), 298–302.
83. Peters, D. Use of Sweet Potato in Pig Production in Asia: Agricultural and Socio-Economic Aspects. *Pig News Inf.* **2004,** *25* (1), 10–15.
84. Ravi, V., Aked, J.; Balagopalan, C. Review on Tropical Root and Tuber Crops, Part I: Storage Methods and Quality Changes. *Crit. Rev. Food Sci. Nutr.* **1996,** *36* (7), 661–709.
85. Scazzina, F.; Siebenhandl-Ehn, S.; Pellegrini, N. The Effect of Dietary Fibre on Reducing the Glycaemic Index of Bread. *Br. J. Nutr.* **2013,** *109* (7), 1163–1174.
86. Sefa-Dedeh, S.; Agyir-Sackey, E. K. Chemical Composition and the Effect of Processing on Oxalate Content of Cocoyam *Xanthosoma sagittifolium* and *Colocasia esculenta* Cormels. *Food Chem.* **2004,** *85* (4), 479–487.
87. Simsek, S.; El, S. N. In Vitro Starch Digestibility, Estimated Glycemic Index and Antioxidant Potential of Taro (*Colocasia esculenta* L. Schott) Corm. *Food Chem.* **2015,** *168* (1), 257–261.
88. Singh, S.; Riar, C. S.; Saxena, D. C. Effect of Incorporating Sweet Potato Flour to Wheat Flour on the Quality Characteristics of Cookies. *Afr. J. Food Sci.* **2008,** *2* (6), 65–72.
89. Thomas, D.; Elliott, E. J.; Baur, L. Low Glycaemic Index or Low Glycaemic Load Diets for Overweight and Obesity. *Cochrane Database System. Rev.* **2007,** *34* (3), 1–7.
90. Stathers, T.; Carey, T.; Mwanga, R.; Njoku, J.; Malinga, J.; Njoku, A.; Gibson, R.; Namanda, S. Everything You Ever Wanted to Know About Sweet Potato: Reaching Agents of Change ToT Manual 4: Sweet Potato Production and Management; Sweet Potato Pest and Disease Management. Nairobi (Kenya): International Potato Center (CIP); ISBN 978–92–9060–427–3, 2013; *4*, 188; https://doi.org/10.4160/9789290604273.v4 (accessed on Dec 30, 2021).
91. Tortoe, C.; Ofori, H.; Akonor, P.; Oduro-Obeng, H. Trace Metal Concentrations in Three Pastry Products Prepared from Root and Tuber and Cereal Crops Composite Flours. *Cogent Chem.* **2018,** *4* (1), 142–157.
92. Ukozor, A. U.; Okwulehie, F. C.; Achinihu, G. A.; Mbah, C. A.; Olubiyi, E. O.; Opara, H. C., Glycemic Responses of Some Varieties of Cocoyam in Nigeria Using Normo-Glycemic Subjects. *Nigerian J. Nutr. Sci.* **2019,** *40* (2), 107–114.
93. Venn, B. J.; Green, T. J. Glycemic Index and Glycemic Load: Measurement Issues and Their Effect on Diet–Disease Relationships. *Eur. J. Clin. Nutr.* **2007,** *61* (S1), S122–S127.
94. Woldegiorgis, G.; Tadesse, T.; Gurmu, F.; Chindi, A.; Seid, Al. Root and Tuber Crops Research in Ethiopia: Achievements and Future Prospects. *Agric. Res. Ethiop. Renaissance* **2016,** *23* (2), 1–8.
95. Wolever, T. M. S.; Jenkins, D. J. A.; Jenkins, A. L.; Josse, R. G. The Glycemic Index: Methodology and Clinical Implications. *Am. J. Clin. Nutr.* **1991,** *54* (5), 846–854.
96. Wormenor, D. *Determination of the Glycemic Index of Local Staples in Ghana and the Effect of Processing on Them*; Doctoral dissertation; Department of Food Science and Technology, Kwame Nkrumah University of Science and Technology, Kumasi, Ghana, 2015; pp 46–48.

CHAPTER 8

TRADITIONAL MEDICINAL HERBS IN NORTH-WESTERN REGION OF HIMALAYAS: CASE STUDY

DEEPIKA DIMRI and AMBIKA KUMAR

ABSTRACT

Man has been long associated with plant kingdom from the time of his existence. Ethno-botany has emerged as an important branch in advanced biotechnology. Ethno-botany concentrates on the efficacy of various herbal species and their applications in medicine, food, and other purposes. These medicinal plants are not only cost efficient, but also safe and nearly free from severe side effects. Over the last few decades, the traditional knowledge on the utility of medicinal plants has been widely acknowledged across the globe. In this context, Indian village farmers, elders have notable knowledge toward health issues and corresponding remedial capabilities of medicinal herbs. In India, Uttarakhand is located in the western Himalayan region and has traditionally been famed for various medicinal plants. Realizing their importance, this review chapter emphasizes on significant features of traditional utilizations of medicinal plants for curing various diseases among communities in different districts in the north-western region of Himalayas, India. This chapter also represented about marketing of medicinal plants and herbal medicines, originated from North-Western Region of Himalayas, India. The study has been supported by our field study which has been carried out in various villages of different districts in the Garhwal and Kumaoun regions of Uttarakhand, part of north-western region of Himalayas, India. The present study records 74 species of ethno-medicinal plants along with their families. It may believe that, however, this chapter describes the

Sustainable and Functional Foods from Plants: Health Impact, Bioactive Compounds, and Production Technologies. Megh R. Goyal, Arijit Nath, and Zoltán Kovács (Eds.)
© 2024 Apple Academic Press, Inc. Co-published with CRC Press (Taylor & Francis)

traditional knowledge about medicinal plants in the north-western region of Himalayas, India. It will play a great role in herbal medicine and advanced plant biotechnology.

8.1 INTRODUCTION

Nature plays a great role in the existence of human society, their behavior, lifestyle, modernization, and so on. From early stage, human civilization has depended on nature in several purposes, such as sources of medicines, shelters, food stuffs, clothing, flavors, fragrances, fertilizers, and transportation.[110] Application and consciousness toward medicinal plants to prepare medicine and food materials have been realized through series of trials and errors. Accordingly, with time they explored their successive experiences to advanced level of utility of these plants. Conventional herbal medicines are naturally occurring, which requires almost no industrial/chemical processing prior to treating illness within local or regional curative practices.[147] Owing to these striking features, traditional herbal medicines are getting momentous awareness in global health discussions. In this context, information associated with medicinal plants has long been communicated gradually from generation to generation.[65]

Medicinal plants are utilized as a vital medicinal resource in almost all civilizations. Ensuring the safety, efficiency, quality, and accessibility of medicinal plants, this resource for medical treatment became a striking feature in modern industrialized and developing countries. These medicinal plants are enriched with bioactive compounds; those are exploited for the treatment of many chronic and internal diseases and are sometimes considered complicated or even not possible to cure.[36]

However, facts indicate that medicinal herbs have been utilized as medicines around 60,000 years ago.[134] It is very difficult to find out the exact time when plant became medicine in human civilization. The ancient academicians only believed that medicinal plants are only solutions to treat a number of health-related issues and diseases. They continued their systematic studies regarding various aspects of these plants and conducted different experiments, to utilize these natural resources for medicinal purposes. The experimental evidence made them achieve at accurate conclusions regarding the efficiency of different medicinal plants that have great medicinal value. After having this successive experimental evidence, most of drugs synthesized from these medicinal plants were observed to be almost free from adverse effects.[63,64] Owing to these medicinal values, herbal management is growing in popularity across the globe.

Nowadays, medicinal plant plays a significant role in the healthcare system and a large section of world's population, including advanced and developing countries, have placed their great attention on herbal medicine.[144] WHO has mentioned in a report that more than 80% of the world's population depend more often on traditional drugs, mostly medicinal plants, helping as the main foundation of health care.[42] The corresponding data includes not only a massive population of India, China, and all the developing countries but it also includes many advanced countries.[45] In many advanced countries, like UK, rest of Europe, North America, and Australia, complementary and alternative medicines (CAMs) are now becoming conventional in the use.[8,25,26,33] In these developed countries, the most important reason for adopting herbal treatment is the belief that it will facilitate healthier life without any adverse effects. Herbal medicines are, hence, often viewed as a moderate and balanced approach for healing. Individuals, who use these herbs for home remedies as replacement of synthetic chemical-based drugs, may save sufficient amount of money.[33] Based on high potentiality of herbal medicine without side effects and low cost, herbal medicines have become popular and can improve global drug market in a substantial way.[13,22,23,68,85,102,145] These medicinal resources derived from the medicinal plants are now easily available in food stores and supermarkets along with traditional drug stores. The use of medicinal plants is a valuable resource and necessity, which offers a substitute in primary health care systems.[79]

This review chapter presents a case study on medicinal plants in North-Western Region of Himalayas, India. Also the present chapter documents the comprehensive information about traditional knowledge related with medicinal herbs, utilization of various medicinal plants and its parts for curing different ailments among different communities in distinct parts of Uttarakhand state, north-western region of Himalayas, India, and their contribution in economy. Marketing aspects of medicinal plants and herbal medicines, originated from the North-Western Region of Himalayas, India is represented in this chapter. Furthermore, this chapter also includes the indigenous and traditional knowledge of medicinal herbs, used by local dwellers in Garhwal and Kumaoun regions, a part of Uttarakhand, India, collected by authors through survey basis.

8.2 MEDICINAL PLANTS AND HERBAL MEDICINES IN INDIAN SCENARIO

The detailed contents about medicinal herbs came into existence almost 5000 years ago in India, China, and Egypt, and at least 2500 years in Central Asia and Greece.[6] The medicinal system, developed in India or the medicinal system, which have arrived to India from outside and became incorporated into Indian culture, is recognized as Indian Systems of Medicine. India has the exclusive and well-recognized six division medicinal strategies and these have been categorized as Ayurveda, Unani, Siddha, Yoga, Naturopathy, and Homoeopathy. However, Homoeopathy had been adopted in Indian culture in the 18th Century, which has been completely assimilated in Indian medical system and considered as part of Indian Systems of Medicine.[93] These curative strategies are also mentioned in the ancient scriptures like Vedas. The concept of Ayurveda was observed and began in India between 2500 and 500 BC.[136]

In an event organized in New Delhi (India) recently (January, 2018), it was declared that India is the second leading exporter of medicinal plants in the world. India and China represent two of the major countries in Asia, demonstrating best collections of registered and relatively popular medicinal herbs producer.[99] With 6,600 medicinal plants, India is in second position in this ranking and along with China. Both the nations fabricate over 70% of the herbal medicine's requirement across the globe. In India, Himalayan Region (IHR) exhibits over 1748 plant species comprising well-known medicinal value, which indicate native endemic (31%), threatened species (15.5%), Red Data plant species (14%), and 3.5% of total medicinal plants in different threat categories in Conservation Assessment and Management Plan (CAMP). Numerous species are from the Indian Himalayan region and many of these are found in the Uttarakhand state of India, which has been traditionally recognized as a "Gold mine of medicinal herbs and aromatic" plants in the country.[66] Approximately, there are 320 well-known species possessing medicinal value out of thousands of species in the Uttarakhand state. In this context, the State Government of Uttarakhand has declared it as "Herbal State" during early 2003 as majority of the people have dependency on medicinal plants, which grow in and around of the forest to generate income for their livelihood.[50] The communities living in far-flung regions of Uttarakhand considerably rely on local plant species to find a variety of raw materials.

Conventional health-care system is the single amicable system for the majority of the population residing in the high altitudes in terms of logistically and economically. The primary component of local medicines is the herbs and hence they have great significance in traditional health-care.[14] The use of medicinal plants and its products as medicine can be dated back to the onset of human civilization in India. The earliest recorded evidence was found in the Rigveda regarding the use of medicinal plant in the Himalayan region, which was written between 4500 BC and 1600 BC and described 67 medicinal plants.[94]

There is a vast taxonomical-cum-ethnobotanical research and documentation work presented about Indian herbs. Specific studies on the Himalayan region portray almost 1,200 medicinal plants species in the Himalayan Mountains.[10,32,52,95,130,140] Habit-wise, medicinal plants comprise 23, 22, and 58% species as trees, shrubs, and herbs, respectively,[103] while Nayar and Sastry[83] have reported 73 plant species are rare, vulnerable, and endangered in Uttarakhand. The human population residing in the Himalayan rural villages uses the medicinal plants, available in the forests and also bears a heritage of the knowledge of indigenous health care remedies. The recent trends reported by the researcher exhibit a decline in the traditional herbal-based health-care practices.[1,20] The research concludes that two main reasons are accountable for this declining trend: (a) the younger generation is not showing their interest and leaving this ancient health care practices owing to the proliferation of modern technologies; and (b) over exploitation of medicinal plant raw material from the forest has also resulted in extensive reduction of the number of herbal plant species.[37]

The documentation of traditional knowledge supports in the protection of this old-age practice by recognizing threatened species and contributes to the conservation and management of plant diversity. The valuable indigenous knowledge when accompanied and authenticated by the scientific research can explore new universal models of sustainable development, which are particularly environmentally benign, economically viable, and socially acceptable.[128]

8.3 MEDICINAL PLANTS AND HERBAL MEDICINE IN NORTH-WESTERN REGION OF HIMALAYAS, INDIA

8.3.1 STATE OF ART

Extensive literature review was performed about knowledge of medicinal plant in Uttarakhand, north-western Himalayan region of India. Articles, published during 2010–2018 in national and international journals about ethno-botanical studies, are considered in this context. The majority of these papers comprise English and scientific names of plants and herbs, parts of plants and herbs, and medicinal items from them. Table 8.1 provides information about valuable Indian medicinal plants, their parts and biological activity.

Phondani et al.[89–92] reported that the traditional communities (more than 80%) inhabiting in the villages of eight valleys (Urgam, Niti, Pinder, Berahi, Mandakini, Nandakini, Bhilangana, and Binsar) in Alaknanda catchment and its neighboring areas of Uttarakhand still depend on plant-based domestic treatments for curing veterinary ailments. In this context, old-aged people were observed to have better experience and respective knowledge especially in the areas for curing veterinary ailments. In total, 34 diseases were normally observed in nine different classes of livestock and their treatment with 73 herbal species belonging to 70 genera and 45 families; those grow in forests as well as in close vicinity of the rural areas.[91]

Sharma et al. provided data on ethnomedicinal plants used to treat jaundice by three main indigenous communities, that is, nomadic Tharu, Gujjars, and Bhoxa. Their survey report has documented plants exhibiting relation to pharmacological studies and hepatoprotective phytoconstituents present in them.[112] The indigenous knowledge of medicinal herbs and their respective remedies in various indigenous communities, communicated orally for centuries, is now becoming extinct, due to deterioration in traditional culture and introduction of modern technologies.[46]

Bisht et al. specify the urgent requirement of documentation of the medicinal and aromatic plant species. In total, 124 species belonging to 59 families and 108 genera were used by the communities for the treatment of 39 diseases and it has been reported from the Chamoli district of Uttarakhand. The study also reported that roots/rhizomes (38%), leaves (28%), fruits/seeds (10%), and whole plant (6%) of the documented plant species were used. It has been reported that local healers are not interested to share the knowledge to other people. Inappropriate use of the plant parts as a medicine, concern related to over exploitation of plant species, and fear of losing the position of local healers in the community are the reasons in this context.[21]

Bhat et al. reported the traditional knowledge and ecological status of medicinal plants in a protected area of Garhwal Himalaya. They reported that most frequently used parts of medicinal plants were leaves (32%), roots (24%), entire parts or plant (13%), fruits (9%), and seeds and flowers (6% each).[15]

TABLE 8.1 Indian Medicinal Plants: Their Parts and Biological Activity.

Part of plant	Scientific name	Biological activities
Root	*Asparagus racemosus* Wild (Alliaceae)	Adaptogen, galactogogue[34,51]
	Berberis aristata DC (Berberidaceae)	Anti-protozoal, hypoglycemic, antitrachoma[40,125]
	Boerhavia diffusa L. (Nyctaginaceae)	Diuretic, anti-inflammatory, antiarthritic[54,127]
	Calotropis gigantea (Linn.) R. Br. (Asclepiadaceae)	Anti-inflammatory, spasmolytic, treatment of asthma[126]
	Callicarpa macrophylla Vahl. (Verbenaceae)	Treatment of uterine disorder[135]
	Inula racemosa Hk.f (Asteraceae; Compositae)	Recovery from gastro-intestinal disorder, diuretic, expectorant and allergic disorders, etc.[77]
	Chlorophytum boriavillianum Santapau & RR Fernandus (Alliaceae)	Aphrodisiac[43]
	Curculigo orchioides Gaertn. (Amaryllidaceae)	Spermatogenesis enhancer[60]
	Desmodium gangeticum (Linn.) DC (Papillionaceae)	Antioxidant and antirheumatic[60,117]
	Leptadenia reticulata (Retz.) Wt. & Arn. (Asclepiadaceae)	Galactogogue, vasodilator, anabolic [7]
	Gymnema sylvestre R. Br. (Asclepiadaceae)	Antidiabetic, antihyperglycemic [81]
	Rubia cordifolia L (Rubiaceae)	Anti-inflammatory, antitumor, hypoglycemic[113]
	Rauvolfia serpentine Benth (Apocynaceae)	Hypertension activity, treatment of mental disorders[30,67]
	Oroxylum indicum (Linn.) Vent (Bignoniaceae)	Anti-inflammatory, diuretic[50]
	Vetiveria zizanioides (L.) Nash (Poaceae)	Application of vetiver oil in cosmetics[72]
	Vitex negundo Linn. (Verbenaceae)	Anti-inflammatory, antiarthritic, immunomodulator[80]
	Withania somnifera (Linn.) Dunal (Solanaceae)	Adaptogen, anti-rheumatic[131]
	Saussurea lappa (Decne,) Sch. Bip (Asteraceae)	Analgesic, aphrodisiac, treatment of Asthma[31]
	Momordica charantia Linn. (Cucurbitaceae)	Antidiabetic[3]
	Clerodendrum serratum (Linn.) Moon (Verbenaceae)	Treatment of malaria, antiasthmatic, antiallergic[53,133]

TABLE 8.1 *(Continued)*

Part of plant	Scientific name	Biological activities
Fruit	*Aegle marmelos* (L.) Corr. (Rutaceae)	Hypoglycemic, chemo preventive[38,142]
	Emblica officinalis Gaertn. (Euphorbiaceae)	Adaptogen, antioxidant[96,143]
	Hippophae rhamnoides L. (Elaeagnaceae)	Treatment of circulatory disorders, wound healing enhancer, duodenal ulcer[9]
	Piper longum Linn. (Piperaceae)	Treatment of cough, asthma, fever[67,107]
	Plumbago zeylanica Linn. (Plumbaginaceae)	Antipyretic, anticancer, anticoagulant, cytotoxic[70,125]
Bark	*Ficus religiosa* Linn. (Urticaeae)	Antiulcer (gastric ulcer), anti-inflammatory, hypoglycemic agent[4,116]
	Butea monosperma (Lam.) Taub (Fabaceae)	Adaptogen, abortifacient, antioestrogenic, antigout, antiovulatory[124]
	Tecomella undulata (Sm.) Seem. (Bignoniaceae)	Antibacterial, hypoglycemic, hepatoprotective[16]
	Terminaliaarjuna (Roxb.) Wt. & Arn. (Combretaceae)	Heart diseases[47]
	Pterocarpus marsupium Roxb. (Fabaceae)	Hypoglycemic, antifungal[86,106]
	Terminalia arjuna (Roxb.) Wt. & Arn. (Combretaceae)	Treatment of heart diseases[47]
	Symplocos racemosa Roxb. (Symplocaceae)	Antidiarrheal[120]
	Berberis aristata DC (Berberidaceae)	Anti-protozoal, hypoglycemic, antitrachoma[40,123]
Whole plant	*Andrographis paniculata* (Burm.f.) Wallich ex Nees (Acantahceae)	Treatment of cold, flu, liver disease[69,119]
	Celastrus paniculatus Wild (Celastraceae)	Brain tonic, memory enhancer, treatment of depression[39,57]
	Centella asiatica (Linn.) Urban (Umbelliferae)	Tranquilizer, memory enhancer, wound healing[121,137]
	Cissus quadrangularis L (Vitaceae)	Treatment of bone fracture, inflammation[35,139]]
	Tribulus terrestris Linn. (Zygophyllaceae)	Diuretic, anti-urolithiatic, cytoprotective[27,104]
	Bacopa monnieri (L.) Pennel (Scorphulariaceae)	Antioxidant, memory enhancing[129]
	Eclipta alba (Linn.) Hask (Compositae)	Hepatoprotective/promotes hair growth[28]

TABLE 8.1 *(Continued)*

Part of plant	Scientific name	Biological activities
	Solanum xanthocarpum Sch. And Wendl. Syn *S. Virginianum* Linn. (Solanaceae)	Treatment of asthma and related respiratory disorders[132]
	Swertia chirata Buch-Ham (Gentianaceae)	Antimalarial, hypoglycemic, Febrifuge[38,48]
	Phyllanthus amarus Schum. And Thonn. (Euphorbiaceae)	Hepatoprotective[11]
	Ocimum sanctum Linn.(Lamiaceae)	Adaptogen, antioxidant, hypoglycemic, immunomodulator, radio-protector[84]
Resin	*Boswellia serrata* Roxb.(Burseraceae)	Anti-rheumatic, anticolitis, anti-inflammatory, anticancer[118]
	Cassia fistula Linn.(Leguminosae)	Laxative, anti-pyretic, worm infestation[59]
	Commiphora muku l(Hooker Stedor) Engl. (Burseraceae)	Hypolipidemic, antiobesity, treatment of rheumatoid arthritis[109]
Stem	*Crateva nurvala* Buch-Ham (Capparidaceae)	Recovery from urinary disorders including kidney stones[5]
	Glycyrrhiza glabra Linn. (Papillionaceae)	Expectorant, treatment of peptic ulcer[78]
	Tinospora cordifolia (Willd.) Hook.f. and Thoms. (Menispermaceae)	Adaptogenator, immunomodulator[34,138]
	Saracaasoca (Roxb.) de Wilde (Caesalpiniaceae)	Treatment of postmenopausal syndrome and gynecological disorders[82,141]
	Holarrhena antidysenterica(Linn.) Wall ex DC (Apocynaceae)	Antispasmodic, anticolitis, hypoglycemic[78]
	Mucuna pruriens (Linn.) DC(Fabaceae; Papilionaceae)	Treatment of Parkinson's disorder, male sexual disorders[105]
	Myristicafragrans Houtt (Myristicaceae)	Aphrodisiac, hypolipidemic, anti-inflammatory[114]
Stigma	*Crocus sativus* Linn. (Iridaceae)	Aphrodisiac, antistress, antioxidant[18]

TABLE 8.1 *(Continued)*

Part of plant	Scientific name	Biological activities
Rhizome	*Acoruscalamus* Linn. (Araceae)	Nervine tonic, antispasmodic[24,108]
	Curcuma longa Linn. (Zingiberaceae)	Anti-inflammatory, wound healing enhancer; chemo preventive agent, antioxidant, anticancer[81]
	Hedychium spicatum Buch-Ham.Ex. Smith (Zingiberaceae)	Soothening, expectorant, antitussive antiasthmatic[29]
	Zingiber officinale Rosc (Zingiberaceae)	Treatment of fever, cough and asthma, antiemetic[121]
Bulb	*Allium sativum* Linn. (Alliaceae)	Anti-inflammatory, antihyperlipidemic, antifibrinolytic[38]
Fruits	*Terminalia chebula* Retz., and *Terminalia bellerica* Roxb. (Combretaceae)	Laxative, antioxidants[81]
Gel	*Aloe barbadensis* Mill., and *Aloe vera*Tourn. Ex Linn. (Alliaceae)	Treatment of skin diseases, sunburn, frostbite, scalds, wound healing[12]
Seed	*Eugenia jambolana* Lam. (Myrtaceae)	Hypoglycemic, anti-inflammatory, antidiarrheal, antipyretic[122]
Tubor	*Gloriosa superba* Linn.(Liliaceae)	Spasmolytic, oxytocic[115]
	Picrorhiza kurroa Royle ex. Benth (Scorphulariaceae)	Hepatoprotective, adaptogen[81]
	Pueraria tuberosa (Roxb. Ex Willd). DC (Fabaceae)	Anti-implantation, estrogenic, anti-inflammatory, dysmenorrhea, treatment of dysfunctional uterine bleeding (DUB)[17]
Source of taxol	*Taxusbaccata* Linn. (Taxaceae)	Used for treatment of metastatic breast cancer[30]

Phondani with coworkers[89] reported a detailed information about the ethnomedicinal applications of 54 Medicinal and Aromatic Plants (MAPs), their botanical and vernacular names, respective family, collection season, purpose of collection, threat status, quantity, conservation practices, habit, habitat, market potential, and various parts used in the traditional medicinal system by the traditional communities in the region of Binsar Wildlife Sanctuary located in the Uttarakhand state. The communities of this region have been practicing the traditional culture of curing several illness for the past hundreds of years. The reported species belong to 38 families and are applied to treat more than 47 different kinds of diseases. It has been concluded that approximately 70% of population rely on herbal medicinal systems for their treatments and often visit to Vaidyas (local doctors) for curing a variety of ailments while modern health care amenities (Allopathy medicine) were either rare or in very deprived conditions.[90]

According to Sharma et al. (2013), 109 medicinal plants belonging to 57 families and 102 genera are being used by the Gujjars tribe of Sub-Himalayan region of Uttarakhand for curing different skin ailments such as allergy, chilblain, eczema, ringworms, blisters, boils, cracked feet, cuts, leprosy, leukoderma, sore, and wounds.[112]

It has been highlighted by Joshi and Joshi (2014) that the important aspect of medicinal plants in Uttarakhand state is deeply associated with the local's livelihood. The authors studied the local perceptions of the use and the cultivation of medicinal plants.[58]

Mehra et al. (2014) reported the ethno-medicinal application and respective sacred value of important medicinal plants in the Kumaun region. Under this study, the authors reported 58 species of ethno-medicinal plants, which were belonging to 38 families and their applications in the indigenous-medicine system. Apart from this, study has been also conducted on 15 sacred plants documented from 13 families having high sacred significance in the Himalayan region and applied in various rituals.[76]

Parihar et al. (2014) estimated information with the help of semistructured questionnaires on the diversity and utilization of medicinally significant plants in the existing traditional agro forestry systems in the Kumaoun region of Uttarakhand. From the respective region, the authors also reported that around 68 plant species belong to 38 families and 63 different genera from the agro forestry systems. Distinct plant parts, such as roots, rhizomes, bulbs, leaves, bark, fruits, seeds, flowers, stem, and whole plant, were adopted by the local inhabitants for the treatment of several diseases. The study revealed that harmful effect of the allopathic drugs may boost the people toward medicinal plants for curing diseases. Furthermore, it has been concluded that many of these plants, used by the local inhabitants, are uncommon and not familiar in people outside. Hence, there is a crucial requirement to preserve the indigenous knowledge which is associated with managing and utilizing plants for various treatments.[87]

Sharma et al. (2014) performed an ethno-pharmacological survey and documented medicinal plants used for healing fractures. The investigation was performed by interview among common villagers and herbal practitioners in folk tradition of Kumaon region of Uttarakhand. The inventors documented a list of plants with 15 genera belonging to 13 families; those were used for the healing of bone fractures. Popular and effective plant species were *Pholidota articulata* Lindley and *Coelogyne cristata* Lindley, belonging to Orchdaceae family.[121]

The survey studies documented by Kala, C.P. (2015) revealed that in total 84 plant species including medicinal, aromatic, and spice were adopted in the villages beside tons watershed in the Uttarkashi district of Uttarakhand, for curing diseases and its applicability as perfumes. Approximately, 17% of his recorded plant species, including *Arnebia benthamii* (D. Don) Johnston, *Dactylorhiza hatagirea* (D.Don) Soo, *Picrorhiza kurrooa* Benth, *Podophyllum hexandrum* Royle, *Rheum australe* D. Don, and *Angelica glauca* Edgew, have become on the verge of being endangered due to amplified natural and anthropogenic activities. Sustainable application of these valuable plant species has been suggested by the author.[61]

Ratha et al. (2015) reported that the bhotiya tribe living in alpine and subalpine cold desert areas in Niti valley under Nandadevi Biosphere Reserve of Uttarakhand use high-altitude medicinal herbs in the conventional health care system. Unfortunately, the ethno-medicinal knowledge among this tribe (especially the younger generation) is declining gradually. Therefore, it is important for the documentation of the plants and introduction of modern techniques for *ex-situ* cultivation of commercially viable species. This will benefit the younger generation in terms of income and will also prevent the migration of the young generation to other cities in search of jobs for income.[97]

Rathore et al. recognized 79 medicinal plant species, which were belonging to 75 genera and 32 families; those are conventionally used in health care practices by local residents of Dhundsir Gad watershed area of Garhwal Himalaya. The commonly used parts of medicinal plants have been reported as leaves (37%), roots (24%), whole plants (23%), seeds (8%), fruit (6%), and flowers (2%). The dwellers preferred to use herbal drugs as an alternative source for their primary health care, even the health facilities are available nearby. The authors also noted that conservation and protection of the valuable plant species are serious concern due to the decline in diversity of important plant species, which may finally affect the traditional system associated with medicine, spice, and food.[98]

Sharma et al. (2014) represented some preparations and formulations with medicinal plants and herbs used by the Tharu community in Uttarakhand state to cure various skin diseases, such as wounds (38 species), boils (32 species), cuts (18 species), leprosy (11 species), eczema (10 species), itching (7 species), ringworm (5 species), burns (4 species), leucoderma (4 species), cracked-heels (2 species), dandruff (3 species), body-infection (2 species), chilblains (2 species), hair fall (2 species), and toes infection (2 species).[111]

Phondani and coauthors (2016) discussed the advancement of a participatory approach to encourage Medicinal and Aromatic Plant (MAP) farming for the conservation of biodiversity as well as livelihood development in Champawat district of Uttarakhand state. The authors also developed propagation protocols for 11 selected MAPs from the study sites and also performed cost-benefit analysis.[89]

Khajuria and Bisht (2017) pointed out 30 medicinal plants from 24 families and found out by survey basis. These plants were found in the temperate forest of Pauri, Garhwal of Uttarakhand. These medicinal plants were used in the curing of Nephrolithiasis disease. The study portrayed that the knowledge of medicinal plants was restricted to few elder people in villages. The authors also suggested the phytochemical screening from these plants may be useful for the preparation of some new drug for Nephrolithiasis treatment.[62]

It has been observed by Bisht and Adhikari (2018) that ethnobotanical knowledge plays a major role in curing various ailments in indigenous people of the Indian Himalayan Region. For the treatment of 31 human ailments, 70 plant species belong to 41 families and 64 genera were reported. Their study found that the leaves and roots representing (23%) were most commonly plant part used to cure diseases by the indigenous communities followed by bark and latex representing (13%). It was also recommended by the authors to further investigate the efficiency and safety of reported ethnomedicinal plants prior to treatment for various ailments.[20]

According to Bisht et al. (2018) the forest fringe communities belonging to Lwali village in Pauri Garhwal of Uttarakhand utilized 35 medicinal plant species in various forms, mostly as medicines to cure different ailments. It has been observed that decreasing dependency on forests has resulted in the rapid loss of traditional knowledge of respective communities. Moreover, due to easily availability of allopathic medicines, the conventional healing methods are losing their place among the communities.[19]

Further, Jan et al. (2018) recorded 50 ethno medicinal plant species belonging to 33 families in Karbari Grant village located in Dehradun, Uttarakhand. Their study was followed by extensive field visits, interviews, and discussions with local tribal communities for the collection of information of the particular local plants. The study emphasized regarding conservation of these plants and their natural habitats for the benefit of mankind.[56] In other words, appropriate implementation of protection and management of invasive plants as well as proper control of destructive land are essential for securing and management of native endemic species and their habitats.[101] Moreover, promoting education and creating biodiversity register of medicinal plants in the Himalayan regions could act as elemental cornerstone in this reform.[71,92,100]

8.3.2 MARKETING STRATEGIES OF MEDICINAL PLANTS/HERBAL MEDICINES IN UTTARAKHAND STATE

The state of Uttarakhand has different geography as well as diversification in agro-climatic regions. The state is home to numerous vital Medicinal and Aromatic Plants (MAPs). In India, approximate marketing value of medicinal plants is INR 3.5 billion/year and this amount is growing at the rate of 7% per year. There are also inadequate data on the quantity of MAPs being supplied from the Uttarakhand region. Hence, the contribution in terms of monetary to India's total MAPs trade is not well cleared.[88] The history about gathering and marketing of indigenous medicinal and aromatic plants (MAPs) from the Uttarakhand region was started in the year 1949. Several organizations have taken initiative in the sector of medicinal herbal plants in this region. For the first time, a program "Indo-Tibet border medicinal plants scheme" for the capitalization of medicinal plants from this region was began in 1958.

In the year 1960, Herbal Pharmaceutical Development Unit (HPDU), a government-owned department, was established in Ranikhet of Almora District in Uttarakhand. Afterward, collection and sale cooperatives known as Janpadiya Bhesaj Kraya and Vikraya Sahkari Sanghs Limited, India at districts level were established under (HPDU) for harvesting of MAPs and selling it to herbal pharmaceuticals, such as Indian Medicines and Pharmaceuticals Limited (IMPCL), Mohand-Ramnagar and Cooperative Drug Factory (CDF), and Ranikhet.[2] In general, trade of Himalayan medicinal plants relies on harvesting from the natural habitats.[74,75] Till now, many institutions have been established by the government support in this hilly region. These are Herbal Pharmaceutical Development Unit (HPDU), (Janpadiya

Bhesaj Krayaevam Vikraya Sahkari Sanghs, which is a government-approved collection and sales cooperatives), Herbal Research and Development Institute (HRDI), Centre for Aromatic Plants (CAP), and Sate Medicinal Plants Board (SMPB) which are working on policies, legal issues, and facilitation for conservation, collection, cultivation, and marketing of medicinal and aromatic plants (MAPs). Different centers for the collection of plants at state and local levels are established in Uttarakhand to simplify the marketing of MAPs. Some cultivated MAPs have also fixed minimum support prices. In total, 161 nurseries were developed for developing planting materials of important MAPs and approximately 36.444 million plants were cultivated. It is important to mention that *Allium stracheyi*, *Cinnamomum tamala*, *Asparagus racemosus*, *Saussurea lappa*, *Picrorhiza kurrooa*, *Sapindus mukorossi*, *Rauvolfia serpentina*, *Emblica officinalis*, *Terminalia chebula*, and *Terminalia bellerica* are traded from cultivation-associated sources. In Table 8.2, the fixed minimum support prices for various aromatic oils extracted from twelve aromatic plants are given.[73]

TABLE 8.2 Medicinal and Aromatic Plants and Their Minimum Support Prices: Self-Developed with Information Were Collected from Kuniyal et al. (2015).[73]

MAPs (Oil derived plants)	Minimum support prices (MSP), per 10^5		
	Indian Rupee	**US $**	**Euro**
Artimisia	2000	28	26
Basil	700.00	9.8	9.1
Caraway	500	70	65
Chamomile	36,000	504	468
Citronella	600.00	8.4	7.8
Costus	50,000	700	650
Geranium	6000	84	78
Indian bay leaf	1500	21	19.5
Japanese mint	600.00	8.4	7.8
Lemongrass	550.00	7.7	7.15
Marigold	4000	56	52
Palmarosa	1000.00	14	13
Picrorhiza kurrooa (dried roots and stolen)	450.00/kg	6.3/kg	5.85/kg
Saussurea lappa (dried roots)	150.00/kg	2.1/kg	1.95/kg

One Indian Rupee ≈ 0.014 United States $;

One Indian Rupee ≈ 0.013 Euro

According to State Medicinal Plant Board, three species including Reetha (*Sapindus mukrossi*), Tejpatta (*Cinnamomum tamala*), and Jhula (*Parmelia* spp.) are gathered in sufficient amount from different parts of the state to profit local communities. Two institutions such as "Garhwal Mandal Vikas Nigam" and the "Bhesaj Sangh" have the right of collection and obtaining of these medicinal plant species. The income generated by the approved procurement agencies is retained by them and it is transferred to the state revenue. The medicinal plants are sold through the auctions conducted at three places in Uttarakhand, such as Aamdanda, Tanakpur, and Bibiwala on every 2nd and 16th dates in a month. (http://www.smpbuk.org/; http://www.smpbuk.org/detailsphp?pgID=farmer_13n)

8.3.3 INVESTIGATION AND SURVEY ABOUT TRADITIONAL KNOWLEDGE AND MEDICINAL PLANTS

The present chapter documents the traditional knowledge of utilization of various medicinal plants and its parts for curing different ailments by communities in diverse regions of Uttarakhand state. The investigation (survey) was performed in the Garhwal and Kumaoun regions in Uttarakhand, India. The survey was conducted in various villages in Uttarakhand, namely in Pauri district, Almora district, Bageshwar district, and Nainital district. In this context, elderly women in the households were contacted for the detailed knowledge of the use of medicinal plants. Extensive field surveys of medicinal herbs were carried out in different months/weather conditions.

TABLE 8.3 Based on Information Collected from the Survey, Important Medicinal Plants, Their Local Names, Families, Parts Used, and Ethno-Medicinal Uses in Uttarakhand (India).

Local name	Botanical name	Family	Parts of plants and their biological activities
Adrak	*Zingiber officinale*	Zinziberaceae	Rhizome used in common cold and cough
Ajwaen	*Carumcopticum*	Apiaceae	Seed is used with water for treatment of gastric problems
Akhroat	*Juglansregia*	Juglandaceae	Leaves are used for treatment of toothache
Alu	*Solanum tuberosum*	Solanaceae	Sliced tubers are used in burning
Amal tas	*Cassia fistula*	Caesalpiniaceae	Seed is used in constipation
Amla	*Emblica officinales*	Euphorbiaceae	Used for making a kind of medicine used in curing constipation

TABLE 8.3 *(Continued)*

Local name	Botanical name	Family	Parts of plants and their biological activities
Arandi	*Ricinuscommunis*	Euphorbiaceae	Oil and leaves are applied for wound healing
Banana	*Musa paradisiacal*	Musaceae	Fruit with milk is used for body-weight increase.
Basinga	*Adhatoda vasica*	Acanthaceae	Stem is used for cleaning teeth, juice is used for gastric and diabetic patient
Bedu	*Ficus palmate*	Urticaceae	The secretion is used to cure wounds
Bedu	*Ficus palmate*	Moraceae	Plant leaves are used for treatment of spine stinging
Bel patri	*Aegle marmelos*	Rutaceae	Fruit is used to cure dysentery
Ber	*Zizyphusjujuba*	Rhamnaceae	Fruit is used for eating and bark is used for medicine
Bhang	*Cannabis sativa*	Cannabinaceae	Product of bhang is locally called as *Attar*, which is used for healing wounds
Bheemal	*Grewiaoptiva*	Tiliaceae	Barks and leaves are used in scalp problems
Chir	*Pinusrouxburghii*	Pinaceae	Bark is used for ear disease
Chir	*Pinusrouxburghii*	Pinaceae	Ash of bark is used in ear and eye diseases
Chora	*Angelica gluca*	Umbelliferae	Roots are used as spices
Daak	*Butea monosperma*	Papilionaceae	Dried flowers are used for treatment of stomach disease and seeds are used for destroy pin worm
Dadim	*Punicagranatum Linn.*	Punicaceae	Used for treatment of weakness and vomiting
Daikan	*Melia azedarach*	Meliaceae	Leaf paste is used to fight pimples and boils
Daikan	*Melia azedarach*	Meliaceae	Leaves are used in bathing water to prevent skin disease
Dayya	*Callicarpamacrophylla*	Verbenaceae	Fruits is used in mouth ulcer
Deodar	*Cedrus deodara*	Pinaceae	Used as an against of snakes and other poisonous organisms
Dhak	*Butea monosperma*	Papilionaceae	Flowers are used for making medicine for infants.
Dhaniya	*Corariandumsativum*	Umbelliferae	Seed is used in vomiting
Doobghaas	*Cynodondactylon*	Poaceae	Leaf is used for the treatment of jaundice
Gehat	*Dolichosbilflorus*	Papilionaceae	Seeds is used for kidney stone

TABLE 8.3 *(Continued)*

Local name	Botanical name	Family	Parts of plants and their biological activities
Gehu	*Triticumaestivum*	Poaceae	Seeds are used for digestive problems, healing bounds.
Gethi	*Dioscorea*	Dioscoreaceae	Fruits roasted in coals used for treatment cough and cold
Haldi	*Curcuma domestica*	Zinziberaceae	Rhizome is used in common cold, cough, fever, and bone fracture
Harad	*Terminalachebula*	Combretaceae	Fruit is used in cough and in making a kind of powder (locally called as *triphala-churna)* to cure constipation
Jamun	*Syzygiumcumini*	Myrtaceae	Seed is used for treatment of diabetes
Jow	*Hordeum vulgare*	Poaceae	Juice of saplings (A young tree with a slender trunk is known as a sapling) and plant leaves are used as medicine for treatment of vomiting of dog
Kaala jeera	*Carumcarvii*	Umbelliferae	Used for treatment of diabetes
Kaddukabhina	*Cucurbita maxima*	Cucurbitaceae	Stalk of pumpkin is used for treatment in skin burning
Kala-bansa/kala khaud	*Barleriacristata*	Acanthaceae	Leaf juice is used to stop bleeding and filling the wounds
Kandali (Bicchughaas)	*Urticadioica*	Urticaceae	Used for treatment backache and headache
Kantkari	*Solanum indicum*	Solanaceae	Seed is used for treatment of toothache
Karela	*Momordica charantia*	Cucurbitaceae	Juice of leaves is used in headache
Kari patta	*Murrayakoenigii*	Rutaceae	Leaf paste is used in acne
Kingoda	*Berberis spp.*	Barberidaceae	Root is used for treatment of eye flu
Kunj	*Artemisia wallichians*	Asteraceae	Leaf juice is used to kill worms in ear (centipede)
Kveral	*Bauhinia variegate*	Caesalpiniaceae	Floral bud and bark are used in stomach disease
Lehsun	*Allium sativum*	Lilliaceae	Bulb is used in joint pain and diabetes
Mandook parni	*Cenella asiatica*	Umbelliferae	Leaves are used for sharpen memory
Manjistha/ Chatkura	*Rubus cordifolia*	Rubiaceae	Leaves pastes are used for healing wounds
Maradu	*Polygala chinensis*	Polygalaceae	Applied for treatment in insect bite
Masoor	*Lens Culinaris*	Leguminosae	Paste with milk is used for treatment of skin disease

TABLE 8.3 *(Continued)*

Local name	Botanical name	Family	Parts of plants and their biological activities
Mehal	*Prunusphasia*	Rosaceae	Leaf paste is applied on fungal infection in rainy seasons
Methi	*Trigonellafoenum-graecum*	Fabaceae	Fenugreek is used for digestive problems, such as loss of appetite, stomach upset, constipation, and stomach inflammation (gastritis)
Muli	*Raphanus sativus*	Brassicaceae	Whole plant is used for treatment of jaundice
Paddy	*Oryza sativa*	Gramineae	Paste of turmeric together with rice is used for treatment of ankle sprain
Pavand	*Cassia tora*	Caesalpiniaceae	Leaves are used in skin disease
Pudina	*Menthalongifolia*	Labiatae	Protection against heat stock as juice
Pyaj	*Allium cepa*	Lilliaceae	Bulb is used in vomiting and toothache
Reetha	*Sapindusmukorossi*	Sapindaceae	Used in hair problems (lees, dandruff)
Runi	*Citurshistrus*	Rutaceae	Used for preparing medicine for infants
Sandan	*Ougeiniadalbergioides*	Papilionaceae	Bark is used in urinary disease
Sarson	*Brassica compestris*	Brassicaceae	Oil from seed is used for treatment of joint pain, skin burns, and jaundice
Satavar	*Asparagus racemosus*	Liliaceae	Powder and juice of root are used in piles
Silfoda	*Bergenialigulata*	Saxifragaceae	Roots are used in urinary disease
Snuhi	*Euphorbia nivulia*	Euphorbiaceae	Milky secretion is used for treatment of piles
Tamatar	*Lycoperciumindicum*	Solanaceae	Fruit is eaten in the early morning with empty stomach to avoid pin worms in the children
Tejpatta	*Cinnamomumtamala*	Lauraceae	Leaves and bark are used as spices
Til	*Sesamumindicum*	Pedaliaceae	Paste of roots are used in muscular diseases
Timur	*Xanthoxylum piperitum*	Rutaceae	Seeds are used in toothache and stems are used as toothbrush
Tulsi	*Ocimum sanctum*	Labiatae	Leaves and inflorescence are used in cough and cold
Vasa	*Adhatoda vasica*	Acanthaceae	Leaves are useful for diabetic patients
Zameer	*Citurshistrus*	Rutaceae	Fruit is used in stomachache, vomiting during pregnancy

Our research study documented around 74 ethno medicinal plant species belonging to total 47 families. The complete list of the recorded plant species with ethnomedicinal properties is provided in Table 8.3. In terms of number of species used, *Umbelliferae* (4 species), *Liliaceae* (4 species), *Acanthaceae* (3 species), *Caesalpiniaceae* (3 species), *Rutaceae* (3 species), *Solaceace* (3 species), *Euphorbiaceae* (3 species), *Cucurbitaceae* (2 species), *Brassicaceae* (2 species), *Poaceace* (2 species), and *Papilionaceae* (2 species) are found as the most used family of the herbal plant species. The communities in the study area used different plant parts for medicinal uses, such as roots, leaves, stem, seeds, and whole plants. The plants reported in Table 8.3 are extremely precious herbs for medicinal purposes in various ailments, like liver complaints, ulcer, stomach complaints, tumor, arthritis, wound, cholera, body swelling, malaria, rheumatic pain, skin ailments, swelling, brain tonic, eye diseases, urine complaints, bronchitis, etc.

According to the survey by the authors of this chapter, people from these rural areas do not use only medicines during illness, but they also work with mind and body to cure an illness. Due to improper medical facilities in the rural villages, traditional knowledge must be promoted, which can be applied as a first aid in the worst situation. However, old-generation people in the survey area like to use herbal medicine; at present, people residing in those villages are heavily relying on allopathic medicines for immediate cure and for minimizing the death risk. Owing to the lack of sufficient knowledge, villagers are not aware about the values and costs of these herbal plants. Thus, some outsider companies in studied sites are exploiting medicinal plants through the villagers at a very low cost. Eventually, this will diminish their traditional knowledge among these villagers. Therefore, it is necessary to combine traditional knowledge of medicinal plants with their livelihood on a sustainable manner. As per the perceptions of the communities, numbers of herbal medicinal plants were quite abundant before but gradually their natural habitat has been deteriorated due to the climate change, forest fire, and invasive alien species, that is, *Lantana camara* in Pauri district, *Parthenium hysterophorus* in Nainital district, and *Pinus wallichiana/ roxburgii* in Almora district. Thus, rights of practicing local knowledge about herbal medicine and medicinal plants among local people need to be protected. The ethno-medicinal wisdom of the local community can be recognized and preserved by the management authorities along with their role in the conservation and management of biodiversity.

8.4 SUMMARY

In India, medicinal herbs have been utilized to cure various ailments for thousands of years. In this regard, it has been observed that Indian citizens particularly inhabitants of mountain or forest area have remarkable knowledge about reduction of health hazards with medicinal plant-based remedies. Among several regions in India, Uttarakhand state has traditionally been famed for various medicinal plants. This review chapter is undertaken to emphasis on significant features of traditional utilizations of medicinal herbs for curing various ailments among traditional communities in Uttarakhand, India. Furthermore, this chapter represents about economic issues related to herbal medicines and medicinal plants, developed in the north-western region of Himalayas, India. Besides the representation of the state of the art, the present chapter also represents the survey results of local perceptions about utilization of medicinal plants. The field survey studies have been conducted in various villages of different districts in the Garhwal and Kumaoun regions of Uttarakhand. Information on medicinal plants was reported through questionnaire survey, formal and informal conversations with local inhabitants. In the present report, we recorded almost 74 categories of ethno-medicinal plants, belonging to total 47 families. The economy and daily needs of local people in those regions hinge on the agriculture and forest products for their livelihood. The present studies can serve as an inclusive nonredundant catalogue for future investigation, related to accessibility of medicinal herbs in the Uttarakhand region of the Himalayas. It is expected that this chapter will facilitate to formulation of herbal medicines and receive a great attention of ethno-botanists and computational bioinformaticians/biologist, who are aiming more sustainable (effective against health risk without side effects) use of medicinal herbs to treat various ailments.

ACKNOWLEDGMENTS

The authors are highly thankful to all local residents, traditional healers, and specialists of the surveyed area for their massive cooperation and active participation. The authors would also like to show their gratitude to the faculty members of Doon University Dehradun (India), Bhagalpur National College Bhagalpur, Bihar (India), and The Elite Society Tajpur Samastipur, Bihar (India) for sharing suggestions and knowledge with us during the course of this review work.

KEYWORDS

- **marketing of medicinal herbs**
- **medicinal herbs**
- **north-western region of Himalayas**
- **traditional knowledge**
- **Uttarakhand**

REFERENCES

1. Agarwal, V. S., Ed. *Drug Plants of India*; Kalyani Publishers: New Delhi, India, 1997; pp 57–318.
2. Agriculture Finance Corporation (AFC). *Action Plan for Procurement of Medicinal Plants and Revival and Strengthening of Bhesaj Sanghs in Uttarakhand* (Unpublished); Agriculture Finance Corporation (AFC) Ltd.: Mumbai, 2003; p 75.
3. Ahmad, I.; Lakhani, M. S.; Gillet, M.; John, A.; Raza, H. Hypotriglyceridemic and Hypocholesterolemia Effects of Anti-Diabetic *Momordica charantia* (Karela) Fruit Extract in Streptozotocin Diabetic Rats. *Diab. Res. Clin. Pract.* **2001,** *51*, 155–161.
4. Ambike, S. H.; Rao, M. R. R. Studies on a Phytosterolin from the Bark of *Ficus religiosa*, Part-I. *Indian J. Pharmacol.* **1967,** *29*, 91–95.
5. Anand, R.; Patnaik, G. K.; Roy, K.; Bhaduri, A. P. Anti-Oxaluric and Anticalciuric Activity of Lupeol Derivatives. *Indian J. Pharmacol.* **1995,** *27*, 265–268.
6. Ang-Lee, M. K.; Moss, J.; Yuan, C. S. Herbal Medicines and Perioperative Care. *J. Am. Med. Assoc.* **2001,** *286* (2), 208–216.
7. Anjaria, J. V.; Varia, M. R.; Janakiraman, K.; Gulati, O. D. Studies on *Leptadenia reticulata*: Lactogenic Effect on Rats. *Indian J. Exp. Biol.* **1975,** *13* (5), 448–449.
8. Anquez-Traxler, C. The Legal and Regulatory Framework of Herbal Medicinal Products in the European Union: A Focus on the Traditional Herbal Medicines Category. *Drug Inf. J.* **2011,** *45* (1), 15–23.
9. Arora, R. K.; Chawla, R.; Shikha, R.; Kumar, R.; Sharma, A.; Puri, S. C.; Sinha, A. K.; Tripathi, R. M.; Sharma, R. Radioprotection by Himalayan High-Altitude Region Plants. In: *Herbal Drugs: A Twenty First century Perspective*; Jaypee Brothers: New Delhi, 2006; pp 301–325.
10. Badola, H. K.; S. S., Dhar, U.; Palni, L. M. S., Eds. *Himalayan Medicinal Plants: Potential and Prospects*, Vol. 435; Himavikas Occasional Publ. 14, Gyanodaya Prakashan, Nainital, 2001; pp 87–116.
11. Balachandran, P.; Govindarajan, R. Hepatic Disorders. In: *Scientific Basis for Ayurvedic Therapies*; Mishra, L. C., Ed.; CRC Press: Boca Raton, 2004; pp 232–250.
12. Baliga, M. S. Mechanisms and Pre-Clinical Efficacy of Plants in Preventing UV-Induced Skin Damage: Current Status and Future Prospects. In: *Herbal Drugs: A Twenty First*

Century Perspective; Sharma, R. K., Arora, R., Eds.; Jaypee Brothers: New Delhi, 2006; pp 497–521.
13. Bandaranayake, W. M. Quality Control, Screening, Toxicity, and Regulation of Herbal Drugs. *Modern Phytomed.* **2006,** *2006*, 25–57.
14. Baptista, M. M.; Ramos, M. A.; Albuquerque, U. P.; Coelho-de-Souza, G.; Ritter, M. R. Traditional Botanical Knowledge of Artisanal Fishers in Southern Brazil. *J. Ethnobiol. Ethnomed.* **2013,** *9*, 54–58.
15. Bhat, J. A.; Kumar, M.; Bussmann, R. W. Ecological Status and Traditional Knowledge of Medicinal Plants in Kedarnath Wildlife Sanctuary of Garhwal Himalaya, India. *J. Ethnobiol. Ethnomed.* **2013,** *9* (1), 1–18.
16. Billore, K. V.; Yelne, M. B.; Dennis, T. J.; Chaudhari, B. G. Rohitaka (*Tecomella undulata*). In: *Database on Medicinal Plants Used in Ayurveda*; Central Council for Research in Ayurveda and Siddha: New Delhi, India, 2004; pp 321–329.
17. Billore, K. V.; Yelne, M. B.; Dennis, T. J.; Chaudhari, B. G. Vidari (*Pueraria tuberosa*). In: *Database on Medicinal Plants Used in Ayurveda*; Central Council for Research in Ayurveda and Siddha: New Delhi, India, 2004; pp 441–451.
18. Billore, K. V.; Yelne, M. B.; Dennis, T.J; Chaudhari, B. G. Kumkuma (*Crocus sativus*). In: *Database on Medicinal Plants Used in Ayurveda.* Central Council for Research in Ayurveda and Siddha: New Delhi, India, 2004; pp 110–132.
19. Bisht, N.; Verma, P. K.; Negi, R.; Chandra, A. An Ethnobotanical Study of Plants Used by Forest Fringe Communities of Lwali Village (Pauri Garhwal, Uttarakhand). *Plant Sci. Today* **2018,** *5* (2), 39–43.
20. Bisht, S.; Adhikari, B. S. Ethnobotanical Study of Traditional Medicinal Plants Used by *Banraji* Community in Uttarakhand, West Himalaya. *J. Ethnobiol. Tradit. Med.* **2018,** *129*, 1426–1441.
21. Bisht, V. K.; Kandari, L. S.; Negi, J. S.; Bhandari, A. K.; Sundriyal, R. C. Traditional Use of Medicinal Plants in District Chamoli, Uttarakhand, India. *J. Med. Plants Res.* **2013,** *15*, 918–929.
22. Blumenthal, M.; Brusse, W. R.; Goldberg, A.; Gruenwald, J.; Hall, T.; Riggins, C. W., Eds. *The Complete German Commission—E Monographs: Therapeutic Guide to Herbal Medicines*; The American Botanical Council: Austin, 1998; p 685.
23. Bodeker, G.; Ong, C. K.; Grundy, C.; Burford, G.; Shein, K., Eds. *WHO Global Atlas of Traditional, Complementary and Alternative Medicine*, Vol. 1; WHO Centre for Health Development: Japan, 2005; p 347.
24. Bose, B. C.; Vijayavargiya, R.; Safi, A. Q.; Sharma, S. K. Some Aspects of Chemical and Pharmacological Studies of *Acorus calamus* Linn. *J. Am. Pharma. Assoc.* **1960,** *49*, 32–34.
25. Braun, L. A.; Tiralongo, E.; Wilkinson, J. M.; Spitzer, O.; Bailey, M.; Poole, S. Perceptions, Use and Attitudes of Pharmacy Customers on Complementary Medicines and Pharmacy Practice. *Complem. Altern. Med. 2010, 10*, 38–44.
26. Calapai, G. European Legislation on Herbal Medicines: A Look into the Future. *Drug Saf.* **2008,** *31*, 428–431.
27. Chakraborty, B.; Neogi, N. C. Pharmacological Properties of *Tribulus terrestris* L. *Indian J. Pharmacol. Sci.* **1978,** *40* (2), 50–52.
28. Chandra, T.; Sadique, J.; Somasundaram, S. Effect of *Eclipta alba*on Inflammation and Liver Injury. *Fitoterapia* **1987,** *58* (1), 23–32.

29. Chaturvedi, G. N.; Sharma, B. D. Clinical Studies on *Hedychium spicatum* (Shati): An Anti-Asthmatic Drug. *Indian J. Med. Res.* **1975,** *10* (2), 941–945.
30. Chauhan, S. M. S.; Bisht, A. T.; Singh, P. P. Bio-Active Compounds from Himalayan Medicinal Plants. In: *Herbal Drugs: A Twenty First Century Perspective*; Sharma, R. K., Arora, R., Eds.; Jaypee Brothers: New Delhi, India, 2006; pp 190–199.
31. Chaurasia, O. P.; Parimelazhagan, T.; Ahmed, Z. High Altitude Trans-Himalayan Medicinal Plants of Defense Importance. In: *Herbal Drugs: A Twenty First Century Perspective*; Sharma, R. K., Arora, R., Eds.; Jaypee Brothers: New Delhi, India; 2006; pp 182–189.
32. Chowdhery, H. J.; Wadhwa, B. M., Eds. *Flora of Himachal Pradesh*, Vol. 1; Botanical Survey of India Calcutta, 1984; p 242.
33. Committee on the Use of Complementary and Alternative Medicine by the American Public Board on Health Promotion Disease Prevention Institute of Medicine. *Complementary and Alternative Medicine in the United States*; The National Academies Press: Washington, DC, 2005; p 43.
34. Dahanukar, S. A.; Rege, N. N.; Thatte, U. Adaptogens: Medicinal Plants, Their Bioactivity, Screening and Evaluation. *Proc. Int. Workshop* **1997,** *2* (5), 143–163.
35. Deka, D. K.; Lahon, L. C.; Saikia, J.; Mukit, A. Effect of *Cissus quadra angularis* in Accelerating Healing Process of Experimentally Fractured Radius-Ulna of Dog: A Preliminary Study. *Indian J. Pharmacol.* **1994,** *26*, 44–45.
36. Dhanik, J.; Verma, A.; Arya, N.; Verma, A.; Nand, V. Brief Review on Some Medicinal Plants of Uttarakhand. *J. Pharmacogn. Phytochem.* **2017,** *6* (6), 2497–2506.
37. Dhar, U.; Manjkhola, S.; Joshi, M.; Bhatt, A.; Bisht. A. K.; Joshi, M. Current Status and Future Strategy for Development of Medicinal Plants Sector in Uttaranchal, India. *Curr. Sci.* **2002,** *83*, 8–25.
38. Dixit, P. P.; Londhe, J. S.; Ghaskadbi, S. K.; Devasagayam, T. P. A. Anti-Diabetic and Related Beneficial Properties of Indian Medicinal Plants. In: *Herbal Drugs: A Twenty First Century Perspective*; Sharma, R. K., Arora, R., Eds.; Jaypee Brothers: New Delhi, India, 2006; pp 377–395.
39. Doshi, T. The Study on the *Medhyarasayna* Drugs (Vacha and Jyotishmati) with Special Reference to Their Effect on Depression. M. D. (Ayu) Dissertation Submitted to Gujarat Ayurved University, Jamnagar, India, 1991; p 132.
40. Dutta, N. K.; Iyer, N. S. Anti-Amoebic Value of Berberine and Kurchi Alkaloids. *J. Indian Med. Assoc.* **1968,** *50*, 349–354.
41. Ekor, M. The Growing Use of Herbal Medicines: Issues Relating to Adverse Reactions and Challenges in Monitoring Safety. Front. Pharmacol. *2*0**13,** *4*, 177–182.
42. Farnsworth, N. R.; Akerele, O.; Bingel, A. S.; Soejarto, D. D.; Guo, Z. Medicinal Plants in Therapy. *Bull. World Health Org.* **1985,** *63* (6), 965–981.
43. Farooqi, A. A.; Khan, M. M.; Vasundhara, M. Production Technology of Medicinal and Aromatic Crops. *Nat. Remed.* **2001,** *2001*, 90–91.
44. Final Report on Pilot Study on Mechanism for Sustainable Development & Promotion of Herbal Medicinal Plants (H & MP) in the State of Uttarakhand (India); Uttarakhand State Government, 2004; p 86.
45. Ganesan, A. The Impact of Natural Products Upon Modern Drug Discovery. *Curr. Opin. Chem. Biol.* **2008,** *12* (3), 306–317.
46. Gangwar, K. K..; Deepali, G. R. S.; Gangwar, R. S. Ethnomedicinal Plant Diversity in Kumaun Himalaya of Uttarakhand, India. *Nat. Sci.* **2010,** *8* (5), 66–78.

47. Gauthaman, K.; Mishra, L. C. Ischemic Heart Diseases. In: *Scientific Basis for Ayurvedic Therapies*; Mishra, L. C., Ed.; CRC Press: Boca Raton, 2004; pp 512–531.
48. Gopal, H.; Sukumar, S.; Purushotaman, K. K. Anti-Malarial from Indian Medicinal Plants. *J. Res. Ayurveda Siddha* **1981,** *11* (3), 286–295.
49. Govindarajan, R.; Vijayakumar, M. Anti-Oxidant Approach to Diseases Management with Special Emphasis on Herbal Drugs. In: *Herbal Drugs: A Twenty First Century Perspective*; Sharma, R. K., Arora, R., Eds., Jaypee Brothers: New Delhi, India, 2006; pp 421–431.
50. Gujral, M. L.; Saxena, P. N.; Mishra, S. S. Experimental Study of the Comparative Activity of Indigenous Diuretics. *J. Indian Med. Assoc.* **1955,** *25* (2), 49–51.
51. Gupta, A. K.; Mishra, N. Ancient Indian Medicine and Medicinal Plants. In: *Herbal Drugs: A Twenty First Century Perspective*; Sharma, R. K., Arora, R., Eds.; Jaypee Brothers: New Delhi, India, 2006; pp 571–577.
52. Gupta, R. Survey Record of Medicinal and Aromatic Plants of Chamba Forest Division, Himachal Pradesh. *Indian Foresters* **1964,** *90*, 454–454.
53. Gupta, S. S.; Gupta, N. K. Effect of *Solanum xanthocarpum* and *Clerodendron serratum* on Histamine Release from Tissue. *Indian J. Med. Sci.* **1967,** *21*, 795–799.
54. Harvey, S. K. Preliminary Experimental Study of the Diuretic Activity of Some Indigenous Drugs. *Indian J. Med. Res.* **1966,** *54* (8), 774–778.
55. Jalal, J. S.; Kumar, P.; Pangtey, Y. P. S. Ethnomedicinal Orchids of Uttarakhand, Western Himalaya. *Ethnobot. Leaflets* **2008,** *1*, 164–169.
56. Jan, M.; Mir, T. A.; Dhyani, S. Ethnobotanical Study of Plants Used by Ethnic People of Karbari Grant Village Dehradun, Uttarakhand. *Int. J. Multidisc. Sci.* **2018,** *1* (3), 1–3.
57. Joglekar, G. V.; Balwani, J. H. Certain Central Nervous System Effect of Polyester of *Celastrus paniculatus (*Malkanguni) Oil. *Indian J. Med. Res.* **1967,** *1*, 190–195.
58. Joshi, B. C.; Joshi, R. K. The Role of Medicinal Plants in Livelihood Improvement in Uttarakhand. *Int. J. Herb. Med.* **2014,** *1* (6), 55–58.
59. Joshi, P. K. Pharmaco-Clinical Study of Argavadha with Special Reference to Vicharchika. M. D. (Ayu) Dissertation Submitted to Gujarat Ayurved University, Jamnagar, India, 1998; p 135.
60. Joshi, P. K. A Comparative Pharmacognostical, Phytochemical and Pharmacotherapeutic Study of Sweta Musali (*Curculigo orchioides*) with Special Reference to Vrsya Karma. Ph.D. (Ayu) Thesis Submitted to Gujarat Ayurved University, Jamnagar, India, 2005; p 236.
61. Kala, C. P. Medicinal and Aromatic Plants of Tons Watershed in Uttarakhand Himalaya. *Appl. Ecol. Environ. Sci.* **2015,** *3* (1), 16–21.
62. Khajuria, A. K.; Bisht, N. S. Ethnomedicinal Plants Used to Treat Nephrolithiasis: A Case Study Pauri (Pauri—Garhwal), Uttarakhand. *Int. J. Herb. Med.* **2017,** *5* (1), 10–13.
63. Karimi, A.; Majlesi, M., Rafieian-Kopaei, M. Herbal Versus Synthetic Drugs; Beliefs and Facts. *J. Nephropharmacol.* **2015,** *4* (1), 27–30.
64. Kazemipoor, M.; Jasimah, C. W.; Radzi, W. M.; Geoffrey A. Cordell; Yaze, I. *Int. J. Chem. Eng. Appl.* **2012,** *3*, 4–11.
65. Kia, F. J.; Lorigooini, Z.; Khoei, H. A. Medicinal Plants: Past History and Future Perspective. *J.Herbmed Pharmacol.* **2018,** *7* (1), 1–7.
66. Kirtikar, K. R.; Basu, B. D., Eds. *Indian Medicinal Plants, Volumes 1 to 4*; Allahabad Lalit Mohan Basu Publisher: Allahabad, India; and Archaeological Survey of India/

Government of India, 1935. https://www.indianculture.gov.in/ebooks/indian-medicinal-plants (accessed on May 10, 2022).
67. Kohli, K.; Aiman, S. Convention on Herbal Bio-Diversity: Quantitative Aspects of Research. In: *Herbal Drugs: A Twenty First Century Perspective*; Sharma, R. K., Arora, R., Eds.; Jaypee Brothers: New Delhi, India, 2006; pp 45–49.
68. Kon, J. M.; Goh, N. K.; Chia, L. S.; Chia, T. F. Recent Advances in Traditional Plant Drugs and Orchids. *Acta Pharma. Sinica* **2003,** *24*, 7–21.
69. Koul, I. B.; Kapil, A. Effect of Diterpenes from *Andrographis paniculata* on Anti-Oxidant Defense System and Lipid Peroxidation. *Indian J. Pharmacol.* **1994,** *26*, 296–300.
70. Krishnaswamy, M.; Purushothaman, K. K. Plumbagin: A Study of Its Anti-Cancer, Antibacterial and Antifungal Properties. *Indian J. Exp. Biol.* **1980,** *18* (8), 876–877.
71. Kumar A.; Mitra M.; Adhikari, B. S.; Rawat, G. S. Depleting Indigenous Knowledge of Medicinal Plants in Cold-Arid Region of Nanda Devi Biosphere Reserve, Western Himalaya. *Med. Aromatic Plants* **2015,** *4* (3), 1–6.
72. Kumar, S.; Singh, J.; Shah, N. C., Ranjan, V.; Hasan, S. A., Eds. *Marketing Directory*; Central Institute of Medicinal and Aromatic Plants: Lucknow, India, 1997; p 49.
73. Kuniyal, C. P.; Bisht, V. K.; Negi, J. S.; Bhatt, V. P.; Bisht, D. S.; Butola, J. S.; Sundriyal, R. C.; Singh, S. K. Progress and Prospect in the Integrated Development of Medicinal and Aromatic Plants (MAPs) Sector in Uttarakhand, Western Himalaya. *Environ. Dev. Sustain.* **2015,** *17* (5), 1141–1162.
74. Kuniyal, C. P.; Kuniyal, P. C.; Butola, J. S.; Sundriyal, R. C. Trends in the Marketing of Some Important Medicinal Plants in Uttarakhand, India. *Int. J. Biodiversity Sci. Ecosyst. Serv. Manage.* **2013,** *9* (4), 324–329.
75. Kuniyal, C. P.; Rawat, Y. S.; Oinam, S. S.; Kuniyal, J. C.; Vishvakarma, S. C. R. Kuth (*Saussurea lappa*) Cultivation in the Cold Desert Environment of Lahaul Valley: Arising Threats and Need to Revive Socio-Economic Values. *Biodiversity Conserv.* **2005,** *14*, 1035–1045.
76. Mehra, A.; Bajpai, O.; Joshi, H. Diversity, Utilization and Sacred Values of Ethno-Medicinal Plants of Kumaun Himalaya. *Trop. Plant Res.* **2014,** *1* (3), 80–86.
77. Mishra, L. C., Ed. *Scientific Basis for Ayurvedic Therapies*; CRC Press: Boca Raton, 2004, pp 203–207.
78. Mitra, S. K.; Rangesh, P. R. Irritable Colon (Grahani). In: *Scientific Basis for Ayurvedic Therapies*; Mishra, L. C., Ed.; CRC Press: Boca Raton, 2004, pp 356–368.
79. Mukemre, M.; Behçe, L.; Çakılcıoglu, U. Ethnobotanical Study on Medicinal Plants in Villages of Çatak (Van-Turkey). *J. Ethnopharmacol.* **2015,** *166*, 361–374.
80. Nair, A. M.; Saraf, M. N. Inhibition of Antigen and Compound 48/80 Induced Contractions of Guinea Pig Trachea by the Ethanolic Extract of the Leaves of *Vitex negundo Linn. Indian J. Pharmacol.* **1995,** *27*, 230–233.
81. Narasimhan, S.; Anitha, G.; Illango, K.; Kumar, M. R. Bioassay Guided Isolation of Active Principles from Medicinally Important Plants. In: *Herbal Drugs: A Twenty First Century Perspective*; Sharma, R. K., Arora, R., Eds.; Jaypee Brothers: New Delhi, India, 2006; pp 70–76.
82. Narasimhan, S.; Anitha, G.; Illango, K.; Kumar, M. R. Bio-Assay Guided Isolation of Active Principles from Medicinally Important Plants. In: *Herbal Drugs: A Twenty First Century Perspective*; Sharma, R. K., Arora, R., Eds.; Jaypee Brothers: New Delhi, India, 2006; pp 421–423.

83. Nayar, M. P.; Sastry, A. R. K. *Red Data Book of Indian Plants*, Vol. I–III; Botanical Survey of India: Calcutta, 1987, 1988, 1990.
84. P, U. D. Development of a Radioprotective Drug from *Ocimum sanctum*. In: *Herbal Drugs: A Twenty First Century Perspective*; Sharma, R. K., Arora, R., Eds.; Jaypee Brothers: New Delhi, India, 2006; pp 291–300.
85. Pal, S. K.; Shukla, Y. Herbal Medicine: Current Status and the Future. *Asian Pacific J. Cancer Prev.* **2003,** *4*, 281–288.
86. Pandey, M. C.; Sharma, P. V. Hypoglycemic Effect of *Pterocarpus marsupium* Roxb. (Bijaka): A Clinical Study. *Med. Surg.* **1975,** *25* (11), 21–30.
87. Parihaar, R. S.; Bargali, K.; Bargali, S. S. Diversity and Uses of Ethno-Medicinal Plants Associated with Traditional Agroforestry Systems in Kumaun Himalaya. *Indian J. Agric. Sci.* **2014,** *84* (12), 1470–1476.
88. Pauls, T.; Franz, M. Trading in the Dark: Medicinal Plants Production Network in Uttarakhand. *Singapore J. Trop. Geog.* **2013,** *34* (2), 229–243.
89. Phondani, P. C.; Bhatt, I. D.; Negi, V. S.; Kothyari, B. P.; Bhatt, A.; Maikhuri, R. K. Promoting Medicinal Plants Cultivation as a Tool for Biodiversity Conservation and Livelihood Enhancement in Indian Himalaya. *J. Asia-Pacific Biodiversity* **2016,** *9*, 39–46.
90. Phondani, P. C.; Maikhuri, R. K.; Bisht, N. S. Endorsement of Ethnomedicinal Knowledge Towards Conservation in the Context of Changing Socio-Economic and Cultural Values of Traditional Communities Around Binsar Wildlife Sanctuary in Uttarakhand, India. *J. Agric. Environ. Ethics* **2013,** *26*, 573–600.
91. Phondani, P. C.; Maikhuri, R. K..; Kala, C. P. Ethnoveterinary Uses of Medicinal Plants Among Traditional Herbal Healers in Alaknanda Catchment of Uttarakhand, India. *Afr. J. Tradit. Complem. Altern. Med.* **2010,** *7* (3), 195–206.
92. Phondani, P. C.; Maikhuri, R. K.; Rawat, L. S. Ethnobotanical Uses of Plants Among the Bhotiya Tribal Communities of Niti Valley in Central Himalaya, India. *Ethnobot. Res. Appl.* **2010,** *8*, 233–244.
93. Prasad, L. V. *Indian System of Medicine and Homoeopathy Traditional Medicine in Asia*; WHO-Regional Office for South East Asia: New Delhi, 2002; pp 283–286.
94. Rahul, C.; Pankaj, P.; Sarwan, S. K., Mahesh, J. K. Phytochemical Screening and Antimicrobial Activity of *Albizzia lebbeck*. *J. Chem. Pharma. Res.* **2010,** *2* (5), 476–484.
95. Rana, M. S.; Samant, S. S. Diversity, Indigenous Uses and Conservation Status of Medicinal Plants in Manali Wildlife Sanctuary, North Western Himalaya. *Indian J. Tradit. Knowl.* **2011,** *10* (3), 439–459.
96. Rao, M. R. R.; Siddiqui, H. H. Pharmacological Action of *Emblica officinalis*. *Indian J. Exp. Biol.* **1964,** *2*, 29–33.
97. Ratha, K. K.; Joshi, G. C.; Rungsung, W.; Hazra, J. Use Pattern of High-Altitude Medicinal Plants by Bhotiya Tribe of Niti Valley, Uttarakhand. *World J. Pharm. Pharma. Sci.* **2015,** *4 (6)*, 1042–1061.
98. Rathore, S.; Tiwari J. K.; Malik, Z. A. Ethnomedicinal Survey of Herbaceous Flora Traditionally Used in Health Care Practices by Inhabitants of Dhundsir Gad Watershed of Garhwal Himalaya, India. *Global J. Res. Med. Plants Indigen. Med.* **2015,** *4* (4), 65–78.
99. Raven, P. H. Medicinal Plants: A Global Heritage. In: *Proceedings of the International Conference on Medicinal Plants for Survival*; International Development Research Center: New Delhi, 1998; pp 14–18.

100. Rawal, R. S.; Rawat, B. Nanda Devi Biosphere Reserve, Western Himalaya, India. In: *Compendium on Indian Biosphere Reserves. Progression During Two Decades of Conservation*; Palni, L. M. S., Rawal R. S., Eds.; G. B. Pant National Institute of Himalayan Environment & Development, Government of India, 2012; pp 42–55.
101. Rawat, B.; Gairola, S.; Rawal, R. S. Assessing Conservation Values of Forest Communities in Nanda Devi Biosphere Reserve: Plant Diversity, Species Distribution and Endemicity. *J. Mount. Sci.* **2015,** *12* (4), 878–890.
102. Roberts, J. E.; Tyler, V. E. *Tyler's Herbs of Choice*: *The Therapeutic Use of Phytomedicinals;* The Haworth Press: New York, 1997; p 135.
103. Samant, S. S.; Dhar, U.; Palni, L. M. S. *Medicinal Plants of Indian Himalaya: Diversity Distribution Potential Values*; Gyanodaya Prakashan: Nainital, India, 1998; p 163.
104. Sangeeta, D.; Sidhu, H.; Thind, S. K.; Nath, R.; Vaidyanathan, S. Therapeutic Response of *Tribulus terrestris* (Gokhru) Aqueous Extract on Hyperoxaluria in Male Adult Rats. *Photother. Res.* **1993,** *7* (2), 116–119.
105. Satyavati, G. V.; Gupta, A. K..; Tandon, N., Eds. *Mucuna Adans. Medicinal Plants of India*, Vol. 1; Indian Council of Medical Research: New Delhi, 1987; pp 282–289.
106. Satyavati, G. V.; Gupta, A. K.; Tandon, N., Eds. *Pterocarpus Jacq. Medicinal Plants of India Volume 2*; Indian Council of Medical Research: New Delhi, 1987; pp 530–539.
107. Satyavati, G. V.; Gupta, A. K.; Tandon, N., Eds. *Piper Linn. Medicinal Plants of India*, Vol. 2; Indian Council of Medical Research: New Delhi, 1987; pp 426–456.
108. Satyavati, G. V.; Raina, M. K.; Sharma, M., Eds. *Medicinal Plants of India*; Indian Council of Medical Research: New Delhi, 1976; pp 1–20.
109. Satyavati, G. V. Guggulipid: A Promising Hypolipidemic Agent from Guggul (*Commiphorawightii*)*:* Economic and Medicinal Plant Research. *Plants Tradit. Med.* **1991,** *5*, 47–53.
110. Seymour, V. The Human-Nature Relationship and Its Impact on Health: A Critical Review. *Front. Public Health* **2016,** *4*, 260–269.
111. Sharma, J.; Gairola, S.; Sharma. Y. P.; Gaur, R. D. Ethnomedicinal Plants Used to Treat Skin Diseases by Tharu Community of District Udham Singh Nagar, Uttarakhand, India. *J. Ethnopharmacol.* **2014,** *158*, 140–206.
112. Sharma, J.; Gaur, R. D.; Gairola, S.; Painuli, R. M..; Siddiqi, T. O. Traditional Herbal Medicines Used for the Treatment of Skin Disorders by the Gujjars Tribe of Sub-Himalayan Tract, Uttarakhand. *Indian J. Tradit. Knowl.* **2014,** *12* (4), 736–746.
113. Sharma, P. C.; Yelne, M. B.; Dennis, T. J., Eds. Manjishtha (*Rubia cordifolia*). In: *Data Base on Medicinal Plants Used in Ayurveda*; Central Council for Research in Ayurveda and Siddha: New Delhi, 2002; pp 171–186.
114. Sharma, P. C.; Yelne, M. B.; Dennis, T. J., Eds. Jatiphala (*Myristicafragrans*). In: *Data Base on Medicinal Plants Used in Ayurveda*; Central Council for Research in Ayurveda and Siddha: New Delhi, 2002; pp 213–241.
115. Sharma, P. C.; Yelne, M. B.; Dennis, T. J., Eds. Langli (*Gloriosasuperba*). In: *Data Base on Medicinal Plants used in Ayurveda*; Central Council for Research in Ayurveda and Siddha: New Delhi, 2002; pp 341–357.
116. Sharma, P. C.; Yelne, M. B.; Dennis, T. J., Eds. Aswatha (*Ficus religiosa*). In: *Data Base on Medicinal Plants Used in Ayurveda*; Central Council for Research in Ayurveda and Siddha: New Delhi, 2002; pp 130–139.
117. Sharma, P. C.; Yelne, M. B.; Dennis, T. J., Eds. Shalaparni (*Desmodium-gangeticum*): In: *Data Base on Medicinal Plants used in Ayurveda*; Central Council for Research in Ayurveda and Siddha: New Delhi, 2002; pp 472–480.

118. Sharma, P. C.; Yelne, M. B.; Dennis, T. J., Eds. Shallaki (*Boswellia serrata*). In: *Data Base on Medicinal Plants Used in Ayurveda*; Central Council for Research in Ayurveda and Siddha: New Delhi, 2002; pp 404–417.
119. Sharma, P. C.; Yelne, M. B.; Dennis, T. J., Eds. Mandookaparni (*Centella asiatica*). In: *Data Base on Medicinal Plants Used in Ayurveda;* Central Council for Research in Ayurveda and Siddha: New Delhi, 2002; pp 264–279.
120. Sharma, P. C.; Yelne, M. B.; Dennis, T. J., Eds. Lodhra (*Symplocos racemosa*). In: *Data Base on Medicinal Plants Used in Ayurveda*; Central Council for Research in Ayurveda and Siddha: New Delhi, 2002; pp 164–170.
121. Sharma, P. C.; Yelne, M. B.; Dennis, T. J., Eds. Shunthi (*Zingiber officinale Rosc*). In: *Data Base on Medicinal Plants Used in Ayurveda*; Central Council for Research in Ayurveda and Siddha: New Delhi, 2002; pp 315–390.
122. Sharma, P. C.; Yelne, M. B.; Dennis, T. J., Eds. Jambu (*Syzigiumcumini*). In: *Data Base on Medicinal Plants Used in Ayurveda*; Central Council for Research in Ayurveda and Siddha: New Delhi, 2002; pp 314–331.
123. Sharma, P. C.; Yelne, M. B.; Dennis, T. J., Eds. Daruharidra (*Beriberis aristata*). In: *Data Base on Medicinal Plants Used in Ayurveda*; Central Council for Research in Ayurveda and Siddha: New Delhi; 2002; pp 120–126.
124. Sharma, P. C.; Yelne, M. B.; Dennis, T. J., Eds. Palasha (*Butea monosperma*). In: *Data Base on Medicinal Plants Used in Ayurveda*; Central Council for Research in Ayurveda and Siddha: New Delhi; 2002; pp 336–347.
125. Sharma, P. C.; Yelne, M. B.; Dennis, T. J., Eds. Chitraka (*Plumbago zeylanica*). In: *Data Base on Medicinal Plants Used in Ayurveda*; Central Council for Research in Ayurveda and Siddha: New Delhi, 2002; pp 102–113.
126. Sharma, P. C.; Yelne, M. B.; Dennis, T. J., Eds. Alarka (*Calotropis gigantea*). In: *Data Base on Medicinal Plants Used in Ayurveda*; Central Council for Research in Ayurveda and Siddha: New Delhi, 2002; pp 1–10.
127. Sharma, P. C.; Yelne, M. B.; Dennis, T. J., Eds. Raktapunarnava (*Boerhavia diffusa*). In: *Data Base on Medicinal Plants Used in Ayurveda*; Central Council for Research in Ayurveda and Siddha: New Delhi, 2002; pp 360–377.
128. Shinwari, S. K.; Gilani, S. S. Sustainable Harvest of Medicinal Plants at Bulashbar Nullah, Astore (Pakistan). *J. Ethnopharmacol.* **2003,** *84*, 289–298.
129. Singh, H. K.; Dhawan, B. N. Neuropsychopharmacological Effects of Ayurvedic Nootropic *Bacopa monniera* Linn. (Brahmi). *Indian J. Pharmacol.* **1997,** *29*, 359–365.
130. Singh, P. B.; Aswal, B. S. Medicinal Plants of Himachal Pradesh Used in Indian Pharmaceutical Industry. *Bull. Medico-Ethnobot. Res.* **1992,** *13*, 172–208.
131. Singh, S.; Kumar, S. *Withania Somnifera—The Indian Ginseng-Ashwagandha*; Central Institute of Medicinal and Aromatic Plants (CIMAP): Lucknow, India, 1998; p 43.
132. Sinha, A.; Yadav, S. K.; Khan, S., Kothari, A.; Sharma, A.; Kothari, S. L. Biotechnologically Important Solanaceous Plants. In: *Herbal Drugs: A Twenty First Century Perspective*; Sharma, R. K., Arora, R., Eds.; Jaypee Brothers: New Delhi, India, 2006; pp 117–127.
133. Sivarajan, V. V.; Balachandran, I., Eds. *Ayurvedic Drugs and Their Plant Sources.* Oxford and IBH Publishing Co., Pvt., Ltd.: New Delhi, 1999; p 237.
134. Solecki, R.; Shanidar, IV. A Neanderthal Flower Burial in Northern Iraq. *Science* **1975,** *190*, 880–889.

135. Sood, S. *Identification of Priyangu Through Pharmaco-Therapeutic Studies with Special Reference to Its Sonitasthapan Activity.* M. D. (Ayu) Dissertation Submitted to Gujarat Ayurved University, Jamnagar, India, 1995; p 163.
136. Subhose, V. *Bull. Indian Inst. Hist. Med. Hyderabad* **2005,** *35* (83), 32.
137. Suguna, L.; Sukumar, P.; Chandrakasan, G. Effects of *Centella asiatica* Extract on Dermal Wound Healing in Rats. *Indian J. Exp. Biol.* **1996,** *34*, 1208–1211.
138. Thatte, U. M.; Rao, S. G.; Dahanukar, S. A. *Tinospora cordifolia* Induces Colony Stimulating Activity in Serum. *J. Postgraduate Med.* **1994,** *40*, 202–203.
139. Udupa, K. N.; Prasad, G. C. Biochemical and Ca^{45} Studies on the Effect of *Cissus quadrangularis*in Fracture Healing. *Indian J. Med. Res.* **1964,** *52*, 480–487.
140. Uniyal, M. R.; Chauhan, N. S. Medicinal Plants of Uhal Valley in Kangra Forest Division. H. P. *J. Res. Indian Med.* **1971,** *6* (3), 287–299.
141. Vinjamury, M.; Bagashayana, N.; Sivarama, P.; Singh, B. B. Gynecological Diseases. In: *Scientific Basis for Ayurvedic Therapies*; Mishra, L. C., Ed.; CRC Press: Boca Raton, 2004; pp 552–571.
142. Vyas, D. S.; Sharma, V. N.; Sharma, H. K.; Khanna, N. K. Preliminary Study of Antidiabetic Properties of *Aegle marmelos* and *Enicostemma littorale*. *Res. Indian Med. Yoga Homeopathy* **1979,** *14*, 63–65.
143. Vyas, M. K.; Apte, B. K. Amalaki. *Aushadhi* **1977,** *7* (1), 1–17.
144. Wachtel-Galor S.; Benzie I. F. F. Herbal Medicine: An Introduction to Its History, Usage, Regulation, Current Trends, and Research Needs. In: *Herbal Medicine*; CRC Press/Taylor & Francis: Boca Raton, 2011; pp 2–28.
145. World Health Organization (WHO) Monograph on Selected Medicinal Plants, Vol. 2; World Health Organization, 2002; p 85.
146. Yuan, X.; Chapman, R. L.; Wu, Z. Analytical Methods for Heavy Metals in Herbal Medicines. *Phytochem. Anal.* **2011,** *22* (3), 189–198.
147. Zhang, J.; Onakpoya, I. J.; Posadzki, P.; Eddouks, M. The Safety of Herbal Medicine: From Prejudice to Evidence. *Evid. Based Complem. Altern. Med.* **2015,** *2015*, 1–3.

CHAPTER 9

XANTHOHUMOL (*Humulus lupulus* L.): POTENTIAL OF BIOACTIVE POLYPHENOLS IN BEER

ANNA JÚLIA ÉLIÁS, ZSANETT BODOR, and CSILLA BENEDEK

ABSTRACT

Beer is a highly popular fermented alcoholic beverage known for its high content of bioactive compounds, including phenolic constituents, bitter acids, vitamins, melanoidins, and essential oils. They are derived from both malted barley (70–80%) and hops (20–30%). Many of the original compounds undergo substantial transformations during brewing and fermentation. As main phytochemicals in beer, polyphenols cover structurally extremely diverse compounds: various types of flavonoids, including prenylflavonoids, proanthocyanidins, chalcones, and non-flavonoids, mainly phenolic acids. Scientific evidence shows that many of these constituents, especially those derived from hops, have the potential in preventing cancer. Xanthohumol is a hop-derived flavonoid found in beer, which has been widely examined for its health-protecting role. It has been found to be effective against many lifestyle-related illnesses, and its antitumorigenic protective effect was proven by numerous in vitro and in vivo experiments. The aim of this chapter is to give a general presentation of the major phytochemicals in beer, especially polyphenols, and xanthohumol. Another objective is to summarize existing information on the role of these compounds within nutritional benefits and health-related aspects of beer, that is, prevention of cancer, diabetes, osteoporosis, cardiovascular, and neurodegenerative diseases. Risks of excessive beer consumption are also shortly outlined.

Sustainable and Functional Foods from Plants: Health Impact, Bioactive Compounds, and Production Technologies. Megh R. Goyal, Arijit Nath, and Zoltán Kovács (Eds.)
© 2024 Apple Academic Press, Inc. Co-published with CRC Press (Taylor & Francis)

9.1 INTRODUCTION

Beer is the most preferred alcoholic beverage worldwide, part of many people's lifestyle, with a consumption of about 200 billion liters per year.[15]

Its rich nutrient content includes both macro- (carbohydrates and amino acids) and micronutrients (vitamins, minerals, bitter acids, and phenolic compounds). Since medieval times, beer is traditionally manufactured according to the German Beer Purity Law, which allows only barley, hop, water, and yeast as raw materials. The type of barley and hops, variety of the malting procedure, pH and temperature during the different technological steps, and the yeast fermentation, all have a major impact on beer type and quality.

Phytochemicals in beer originate from both barley and hop or are formed during technology (e.g., melanoidins, isohumulones). Polyphenols are secondary plant metabolites present in both barley and hop; however, most of the health-related potential attributed to beer is related to the latter. Phenolic compounds were shown to have potential health benefits that are strongly depending on their structure. They are usually categorized as flavonoids and non-flavonoids, the latter comprising the subclasses of phenolic acids, coumarins, stilbenes, and lignans. Flavonoids typically contain a 1,4-benzo-pyrone backbone with 1-, 2-, or 3-phenyl substituents. Their subfamilies differ in the oxidation state of the carbon atoms and the substituents on the benzopyrone ring.[20]

They occur naturally mainly as the aglycone fragment of various glycosides, in bond with different sugars. Phenolic acids are derived from hydroxybenzoic or hydroxycinnamic acids; representing an important family of phenolic compounds.[5]

The perspective of health effects and the abundance of polyphenols in beer have attracted many research communities to explore beer as a potent and attractive food and a good resource of valuable phytochemicals. Nevertheless, its alcohol content still remains one of the risk factors that hampers the overall positive assessment of beer and seriously restricts its consumption.

9.2 CLASSIFICATION OF POLYPHENOLS IN BEER

9.2.1 FLAVONOIDS

Beer contains compounds from seven groups of flavonoids (including tannins) that have been reported in beer in an extended concentration scale, ranging from 0.001 to 20 mg/L. This is attributed to differences in raw materials (barley and hops), brewing and stabilization procedures, and storage time. The main flavonoids in beer are catechin and epicatechin (also present as gallates), isoxanthohumol, quercetin, kaempferol, and anthocyanidins, mainly procyanidin B3, and prodelphidins B3 and B9.[16]

The major flavonoid compounds reported in beer with their origin are presented in Table 9.1.

9.2.2 NON-FLAVONOID PHENOLIC COMPOUNDS

The non-flavonoid polyphenols in beer are shown in Table 9.2. These are mainly phenolic acids originating from barley, including p-coumaric, gallic, syringic, ferulic, and caffeic acids. Phenolic acids are important not only as bioactive molecules but also as flavoring constituents. All of them are reported to be bitter, most of them being also astringent; other typical notes are sour, harsh, medicinal (p-coumaric, vanillic acids), bitter-sweet (chlorogenic, ferulic, vanillic acids), dry (gallic, sinapic acids) or even vanilla-like (ferulic acid).[9]

9.2.3 OTHER PHYTOCHEMICALS IN BEER

9.2.3.1 ISOHUMULONES

The main phytochemicals in hops are hop acids, classified as alpha-acids (humulones) and beta-acids (lupulones), which are differentiated by their side chain. Hop cones contain up to 25% or even more hop acids, the individual constituents are determined by the hop variety and the conditions of growing, the major alpha-acid being generally humulone. During beer brewing alpha-acids suffer thermal isomerization, yielding iso-alpha-acids (isohumulones) as shown in Figure 9.1.

TABLE 9.1 Main Flavonoids and Their Origin in Beer: Self-Developed with Concepts from Ref. [21,22,39,41]

Flavonoid subclass	General structure	Name of the specific flavonoid in beer	Structure of specific flavonoid in beer	Source
Flavonols	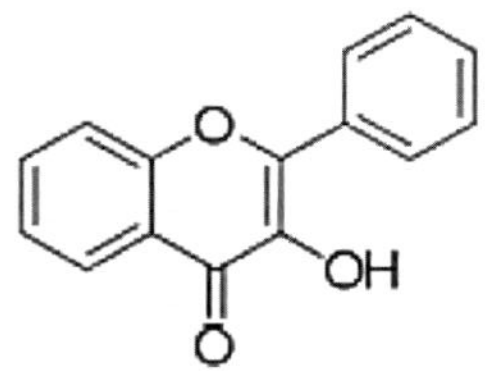	**Kaempferol**	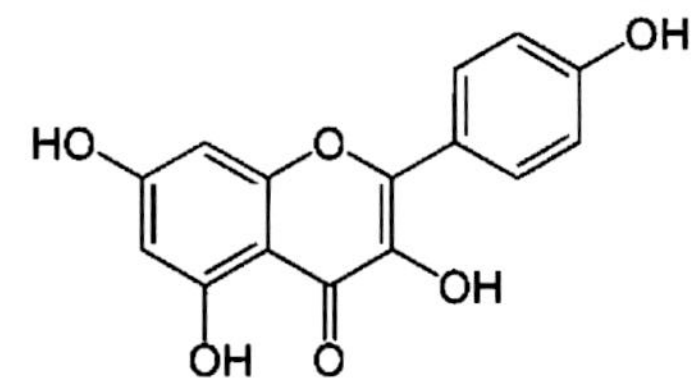	Barley [16,29]
		Quercetin	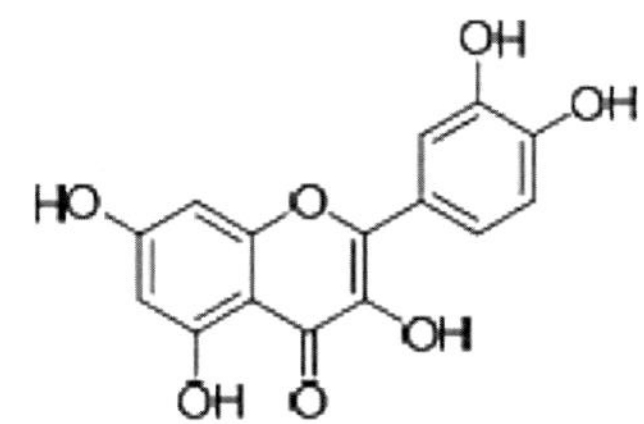	Barley [16,21,29,39,63]
Isoflavones	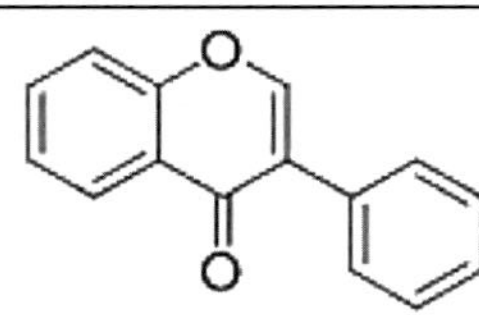	**Genistein**	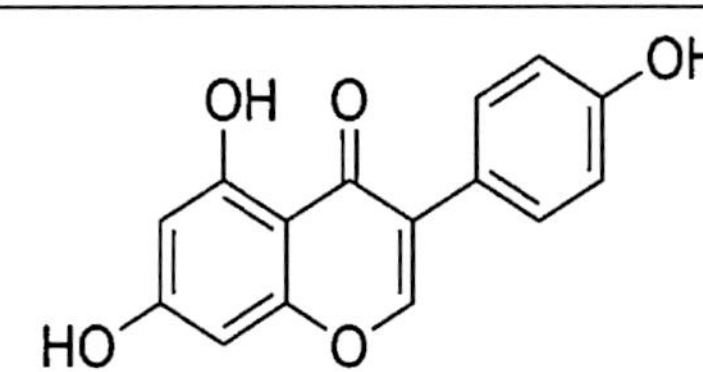	Hops [18,45]
		Daidzein	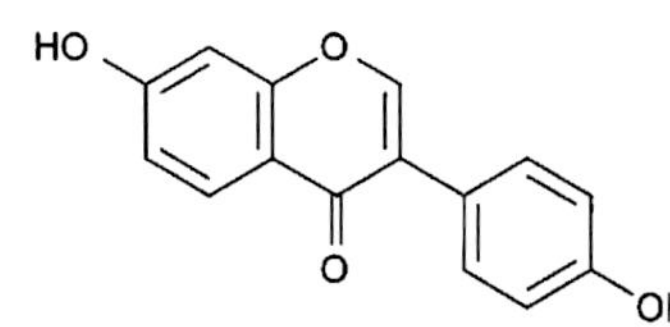	Hops [18,45]

TABLE 9.1 *(Continued)*

Flavonoid subclass	General structure	Name of the specific flavonoid in beer	Structure of specific flavonoid in beer	Source
		Formononetin	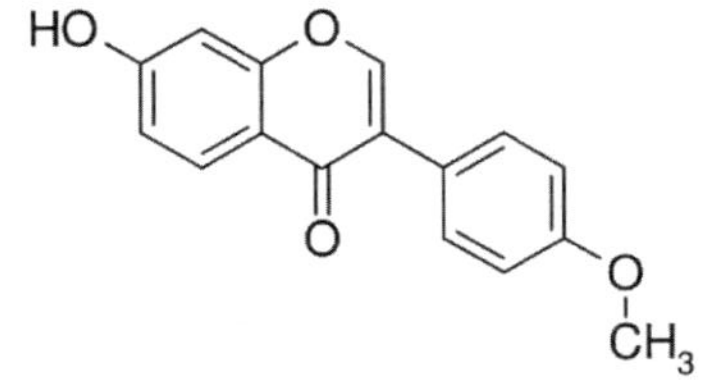	Hops [45]
Antho-cyanidins	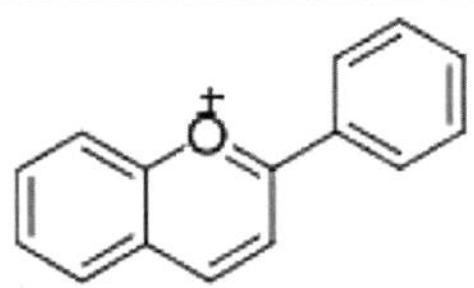	**Proantho-cyanidin B3**	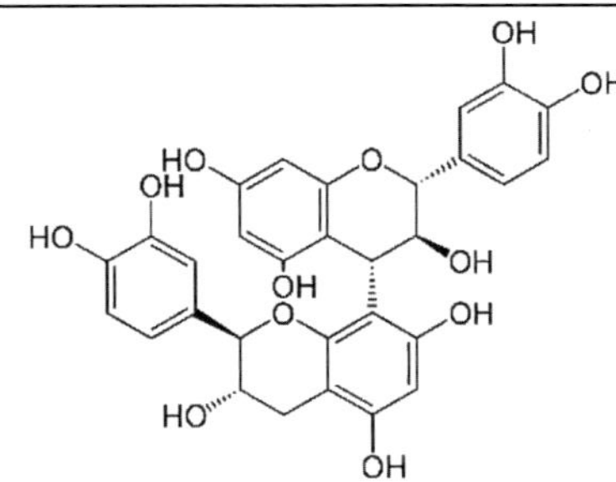	Hops, barley [16,29]
		Pro-delphinidin B3	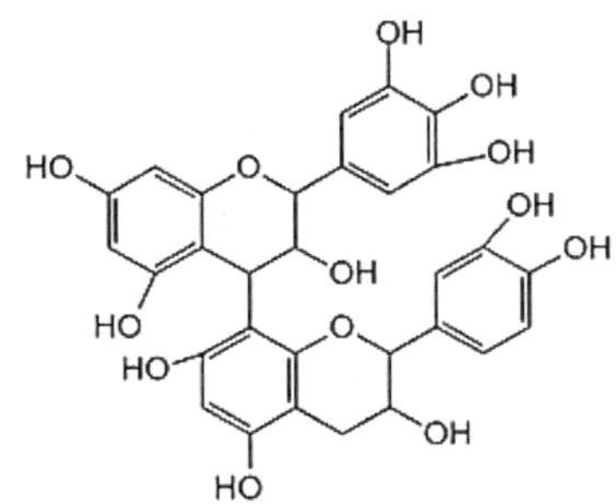	Hops, barley [16,29]

TABLE 9.1 *(Continued)*

Flavonoid subclass	General structure	Name of the specific flavonoid in beer	Structure of specific flavonoid in beer	Source
		Pro-delphinidin B9	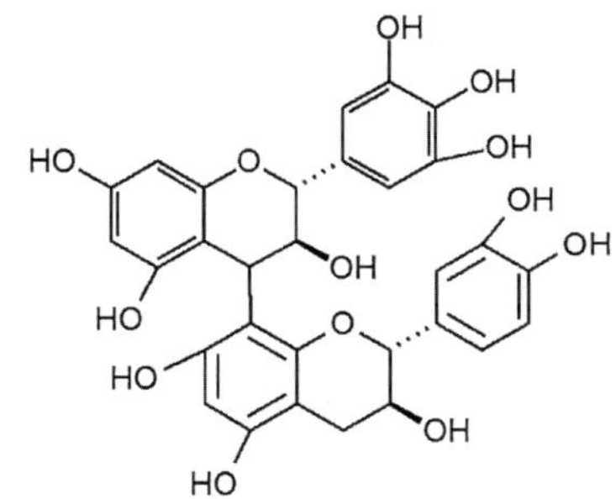	Hops [16,29]
Flavanones	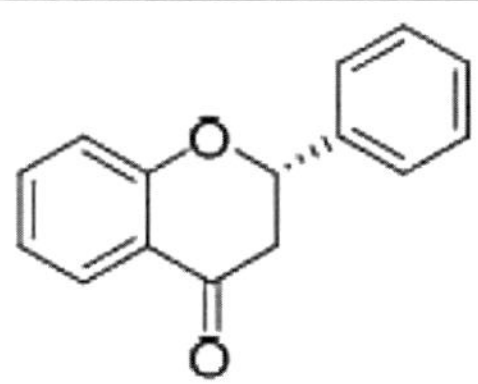	**Iso-xanthohumol**	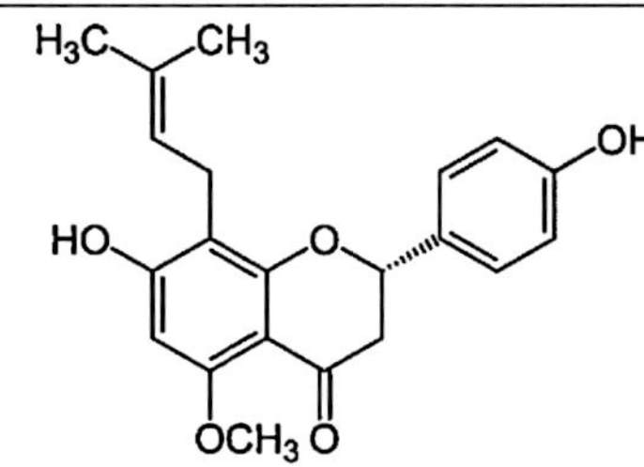	Hops (isomerization) [16,29,50,80]
		8-Prenyl-naringenin	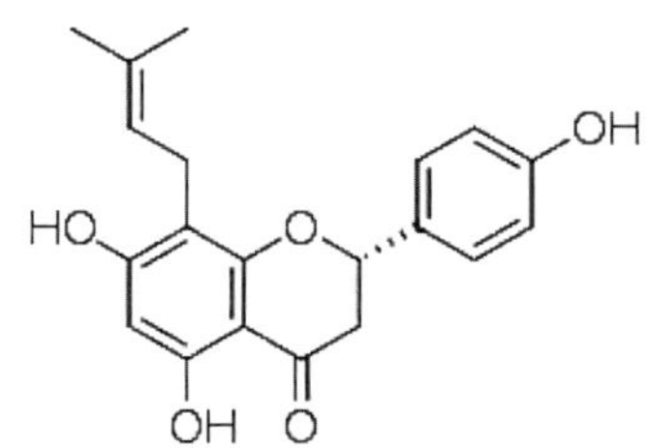	Hops [18,29,80]

TABLE 9.1 *(Continued)*

Flavonoid subclass	General structure	Name of the specific flavonoid in beer	Structure of specific flavonoid in beer	Source
		6-Prenyl-naringenin	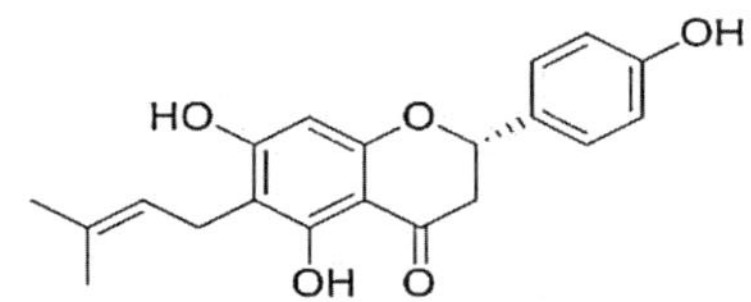	Hops [18,29,80]
		6-Geranyl-naringenin	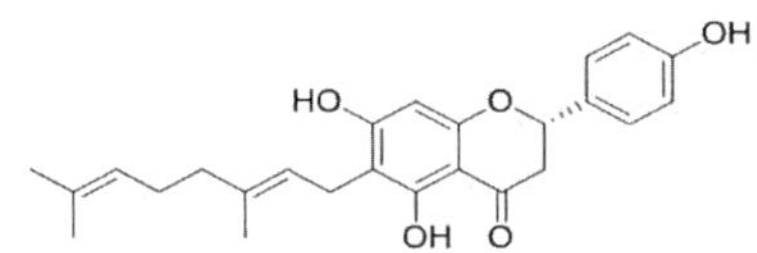	Hops [15,80]
Flavan-3-ols	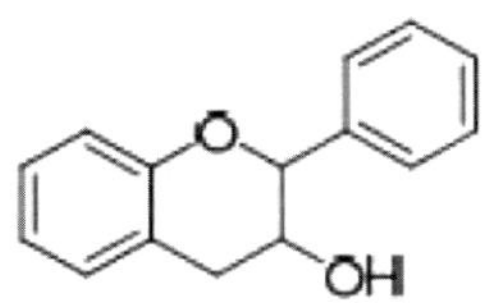	**Catechin**	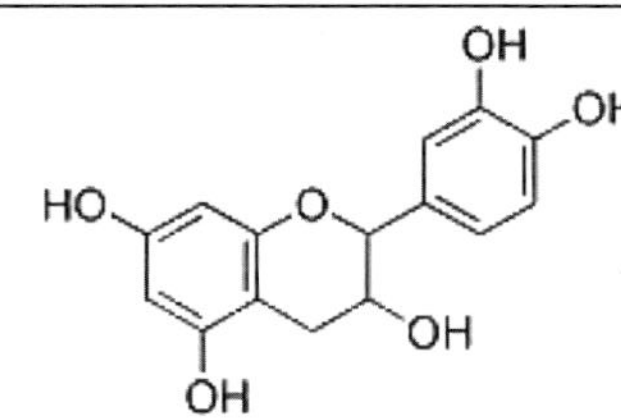	Barley (also as gallate) [12,16,21,29,39,57]
		Epicatechin	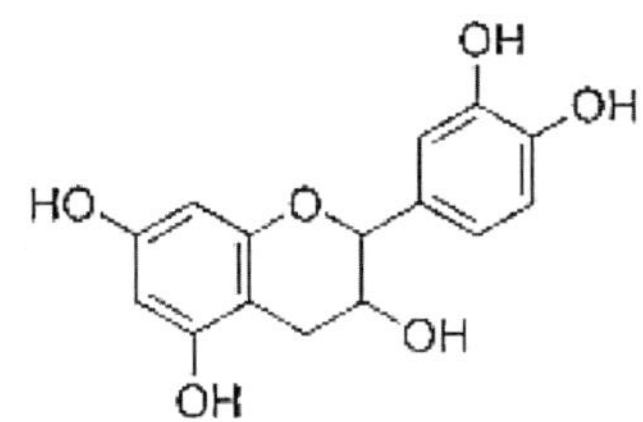	Barley (also as gallate) [12,16,21,29,39,57]

TABLE 9.1 *(Continued)*

Flavonoid subclass	General structure	Name of the specific flavonoid in beer	Structure of specific flavonoid in beer	Source
Chalcones	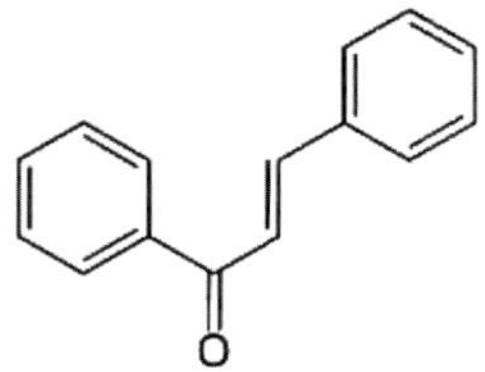	**Xanthohumol**	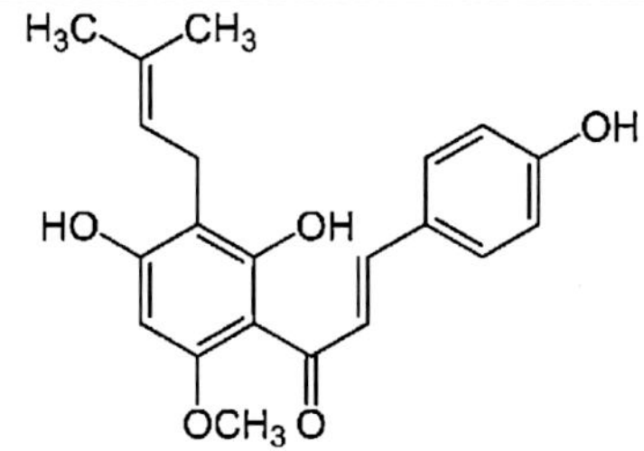	Hops [29,50,80]
Tannin derivatives		**Tannic acid**	$C_{76}H_{52}O_{46}$	Barley malt, hops (also: wood barrels during aging) [63]
		Ellagic acid	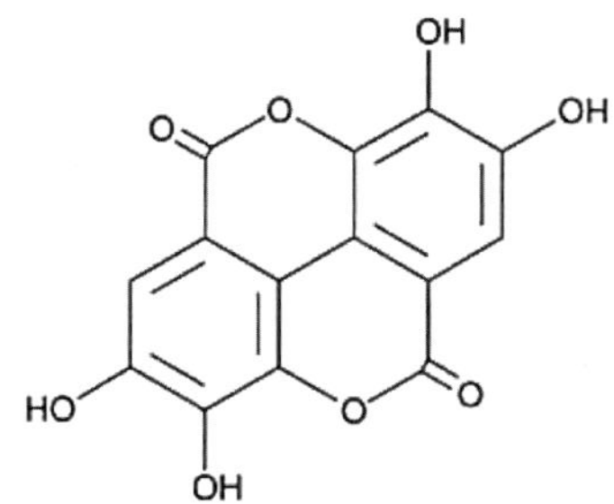	Hops [9]

TABLE 9.2 Main Non-flavonoid Phenolics in Beer and their Source: Self-Developed with Concepts from Ref. [21,22,39,41]

Non-flavonoid polyphenols	General structure	Name of the specific phenolic acid in beer	Structure of specific phenolic compound in beer	Source
Phenolic acids/ benzoic acid derivatives	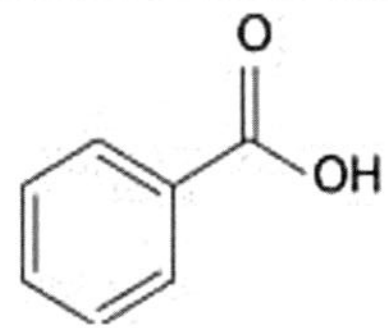	Syringic acid	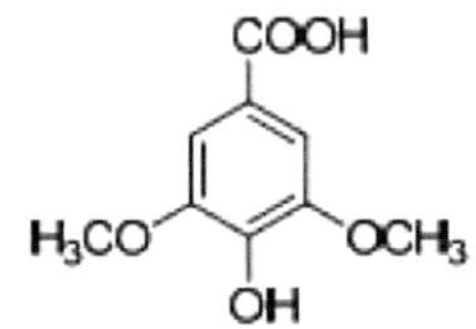	Barley [29,57,77]
		4-Hydroxy-benzoic acid	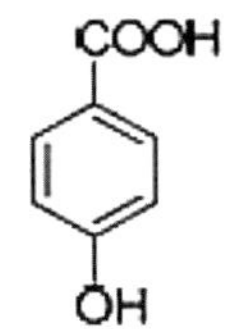	Barley [29,57]
		Salicylic acid	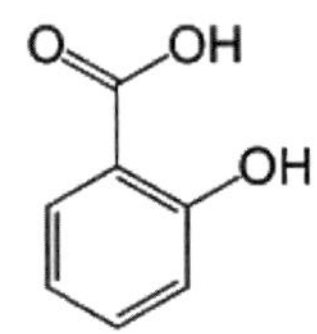	[21,57]
		Protocatechuic acid	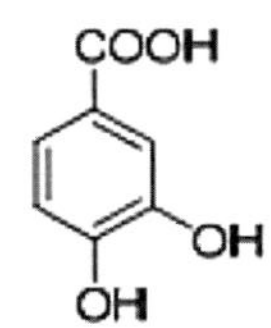	Barley [12,21,29,39,57]

TABLE 9.2 *(Continued)*

Non-flavonoid polyphenols	General structure	Name of the specific phenolic acid in beer	Structure of specific phenolic compound in beer	Source
		Gallic acid	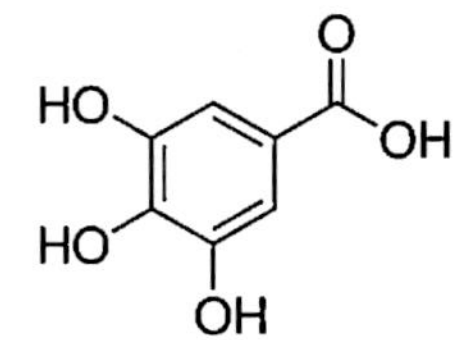	Barley, hops [12,21,23,29,39, 57,63]
		Vanillic acid	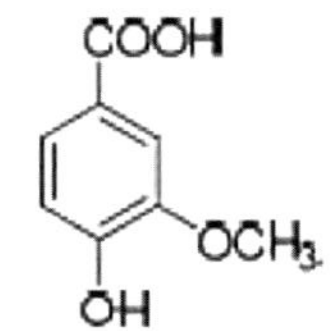	Barley malt, hops [12,23,29,39,57,77]
		Vanillin	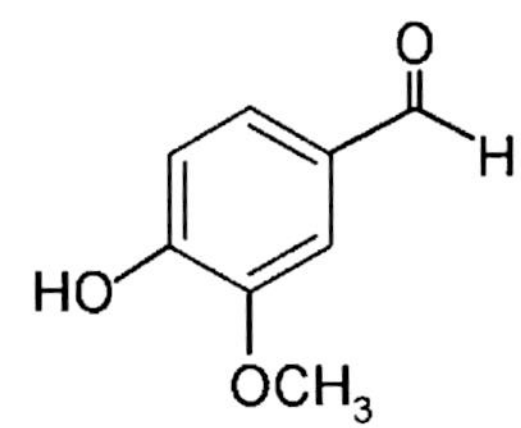	Barley malt [39,63]
		Gentisic acid	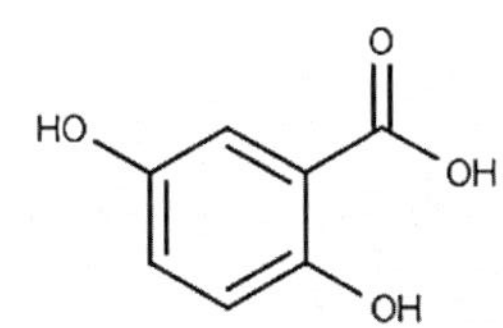	Barley [21,57]

TABLE 9.2 *(Continued)*

Non-flavonoid polyphenols	General structure	Name of the specific phenolic acid in beer	Structure of specific phenolic compound in beer	Source
Phenolic acids/4-Hydroxy-phenylacetic acid derivatives	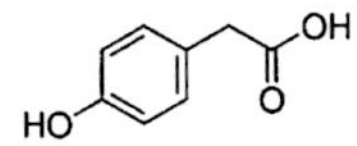	4-Hydroxy-phenylacetic acid	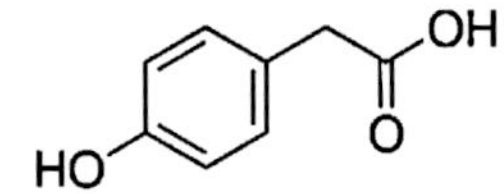	Barley [39]
Phenolic acids/cinnamic acid derivatives	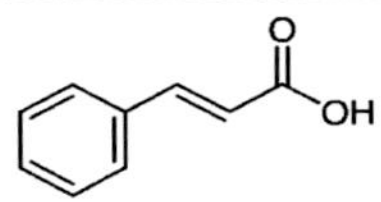	Sinapic acid	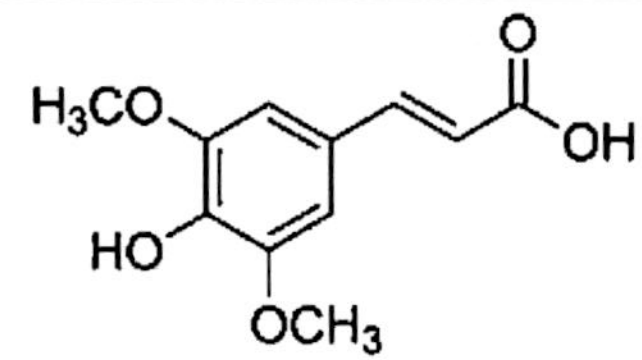	Barley [12,29,39,57,77]
		p-Coumaric acid	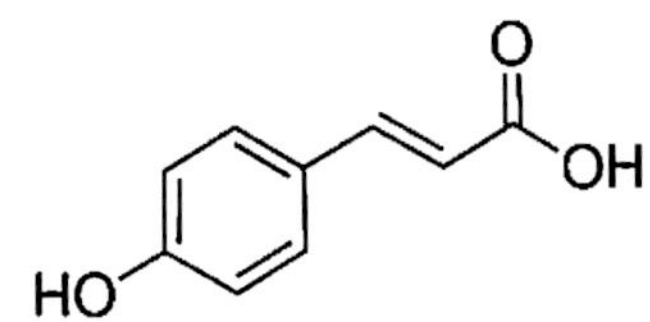	Barley [12,21,29,57,63]
		o-Coumaric acid	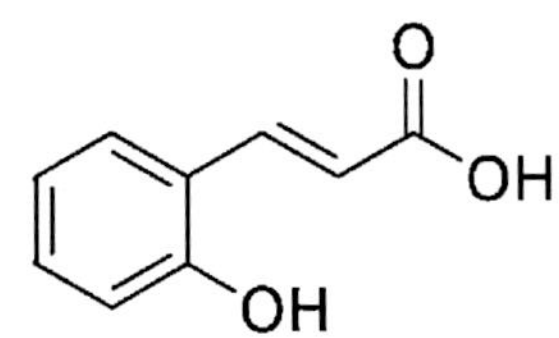	Barley [77]
		Caffeic acid	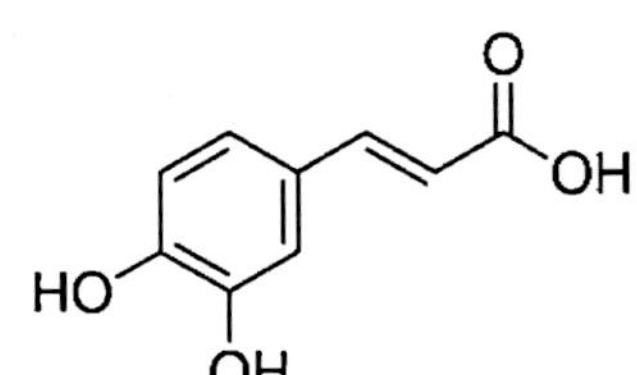	Barley [12,21,29,39,57,63,77]

TABLE 9.2 *(Continued)*

Non-flavonoid polyphenols	General structure	Name of the specific phenolic acid in beer	Structure of specific phenolic compound in beer	Source
		Ferulic acid	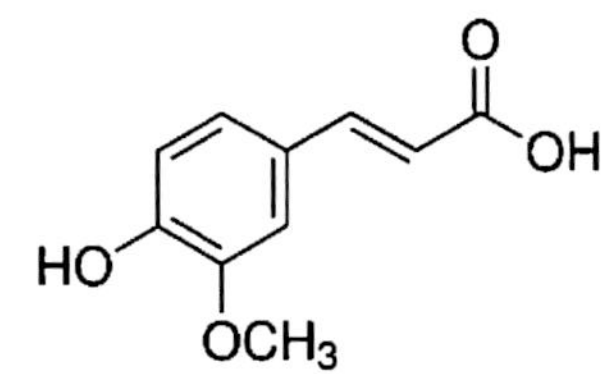	Barley malt [12,21,29,39, 57,77]
		Chlorogenic acid	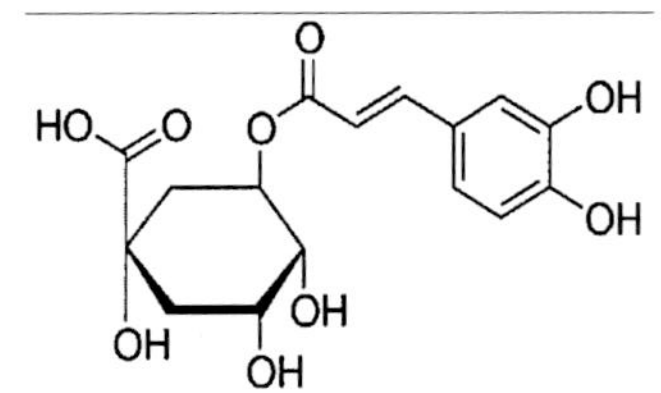	Barley, hops [23,39,77]
		4-Hydroxy-phenylactic acid	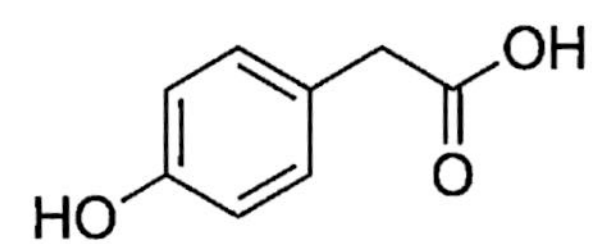	Barley [39]
Other phenolic compounds		Catechol	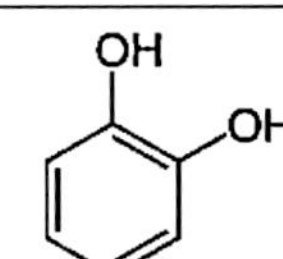	Barley [63]

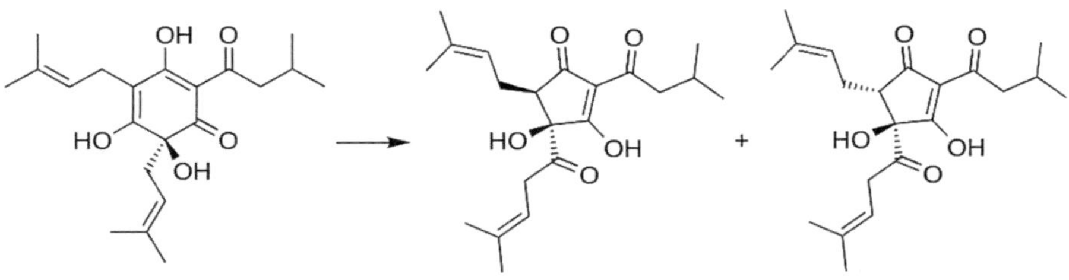

FIGURE 9.1 Thermal isomerization process of humulone to isohumulones.

Each humulone yields two epimeric isohumulones, that is, cis- and trans-isohumulones, determined by the relative position of the OH moiety at the carbon atom (4) and the prenyl sidechain at the carbon atom (5). As a result, six main iso-alpha-acids (cis- and trans-isohumulones, isocohumulones, and isoadhumulones) are formed due to the conversion of the three key alpha-acids (humulone, cohumulone, and adhumulone).[39] About 70% of bitterness in beer is determined by the presence of isohumulones, present in amounts of 20–60 mg/L.[9]

In addition to bitter taste, isohumulones prevent growing of Gram-positive bacteria, ensuring thus the protection of the beer against these.[41]

Iso-α-acids were recently discovered to have a significant potential in the prevention of Alzheimer disease.[15]

9.2.3.2 HOP ESSENTIAL OIL

Most of the flavor in beer is added by the essential oils of hops. It is composed of hydrocarbons, mainly terpenes (mono- and sesquiterpenes) and oxygenated compounds (alcohols, aldehydes, ketones, esters). The oxygenated fraction is less volatile and more aromatic. The main terpenes in beer include myrcene, β-caryophyllene, and α-humulene. Linalool and geraniol are also important aroma compounds that contribute to the floral character of beer, while β-ionone, β-damascenone, neral, geranial, 1,3(E),5(Z)-undecatriene, trans-4,5-epoxy-(E)-2-decenal, 1,3(E),5(Z), 9-undecatetrene, and oxygenated compounds as methyl, ethyl, and propyl 2-methylbutanoate, nonanal, (Z)-1,5-octadien-3-one, and isovaleric acid are also potent odor sources in hop essential oil.[27]

9.2.3.3 MELANOIDINS

Melanoidins cover a diversity of colored compounds formed in the nonenzymatic browning process from multi-step reactions between reducing carbohydrates and molecules possessing free amino moieties. They are formed first during malt kilning and then during wort boiling. Their molecular weights can rise up to 100,000 Da with darker products having higher molecular masses than lighter ones. Some of the compounds present in melanoidins have nitrogen in their molecules; others contain only carbon, hydrogen, and oxygen, however, their exact structure is highly unknown.[6]

Together with the polyphenols, they act as antioxidants and were confirmed to prevent DNA damage induced by oxygen-containing free radicals. This effect was stronger for dark beers compared with blond ones, the latter containing less melanoidins. The antioxidant effect of melanoidins is manifested in scavenging free radicals and chelating metal cations, especially iron. Besides their tumor growth-inhibiting effects, they also exhibit antimicrobial action against many pathogens including *Helicobacter pylori*, *Escherichia coli*, and *Salmonella typhimurium*.[43]

9.2.3.4 VITAMINS

Beer is a widely known source of vitamins from group B (Table 9.3). As one of the vitamin B compounds present in beer, riboflavin (just like other flavins), in addition to its evident nutritional value, is supposed to generate reactive oxygen species that can develop off-flavors in beer.[9,70]

TABLE 9.3 Vitamin B Content in Beer.

Vitamin B	Concentration (mg/L)
Thiamin (B_1)	0.08
Riboflavin (B_2)	0.02–0.1
Nicotinic acid, niacin (B_3)	0.3–5
Pantothenic acid (B_5)	0.04–2
Pyridoxine (B_6)	0.07–1.7
Biotin (B_7)	<0.015
Folic acid (B_9)	0.04–0.6
Cyanocobalamin (B_{12})	<0.03

9.2.4 PRENYLATED FLAVONOIDS IN BEER

Prenylated flavonoids and isoflavonoids, containing specifically one or more prenyl moieties bound to the flavonoid skeleton, are known for their unique bioactivities, which are significantly different from the activities of the unmodified parent compounds. This higher bioactivity results from the presence of the prenyl moiety, a lipophilic group that facilitates the transmembrane transport of these compounds.[69]

Studies on animals suggest that the presence of the prenyl group has a significant role in the physiological impact of dietary flavonoids. However, it is not exactly known how prenylation can enhance the biological function of food flavonoids in the organism.[51]

As the number of prenyl moieties rises, the lipophilic character, the membrane attachment, and transmembrane transport of the molecule increases, enhancing thus biological activity. Due to this latter, interest toward prenylated flavonoids is continuously increasing. Some of them are reported as phytoestrogens and many possess antioxidant, anticoagulant, anti-inflammatory, hypoglycemic, antihyperlipidemia, antimicrobial, antiviral, and anthelmintic activities. However, their most remarkable property is their anticancer activity which has been confirmed in vitro in numerous human cancer cell lines and in vivo in animal experiments.[46,80]

Prenylflavonoids are grouped into several subclasses, according to their chemical structure; prenylflavones, prenylflavanones, and prenylflavonols are the main groups. According to the new classification of phenolic compounds, chalcones possessing an open C-ring system are a subclass of flavonoids.

The 1100 known prenylated compounds are found only in a restricted number of plants. Among these, the most elevated amounts of prenylflavonoids are available in hops (*Humulus lupulus*).[80]

The dominant prenylflavonoid in hops is a chalcone, xanthohumol that is produced in every part of the plant, its content rising up to 0.1–1% (in the dry weight) in the cone-like female inflorescences used for beer. It can be isolated by chromatography, most recently by high-speed counter-current chromatography. Even though xanthohumol contributes in a lesser extent to the antioxidant capacity of beer, due to its highly lipophilic character, it is possibly more bioactive than other polyphenols in beer.[80]

Although beer is the main nutritional source of xanthohumol and associated prenylated flavonoids, xanthohumol is a minor phytochemical in beer. This is attributed to several factors, starting from its inefficient extraction yield from the hops into the wort, its binding to the insoluble proteins originated from malt, and further binding to the yeast cells present during fermentation. Furthermore, xanthohumol undergoes its thermal isomerization (cyclization) into the flavanone isoxanthohumol (Fig. 9.2).

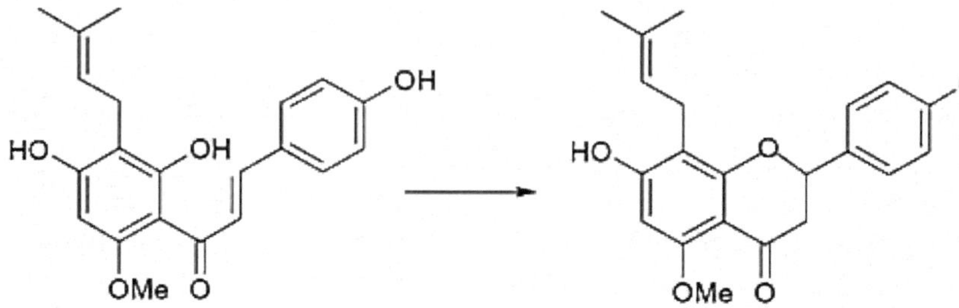

FIGURE 9.2 Isomerization of xanthohumol (chalcone) to isoxanthohumol (flavanone).[22,46]

Thus, on average, less than 0.2 mg/L xanthohumol is found in commercial beers, between 0.10 ± 0.07 mg/L in Ale-style beers and 0.3 ± 0.03 mg/L in dark beers. This depends on the hop varieties, breeding conditions and age, as well as the brewing process. New brewing technologies were implemented to increase the xanthohumol content in beer, for example, the use of hop products enriched in xanthohumol, special malts, addition of hops to the boiling worts at a later stage, temperature control during the addition of xanthohumol-containing products to sweet worts and special methods like hydrodynamic cavitation. Pharmacologically relevant effects can only be achieved by oral administration of doses starting from 16.9 mg/kg (impact on glucose metabolism), up to 1 g/kg (antiangiogenic effect), without any relevant toxicity.[15]

The other prenylated chalcone-type flavonoid in hops, desmethylxanthohumol is practically consumed during thermally induced cyclization, yielding a combination of 6-prenylnaringenin and 8-prenylnaringenin (two phytoestrogenic compounds) as shown in Figure 9.3.

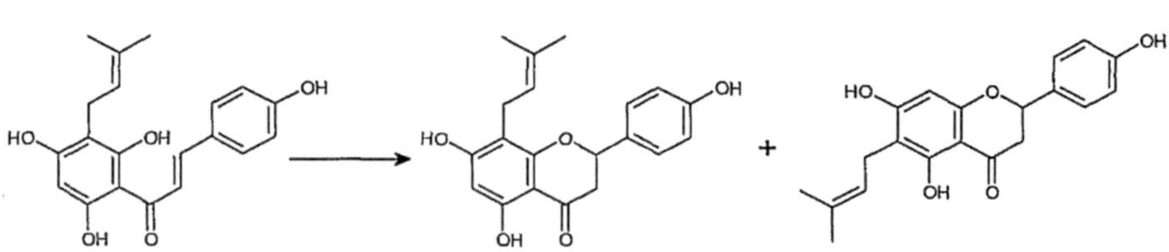

FIGURE 9.3 Isomerization of desmethylxanthohumol (chalcone) to 8-prenylnaringenin and 6-prenylnaringenin (flavanones): Self-developed with concepts from Ref. [24,42,72]

Desmethylxanthohumol (Fig. 9.3) is practically absent in beer, while its thermal conversion product, 8-prenylnaringenin, is at the same time a post-ingestion metabolic product of isoxanthohumol in the human liver through the CYP450 reaction, respectively in gut under the action of the microbiota.[18,51]

9.3 ROLE OF FLAVONOIDS IN CANCER RESEARCH

Cancer is a major burden on society today, occupying a prominent position in the disease and death statistics in the world; and about 17 million new cases were registered globally in 2018. The most frequent types are lung, female breast, bowel, and prostate cancer. By 2030, we are expected to have 23.6 million new cases worldwide. Beyond prevalence, mortality data are not encouraging either. In 2018, 9.6 million people were killed by various cancers; lung, liver, stomach, and intestinal cancers accounted for nearly half of the cases.[14]

Risk factors are similar all over the world: smoking, inadequate moving, unhealthy diet, immoderate alcohol consumption, overweight, and infections. Treatments are often painful and have several side effects, in addition, some patients can be resistant to the therapies applied. Therefore, new treatment alternatives are constantly being researched. Among these, natural components in plants are of increasing scientific interest. This is mainly due to their high bioavailability and reduced side effects. In particular, flavonoids of the polyphenol family have gained great importance as potential anticancer agents that promote the destruction of tumor cells.[4]

From a nutritional point of view, it is important to underline that these components can naturally be part of our diet, and thus play a role in prevention.

As mentioned above, flavonoids are of great interest in cancer research as chemopreventive agents that have a promising potential due to their cytotoxic effects on tumor cells. Flavonoids, such as silymarin, genistein, quercetin, morusin, pomiferin, and cudraflavon B were described among others as potential chemopreventive agents. In addition to the main anticarcinogenic substance in hops, that is, xanthohumol, other hop-derived components were also found to be effective: isoxanthohumol, 8-prenylaringerin, 6-prenylaringerin, and xanthogalenol were investigated.[80]

Their effects have been proven on tumors including prostate, colon/rectal, breast, thyroid, lung, and ovarian. One of the great advantages of these compounds is their extremely low toxicity. Examining the safety of xanthohumol, an oral test on 70 female C3H laboratory mice was carried out. Mice were divided into three groups. As controls ($n = 10$), untreated plain water subjects were studied. 30-30 mice were enrolled in the other two groups, one of which received 0.1% ethanol (control group) and the other one 5×10^{-4} M xanthohumol dissolved in drinking water. During the study, members of each group consumed approximately the same amount of fluid and feed, and their body weight did not change or differ. After 4 weeks, blood samples were taken and analyzed for hematological and biochemical parameters, and multiple organs and tissues were investigated. The results showed that xanthohumol had no effect on hematological and biochemical parameters, nor did it alter the function of bone marrow, kidney, exocrine pancreas, liver, muscle, thyroid, ovary, and adrenal cortex. There was no change in fat, protein, or carbohydrate metabolism either. The daily amount of xanthohumol injected in this study into mice would correspond to an amount of approximately 1.6 g in humans, which would be unlikely to be consumed, even supposing extreme beer consumption. Based on these, xanthohumol can be considered safe for use in research.[83]

The anticancer effect of flavonoids is achieved through the following mechanisms:

- Some flavonoids reduce multiple drug resistance, which is a very important factor in the successful treatment of cancer patients.[7,68]
- They inhibit angiogenesis driven by vascular endothelial growth factor (VEGF) and bFGF (basic fibroblast growth factor).[54,60]
- They inhibit the development of new tumor cells and induce apoptosis (programmed cell death). They can influence the cell cycle process that occurs between two divisions in a eukaryotic cell. It consists of interphase, mitosis, and cell division. The cell cycle is co-regulated by cyclin proteins and cyclin-dependent kinases. Flavonoids interfere with the cell cycle process, and by their interference with the function of certain regulators, cells become arrested in cell cycle phases G1 or G2/M. Phase G1 is the first phase of growth. Phase G2 is the second stage of growth, preparing the cell for division and phase M or mitosis and cytokinesis, the division of a cell into two progeny cells. By influencing this process, flavonoids can prevent the division of tumorous cells.[13,17,95]
- They inhibit the enzymes involved in the metabolism of certain substances, thereby preventing the activation of certain carcinogenic compounds.[78,85]

However, there are also limitations to the applicability of these compounds. On the one hand, it is difficult to isolate and purify the active ingredients. Many plant sources are limited and the amount of flavonoids that can be extracted is very low. The extraction and purification procedures are expensive and time-consuming.

Pharmacokinetic phenomena also cause considerable difficulties. Poor solubility and structural instability, as well as interactions with microflora in the body and other drugs, all influence the usability and function of flavonoids.

Nevertheless, it is not easy to clearly confirm the positive effects of these compounds. The epidemiological studies required for this are expensive, time-consuming, and the results are often unclear. Because of the heterogeneity of society and many other influencing factors, it is difficult to draw reliable conclusions from each study.[1,66]

However, flavonoids, including xanthohumol, deserve attention in research of cancer prevention and treatment.

9.4 XANTHOHUMOL AND CANCER RESEARCH: RESULTS OF IN VITRO EXPERIMENTS

The most interesting aspects of xanthohumol are its antiproliferative and anticarcinogenic effects.

Based on the studies available to the date, xanthohumol has been shown to be chemoprotective against breast, colon, ovarian, and prostate cancer cells. It exerts its effect on initiation, promotion, and progression of tumors as well. By neutralizing reactive molecules, it reduces the degree of DNA damage and also inhibits members of the so-called cytochrome P450 enzyme system, which would otherwise metabolically activate pro-carcinogens. In addition, its anti-inflammatory and antiangiogenesis effects are important in inhibiting the development and growth of tumors.[50]

There have been many studies aiming the exploration of the antitumorigenic effects of xanthohumol, and their enhancement. There are many in vitro studies that have proven this effect on human tumorous cell lines.

9.4.1 GASTRIC CANCER

Xanthohumol can reduce the viability of tumorous gastric cells (SGC-7901, SNU216, SNU668 cell types), but does not have or has just mild toxic effects on normal cells (concentration 1–100 μM). It prevents the multiplication of cancer cells and increases the number of apoptotic cells. This mechanism was associated with the reduced expression of proteins Bcl-XL and Bcl-2 with antiapoptotic properties, while the expression of pro-apoptotic proteins

(Bax, PARP, caspase-3) increased. Besides, xanthohumol decreased the production of NF-kB in cancer cells. NF-kB is able to enhance angiogenesis, thereby increasing the oxygen and nutrients-supply of tumors. These factors can make xanthohumol capable of defense against gastric cancer.[32,90]

9.4.2 PANCREAS CANCER

The antitumorigenic effect of xanthohumol has been investigated against pancreas cancer cells as well. Xanthohumol treatment significantly reduced the proliferation of tumor cells in vitro. At the same time, NF-kB (nuclear factor kappa-light-chain-enhancer of activated B cells) production decreased too. As an additional positive effect, VEGF and IL-8 expression decreased both in mRNA and protein levels.[73]

9.4.3 LARYNX CANCER

The survival of tumorous human larynx cells (RK33 and RK45 types) and normal cells (HSF, OLN-93 types) was compared after xanthohumol treatment in an in vitro experiment. As a result, the survival of cancer cells decreased, while the normal cells barely suffered from cytotoxic effect. The reason is probably the enhancement of apoptosis. Tumor cells stagnated in the G1 phase of cell cycle. More factors can be responsible for this effect. For instance, the enhanced expression of p53, p21/WAF1 enzymes, the reduction of cyclin D1 and Bcl-2 production, and the activation of caspase 3-, -8, -9. Another potential antitumorigenic mechanism is the inhibited phosphorylation via the ERK1/2 pathway.[71]

9.4.4 CERVIX CANCER

The effect of xanthohumol on cervix cancer has been investigated on Ca Ski human cancer cells. The inhibition of cell proliferation was observed in this case as well. In addition, morphologic changes occurred both in cell and nucleus levels. After 48 h of xanthohumol treatment, changes in mitochondrial membrane potential and DNA disruptions were registered. Xanthohumol was able to keep cancer cells in the cell cycle S-phase and activate caspase 3-, -8, -9. Xanthohumol reduced the Bcl-2 and XIAP levels, while increased the amount of PARP, p53, and AIF depending on the dose applied. As a result of all these effects, the number of apoptotic cells increased in the tumorous cell line.[92]

9.4.5 MAMMARY CARCINOMA

The antitumorigenic effect of xanthohumol has been detected for breast cancers as well. Shortened survival time induction of apoptotic progress was observed on MDA-MB-231 and MCF-7 cell lines. Xanthohumol reduced the activity of Bcl-2 and Bcl-XL, and the cells stagnated in G0/G1 cell phase.[76]

9.4.6 CHOLANGIOCARCINOMA

Xanthohumol has been proven to be effective against human cholangiocarcinoma cells in vitro. Decreased cell proliferation was observed after xanthohumol treatment in three types of cell lines (CCLP1, SG-231, CC-SW-1). Xanthohumol was capable to stop the cell cycle and induce apoptosis. In every case, the regulation of pro-apoptotic (PARP and caspase 3) and anti-apoptotic (XIAP) factors played a huge role. Reduced expression of Notch-1 and AKT were described at molecular levels in a time-dependent manner.[87]

9.4.7 HEPATIC CANCER

The effect of xanthohumol against HepG2 hepatic cancer cells has been detected too. The results of a 48-h multiple-dose xanthohumol treatment inhibited proliferation and increased apoptosis in the tumorous cells. As the mechanism of action of the observed time- and dose-dependent effects of xanthohumol, the reduced function of XIAP, Bcl-2/Bax, and NF-kB, and the increased activation of p53, PARP, AIF, and cytochrome C were described. It means that longer exposure time and higher dosage were more effective against these cancer cells.[96]

9.5 XANTHOHUMOL AND CANCER RESEARCH: RESULTS OF IN VIVO EXPERIMENTS

9.5.1 CHOLANGIOCARCINOMA

The efficiency of xanthohumol against biliary cancer has been proven in vivo as well. During the experiment, BALB/c mice with reduced immunity on account of lacking thymus were observed. CCLP-1 (Cancer Cell Line Profiler) or SG-231 cholangiocarcinoma cell lines (each 1 × 10^6 pieces of cells) were administered subcutaneously in the flank of the 6-week-old rodents. After the tumors had grown palpable, they started the xanthohumol treatment, which consisted of intraperitoneally injected solution every other day. Xanthohumol was solved in physiological saline. The control group was treated with blank saline during the experiment. Each mouse was administered a dose of 125 μg xanthohumol on treatment days, which is equivalent to 5 mg/kg (based on the weight of a 25 g mouse). Each group consisted of 5-5 animals. Tumor sizes were measured and documented every 4 days.

Examining the SG-231 cell line, in the xanthohumol-treated group, the tumor sizes grew by 153%, while in the control group, it was 427% on the 8th day. On the 16th day, these groups showed 256% and 1308% growth. The inhibitory effect of xanthohumol on tumor growth was clearly visible.

Even more convincing results were described in the CCLP-1 cell line groups. After 8 days, 381% growth was observed in the control group, while the xanthohumol-treated group achieved 13% decrease in tumor sizes. On the 16th day, 853% growth in the control, and 78% decrease in the treated group were registered.

Based on these results, xanthohumol is not only able to reduce tumor growth but also to turn it back.[87]

In a further study, 2×10^6 pieces of cells of the KKU-M214 cholangiocarcinoma cell line were injected in the flanks of 6-week-old BALB/c nude mice. Animals were kept in pathogen-free environment. After 1 week, visible tumors had developed. The mice were divided into three groups. Xanthohumol was mixed in their drinking water. Treated groups drank 20 μM or 50 μM xanthohumol solutions orally for 30 days. The control group received 0.5% ethanol instead. The solution was kept in a closed bottle and the required amount was replaced every day. Tumor and body weights were measured twice a week, and the fluid consumption was also registered. During the experiment, no side effects or toxic effects on organs (liver, kidney, and spleen) were observed. All the three groups consumed nearly the equal amount of fluid, and their body weights remained almost unchanged, relatively similar to each other. The tumor growth inhibitory effect of xanthohumol was significantly visible between the control and the 50 μM groups. However, this effect was not registered in the 20 μM xanthohumol group.[23]

9.5.2 COLORECTAL CANCER

The 6-week-old BALB/c mice were tested to investigate the impact of xanthohumol on colorectal cancer. HCT-116 or SW-480 cell lines were injected subcutaneously in immunosuppressed nude mice. Some 10^7 pieces of cells were implanted in the link flank of the animals. The treatment has been started 1 week after. By then, the tumors had grown to 150 mm^3. The experiment lasted for 17 days. Xanthohumol was dissolved in DMSO (xanthohumol concentration of 100 μM). This basic solution was diluted with PBS according to each treatment group. The control group has been treated with 0.85% DMSO. The tested groups got 1 mg/kg, 15 mg/kg, and 30 mg/kg xanthohumol intraperitoneally every day. Tumor sizes were followed-up with caliper. After 17 days, the animals were sacrificed. Tumors, heart, liver, spleen, lungs, and pancreas have been removed from the bodies. There were no visible morphological changes in organs, and animals body weights have not decreased either. Inhibitory in tumor growth was described in a dose-dependent manner.[75]

9.5.3 MAMMARY CARCINOMA

An experiment was carried out on 8 weeks old BALB/c mice. The animal weights were 17–18 g on average. Their living conditions were standardized, they could eat and drink ad libitum. The 2×10^5 pieces of cells from the 4T1 breast cancer cell line were injected in the right flank. Animals were divided into three groups randomly. The members of the first group ($n = 8$) were administered 100 mg/kg xanthohumol, while the second group ($n = 8$) received 200 mg/kg xanthohumol every other day intragastrically for 2 weeks. The control group was treated with a placebo. Each animal received 200 μL of placebo or XN-solution every day. The weights of the mice were measured every day. After 2 weeks the animals were sacrificed. In terms of body weight, significant differences were described between the group treated with 200 mg/kg xanthohumol and the control group. The treated animals maintained higher body weights, and their tumor sizes were smaller. Examining the tumor weights, xanthohumol showed a dose-dependent effect. After 2 weeks, the tumor weights were lower by 27.22% in the 100 mg/kg xanthohumol group and by 46.79% at the 200 mg/kg xanthohumol group, in comparison with the control group.[76]

9.5.4 PROSTATE CANCER

TRAMP C57BL/6 (transgenic adenocarcinoma of the mouse prostate) modified genome mice were tested as models for investigation of the effect of xanthohumol on prostate cancer. The experiment was started on 4 weeks old animals and lasted for 20 weeks. Xanthohumol was dissolved in ethanol (2.5 mg/mL), which was then diluted. The mice were fed orally with 50 μg xanthohumol solved in 0.2 mL polyvinylpyrrolidone (1%) and Tween 20 (0.01%) every other day. Each group consisted of 2 × 10 mice. During the experiment, rodents could drink water and eat ad libitum. Body weights were registered once a week. After their death, the urogenital tract, bladder, ureter, prostate, and testicles were extracted from the animals. There were no observed side effects, and in the mice treated with xanthohumol decreased urogenital tract weights were measured compared with the control group.[81]

9.5.5 PANCREAS CANCER

The effect of xanthohumol on pancreas tumor cells in vivo has been observed. 5×10^6 BxPC-3 human cancer cells were injected in the right flanks of 6-week-old BALB/c nude mice. Tumor sizes were measured every week. After the tumors reached the size of 100 mm^3, the mice were distributed into two groups (5-5 mice). The first group was administered 10 mg/kg xanthohumol intraperitoneally once a week, while the control group was treated with blank DMSO. This part of the experiment lasted 5 weeks. Mice were kept in conditions of constant temperature and regulated light-dark rhythm. Their food was standard autoclaved rodent food. The animals could eat and drink water ad libitum. After 6 weeks, mice were sacrificed and tumors were examined. Xanthohumol-treated animals had significantly lower tumor sizes in comparison with the control group. Meanwhile, there were not any relevant changes in the body weights neither in the control, nor in the treated group, and nor relative to each other.[73]

9.5.6 HUMAN INTERVENTIONAL STUDY

It is important to mention the human interventional study carried out in 2016. Twenty-two people (11 women, 11 men) tested a commercially unavailable drink that contained 12 mg/L xanthohumol. The alcohol content of the drink was not reported. The drink contained 42 mg malic acid, 1 mg roasted malt extract, and 40 g glucose per liter. The placebo drink had the same composition, except that it contained no xanthohumol. Subjects were divided into two groups, and a placebo-controlled study was carried out. Half of the subjects drank xanthohumol-drink and the other half drank a placebo for 14 days, in amounts of 1 L per day. Blood samples were taken on the 0th,

7th, and 14th days and were incubated with dietary carcinogens (NDMA, IQ, B(a)P) and human liver homogenates. It was observed that the degree of DNA damage was significantly decreased in the xanthohumol-consumer group in comparison with the control group. Similar results were described in connection with another study, in which xanthohumol-containing (12 mg/piece) pills were tested (n = 10). According to reports, this effect was attributed to the elevated level of the detoxifier α-GST.[88]

9.5.7 POTENTIAL OF XANTHOHUMOL AS ANTICANCER AGENT

Summing up the available scientific reports and their results, it can be concluded that xanthohumol can represent a promising new therapeutic line. Its potential in cancer research is particularly important. Its antitumorigenic effect has been confirmed many times in vitro and in vivo as well. Laboratory experiments revealed the mechanism by which xanthohumol can act on the cancer cells. Its most important role is the induction of apoptotic progresses in cells. Xanthohumol causes shorter survival time and decreased proliferation of tumorous cells. This effect is due to the activation of pro-apoptotic and the inhibition of antiapoptotic agents. Another important factor is the decrease of angiogenesis via several pathways. Thus, the nutrient and oxygen supply of the cells can be reduced. In vivo studies have confirmed the in vitro results. The results of immunohistochemical analysis performed on the tissues of experimental animals have highlighted the importance of the factors specified in vitro. The other remarkable asset of these studies is that tumor sizes can be made comparable. For the tumorous cell lines examined, significantly smaller tumor weights and sizes have been registered after xanthohumol treatment.

9.5.8 OTHER PHARMACOLOGICAL ASPECTS OF XANTHOHUMOL

In addition to its anticarcinogenic properties, the beneficial effects of xanthohumol have been confirmed in many other health-related aspects as well.

The widespread cardiovascular problems occupy a prominent place in the mortality statistics of developed countries. Studies have shown the anti-arteriosclerotic effect of xanthohumol. On one hand, it inhibits inflammatory fiber growth, which could cause plaque formation on the walls of large and medium blood vessels.[50]

Moreover, it inhibits the cholesteryl ester transfer protein, which can cause an elevated level of HDL cholesterol.[34]

Increasing reverse cholesterol transport is another mechanism to elevate the level of HDL in blood. Xanthohumol can influence this process as well, thus leading to an increased cholesterol efflux in macrophages and resulting in a potential antiarteriosclerotic effect.[35]

Xanthohumol can prevent the ADP-induced platelet reactivity, which enables it a potential in the prophylaxis of cardiovascular disease.[47]

Thus, the affected blood vessels are less likely to develop stenosis and the organs involved are not damaged.

Xanthohumol is capable of neutralizing reactive oxygen and nitrogen radicals (e.g., superoxide). This is due to its antioxidant properties, which are otherwise typical for flavonoids. Research has shown that xanthohumol can inhibit lipid peroxidation in a variety of processes, thereby reducing LDL modification, which could be a potential factor in the progress of cancer or atherosclerosis. In an in vitro experiment applying low micromolecular concentrations, xanthohumol was found to be a more potent antioxidant than α-tocopherol.[59]

Surprisingly, xanthohumol can also be effective against obesity and metabolic syndrome; it has been proven to induce apoptosis in fat cells. In addition, it is able to inhibit their proliferation during the cell differentiation phase.[91]

Another aspect of the effect of xanthohumol is influencing the AMPK signaling pathway. Phosphorylated-AMPK (p-AMPK) reduces anabolic processes and elevates catabolic processes in adipose tissue, which leads to fatty acid degradation.[67]

Dietary xanthohumol-rich hop extract was shown to prevent body weight increase in high-fat-induced obese rats compared with the control group. Furthermore, xanthohumol-rich hop extract may reduce liver weight, and triacylglycerol in the plasma. This is reached by improved hepatic fatty acid metabolism and reduced intestinal fat absorption.[93]

Mice treated with xanthohumol, showed impaired glucose tolerance compared with the controls, in animal models of metabolic syndrome.[55]

Furthermore, a dose-dependent effect was described on plasma glucose levels and body mass of 4-week-old Zucker fa/fa rats (as obesity models) suffering of metabolic syndrome.[48]

Interestingly, micellar solubilization of xanthohumol is able to improve its bioavailability and efficacy. A study was carried out on a model of obesity, diabetes, and non-alcoholic fatty liver disease (NAFLD) induced by a Western-type diet. The positive health effects were registered in a dose that was more than 10 times less than the dose indicated in previous in vivo experiments using native xanthohumol.[58]

Xanthohumol can be a promising biocompound in type 2 diabetes and related illnesses. Studies show that xanthohumol can suppress lipogenesis, prevent weight gain, improve insulin resistance, and glucose tolerance, and positively influence the plasma lipid profile. Moreover, xanthohumol can reduce diabetes-related oxidative stress.[19,22,44]

Xanthohumol decreases the function of the sterol regulatory element-binding protein-1 (SREBP-1), a transcription factor that can prompt the expression of genes playing a crucial role in fatty acid and cholesterol biosynthesis.[53]

The hepatoprotective aspect of xanthohumol has been observed as well. According to the current available data, xanthohumol can protect the liver via several pathways including inhibition of inflammation and fibrosis, anticipating the development of cancer. Moreover, by improving blood lipid parameters, it can be effective against the development of fatty liver disease.[40,89]

Xanthohumol also has broad antimicrobial and antiviral activity. The antiviral properties of its metabolites have been demonstrated in vitro, for example, in malaria, where xanthohumol is regarded as a potential antidote to resistant species.[28,30]

In addition, its anti-HIV effect has been studied. Xanthohumol reduces HIV-derived cell damage and in some lymphocytes, reduces the activity of reverse transcriptase enzyme and viral replication in peripheral mononuclear cells. However, it does not affect the process of the access of the virus into the cells.[86]

Other experiments and studies have also been carried out against herpes, rhinovirus, and cytomegalovirus, for which weak or moderate effects have been demonstrated.[11]

There are also several studies on the antifungicidal effects of xanthohumol. The growth of several pathogenic fungi has been successfully reduced, including *Fusarium, Candida albicans, Trichophyton rubrum*.[52]

Attempts against certain resistant species have been successful as well. For example, for *Streptococcus mutans*, which is primarily responsible for tooth decay, xanthohumol had an antibacterial activity similar to the substances commonly found in toothpastes (e.g., thymol).[10]

Research also reported that xanthohumol, acting selectively on harmful microorganisms, does not affect beneficial flora in the intestine.[49]

Xanthohumol can play a role in the improvement of the neurologic status too. A study showed that xanthohumol treatment decreased the inflammation and apoptosis of aged brains, showing a protective effect on aging-induced damage in mice.[65]

In another experiment, it has been observed that xanthohumol may have a role in the improvement of cognitive flexibility in young mice via adjusting the palmitoylation status of neuronal proteins but seemed to be ineffective in aged ones.[94]

9.6 9-PRENYLNARINGENIN, A PHYTOESTROGEN IN BEER AND ITS POTENTIAL HEALTH BENEFITS

The 8-prenylnaringenin is one of the most potential phytoestrogens. The original concentration of 8-prenylnaringenin in beer is very low, which would mean a negligible effect of this compound on health. However, it is produced in considerable amounts by the liver and gut as a metabolite of isoxanthohumol, thus it should be taken into account as a phytoestrogen.[51]

A study showed that human microbiota is able to activate 4 mg/L isoxanthohumol into 8-prenylnaringenin. Depending on individual transformational potential, the final amount of 8-prenylnaringenin can be 10-folds higher than originally found in beer, thus it may have a biologically active effect.[61]

8-Prenylnaringenin is widely examined due to its estrogenic effect. Studies showed that it can reduce menopausal/postmenopausal symptoms, for example, heat waves, mood fluctuation, and low energy.[2,26,36]

Its administration is considered as a novel potential therapeutic tool in the treatment of symptoms of menopause and postmenopause.[72]

One of the most interesting aspects of these is the antiosteoporotic effect of 8-prenylnaringenin, which is able to prevent the damage of bone mass.[25]

It has been proven that it can promote osteoblast differentiation and at the same time inhibit the activation of osteoclasts.[56]

The advantage of 8-prenylnaringenin over other phytoestrogens is that it binds preferentially to ERα (estrogen receptor alpha), which dominates in the bone tissue.[33]

However, the effect of the long-term use of 8-prenylnaringenin as a potential anti-menopausal agent is still in question and further studies and precise evidence definition of the safety hazards associated with its therapeutic use are strongly needed. Consequently, the health claim stating that hop diminishes symptoms of menopause is waiting for approval in the European Union.[72]

Furthermore, 8-prenylnaringenin proved its benefic effect by inhibiting angiogenesis in vitro and in vivo.[62]

In type 2 diabetes mice, 8-prenylnaringenin was able to reduce lipogenesis, prevent body mass increase and improve glucose tolerance and insulin resistance. Lipid profile improved.[19]

8-Prenylnaringenin showed a protective effect against aggregation and adhesion of human platelets. Due to this mechanism, 8-prenylnaringenin may help prevention of cardiovascular problems, such as atherosclerosis, myocardial infarction, coronary artery disease, and thrombosis.[82]

Certain studies described the potential role of 8-prenylnaringenin in cancers as well. 8-prenylnaringenin was able to act against U-118 MG cancer cell line, due to its cytotoxic effect in vitro.[74]

By inhibiting the proliferation, it was effective against UO.31 renal carcinoma cells and PC-3 prostate cancer cells.[8]

9.7 IS BEER REALLY GOOD FOR OUR HEALTH?: THE CONTROVERSIAL EFFECTS OF ALCOHOL

Whatever the potential benefits of phytochemicals in alcoholic beverages are the influential role of alcohol should not be overlooked. Ethanol was officially classified as an IARC 1 type carcinogen in 2012. Acetaldehyde present in fermented alcoholic beverages is also a known carcinogen too. Both these compounds have been found to be directly connected to tumors of the upper gastrointestinal tract, liver, rectum, colon, breast, larynx, and pharynx. A relationship is suspected between pancreatic and lung tumors.[37,38]

Ethanol degradation products, oxidative stress, aberrant folic acid, retinoid, and estrogen metabolites all play a role in the carcinogenic effect. These are being mapped continuously. The central role of the CYP2E1 enzyme has been repeatedly confirmed (P450 enzyme system). Alcohol load also explains the tumorigenic and metastatic lesions of the immune system. Inappropriate DNA methylations may be important epigenetic factors for pathogenic effects.[64,79,84]

Due to the strong evidence that alcohol itself belongs to the group of proven carcinogens, we cannot be sure that the otherwise positive effects of beer can fully exert their beneficial properties.[38]

This is also highlighted in a study published in 2018, processing data from 195 regions worldwide by age group and gender breakdown. The data collected between 1990 and 2016 covers a very long time interval. Different prospective and retrospective studies comparing the health risks of alcohol have been compared. A new meta-analysis was also developed to determine the relative risk of alcohol-related outcomes for 23 health status outcomes. Among others, the study revealed that overall, alcohol consumption was the 7th highest risk factor in terms of both mortality and life-adjusted life expectancy. Researchers have found that the risk of death, especially the risk of cancerous tumors increases even with minimal alcohol consumption. The authors conclude that the recommended intake to reduce the risk should be set at zero, despite of any phytochemical content of the alcoholic beverages.[31]

However, it should be mentioned that the results of the article above were questioned by other researchers, considering that the study has its limitations. It was highlighted that there are huge society and cultural differences between the observed countries, and the examined 23 health status outcomes cannot be applied on regional and individual levels. There is consensus on the fact that heavy alcohol consumption is harmful, but setting of the recommended alcohol consumption tolerance to zero still needs consideration and evidence.[3]

9.8 SUMMARY

Beer is an appreciated natural source of various phytochemicals with a series of nutritional benefits. Most of these benefits are linked to polyphenols. These biologically active plant-derived compounds play a crucial role in the health impacts attributed to beer, especially those related to its potential antitumorigenic effects. These latter are linked mainly to a prenylated flavonoid, xanthohumol.

Evidences of the physiological impact of beer phytochemicals have been presented in-text, with special emphasis on xanthohumol as a potent antiproliferative and anticarcinogenic agent. Based on the reports available, it has been shown to have a chemoprotective against breast, colon, ovarian, and prostate cancer cells. It also has an efficient impact on the initiation, promotion, and progression of tumors as well. By neutralizing reactive species, it ameliorates DNA damage and also inhibits members of the so-called cytochrome P450 enzyme system, which can metabolically activate procarcinogens. Its anti-inflammatory and antiangiogenesis properties are important in inhibiting the development and growth of tumors. Furthermore, xanthohumol was found to reduce the risks of metabolic syndrome and diabetes, it can improve the immune functions, the neurological status, and contribute to higher HDL cholesterol levels. Some of these promising results have been already proven in vitro, mainly in animal experiments, but more in-depth investigation is still needed.

The pharmacological effects of another bioactive compound in beer, 8-prenylnaringenin have also been shortly summarized. Like in the case of xanthohumol, more objective evidence is awaiting regarding the clearly defined positive impact of flavonoids. Also, the global impact of bioactive phytochemicals present in beer must be considered keeping in mind that beer is an alcoholic beverage, delivering a considerable intake of carcinogenic ethanol that makes the overall evaluation of its physiological effects even more difficult.

KEYWORDS

- **antitumorigenic**
- **beer**
- **flavonoids**
- **hops**
- **polyphenols**
- **xanthohumol**

REFERENCES

1. Amawi, H.; Ashby, C. R.; Tiwari, A. K. Cancer Chemoprevention through Dietary Flavonoids: What's Limiting? *Chinese J. Cancer* **2017,** *36* (1), 1–13.
2. Aghamiri, V.; Mirghafourvand, M.; Mohammad-Alizadeh-Charandabi, S.; Nazemiyeh, H. The effect of Hop (*Humulus lupulus* L.) on Early Menopausal Symptoms and Hot Flashes: A Randomized Placebo-Controlled Trial. *Complement. Ther. Clin. Pract.* **2016,** *23*, 130–135.
3. Astrup, A.; Estruch, R. Alcohol and the Global Burden of Disease. *Lancet* **2019,** *393* (10189), E-article ID: 2390.
4. Abotaleb, M.; Samuel, S. M.; Varghese, E.; Varghese, S.; Kubatka, P.; Liskova, A.; Büsselberg, D. Flavonoids in Cancer and Apoptosis. *Cancers* **2019,** *11* (28), 1–39.
5. Belščak-Cvitanović, A.; Durgo, K.; Huđek, A.; Bačun-Družina, V.; Komes, D. Overview of Polyphenols and their Properties. In *Polyphenols: Properties, Recovery, and Applications*; Galanakis, C., Ed.; Woodhead Publishing: London, 2018; pp 3–44.

6. BeMiller, J. N. Nonenzymic Browning and Formation of Acrylamide and Caramel. In *Carbohydrate Chemistry for Food Scientists*; Ball, M. R., Ed.; Woodhead Publishing: London, 2019; pp 351–370.
7. Blagosklonny, M. V. Treatment with Inhibitors of Caspases, that are Substrates of Drug Transporters, Selectively Permits Chemotherapy-Induced Apoptosis in Multidrug-Resistant Cells but Protects Normal Cells. *Leukemia* **2001,** *15* (6), 936–941.
8. Busch, C.; Noor, S.; Leischner, C.; Burkard, M.; Lauer, U. M.; Venturelli, S. Antiproliferative Aktivität gegenüber humanen Krebszellen von Prenylflavonoiden aus dem Hopfen (Activity Against Human Cancer Cells of Hop Prenylflavonoids). *Wiener Medizinische Wochenschrift*, (Vienna Medical Weekly) **2015,** *165* (11–12), 258–261.
9. Buiatti, S. Beer Composition: An Overview. In *Beer in Health and Disease Prevention*; Preedy, V., Ed.; Elsevier Inc.: San Diego, CA, 2009; pp 213–225.
10. Bhattacharya, S.; Virani, S.; Zavro, M.; Haas, G. J. Inhibition of Streptococcus Mutans and Other Oral Streptococci by Hop (*Humulus lupulus* L.) Constituents. *Econ. Bot.* **2003,** *57* (1), 118–125.
11. Buckwold, V. E.; Wilson, R. J. H.; Nalca, A.; Beer, B. B.; Voss, T. G.; Turpin, J. A.; Buckheit, R. W.; Wei, J. Antiviral Activity of Hop Constituents Against a Series of DNA and RNA Viruses. *Antivir. Res.* **2004,** *61* (1), 57–62.
12. Carvalho, D.; Curto, A.; Guido, L. Determination of Phenolic Content in Different Barley Varieties and Corresponding Malts by Liquid Chromatography-Diode Array Detection-Electrospray Ionization Tandem Mass Spectrometry. *Antioxidants* **2015,** *4* (3), 563–576.
13. Casagrande, F.; Darbon, J. M. Effects of Structurally Related Flavonoids on Cell Cycle Progression of Human Melanoma Cells: Regulation of Cyclin-Dependent Kinases CDK2 and CDK1. *Biochem. Pharmacol.* **2001,** *61* (10), 1205–1215.
14. Cancer Research, UK; Worldwide Cancer Statistics, 2016. https://www.cancerresearchuk.org/health-professional/cancer-statistics/worldwide-cancer
15. (accessed on June 22, 2022).
16. Ciriminna, R.; Albanese, L.; Stefano, V. Di.; Delisi, R.; Avellone, G.; Meneguzzo, F.; Pagliaro, M. Beer Produced via Hydrodynamic Cavitation Retains Higher Amounts of Xanthohumol and Other Hops Prenyl Flavonoids. *LWT Food Sci. Technol.* **2017,** *91*, 160–167.
17. Chen, C. Y.; Blumberg, J. B. Flavonoids in Beer and their Potential Benefit on the Risk of Cardiovascular Disease. In *Beer in Health and Diesease Prevention*; Preedy, V. R., Ed.; Academic Press: New York, 2009; pp 831–843.
18. Choi, J. A.; Kim, J. Y.; Lee, J. Y.; Kang, C. M.; Kwon, H. J.; Yoo, Y. D.; Kim, T. W.; Lee, Y. S. Induction of Cell Cycle Arrest and Apoptosis in Human Breast Cancer Cells by Quercetin. *Int. J. Oncol.* **2001,** *19* (4), 837–844.
19. Clarke, D. B.; Barnes, K. A.; Lloyd, A. S. Determination of Unusual Soya and Non-soya Phytoestrogen Sources in Beer, Fish Products and Other Foods. *Food Addit. Contam.* **2004,** *21* (10), 949–962.
20. Costa, R.; Rodrigues, I.; Guardão, L.; Rocha-Rodrigues, S.; Silva, C.; Magalhães, J.; Ferreira-de-Almeida, M.; Negrão, R. Xanthohumol and 8-prenylnaringenin Ameliorate Diabetic-Related Metabolic Dysfunctions in Mice. *J. Nutr. Biochem.* **2017,** *45*, 39–47.
21. Ciulu, M.; Spano, N.; Pilo, M. I.; Sanna, G. Recent Advances in the Analysis of Phenolic Compounds in Unifloral Honeys. *Molecules* **2016,** *21* (4), 1–32.

22. Dvořáková, M.; Hulín, P.; Karabín, M.; Dostálek, P. Determination of Polyphenols in Beer by an Effective Method Based on Solid-Phase Extraction and High-Performance Liquid Chromatography with Diode-Array Detection. *Czech J. Food Sci.* **2007,** *25* (4), 182–188.
23. Dostálek, P.; Karabín, M.; Jelínek, L. Hop Phytochemicals and their Potential Role in Metabolic Syndrome Prevention and Therapy. *Molecules* **2017,** 22 (10), 1761.
24. Dokduang, H.; Yongvanit, P.; Namwat, N. Xanthohumol Inhibits STAT3 Activation Pathway Leading to Growth Suppression and Apoptosis Induction in Human Cholangiocarcinoma Cells. *Oncol. Rep.* **2016,** *35* (4), 2065–2072.
25. Doller, R. A.; Riepl, H. M.; Rose, O.; Frias, C.; Henze, G.; Prokop, A. Desmethylxanthohumol from Hops, Chemistry and Biological Effects. In *Beer in Health and Disease Prevention*; Preedy, V., Ed.; Elsevier Inc.: San Diego, CA, 2008; pp 703–709.
26. Effenberger, K. E.; Johnsen, S. A.; Monroe, D. G.; Spelsberg, T. C.; Westendorf, J. J. Regulation of Osteoblastic Phenotype and Gene Expression by Hop-Derived Phytoestrogens. *J. Steroid Biochemi. Mol. Biol.* **2005,** *96* (5), 387–399.
27. Erkkola, R.; Vervarcke, S.; Vansteelandt, S.; Rompotti, P.; De Keukeleire, D.; Heyerick, A. Randomized, Double-Blind, Placebo-Controlled, Cross-Over Pilot Study on the use of a Standardized Hop Extract to Alleviate Menopausal Discomforts. *Phytomedicine* **2010,** *17* (6), 389–396.
28. Eyres, G.; Dufour, J. P. Hop Essential Oil: Analysis, Chemical Composition and Odor Characteristics. In *Beer in Health and Disease Prevention*; Preedy, V., Ed.; Elsevier Inc.: San Diego, CA, 2008; pp 239–254.
29. Frölich, S.; Schubert, C.; Bienzle, U.; Jenett-Siems, K. *In vitro* Antiplasmodial Activity of Prenylated Chalcone Derivatives of Hops (*Humulus lupulus* L.) and their Interaction with Haemin. *J. Antimicrob. Chemother.* **2005,** *55* (6), 883–887.
30. Gerhäuser, C. Beer Constituents as Potential Cancer Chemopreventive Agents. *Eur. J. Cancer* **2005,** *41* (13), 1941–1954.
31. Gerhäuser, C. Broad Spectrum Antiinfective Potential of Xanthohumol from Hop (*Humulus lupulus* L.) in Comparison with Activities of Other Hop Constituents and Xanthohumol Metabolites. *Mol. Nutr. Food Res.* **2005,** *49* (9), 827–831.
32. Griswold, M. G.; Fullman, N.; Hawley, C.; Arian, N.; Zimsen, S. R. M.; Tymeson, H. D.; Venkateswaran, V.; Tapp, A. D. Alcohol use and Burden for 195 Countries and Territories, 1990–2016: A Systematic Analysis for the Global Burden of Disease Study 2016. *Lancet* **2018,** *392* (10152), 1015–1035.
33. Guo, D.; Zhang, B.; Liu, S.; Jin, M. Xanthohumol Induces Apoptosis via Caspase Activation, Regulation of Bcl-2, and Inhibition of PI3K/ Akt/ mTOR-Kinase in Human Gastric Cancer Cells. *Biomed. Pharmacother.* **2018,** *106*, 1300–1306.
34. Hümpel, M.; Isaksson, P.; Schaefer, O.; Kaufmann, U.; Ciana, P.; Maggi, A.; Schleuning, W. D. Tissue Specificity of 8-prenylnaringenin: Protection from Ovariectomy Induced Bone Loss with Minimal Trophic Effects on the Uterus. *J. Steroid Biochem. Mol. Biol.* **2005,** *97* (3), 299–305.
35. Hirata, H.; Takazumi, K.; Segawa, S.; Okada, Y.; Kobayashi, N.; Shigyo, T.; Chiba, H. Xanthohumol, a Prenylated Chalcone from *Humulus Lupulus* L.; Inhibits Cholesteryl ester Transfer Protein. *Food Chem.* **2012,** *134* (3), 1432–1437.
36. Hirata, H.; Uto-kondo, H.; Ogura, M.; Ayaori, M.; Shiotani, K.; Ota, A.; Tsuchiya, Y.; Ikewaki, K. ScienceDirect Cholesterol Transport. *J. Nutr. Biochem.* **2017,** *47*, 29–34.

37. Heyerick, A.; Vervarcke, S.; Depypere, H.; Bracke, M.; Keukeleire, D. De A First Prospective, Randomized, Double-Blind, Placebo-Controlled Study on the use of a Standardized Hop Extract to Alleviate Menopausal Discomforts. *Maturitas* **2006,** *54* (2), 164–175.
38. International Agency for Research on Cancer (IARC). Agents Classified by the IARC Monographs. *Igarss 2014* **2012,** *1–105* (1), 1–5.
39. International Agency for Research on Cancer (IARC). List of Classifications by Cancer Sites with Sufficient or Limited Evidence in Humans. *IARC* **2018,** *1*, 1–12.
40. Jandera, P. Methods for the HPLC Analysis of Phenolic Compounds and Flavonoids in Beer. In *Beer in Health and Disease Prevention*; Preedy, V., Ed.; Elsevier Inc.: San Diego, CA, 2009; pp 1003–1014.
41. Keukeleire, D. Fundamentals of Beer and Hop Chemistry. *Química Nova* **2000,** *23* (1), 108–112.
42. Karabín, M.; Haimannová, T.; Fialová, K.; Jelínek, L.; Dostálek, P. Preparation of Hop Estrogen-Active Material for Production of Food Supplements. *Molecules* **2021,** *26* (19)
43. Langner, E.; Rzeski, W. Biological Properties of Melanoidins: A Review. *Int. J. Food Prop.* **2014,** *17* (2), 344–353.
44. Luís, C.; Costa, R.; Rodrigues, I.; Castela, Â.; Coelho, P.; Guerreiro, S.; Gomes, J.; Reis, C. Xanthohumol and 8-prenylnaringenin reduce Type 2 Diabetes–Associated Oxidative Stress by Downregulating Galectin-3. *Porto Biomed. J.* **2019,** *4* (1), E-article ID: 23.
45. Lapcík, O.; Hill, M.; Hampl, R.; Wähälä, K.; Adlercreutz, H. Identification of Isoflavonoids in Beer. *Steroids* **1998,** *63* (1), 14–20.
46. Liu, M.; Hansen, P. E.; Wang, G.; Qiu, L.; Dong, J.; Yin, H.; Qian, Z.; Yang, M. Pharmacological Profile of Xanthohumol, a Prenylated Flavonoid from Hops (*Humulus lupulus*). *Molecules* **2015,** *20* (1), 754–779.
47. Luzak, B.; Kassassir, H.; Rój, E.; Stanczyk, L.; Watala, C.; Golanski, J. Xanthohumol from Hop Cones (*Humulus lupulus* L.) Prevents ADP-Induced Platelet Reactivity. *Arch. Physiol. Biochem.* **2017,** *123* (1), 54–60.
48. Legette, L. L.; Moreno Luna, A. Y.; Reed, R. L.; Miranda, C. L.; Bobe, G.; Proteau, R. R.; Stevens, J. F. Xanthohumol Lowers Body Weight and Fasting Plasma Glucose in Obese Male Zucker fa/fa Rats. *Phytochemistry* **2013,** *91*, 236–241.
49. Magalhães, P. J.; Carvalho, D. O.; Cruz, J. M.; Guido, L. F.; Barros, A. A. Fundamentals and Health Benefits of Xanthohumol, a Natural Product Derived from Hops and Beer. *Nat. Prod. Commun.* **2009,** *4* (5), 591–610.
50. Magalhães, P. J.; Guido, L. F.; Cruz, J. M.; Barros, A. A. Analysis of Xanthohumol and Isoxanthohumol in Different Hop Products by Liquid Chromatography-Diode Array Detection-Electrospray Ionization Tandem Mass Spectrometry. *J. Chromatogr. A* **2007,** *1150* (1–2), 295–301.
51. Mahli, A.; Seitz, T.; Freese, K.; Frank, J.; Weiskirchen, R.; Abdel-tawab, M.; Behnam, D.; Hellerbrand, C. Therapeutic Application of Micellar Solubilized Liver Disease. *Cells* **2019,** *8* (4), 1–13.
52. Mikai, R.; Horikawa, H.; Fujikura, Y.; Kawamura, T.; Nemoto, H.; Nikawa, T.; Terao, J. Prevention of Disuse Muscle Atrophy by Dietary Ingestion of 8-prenylnaringenin in Denervated Mice. *PLoS One* **2012,** *7* (9), 1–11.
53. Ming, L. G.; Lv, X.; Ma, X. N.; Ge, B. F.; Zhen, P.; Song, P.; Zhou, J.; Ma, H. P. The Prenyl Group Contributes to Activities of Phytoestrogen 8-prenynaringenin in

Enhancing Bone Formation and Inhibiting Bone Resorption in Vitro. *Endocrinology* **2013,** *154* (3), 1202–1214.
54. Miranda, C. L.; Johnson, L. A.; De Montgolfier, O.; Elias, V. D.; Ullrich, L. S.; Hay, J. J.; Paraiso, I. L.; Choi, J. Non-estrogenic Xanthohumol Derivatives Mitigate Insulin Resistance and Cognitive Impairment in High-Fat Diet-Induced Obese Mice. *Sci. Rep.* **2018,** *8* (1), 1–17.
55. Miranda, C. L.; Stevens, J. F.; Ivanov, V.; McCall, M.; Frei, B.; Deinzer, M. L.; Buhler, D. R. Antioxidant and Prooxidant Actions of Prenylated and Nonprenylated Chalcones and Flavanones In Vitro. *J. Agric. Food Chem.* **2000,** *48* (9), 3876–3884.
56. Mirassay, L.; Varinská, L.; Mojžiš, J. Antiangiogenic Effect of Flavonoids and Chalcones: An Update. *Int. J. Mol. Sci.* **2018,** *19* (1), 27–35.
57. Mitić, S. S.; Paunović, D. D.; Pavlović, A. N.; Tošić, S. B.; Stojković, M. B. Phenolic Profiles and Total Antioxidant Capacity of Marketed Beers in Serbia. *Int. J. Food Prop.* **2014,** *17* (4), 370–377.
58. Miyata, S.; Inoue, J.; Shimizu, M.; Sato, R. Xanthohumol Improves Diet-Induced Obesity and Fatty Liver by Suppressing Sterol Regulatory Element-Binding Protein (SREBP) Activation. *J. Biol. Chem.* **2015,** *290* (33), 20565–20579.
59. Mizobuchi, S.; Sato, Y. A New Flavanone with Antifungal Activity Isolated from Hops. *Agric. Biol. Chem.* **1984,** *48* (11), 2771–2775.
60. Pai, V. C.; Bertoni, A.; Nalin, M.; Sampietro, S.; Zanfa, M.; Sinigaglia, F. The Phytoestrogen 8-prenylnaringenin Inhibits Agonist-Dependent Activation of Human Platelets. *Biochim. Biophys. Acta Gen. Subj.* **2012,** *1820* (11), 1724–1733.
61. Pai, T. V.; Sawant, S. Y.; Ghatak, A. A.; Chaturvedi, P. A.; Gupte, A. M.; Desai, N. S. Characterization of Indian Beers: Chemical Composition and Antioxidant Potential. *J. Food Sci. Technol.* **2015,** *52* (3), 1414–1423.
62. Paper, D. H. Natural Products as Angiogenesis Inhibitors. *Planta Med.* **1998,** *64* (8), 686–695.
63. Pepper, M. S.; Hazel, S. J.; Hümpel, M.; Schleuning, W. D. 8-Prenylnaringenin, a Novel Phytoestrogen, Inhibits Angiogenesis *In Vitro* and *In Vivo*. *J. Cell. Physiol.* **2004,** *199* (1), 98–107.
64. Possemiers, S.; Bolca, S.; Grootaert, C.; Heyerick, A.; Decroos, K.; Dhooge, W.; De Keukeleire, D.; Rabot, S. The Prenylflavonoid Isoxanthohumol from Hops (*Humulus lupulus* L.) is Activated into the Potent Phytoestrogen 8-prenylnaringenin *In Vitro* and in the Human Intestine. *J. Nutr.* **2006,** *136* (7), 1862–1867.
65. Rancán, L.; Paredes, S. D.; García, I.; Muñoz, P.; García, C.; López de Hontanar, G.; de la Fuente, M.; Vara, E. Protective Effect of Xanthohumol Against Age-Related Brain Damage. *J. Nutr. Biochem.* **2017,** *49*, 133–140.
66. Ratna, A.; Mandrekar, P. Alcohol and Cancer: Mechanisms and Therapies. *Biomolecules* **2017,** *7* (3), 1–20.
67. Ren, W.; Qiao, Z.; Wang, H.; Zhu, L.; Zhang, L. Flavonoids: Promising Anticancer Agents. *Med. Res. Rev.* **2003,** *23* (4), 519–534.
68. Sadzuka, Y.; Sugiyama, T.; Sonobe, T. Efficacies of Tea Components on Doxorubicin Induced Antitumor Activity and Reversal of Multidrug Resistance. *Toxicol. Lett.* **2000,** *114* (1–3), 155–162.
69. Saito, K.; Matsuo, Y.; Imafuji, H.; Okubo, T.; Maeda, Y.; Sato, T.; Shamoto, T.; Tsuboi, K. Xanthohumol Inhibits Angiogenesis by Suppressing Nuclear Factor-κB Activation in Pancreatic Cancer. *Cancer Sci.* **2018,** *109* (1), 132–140.

70. Sales, E. K.; Mohaddes, G.; Alipour, M. R. Chalcones as Putative Hepatoprotective Agents: Preclinical Evidence and Molecular Mechanisms. *Pharmacol. Res.* **2018,** *129*, 177–187.
71. Samuels, J. S.; Shashidharamurthy, R.; Rayalam, S. Novel Anti-Obesity Effects of Beer Hops Compound Xanthohumol: Role of AMPK Signaling Pathway. *Nutr. Metab.* **2018,** *15* (1), 1–11.
72. Shen, G.; Huhman, D.; Lei, Z.; Snyder, J.; Sumner, L. W.; Dixon, R. A. Characterization of an Isoflavonoid-Specific Prenyltransferase from *Lupinus albus*. *Plant Physiol.* **2012,** *159* (1), 70–80.
73. Shikata, Y.; Yoshimaru, T.; Komatsu, M.; Katoh, H.; Sato, R.; Kanagaki, S.; Okazaki, Y.; Toyokuni, S. Protein Kinase-A Inhibition Facilitates the Antitumor Activity of Xanthohumol, a Valosin-Containing Protein Inhibitor. *Cancer Sci.* **2017,** *108* (4), 785–794.
74. Sikorska, E. Analysis of Vitamin B12 using Front-Face Intrinsic Beer Fluorescence. *Eur. Food Res. Technol.* **2007,** *225* (1), 43–48.
75. Slawinska-Brych, A.; Król, S. K.; Dmoszynska-Graniczka, M.; Zdzisinska, B.; Stepulak, A.; Gagos, M. Xanthohumol Inhibits Cell Cycle Progression and Proliferation of Larynx Cancer Cells *In Citro*. *Chem. Biol. Interact.* **2015,** *240*, 110–118.
76. Stompor, M.; Uram, Ł.; Podgórski, R. *In Vitro* Effect of 8-prenylnaringenin and Naringenin on Fibroblasts and Glioblastoma Cells-Cellular Accumulation and Cytotoxicity. *Molecules* **2017,** *22* (7), E-article ID: 1092.
77. Štulíková, K.; Karabín, M.; Nešpor, J.; Dostálek, P. Therapeutic Perspectives of 8-prenylnaringenin, a Potent Phytoestrogen From Hops. *Molecules* **2018,** *23* (3), 660–667.
78. Sun, Z.; Zhou, C.; Liu, F.; Zhang, W.; Chen, J.; Pan, Y.; Ma, L.; Liu, Q. Inhibition of Breast Cancer Cell Survival by Xanthohumol via Modulation of the Notch Signaling Pathway In Vivo and In Vitro. *Oncol. Lett.* **2018,** *15* (1), 908–916.
79. Szwajgier, D. Content of Individual Phenolic Acids in Worts and Beers and their Possible Contribution to the Antiradical Activity of Beer. *J. Inst. Brew.* **2009,** *115* (3), 243–252.
80. Tabrez, S.; Priyadarshini, M.; Urooj, M.; Shakil, S.; Ashraf, G. M.; Khan, M. S.; Kamal, M. A.; Alam, Q.; et al. Cancer Chemoprevention by Polyphenols and their Potential Application as Nanomedicine. *J. Environ. Sci. Health C Environ. Carcinog. Ecotoxicol. Rev.* **2013,** *31* (1), 67–98.
81. Theodoratou, E.; Timofeeva, M.; Li, X.; Meng, X.; Ioannidis, J. P. A. Nature, Nurture and Cancer Risks: Genetic and Nutritional Contributions to Cancer. *Annu. Rev. Nutr.* **2017,** *37*, 293–320.
82. Vanhoecke, B. W.; Delporte, F.; Braeckel, E. V. A. V. A. N.; Heyerick, A.; Depypere, H. T.; Nuytinck, M.; Keukeleire, D. D. E.; Bracke, M. E. Safety Study of Oral Tangeretin and Xanthohumol Administration to Laboratory Mice. *In Vivo* **2005,** *19* (1), 103–107.
83. Varela-Rey, M.; Woodhoo, A.; Martinez-Chantor, M. L.; Mato, J. M.; Lu, S. C. Alcohol, DNA Methylation, and Cancer. *Alcohol Res.* **2013,** *35* (1), 25–36.
84. Venè, R.; Benelli, R.; Minghelli, S.; Astigiano, S.; Tosetti, F. Xanthohumol Impairs Human Prostate Cancer Cell Growth and Invasion and Diminishes the Incidence and Progression of Advanced Tumors in TRAMP Mice. *Mol. Med.* **2012,** *18* (1), 1292–1302.
85. Venturelli, S.; Burkard, M.; Biendl, M.; Lauer, U. M.; Frank, J.; Busch, C. Prenylated Chalcones and Flavonoids for the Prevention and Treatment of Cancer. *Nutrition* **2016,** *32* (11–12), 1171–1178.

86. Walden, D.; Kunnimalaiyaan, S.; Sokolowski, K.; Gamblin, T. C.; Kunnimalaiyaan, M. Antiproliferative and Apoptotic Effects of Xanthohumol in Cholangiocarcinoma. *Oncotarget* **2017,** *8* (50), 88069–88078.
87. Waldherr, M.; Mi, M.; Kundi, M.; Nersesyan, A.; Herbacek, I.; Knasmueller, S. Xanthohumol Prevents DNA Damage by Dietary Carcinogens: Results of a Human Intervention Trial. *Cancer Prev. Res.* **2016,** *10* (2), 153–160.
88. Wang, Q.; Ding, Z. H.; Liu, J. K.; Zheng, Y. T. Xanthohumol, A Novel Anti-HIV-1 Agent Purified from *Humulus lupulus*. *Antivir. Res.* **2004,** *64* (3), 189–194.
89. Wattenberg, L. W. Inhibition of Carcinogenesis by Minor Dietary Constituents. *Cancer Res.* **1992,** *52* (7), 2085–2092.
90. Wei, S.; Sun, T.; Du, J. I. E.; Zhang, B.; Xiang, D.; Li, W. Xanthohumol: A Prenylated Flavonoid from Hops, Exerts Anticancer Effects Against Gastric Cancer *In Vitro*. *Oncol. Rep.* **2018,** *40* (6), 3213–3222.
91. Weiskirchen, R.; Mahli, A.; Weiskirchen, S.; Hellerbrand, C. The Hop Constituent Xanthohumol Exhibits Hepatoprotective Effects and Inhibits the Activation of Hepatic Stellate Cells at Different Levels. *Front. Pharmacol.* **2015,** *6*, 140–145.
92. Yang, J. Y.; Della-Fera, M. A.; Rayalam, S.; Baile, C. A. Effect of Xanthohumol and Isoxanthohumol on 3T3-L1 Cell Apoptosis and Adipogenesis. *Apoptosis* **2007,** *12* (11), 1953–1963.
93. Yong, W. K.; Abd Malek, S. N. Xanthohumol Induces Growth Inhibition and Apoptosis in Ca Ski Human Cervical Cancer Cells. *Evid. Based Complement. Alternat. Med.* **2015,** *2015*, 134–139.
94. Yui, K.; Kiyofuji, A.; Osada, K. Effects of Xanthohumol-Rich Extract from the Hop on Fatty Acid Metabolism in Rats Fed a High-Fat Diet. *J. Oleo Sci.* **2014,** *63* (2), 159–168.
95. Zamzow, D. R.; Elias, V.; Legette, L. L.; Choi, J.; Stevens, J. F.; Magnusson, K. R. Xanthohumol Improved Cognitive Flexibility in Young Mice. *Behav. Brain Res.* **2014,** *15* (275), 1–10.
96. Zhao, X.; Jiang, K.; Liang, B.; Huang, X. Anticancer Effect of Xanthohumol Induces Growth Inhibition and Apoptosis of Human Liver Cancer through NF-B/p53 Apoptosis Signaling Pathway. *Oncol. Rep.* **2016,** *35* (2), 669–675.
97. Zi, X.; Feyes, D. K.; Agarwal, R. Anticarcinogenic Effect of a Flavonoid Antioxidant, Silymarin, in Human Breast Cancer Cells MDA-MB 468: Induction of G1 Arrest Through an Increase in Cipl/p21 Concomitant with a Decrease in Kinase Activity of Cyclin- Dependent Kinases and Associated Cycli. *Clin. Cancer Res.* **1998,** *4* (4), 1055–1064.

PART IV

MEASUREMENT TECHNIQUES FOR QUALITY OF NATURAL FOODS

CHAPTER 10

DETECTION METHODS OF ADULTERATION IN NATURAL FOODS

JOHN-LEWIS ZINIA ZAUKUU, ZSANETT BODOR, FLÓRA VITÁLIS, and ZOLTÁN KOVÁCS

ABSTRACT

Food value chains from farm-to-fork entail complex webs of processes that are all risk points for fraudulent activities. These can involve the manipulation of food constituents by addition, removal, or replacement with cheaper but inferior alternatives for selfish gains. This threat to food safety has triggered diverse methods of detection for food quality assurance. This chapter aims to uncover innovative techniques currently being used to adulterate foods produced from plant products and to highlight some effective methods that have been used to combat them. Scientific journals (130) were thoroughly scrutinized and discussed to compare existing and emerging methods of food control. Clearly, food adulteration remains a challenge, but technological advancements have also enabled the development of rapid monitoring systems. Biosensor devices (electronic nose and the electronic tongue) were effective in detecting and predicting advanced adulterations in tea, wine, beer, honey, fruit juices, etc. Spectrometers (handheld and benchtop) have been prominent in the high-accuracy classification and prediction of new adulteration techniques in the mentioned beverages and many other food products including food powders (protein, tamarind, ginger, paprika, tomato, etc.). The adulterants detected ranged from less harmful (inferior food materials) to very hazardous (carcinogenic) ones. The emerging techniques are capable of rapid, high sensitivity, and waste-free analysis with cost efficiency. In combination with chemometrics, they can be applied to

Sustainable and Functional Foods from Plants: Health Impact, Bioactive Compounds, and Production Technologies. Megh R. Goyal, Arijit Nath, and Zoltán Kovács (Eds.)
© 2024 Apple Academic Press, Inc. Co-published with CRC Press (Taylor & Francis)

monitor food adulteration with no risk to the user. Cloud-based systems offer non-invasive online/onsite advantages for handheld spectrometers.

10.1 INTRODUCTION

Plant-based foods have become more sorted in the 21st century due to their purported diverse nutritional advantages. Plant-based foods have been touted to have more antioxidants, minerals, fibers, and carbohydrates than some meat-based foods.[79] Plant-based foods are also reportedly lower in fat and are recommended as a first-choice food for athletes. Plant-based foods have been marked as a major tool for fighting malnutrition.[103] The most common plant-based foods are fruits and vegetables but some parts of certain plants such as the root, stalk, stem, and even leaves are good food sources. Some of these plants provide important culinary benefits and are often referred to as herbs and spices. Herbs and spices have gained fame for their low cost and have been used to fortify foods throughout history as preservatives, flavors, and therapeutic agents.

In spite of the relatively low cost of plant-based foods, some greedy people along the food chain selfishly manipulate the content of these foods for selfish gains. International standards on food properties are crucial tools for enforcing trustworthy trading practices and food safety. Standards generally define the expected quality of a product or its specifications with the aim of harmonizing product trade standards to prevent food manipulation. Product quality criteria can span from the core characteristics of a product or group of products.

Many conventional methods (such as Soxhlet extraction method, sodium dodecyl sulfate–polyacrylamide gel electrophoresis [SDS-page]), Dumas method, chromatography, etc.) are currently being used to monitor the trade and production chain of all foods including plant-based foods. For instance, different types of chromatographic techniques have been used to determine food composition and detect foreign materials. High-performance liquid chromatography (HPLC) has been used to analyze the composition of honey from different origins and detect honey fraud.[3,22,60,128] It was used to determine mycotoxins in cereals.[43,51] Liquid chromatography (LC) has also been used to evaluate different quality parameters in several foods.[10,11,17,80] The use of chromatography however involves the use of reagents which are expensive. The process also requires high technical know-how and can be slow in comparison with some emerging advanced methods. Dumas method has also reportedly been used to determine adulteration of protein powders[30] but is less sensitive to innovative trends of adulteration that can tamper with the nitrogen content of the investigated food. Food adulteration remains a challenge that must be tackled.

Advanced techniques, such as spectroscopy and biosensor devices including the electronic nose and electronic tongue are increasingly gaining attention for their impact on food quality. Spectroscopy involves an interaction (absorption or reflection) between electromagnetic waves and the sample. Electronic tongues and electronic noses contain man-made sensors capable of detecting certain components in food with higher sensitivity than the human senses.

This chapter aims to review and discuss new trends in adulterants and the progress of these instruments in detecting such adulterations.

10.2 FOOD ADULTERATION

Food adulteration can generally refer to any food product that has any of its constituents manipulated by removal, addition, substitution with inferior counterparts, or alliterating the production methods with an aim to deceive consumers for selfish gains. Addition or substitution in food fraud often involves the use of adulterants. An adulterant in a brief description is any material or substance which should not be contained in food or not above its legal limits.[1] Although food adulteration can be unintentional, economically motivated food adulteration or intentional food adulteration is the most dominant in the food industry and it is done for financial gains,[33] at the detriment of the consumer. Unequivocally, illiteracy contributes to the high level of prevalence of food adulteration, but this can be controlled by strict regulations from the appropriate regulatory authorities.

The very essence of food adulteration is that it violates food security. Food security can be defined as a state where everybody has access to safe, quality, nutritious, and affordable food.[104] Food adulteration is also a major hurdle in achieving the United Nations Development Programme (UNDP) sustainable development goal number 2 with a focus on "Zero Hunger" and safe and quality foods.[113] Food adulteration breaks consumer trust in the food supply chain causing doubt in the food regulatory system. Some food adulterants present very serious risks to the health of consumers and can even lead to death. It is therefore crucial that quality assurance of food becomes a priority. Safety of food encompasses an absence of different forms of hazards in food that can cause harm upon consumption.[112] These can include acute, semi-acute, and chronic hazards. Food quality on the other hand describes the attributes of a product that, projects the perceivable characteristics of the product by consumers.[74] This can include discoloration, off-flavors, and odors as a result of spoilage or changes in physical characteristics of the food such as texture and appearance (color) as a result of environmental conditions.

Differentiating between safe and unsafe food is sometimes subjective in the food industry but they are both, influenced by food adulteration. As required by the "Food Law," all foods should be devoid of unacceptable materials and should not be centered on deceiving the consumer.

10.3 FOOD ADULTERATION IN NATURAL FOODS AND THEIR PRODUCTS

Plant-based food products are regarded as high-value products that are required to have the best organoleptic properties without foreign or hazardous materials. However, quality assessment of some herbs by food regulatory authorities in Europe for a 5-year duration showed that many plant-based foods contained chemical hazards, especially spices, and condiments. There were 749 reported cases of chemical hazards in spices and condiments which were all related to food adulteration.[98] In November 2019 alone, arrests were made in India for the adulteration of spices with bran and food color[105] and the adulteration of olive oil with inferior grades.[120] The addition of bran to spices can reduce or eliminate the desired sensory qualities of the spice. Olive oil is also considered as a good source of monosaturated fat and antioxidants which, provide numerous health benefits to consumers.[108]

Adulterating it with lower grade oils can diminish the concentration and viability of these compounds in the oil. In Pakistan, arrests were made in November 2019 for the adulteration of turmeric powder with lead.[68] Turmeric powder is an important spice credited with anti-inflammatory benefits upon consumption.[15] Lead is however considered a potent carcinogen.[107] Carcinogens are substances that are capable of triggering cellular changes and can lead to the formation of tumors or cancerous cells.[31]

Table 10.1 shows trends of adulteration in cereals, flours, spices, fruits, and vegetables. Green vegetables such as green peas and chilies have been reportedly adulterated with very dangerous compounds such as malachite green that present severe health complications to consumers. Fresh vegetables and fruits which are major diets all around the world are often found to contain adulterants as a result of inappropriate fertilizers and pesticides by ambitious farmers who are eager to keep an increased yield at all costs by using these chemicals above their recommended or acceptable limit.[12] These chemicals have been reported as causes of many chronic diseases in humans. Dals and pulses provide affordable, low-fat, protein, minerals, vitamins, and fiber in foods but have also been recently targeted for adulteration.[53] They are mostly legumes and are often recommended as complementary sources of proteins, especially for vegans. Cereals have been adulterated with fungi, body parts of dead insects, and by mixing infected or low-grade cereals with high-quality ones (Table 10.1). Cereals and their products have been acknowledged to play roles in fighting malnutrition and are regarded as staples in many parts of the world.[5] Examples include rice, sorghum, millet, wheat, etc., and they contain important nutrients such as vitamins, minerals, carbohydrates, fats, oils, and proteins.[40]

TABLE 10.1 Some Adulterants in Cereals, Flours, Spices, Fruits, and Vegetables: Self-developed with Concepts from Ref. [5,12,53].

Food item(s)	Adulterant(s)/Hazard(s)	Reason(s)	Negative effect(s)
Cereals	Ergot, insects, soil, infested or low-grade cereals; husks and stems of plants	Increased volume	Stomach disorders Devalues food
Dals and pulses	Sodium bicarbonate, metanil yellow, and coal tar	Imitate freshness for increased patronization	Carcinogenic Throat and stomach irritations
Flours	Chalk, starch powder, milled animal bones, washing soda, and husks	Increased volume Enhance marketable features such as color and appearance	Diarrhea Allergies Eye problems
Green peas, green chilli, and green vegetables	Malachite green oxytocin, saccharin, wax, calcium carbide, copper sulfate, insecticides, and pesticide residues such as arsenic, cadmium, lead, etc.	Enhance marketable features such as color and appearance Mask spoilage	Carcinogenic Tumors Throat and eye infections Reduces food quality Causes allergies Vomiting
Spices and tea	Sawdust, Sudan dyes, coloring agents, fluoride, milled avocado seeds, and Sibutramine	Bright marketable color Increased volume Mask spoilage	Fluoride concentrations above recommended limits can lead to fluorosis Sudan dyes in spices such as paprika can trigger carcinoma and tumors

Spices are reported to constitute the five most manipulated foods in the world due to their high cost as a result of preprocessing.[32] Sudan dyes are cheap lipophilic industrial chemicals often used in oils, textiles, waxes, cosmetics, and polish but have recently been added to spices for increased color.[37] The dyes have been rated as class one carcinogens with mutagenic potentials in humans.[48]

Naturally, most of the adulterants in Table 10.1 can endure harsh processing methods[49] and travel along the food chain until consumption. Therefore, controlling adulteration in raw materials is crucial. Nonetheless, processed foods have also had their fair share of reported adulteration with even more complex substances and innovative techniques. Table 10.2 shows foods that have been processed with plant raw materials and their suspected adulterants.

TABLE 10.2 Adulteration Trends in Food Products with Pant-Based Raw Materials: Self-developed with Concepts from Ref. [13,34,49,56,77,95,130].

Food product	Adulterant(s)/Hazard(s)	Reason(s)	Negative effect(s)
Coffee	Sawdust, husk, chicory, cereal grains, cocoa seeds, mixing with cheaper beans, and roasted corn	Increased volume and patronization	Devalues product nutrition and sensory quality. Health and nutrition risk
Fruit juice	Additives, colorants, water, sweeteners, and using fruits of inferior quality or unripe fruits	Imitate pure juice extracts for increased patronization Additives are used to manipulate the shelf-life of the juice	Additives can cause serious stomach disorders Sweeteners and water devalue the product
Grape seed extracts	Seed, bark, stalk, and skin of crude herbs	Imitate the appearance of pure grape seed extracts for increased sale	Devalues product and presents a health risk
Honey	Sugar syrups, industrial sugars, sulfite–ammonia caramel (E150d)	Imitate high-quality honey For higher patronization Darkening	Devalues product nutrition Exposure to increased sugar levels
Oil	Rancid oils, low-quality spices, production from spoiled or decayed raw materials, and mixing low-grade oils	Increased volume	Devalues the products and can lead to infections and allergies
Protein powders	Melamine, amino acids, bran, and husk	Increased volume and patronization	Health and nutrition risk
Soy sauce	Brine	Increased volume and patronization	Health and nutrition risk
Tomato paste concentrate	Starch, flour, paprika seed powder, salt, and sugar	Increased volume high patronization	Devalues product quality with nutritional risk
Wine	Glycerol, grape concentrate, methanol, sugars and sweeteners, food colorants, and water	Imitate high-quality wine for quicker sales	Devalues the wine and presents health implication

From Table 10.2, the primary motive of food adulterants is a higher throughput of the product irrespective of its quality or consequences. It is certain that food adulteration is on the rise and even new techniques and adulterants are suspected to be used. Immediate actions and control methods are therefore required to monitor this trend and assure food and consumer safety.

10.4 FOOD CONTROL

Food control is described as "the obligatory rules of enforcement by national or local regulatory governing boards to bolster consumer protection and certify that all foods during production, handling, storage, processing, and distribution are safe, free of hazards, and fit for human consumption; conform to safety and quality requirements; and are honestly and accurately labeled as prescribed by law.[122]"

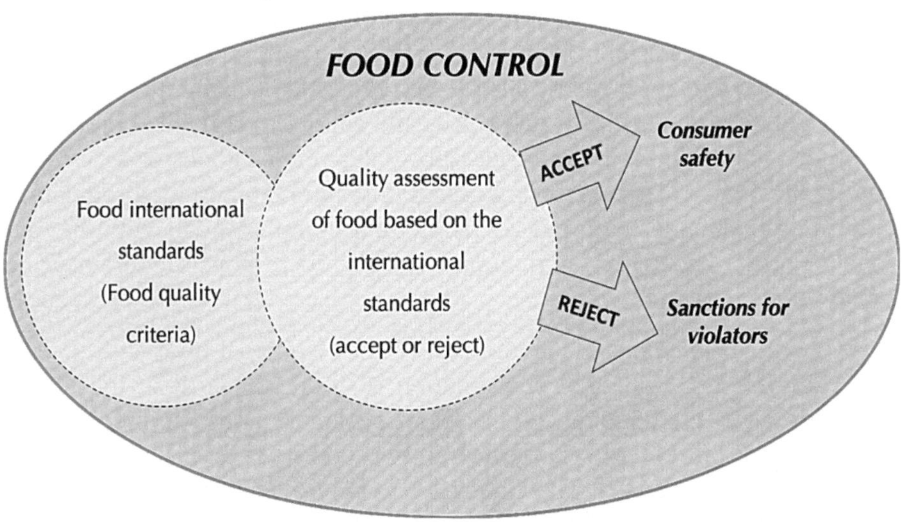

FIGURE 10.1 Food regulatory system for safe and nutritious foods. Source: Self-developed with concepts from *Codex Alimentarius*,[4] ISO directives,[50] and Schaarschmidt.[98]

Food control is important because adulterated food products are often designed to imitate their counterparts such that consumers are unable to detect the differences. Food control and detection of adulterants across the food matrix is of paramount essence to guarantee food quality and consumer safety (Fig. 10.1). Food regulatory authorities are in charge of verifying and confirming the conformations and compliances of accepted standards.

Effective and strict evaluation protocols along the supply chain will expose critical points of attention for controlling food adulteration. Critical points offer opportunities to mitigate the occurrence of these actions and can be implemented to assure optimum human nutrition and lower the health risks associated with inferior food products.

10.5 FOOD REGULATIONS

Food standards and trade have a synergistic effect in guaranteeing quality and safe foods enough to meet the global growing population. For international marketing and trading in the food markets, producers are required to make sure that their products meet the internationally accepted standards as well as, the standards of the local trading markets where they intend to trade. The International Electrotechnical Commission (IEC) and the International Organization for Standardization (ISO) define a standard as a "document that has been agreed upon from a collective view of experts and has been recommended and approved by a legal entity to establish guidelines, requirements, and governing rules for practices and actions in a given context with the objective of high value and trustworthy delivery of goods or services in that given context.[50]" In a growing global population, many food regulatory authorities exist and sometimes cooperate to accomplish goals with international benefits. The major international key players responsible for ensuring global food safety are:

- *Codex Alimentarius* Commission (CAC) is a legal body tasked with the intergovernmental coordination of food standards at the global level for international food safety. Its primary aim is to mediate and roll out fair but strict programs in the food trade to ensure consumer health and safety.[122] The *Codex Alimentarius* Commission is credited for the international harmonization of food quality and safety requirements through the specific regulations of diverse food commodities through uncompromising stipulations and controls that include chemicals and all hazards in foods. The CAC is jointly headed by the FAO and WHO.
- The Food and Agricultural Organization (FAO) is an international body responsible for all aspects of food safety across the food value chain. Mutual cooperation with other recognizable authorities such as

the World Health Organization (WHO) provides a holistic approach in the form of simultaneous objectives, overseeing a series of challenges in order to ensure safe foods for consumers. Its guidelines strengthen international food trade through product quality.[29]

- World Health Organization (WHO) is an international body with the aim of reducing the health and social burden of foodborne diseases through the surveillance of foodborne diseases, better risk assessment, risk communication, new technologies, capacity building, and international cooperation.[119]

The development of strict and reliable food standards is an important criterion for a strong food control and monitoring system. Many countries, especially developing countries do not have very effective standard food legislation systems. This greatly influences food control activities carried out in the country and can affect the international trade reputation of the country. For an optimized food monitoring system, it is important that the appropriate stakeholders use the Codex standards and food safety formulations as guidelines or references. Practical concepts, requirements, experiences, and lessons from other countries should also be taken into consideration. The food adulteration risks are being discovered from diverse backgrounds across the food supply chain in all countries and in both local and imported foods. For food scientists, the most important aspect of food safety is product monitoring and authentication. The methods used in this process of integrity testing are very important as they provide the main support for the standards.

10.6 QUALITY CONTROL METHODS AND INSTRUMENTS

Several techniques have been used to authenticate diverse food types and detect hazards such as synthetic dyes in food. From Table 10.3 chromatography is the most commonly used thus far in detecting adulteration in different foods. It is sometimes used in combination with scientific methods for optimum results. Adulteration of honey with different sugar syrups has been detected with high liquid performance chromatography and mass spectrometry (HPLC-MS), gas chromatography (GC), and HPLC-thin liquid chromatography (HPLC-TLC). Adulterants and hazards in paprika and other spices have also been investigated with HPLC-MS and other equally effective methods. Generally, coffee, milk powder, wine, tea, coconut oil, mastic oil, cochineals, grape seed extracts, wheat, wheat flour, and bread produced from wheat flour have been analyzed for adulterants using the methods in

Table 10.3. Adulterants such as amino acids, sugar syrups, Sudan dyes (I, II, III, and IV), husk, bran, animal fats, coloring agents, sibutramine, sugar, and *Triticum aestivum* have been identified in plant-based foods and their products.

TABLE 10.3 Prominent Quality Control Methods in Use for Plant-Based Food Qualities and Their Weaknesses: Self-developed with Concepts from Ref. [6,26,28,61,65,96,99,123].

Method	Study	Gaps
GC	• Authentication of wine [26] • Honey adulteration with high fructose syrups from inulin [96] • Determination of caffeine in teas, coffees, and eight beverages [99]	• Limited to volatile sample • Injection of the gaseous sample requires high technicality
GC-MS	• Detection of adulteration in mastic oil products [89] • Assessing coconut oil for spiking with animal fats [124] • Mixing processed powdered coffee with processed powdered barley [83]	• Requires expensive extraction and cleanup protocols • Not suitable for nonvolatile samples or samples with high polarity, or heat sensitivity • Chemical transformation through derivation may sometimes be necessary to stimulate heat sensitivities or volatility of the sample
HPLC–isotopic analysis	• Detect red wine adulterated with sugar and water [41]	• Isotopic analysis requires an enriched isotope spike for the analyte of interest which is not always available or sometimes at a very high cost • The process is time-consuming
HPLC-GPC	• Determination of seven illegal dyes in Egyptian spices [28]	• Expensive: requires reagents, waste management, and costly maintenance • Time consuming
HPLC-MS	• Detect adulteration of skimmed milk powder with plant proteins in [65] • Authentication of Greek unifloral kinds of honey [61] • Prediction of Sudan dyes (I, II, II, and IV) in commercial spices [6] • Determination of illegal adulterated aphrodisiac chemical ingredients in health foods [123]	• Expensive: requires reagents, waste management, and • Costly maintenance and difficult cleanup procedures • Slow: high retention time and low sample number per analysis • Sensitivity issues

TABLE 10.3 *(Continued)*

Method	Study	Gaps
HPLC-TLC	• Authentication of honey [69] • Detection of sildenafil adulterated with herbal products [75] • Quantification of sibutramine in spices [92]	• Expensive: requires reagents, waste management, and costly maintenance • Time consuming
LC	• Detection of color adulteration in cochineals [39]	• Expensive, moderate throughput, and requires high technical know-how
MS	• Detection of pesticides in wine [35]	
LC-MS	• Detection of grape seed extracts adulterated with peanut skin [116]	• QuEChERS method was used for extraction and involves the use of expensive chemicals
MIP-TLC	• Determination of Sudan I in paprika powder [37]	• Requires binding assays which, are expensive and time consuming
PCR	• Determine safflower (*Carthamus tinctorius*) adulteration in saffron [115] • Quantification of adulteration in wheat-based composite bread [126] • Detection of honey adulterated with rice molasses [106] • Detect adulteration of wheat flour with *Triticum aestivum* before pasta production [14]	• PCR is prone to errors and can lead to mutations in the fragment that is generated
UPLC-QToF MS	• Discrimination of honey of various floral and geographical origins [52]	• UPLC is characterized by higher back pressures that reduce column life

GC, gas chromatography; GPC, gel permeation chromatography; HPLC, high-performance liquid chromatography; LC, liquid chromatography; MIP, molecularly imprinted polymers; MS, mass spectrometry; PCR, polymerase chain reaction; TLC, thin layer chromatography; UPLC-QToF MS, ultraperformance liquid chromatography-quadrupole time of flight.

Although the methods listed in Table 10.3 have proven to be very capable of detecting the mentioned adulterants, they also have some weaknesses. For instance, GC is limited to volatile compound analysis, UPLC is characterized by higher back pressures that reduce column life, isotopic analysis requires an enriched isotope spike for the analyte of interest which is not always available or sometimes at a very high cost, and requires high technicalities. Although the polymerase chain has high specificity, the specificity is often dependent on how long the targeted deoxyribonucleic acid sequence appears to be copying of the target sequence may not be easy or even if so, may not be very accurate if the sequence of the target is lengthy.[115]

GC-MS is generally not recommended for analytes having low volatility, polarity, or heat resistance unless the analyte is somehow transformed through methods such as derivatization resulting in sample stimulation for optimum results.[90] Generally, most of the methods are expensive due to the need for reagents requiring waste management techniques and costly cleanup techniques. They also require high technicalities and can sometimes be slow. Some of the methods also have low sensitivities that can be challenging for future applications as food adulteration is an evolving concept and novel methods of adulteration continue to be reported day in and day out. With an expanding global market, it is necessary to explore other advanced techniques with equal or better capabilities than the methods in Table 10.3 while considering cost, technicalities, and sensitivities.

10.7 EMERGING INSTRUMENTS

Many advanced instruments have been tested for their capabilities in food control vis-à-vis their efficiency and cost. These include machine vision, ultrasound, sensor-based instruments (electronic tongue, electronic nose, electronic eye, etc.), and spectroscopy. This chapter presents some of the applications of these instruments for plant-based food monitoring with a focus on the electronic tongue, electronic nose, and spectroscopy due to their unique advantages (Table 10.4).

TABLE 10.4 Advantages of the E-Tongue, E-Nose, and Spectroscopy.

Criterion	E-tongue	E-nose	Spectroscopy
Affordable	Yes	Yes	Yes
Bulky	No	No	No
Easy to install and use	Yes	Yes	Yes
High sensitivity	Yes	Yes	Yes
Laborious	No	No	No
Rapid	Yes	Yes	Yes
Requires regents	No	No	No
Sophisticated	No	No	No

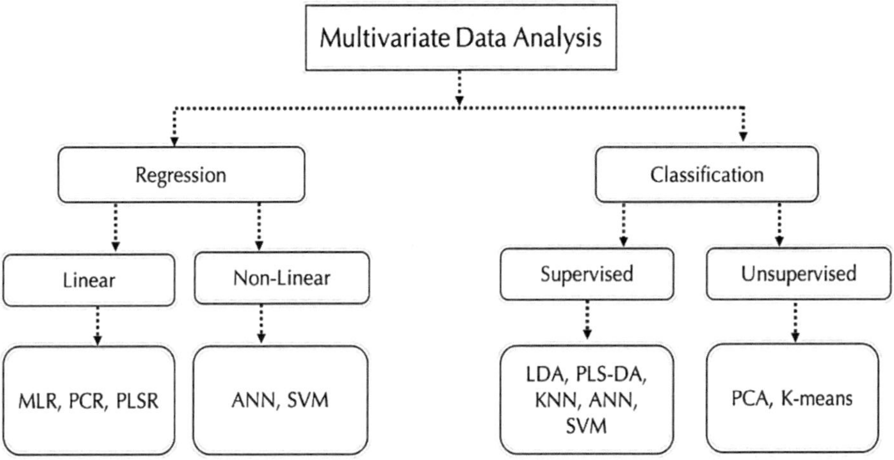

FIGURE 10.2 Multivariate methods associated with using the e-tongue, e-nose, and spectroscopy.

Adopting the e-tongue, e-nose, and spectroscopy for monitoring plant-based foods encompasses multiple variables so multivariate analysis is a necessary option. Multivariate analysis refers to the observation and study of more than one outcome variable with pattern recognition statistical methods which can be supervised or unsupervised.[70] The methods can be used to visualize and observe patterns in the data (principal component analysis) or build supervised or unsupervised models for the variables of interest in the dataset. Depending on the expertise of the researcher and the type of experiment, the methods shown in Figure 10.2 can be used. The advantages and disadvantages of the different types of methods have been explained by Granato et al.[42]

Food processing is a complex stage along the food value chain that is most susceptible to fraudulent malpractices due to the major transformation of foods from raw materials to finished products. Many forms of adulteration may occur, but they are all mostly aimed at disguising the appearance of the food, particularly the color and texture. This is because color and texture are the primary attributes of food that attract consumers. However, manipulating the content of the food also influences the chemical composition which, has a direct relation to its constituents and nutrient composition making these instruments a promising option for quality control.

10.7.1 THE ELECTRONIC TONGUE

The electronic tongue (e-tongue) is a scientific device made up of man-made sensors that are capable of qualitative and quantitative assessment of samples by fingerprinting the compositional patterns of the sample.[121] The sensor array of e-tongues can be said to be equivalent to the human tongue receptors except they are made of either biological components or chemically constituted. Depending on the mode of operation, the potentiometric and voltametric e-tongues have been reported as the most common types of e-tongues. The potentiometric e-tongue operates on the direct application of the Nernst equation and the changes in the electrical potentials of non-polarized electrodes.[71]

In the voltametric e-tongue, the potential is applied on the working electrode, followed by the measurement of the resulting current between the working electrode and the reference electrode. Taste is a subjective attribute due to the sensitivity differences in the human tongue of different consumers. Advanced instruments such as the e-tongue, however, provide reliable non-subjective analysis of taste components in foods as a result of their compositional differences. The instruments require no reagents and they are easy to use (Table 10.4). Taste is the most important sensory parameter for food acceptance and repeated consumption. The components contribute immensely to the food taste. Manipulation of these constituents therefore alters the general qualities of the food including taste. E-tongue has been widely applied in all sectors of the food industry for both liquid and solid food.

The device has been tested for monitoring the characteristics of fruits and many processed plant-based foods based on their taste patterns. The instrument has been applied for the geographical authentication of monofloral honey,[117] Moroccan honey,[25] Hungarian honey,[57] and recently, Romanian honey.[85] Table 10.5 shows the type of e-tongues used to control adulteration in diverse foods and the raw materials of interest. From Table 10.5, e-tongue has been used to detect husks and sticks in coffee. Husk and sticks in coffee do not only project deceptive sensory qualities butalso potent physical hazards that cause throat and stomach irritation in consumers. Adulteration of lower-grade wines with grape concentrate and sucrose has also been reported with the e-tongue with high predictions. The addition of sweeteners in wine is forbidden in some types of wine such as the Tokaj botrytized wines in Hungary. The instrument has also been used to monitor the addition of sweeteners in fruit juices and sugar syrups in honey. The addition of sugars and sweeteners to food products without proper labeling also puts consumers at risk for elevated blood sugar levels more than expected. Ethanolic isomerized hop extract could also be discriminated with the e-tongue for optimum beer quality. Sunflower oil and rancid or winey–vinegary oil have been detected in extra virgin oil using the electronic tongue. Sunflower oil reduces the quality of the extra virgin olive oil. The rancid or winey–vinegary oil affects the shelf-life of the extra virgin olive oil. The e-tongue proved useful in tracking new forms of adulteration in processed foods with plant-based raw materials.

TABLE 10.5 Application of the Electronic Tongue for Monitoring of Plant-Based Foods: Self-developed with Concepts from Ref. [19,24,46,47,54,84,101,131].

Food item	Objective	Type of e-tongue
Beer	Detect ethanolic isomerized hop extract	Potentiometric e-tongue [91]
Coffee	Detect husk, sticks in coffee	Voltametric e-tongue [19]
Cumin	Detect adulterant: Coriander	Potentiometric e-tongue [54]
Extra virgin oil	• Detect adulterant: Sunflower oil • Detect adulterant: Rancid or winey–vinegary oil	• Potentiometric and voltametric e-tongue [7,73] • Potentiometric e-tongue [84]
Fruit juice prepared from orange, apple, pineapple, passion fruit and mango	Predict sucrose in fruit juice	Potentiometric e-tongue [24,46]
Honeydew honey	Detect adulterants: glucose, fructose, inverted sugar, hydrolyzed inulin syrup, and malt wort	Amperometric e-tongue [84]
Mandarin fruit	Predict freshness	Potentiometric e-tongue [94]
Rice	Detect unwanted products of yeast fermentation	Monitor the quality of Korean rice [54]
Sunflower, heather, and orange blossom honey	Detect adulterant: sugar syrups	Voltametric e-tongue [97]
Tomato juice	Detect adulteration of tomato juice with overripe tomato juice	Potentiometric e-tongue [47]
Wine	Predict grape concentrate and sucrose in wine	Potentiometric e-tongue [131]
	Predict lead, cadmium, and iron in wine	Potentiometric e-tongue [101]

Much has not been done with the e-tongue for non-processed plant-based foods (Table 10.5), this may be due to the cumbersome extraction processes involved with the transformation of the food from solid to liquid samples. Studies were only performed on mandarin fruit, rice, and cumin.

10.7.2 ELECTRONIC NOSE

The electronic nose (e-nose) may be referred to as an analytical instrument equipped with electrochemical sensors and artificial algorithm systems that render it capable of sensing gases or volatile compounds. It has been an effective tool for product discrimination, authentication, and quality assurance in general.[110] E-nose systems are characterized by the consistent production of adaptable sensor types for optimum results. Most e-nose systems are equipped with sensing structures for the detection and characterization of gases. Volatile compounds from an analyte can lead to changes in the sensing structures caused by a resistance that results in an elevation or reduction of the charges with respect to the volatile compounds emitted by the analyte.[111] The device is basically designed to determine and discriminate analytes using their odor signals.[110] The smell is an essential parameter for spoilage in the fruit and vegetable industry. Spoilage of food can be a result of uncontrolled circumstances, environmental factors, and microbial activities from the time of harvesting through storage.[88]

During food degradation, the main ingredients of the food will be transformed through enzymatic reactions as a result of microbial presence that often led to the emission of volatile compounds. The proliferation of microbes is a valuable indicator of food freshness and safety because their presence can influence the aroma profile and chemical constituents, a change in appearances such as sliminess or softness. It is therefore important that more reliable methods be used as alternatives or complementary methods for the rather cumbersome ones such as the classical descriptive sensory analysis characterized by panelist olfactory fatigue and subjective responses. Specific volatile compounds can associate with some of these appearances and with the e-nose, it is possible to make correlations that help understand the nature of the microbes present. E-nose provides an advantage of high sensitivity and rapid food product evaluations (Table 10.4) and is a potent tool for the quality assurance of food.[38] It has been used to classify Slovak wines,[59] rice,[87] edible oils,[76] green tea.[129]

A wide group of e-nose gas sensors exist but each is dependent on the objectives of the user and the analytical sample in mind. From Table 10.6, e-noses with different numbers of gas sensors have been used to investigate spoilage in both processed and non-processed plant-based foods. It has been used for the authentication of wine, tomato juice spirits (alcohol), and even microbial assessment of soups, fresh-cut bell pepper, and broccoli. E-nose has been used to predict the freshness of cherry fruit[21] and detect coumarin adulteration in green tea and the adulteration of argan oil with sunflower oil.

TABLE 10.6 Application of the Electronic Nose for Monitoring of Plant-Based Foods: Self-developed Concepts from Ref. [21,27,36,44,47,66,67,127].

Food item	Objective	Type of e-nose
Argan	Sunflower oil	E-nose with five metal oxide gas sensors [13]
Broccoli	Detect *Staphylococcus*, *Salmonella*, and *Shigella* in broccoli	E-nose with six chemiresistive type sensors [27]
Cherry tomato juice	Detect the inclusion of overripe tomato juice	E-nose with 10 metal oxide gas sensors [47]
Chinese spirit	Unwanted flavors	E-nose with 10 metal oxide gas sensors [67]
Fresh-cut green bell pepper (*Capsicum annuum* var. *grossum*)	Predict spoilage bacteria	E-nose with 14 metal oxide gas sensors [21]
Green tea	Detect coumarin in green tea	E-nose with 10 metal oxide gas sensors [127]
Mixed vegetable soup	Predict Enterobacteriaceae in vegetable soup	E-nose with five metal oxide gas sensors [36]
Peach	Detect early freshness as a result of chemicals	E-nose with customized sensors [44]
Wine	Authentication of wines' geographical origin	E-nose with surface acoustic wave (SAW) sensors [66]

10.7.3 SPECTROSCOPY

Spectroscopy is a scientific approach operating on the interactions of analytes in the electromagnetic spectrum of light to give an overview of the compositional structure of the compounds present. Spectrometers are the instruments used and they are majorly composed of a light source, a tool for decomposition of polychromatic light, and a detector.[55] Light can be of different wavelengths, which are represented by the electromagnetic spectrum. In most benchtop spectrometers, grating monochromators or Michelson interferometers are used to decompose light when measuring the full near-infrared spectrum. Instruments may be used in transmittance or reflectance mode.[8] Depending on the energy source and the detectors used,[86] conventional infrared instruments usually operate within defined spectral ranges: in the near-infrared (NIRS), mid-infrared (MIRS), or far-infrared (FIRS) regions (Fig. 10.3). Instruments measuring the MIRS or FIRS regions of the electromagnetic spectrum are mostly used in analytical chemistry where fundamental vibrations of molecules are measured. Depending on the number of bands, they could either be multispectral (3–10 nm) or hyperspectral (10–20 nm). In this chapter, the focus will be on near-infrared spectroscopy (NIRS), Fourier transform near-infrared spectroscopy (FT-NIR), Raman spectroscopy, Ultraviolet–visible spectroscopy (UV-VIS), and hyperspectral imaging.

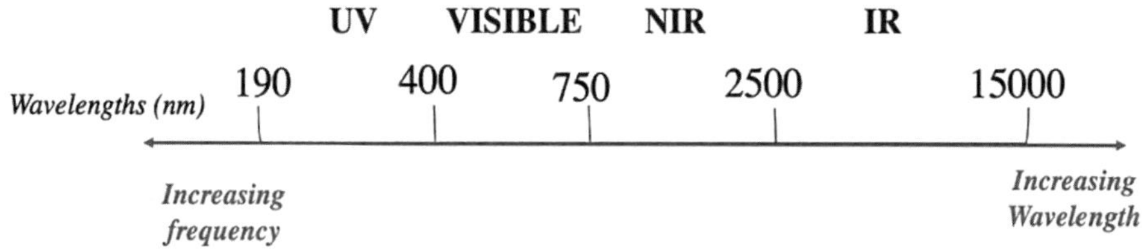

FIGURE 10.3 Electromagnetic spectrum.

The major advantage of spectroscopy is that, it is non-invasive and rapid, therefore, making it suitable for quality control of fruits vegetables, and their products (Table 10.4). It has been used to predict the sugariness and hardness of melons,[109] the non-destructive assessment of fruit and vegetable quality[78] and to investigate the chemical constituent of calçot.[100] The pulp of guava and passion fruit were also assessed for their chemical values with NIRS.[2] There are detailed reports on the application of NIRS to determine diverse coffee quality parameters.[9] From Table 10.7, many advanced forms of food adulteration that use even more complex forms of adulterants have emerged. Some of these adulterants (urea, oxidative compounds, sticks, Sudan dyes, etc.) present serious health consequences to consumers. Nonetheless, NIRS has been effective in tracking these adulterants in some foods.

TABLE 10.7 Application of Spectroscopy for Monitoring of Plant-Based Foods: Self-developed with Concepts from Ref. [20,23,58,62,63,64,93,102,131].

Food item	Adulterant	Type of spectroscopy
Arabica coffee	Peels, sticks, and corn	NIRS [20]
Black pepper powder	Corn flour and starch	NIRS [63]
	Papaya seed	Hyperspectral imaging [82]
Chili pepper powder	Sudan dyes	NIRS and Raman spectroscopy [6,45,72]
Chinese Glutinous Rice Flour (GRF)	Wheat flour, non-GRF, and an illegal food additive, talcum powder	NIRS [125]
Cocoa powder	Carob flour	NIRS [93]
Cumin powder	Corn flour and starch	NIRS [63]
Free-dried Açai	Cornstarch, beet pulp, cassava starch, maltodextrin, and grape juice	NIRS [62]
Herbs: (a) Saffron (*stigmas of Crocus sativus* L.); and (b) safflower	Petals of safflower	2D-NIRS [18]
(flower of *C. tinctorius* L)		
Indonesian black pepper powder	Black pepper husk from India and Vietnam, spent material (two from India), pinheads (two from India), papaya seeds, and chili powder	FT-NIR [118]
Instant green tea	Caffeine	FT-NIR [102]
Irish honey	Beet invert syrup and high fructose corn syrup	NIRS [58]
Oregano (spice)	Olive leaves, myrtle leaves, leaves, sumac leaves, cistus leaves, and hazelnut leaves	FT-IR [10]
Purple sweet potato flour	White sweet potato flour	NIRS [23]
Sunflower, canola, olive vegetable oils	Oxidative compounds	Raman spectroscopy [114]
Tequila spirit	Low-grade spirit	UV–VIS spectroscopy [16]
Whey protein powder	Urea, histidine, and taurine	FT-NIR [64]
Wine	Predict grape concentrate and sucrose in wine	Handheld and bench top NIRS [131]

10.8 SUMMARY

Food adulteration and manipulation is a criminal offense that continues to be on the rise. Substances used in food manipulation range from non-harmful substances to very hazardous ones. Methods such as chromatography, polymerase chain reaction, isotopic analysis, etc. have been effective in controlling food adulteration but they have certain limitations such as high cost, partial sensitivity, and high technicality. They are also relatively slow with lower sample numbers per analysis, long sample extraction protocols, long retention time (e.g., chromatography), etc. Novel instruments that have high sensitivity sensors imbedded in them have proven equally effective in plant-based food control with fewer limitations. The potentiometric and voltametric e-tongue are sensor-equipped devices that have been used to detect adulterants in wine, olive oil, tomato juice, fruit juice, mandarin fruit, honey, beer, etc. E-tongue is, however, better suited for liquid foods than solid foods as solid foods would have to undergo some extraction process before e-tongue analysis. Gas sensor-based e-nose has been used to detect adulteration in spirits, vegetable soups, broccoli, argan oil, fresh-cut bell pepper, etc. E-nose analyses do not need sample preparation and as such are suitable for both solid and liquid plant-based foods as long as they emit volatiles. The duration or waiting period for the analytes in the sensing layers of the e-nose is important in e-nose analysis because it increases the concentration of gases that get to the sensors. Spectroscopy was the most used method among the three instruments in this chapter. It has been used to detect adulteration in powders (cocoa, coffee, sweet potato, whey, and rice), spices (oregano, chili pepper, black pepper, acai, and cumin), herbs, wine, and spirits. It provides a non-invasive rapid real-time analysis advantage. The electronic tongue, electronic nose, and spectroscopy have proven to be worthy in the battle against food adulteration and should be adapted and explored further to guarantee consumer safety.

ACKNOWLEDGMENT

Supported under the (1) ÚNKP-19-3-I-SZIE-71 (Zsanett Bodor) and the ÚNKP-19-4 (Zoltan Kovacs), New National Excellence Program of the Ministry for Innovation and Technology within the framework of the Higher Education Institutional Excellence Program (NKFIH-1159-6/2019) in the scope of plant breeding and plant protection researches of Szent

István University; (2) the Doctoral School of Food Science, Szent István University (John Lewis Zinia Zaukuu, Zsanett Bodor); (3) Janos Bolyai Scholarship of the Hungarian Academy of Sciences; and (4) the European Union. It was co-financed by the European Social Fund (grant agreement no. EFOP-3.6.3-VEKOP-16-2017-00005).

KEYWORDS

- **chemometrics**
- **fraud**
- **sensors**
- **spectroscopy**
- **spice**

REFERENCES

1. Abhirami, S.; Radha, R. Detection of Food Adulteration in Selected Food Items Procured by Homemaker at the Department of Food Service Management and Dietetics, Avinashilingam Institute for Home Science and Higher Education for Women University, Coimbatore. *Int. J. Recent Sci. Res.* **2015,** *6*, 5938–5943.
2. Alamar, P. D.; Caramês, E. T. S.; Poppi, R. J.; Pallone, J. A. L. Quality Evaluation of Frozen Guava and Yellow Passion Fruit Pulps by NIR Spectroscopy and Chemometrics. *Food Res. Int.* **2016,** *85*, 209–214.
3. Alvarez-Suarez, J.; Gasparrini, M.; Forbes-Hernández, T.; Mazzoni, L.; Giampieri, F. The Composition and Biological Activity of Honey: A Focus on Manuka Honey. *Foods* **2014,** *3*, 420–432.
4. *Alimentarius, Codex*. Trade and Food Standards; 2017 [Online]. http://www.fao.org/fao-who-codexalimentarius/en/ (accessed April 23, 2022).
5. Ambrose, A.; Cho, B. K. Technologies for Detection and Measurement of Adulterants in Cereals and Cereal Products. *J. Biosyst. Eng.* **2014,** *39*, 357–365.
6. Anibal, C. V.; Odena, M.; Ruisánchez, I.; Callao, M. P. Determining the Adulteration of Spices with Sudan I-II-II-IV Dyes by UV–Visible Spectroscopy and Multivariate Classification Techniques. *Talanta* **2009,** *79*, 887–892.
7. Apetrei, I. M.; Apetrei, C. Detection of Virgin Olive Oil Adulteration using a Voltametric e-tongue. *Comput. Electron. Agric.* **2014,** *108*, 148–154.
8. Aenugu, H. P. R.; Sathis Kumar, D.; Srisudharson; Parthiban, N.; Ghosh, S. S.; Banji, D. Near Infrared Spectroscopy- An Overview. *Int. J. ChemTech Res.* **2011,** *3*, 825–836.
9. Barbin, D. F.; Felicio, A. L. de S. M.; Sun, D. W.; Nixdorf, S. L.; Hirooka, E. Y. Application of Infrared Spectral Techniques on Quality and Compositional Attributes of Coffee: An Overview. *Food Res. Int.* **2014,** *61*, 23–32.

10. Black, C.; Haughey, S. A.; Chevallier, O. P.; Galvin-King, P.; Elliott, C. T. A Comprehensive Strategy to Detect the Fraudulent Adulteration of Herbs: The Oregano Approach. *Food Chem.* **2016,** *210*, 551–557.
11. Bertoncelj, J.; Polak, T.; Kropf, U.; Korošec, M.; Golob, T. LC-DAD-ESI/ MS Analysis of Flavonoids and Abscisic Acid with Chemometric Approach for the Classification of Slovenian Honey. *Food Chem.* **2011,** *127*, 296–302.
12. Bansal, S.; Singh, A.; Mangal, M.; Mangal, A. K.; Kumar, S. Food Adulteration: Sources, Health Risks, and Detection Methods. *Crit. Rev. Food Sci. Nutr.* **2017,** *57*, 1174–1189.
13. Bougrini, M.; Tahri, K.; Haddi, Z.; Saidi, T.; Bari, N. El; Bouchikhi, B. Detection of Adulteration in Argan Oil by Using an Electronic Nose and a Voltametric Electronic Tongue. *J. Sens.* **2014,** *13*, 26–30.
14. Carloni, E.; Amagliani, G.; Omiccioli, E.; Ceppetelli, V.; Del Mastro, M.; Rotundo, L.; Brandi, G.; Magnani, M. Validation and Application of a Quantitative Real-Time PCR Assay to Detect Common Wheat Adulteration of Durum Wheat for Pasta Production. *Food Chem.* **2017,** *224*, 86–91.
15. Chatzinasiou, L.; Booker, A.; MacLennan, E.; Mackonochie, M.; Heinrich, M. Turmeric (Curcuma longa L.) Products: What Quality Differences Exist? *J. Herb. Med.* **2019,** *24*, 6–11.
16. Contreras, U.; Barbosa-García, O.; Pichardo-Molina, J. L.; Ramos-Ortíz, G.; Maldonado, J. L.; Meneses-Nava, M. A.; Ornelas-Soto, N. E.; López-de-Alba, P. L. Screening Method for Identification of Adulterate and Fake Tequilas by using UV-VIS Spectroscopy and Chemometrics. *Food Res. Int.* **2010,** *43*, 2356–2362.
17. Chu, S.; Metcalfe, C. D. Analysis of Paroxetine, Fluoxetine and Norfluoxetine in Fish Tissues using Pressurized Liquid Extraction, Mixed Mode Solid Phase Extraction Cleanup and Liquid Chromatography-Tandem Mass Spectrometry. *J. Chromatogr. A* **2007,** *1163*, 112–118.
18. Chen, J. bo; Zhou, Q.; Sun, S. Adulteration Screening of Botanical Materials by a Sensitive and Model-Free Approach Using Infrared Spectroscopic Imaging and Two-dimensional Correlation Infrared Spectroscopy. *J. Mol. Struct.* **2016,** *1124,* 262–267.
19. Carpintero, T.; Morais, B. De; Rodrigues, D. R.; Teixeira, U.; Polari, D. C.; Lemos, S. G. A simple Voltametric Electronic Tongue for the Analysis of Coffee Adulteration. *Food Chem.* **2018,** *78*, 1–6.
20. Correia, R. M.; Tosato, F.; Domingos, E.; Rodrigues, R. R. T.; Aquino, L. F. M.; Filgueiras, P. R.; Lacerda, V.; Romão, W. Portable Near Infrared Spectroscopy Applied to Quality Control of Brazilian Coffee. *Talanta* **2018,** *176*, 59–68.
21. Chen, H. Z.; Zhang, M.; Bhandari, B.; Guo, Z. Evaluation of the Freshness of Fresh-Cut Green Bell Pepper (*Capsicum annuum* var. grossum) using Electronic Nose. *LWT Food Sci. Technol.* **2018,** *87*, 77–84.
22. Dimitrova, B.; Gevrenova, R.; Anklam, E. Analysis of Phenolic Acids in Honeys of Different Floral Origin by Solid-Phase Extraction and High-Performance Liquid Chromatography. *Phytochem. Anal.* **2007,** *18*, 24–32.
23. Ding, X.; Ni, Y.; Kokot, S. NIR Spectroscopy and Chemometrics for the Discrimination of Pure, Powdered, Purple Sweet Potatoes and their Samples Adulterated with the White Sweet Potato Flour. *Chemom. Intell. Lab. Syst.* **2015,** *144*, 17–23.
24. Dins, L. G.; Sequeira, C.; Veloso, A. C. A.; Sousa, M. E. B. C.; Peres, A. M. Evaluation of Healthy and Sensory Indexes of Sweetened Beverages using an Electronic Tongue. *Anal. Chem. Acta* **2014,** *848*, 32–42.

25. El Hassani, N.; Elijah A.; Tahri, K.; Llobet, E.; Bouchikhi, B.; Errachid, A.; Zine, N. Emerging Approach for Analytical Characterization and Geographical Classification of Moroccan and French Honeys by Means of a Voltametric Electronic Tongue. *Food Chem.* **2017,** *243*, 36–42.
26. Elisa, J.; Manfroi, V.; Zanus, M.; Lazzarotto, M.; Alcaraz, C. Differentiation of Wines According to Grape Variety using Multivariate Analysis of Comprehensive Two-Dimensional Gas Chromatography with Time-of-Flight Mass Spectrometric Detection Data. *Food Chem.* **2013,** *141*, 3897–3905.
27. Ezhilan, M.; Nesakumar, N.; Babu, K. J.; Srinandan, C. S.; Rayappan, J. B. B. Freshness Assessment of Broccoli using Electronic Nose. *Measurement* **2019,** *145*, 735–743.
28. Erum, J. Van; Dam, D. Van; Deyn, P. P. De Determination of Seven Illegal Dyes in Egyptian Spices by HPLC with Gel Permeation Chromatography Clean up. *Neurosci. Biobehav. Rev.* **2019,** *130*, 35–43.
29. FAO. FAO's Role in Food Safety, 2019 [Online]. http://www.fao.org/food-safety/en/ (accessed April 23, 2022).
30. Field, A.; Field, J. Melamine and Cyanuric Acid do not Interfere with Bradford and Ninhydrin Assays for Protein Determination. *Food Chem.* **2010,** *121*, 912–917.
31. Flytkjær-Virgilsen, L.; Møller, H.; Vedsted, P. Cancer Diagnostic Delays and Travel Distance to Health Services: A Nationwide Cohort Study in Denmark. *Cancer Epidemiol.* **2019,** *59*, 115–122.
32. Forsyth, J. E.; Nurunnahar, S.; Islam, S. S.; Baker, M.; Yeasmin, D.; Islam, M. S.; Rahman, M.; Fendorf, S. Turmeric Means "Yellow" in Bengali: Lead Chromate Pigments Added to Turmeric Threaten Public Health Across Bangladesh. *Environ. Res.* **2019,** *179*, *55*, 15–22.
33. Gao, F.; Hu, Y.; Chen, D.; Li-Chan, E. C. Y.; Grant, E.; Lu, X. Determination of Sudan I in Paprika Powder by Molecularly Imprinted Polymers-Thin Layer Chromatography-Surface Enhanced Raman Spectroscopic Biosensor. *Talanta* **2015,** *143*, 344–352.
34. Geana, E. I.; Popescu, R.; Costinel, D.; Dinca, O. R.; Stefanescu, I.; Ionete, R. E.; Bala, C. Verifying the Red Wines Adulteration through Isotopic and Chromatographic Investigations Coupled with Multivariate Statistic Interpretation of the Data. *Food Control* **2016,** *62*, 1–9.
35. Georgouli, K.; Martinez Del Rincon, J.; Koidis, A. Continuous Statistical Modelling for Rapid Detection of Adulteration of Extra Virgin Olive Oil using Mid Infrared and Raman Spectroscopic Data. *Food Chem.* **2017,** *217*, 735–742.
36. Girish, C.; Meena, R..; Mahima, D.; Mamta, K. Nutritional Properties of Minor Millets: Neglected Cereals with. *Curr. Sci.* **2014,** *107*, 1109–1111.
37. Gliszczy, A. Electronic Nose as a Tool for Monitoring the Authenticity of Food: A Review. *Food Anal. Meth.* **2017,** *10*, 1800–1816.
38. Gobbi, E.; Falasconi, M.; Zambotti, G.; Sberveglieri, V.; Pulvirenti, A.; Sberveglieri, G. Rapid Diagnosis of Enterobacteriaceae in Vegetable Soups by a Metal Oxide Sensor Based Electronic Nose. *Sens. Actuators B Chem.* **2015,** *207*, 1104–1113.
39. Gong, X.; Wang, H.; Zhang, Y.; Sun, J.; Dong, J.; Lin, L.; Yu, J.; Feng, Z. Chromatography Determination of 15 Mycotoxins in Foods and Feeds using High Performance Liquid Chromatography-Tandem Mass Spectrometry with Gel Permeation Chromatography Combined QuEChERS purification. *J. Chromatogr. Sep. Tech.* **2012,** *3*, 120–125.

40. González, M.; Lobo, G. M.; Méndez, J.; Carnero, A. Detection of Colour Adulteration in Cochineals by Spectrophotometric Determination of Yellow and Red Pigment Groups. *Food Control* **2005,** *16*, 105–112.
41. Granato, D.; Putnik, P.; Kovačević, D. B.; Santos, J. S.; Calado, V.; Rocha, R. S.; Cruz, A. G. Da; Jarvis, B. Trends in Chemometrics: Food Authentication, Microbiology, and Effects of Processing. *Compr. Rev. Food Sci. Food Saf.* **2018,** *17*, 663–677.
42. Guo, T.; Fang, P.; Jiang, J.; Zhang, F.; Yong, W.; Liu, J.; Dong, Y. Rapid Screening and Quantification of Residual Pesticides and Illegal Adulterants in Red Wine by Direct Analysis in Real Time Mass Spectrometry. *J. Chromatogr. A* **2016,** *1471*, 27–33.
43. Guohua, H.; Yuling, W.; Dandan, Y.; Wenwen, D.; Linshan, Z.; Lvye, W. Study of Peach Freshness Predictive Method Based on Electronic Nose. *Food Control* **2012,** *28*, 25–32.
44. Haddi, Z.; Mabrouk, S.; Bougrini, M.; Tahri, K.; Sghaier, K.; Barhoumi, H.; El Bari, N.; Maaref, A.; et al. E-Nose and e-Tongue Combination for Improved Recognition of Fruit Juice Samples. *Food Chem.* **2014,** *150*, 246–253.
45. Haughey, S. A.; Galvin-King, P.; Ho, Y. C.; Bell, S. E. J.; Elliott, C. T. The Feasibility of Using Near Infrared and Raman Spectroscopic Techniques to Detect Fraudulent Adulteration of Chili Powders with Sudan Dye. *Food Control* **2015,** *48*, 75–83.
46. Hong, X.; Wang, J. Detection of Adulteration in Cherry Tomato Juices Based on Electronic Nose and Tongue: Comparison of Different Data Fusion Approaches. *J. Food Eng.* **2014,** *126*, 89–97.
47. IARC Aflatoxins. In *Overall Evaluations of Carcinogenicity - An Updating of IARC Monographs;* International Agency for Research on Cancer (WHO): Lyon, France, 1987; pp 1–42.
48. ISO / IEC DirectivesISO / IEC Directives ISO / IEC Directives, Part 2 Rules for the structure and drafting, 2011 [Online]. https://www.iec.ch/members_experts/refdocs/iec/isoiecdir-1%7Bed12.0% (accessed March 13, 2022).
49. Izquierdo, M.; Lastra-mejías, M.; González-flores, E.; Cancilla, J. C.; Pérez, M.; Torrecilla, J. S. Convolutional Decoding of Thermographic Images to Locate and Quantify Honey Adulterations. *Talanta* **2019,** *12*, 20–25.
50. Jandrić, Z.; Frew, R. D.; Fernandez-Cedi, L. N.; Cannavan, A. An Investigative Study on Discrimination of Honey of Various Floral and Geographical Origins using UPLC-QToF MS and Multivariate Data Analysis. *Food Control* **2017,** *72*, 189–197.
51. Janeiro, R. *Intensive-rearing in Brazil incidencia de hongos y aflatoxina B-1 en silos de sorgo tipo trinchera pre y post-fermentados* (Intensive-Rearing in Brazil: Incidence of Fungi and Aflatoxin B-1 in Pre- and Post-fermented Trench-type Sorghum Silos). *Revista Biociencias* **2012,** *2*, 81–91.
52. John, W. S. Food Fraud Prevention. In *Food Microbiology and Food Safety*; John, W. S., Ed.; Springer: New York, NY, 2019; pp 200–225.
53. Joseph, A.; Vikraman, N.; Gelu, A.; Kr, C. Detection of Adulteration in Food Commodities from Selected Outlets. *Int. J. Home Sci.* **2018**, *4*, 257–260.
54. Kang, B. S.; Lee, J. E.; Park, H. J. Electronic Tongue-based Discrimination of Korean Rice Wines (*makgeolli*) Including Prediction of Sensory Evaluation and Instrumental Measurements. *Food Chem.* **2014,** *151*, 317–323.
55. Kagaya, Y.; Miyamoto, S. A Systematic Review of Near-infrared Spectroscopy in Flap Monitoring: Current Basic and Clinical Evidence and Prospects. *J. Plast. Reconstr. Aesthet. Surg.* **2017,** *71*, 246–257.
56. Kamiloglu, S. Authenticity and Traceability in Beverages. *Food Chem.* **2019,** *277*, 12–24.

57. Koncz, F. A.; Bodor, Z.; Rashed, M. S.; Kaszab, T.; Gillay, B.; Kovacs, Z.; Benedek, C. Floral and Geographical Origin Identification of Hungarian Honey with Electronic Tongue and Classical Analytical Methods. *Anal. Tech. Szeged.* **2017,** *11*, 1–9.
58. Kelly, J. D.; Petisco, C.; Downey, G. Potential of Near Infrared Reflectance Spectroscopy to Detect Adulteration of Irish Honey by Beet Invert Syrup and High Fructose Corn Syrup. *J. Near Infrared Spec.* **2006,** *14*, 139–146.
59. Kruzlicova, D.; Mocak, J.; Balla, B.; Petka, J.; Farkova, M.; Havel, J. Classification of Slovak White Wines using Artificial Neural Networks and Discriminant Techniques. *Food Chem.* **2009,** *112*, 1046–1052.
60. Kenjerić, D.; Mandić, M. L.; Primorac, L.; Bubalo, D.; Perl, A. Flavonoid Profile of Robinia Honeys Produced in Croatia. *Food Chem.* **2007,** *102*, 683–690.
61. Karabagias, I. K.; Vavoura, M. V.; Nikolaou, C.; Badeka, A. V.; Kontakos, S.; Kontominas, M. G. Floral Authentication of Greek Unifloral Honeys Based on the Combination of Phenolic Compounds, Physicochemical Parameters and Chemometrics. *Food Res. Int.* **2014,** *62*, 753–760.
62. LC Agilent Systems. Analysis of fumonisin: FB1 and FB2 mycotoxins in corn food and feed samples using the Agilent 1120 Compact LC System coupled to the Agilent 6140 Single Quadrupole LC / MS System Application, [Online]. https://gimitec.com/sites/default/files/5990-6560en1 (accessed March 13, 2020).
63. Liji, T. *Turmeric adulteration: eating your brain*? [Online] 2019. https://www.news-medical.net/news/20190925/Turmeric-adulteration-eating-your-brain.aspx (accessed March 22, 2022).
64. Lima, A. B. S. de; Batista, A. S.; Jesus, J. C. de; Silva, J. de J.; Araújo, A. C. M.; Santos, L. S. Fast Quantitative Detection of Black Pepper and Cumin Adulterations by Near-Infrared Spectroscopy and Multivariate Modeling. *Food Control* **2020,** *107*, E-article: 106802.
65. Liu, M.; Han, X.; Tu, K.; Pan, L.; Tu, J.; Tang, L.; Liu, P.; Zhan, G. Application of Electronic Nose in Chinese Spirits Quality Control and Flavour Assessment. *Food Control* **2012,** *26*, 564–570.
66. Lobato, K. B. de S.; Alamar, P. D.; Caramês, E. T. dos S.; Pallone, J. A. L. Authenticity of Freeze-Dried Açai Pulp by Near-infrared Spectroscopy. *J. Food Eng.* **2018,** *224*, 105–111.
67. Locher, C.; Neumann, J.; Sostaric, T. Authentication of Honeys of Different Floral Origins *via* High-Performance Thin-Layer Chromatographic Fingerprinting. *J. Planar Chromatogr. Mod. TLC* **2017,** *30*, 57–62.
68. Lozano, J.; Fernández, M. J.; Fontecha, J. L.; Aleixandre, M.; Santos, J. P.; Sayago, I.; Arroyo, T.; Cabellos, J. M.; et al. Wine Classification with a Zinc Oxide SAW Sensor Array. *Sens. Actuators B Chem.* **2006,** *120*, 166–171.
69. Lukacs, M.; Bazar, G.; Pollner, B.; Henn, R.; Kirchler, C. G.; Huck, C. W.; Kovacs, Z. Near Infrared Spectroscopy as an Alternative Quick Method for Simultaneous Detection of Multiple Adulterants in Whey Protein-Based Sports Supplement. *Food Control* **2018,** *94*, 331–340.
70. Luykx, D. M. A. M.; Cordewener, J. H. G.; Ferranti, P.; Frankhuizen, R.; Bremer, M. G. E. G.; Hooijerink, H.; America, A. H. P. Identification of Plant Proteins in Adulterated Skimmed Milk Powder by High-Performance Liquid Chromatography-Mass Spectrometry. *J. Chromatogr. A* **2007,** *1164*, 189–197.

71. Magdalena, S.; Wardencki, W. Food Analysis Using Artificial Senses. *J. Agric. Food Chem.* **2015,** *6*, 80–233.
72. Majchrzak, T.; Wojnowski, W.; Dymerski, T.; Gębicki, J.; Namieśnik, J. Electronic Noses in Classification and Quality Control of Edible Oils: A Review. *Food Chem.* **2018,** *246*, 192–201.
73. Meenu, M.; Cai, Q.; Xu, B. A Critical Review on Analytical Techniques to Detect Adulteration of Extra Virgin Olive Oil. *Trends Food Sci. Technol.* **2019,** *91*, 391–408.
74. Minh, D. T. C.; Thi, L. A.; Huyen, N. T. T.; Van Vu, L.; Anh, N. T. K.; Ha, P. T. T. Detection of Sildenafil Adulterated in Herbal Products Using Thin Layer Chromatography Combined with Surface Enhanced Raman Spectroscopy: Double Coffee-Ring Effect Based Enhancement. *J. Pharm. Biomed. Anal.* **2019,** *174*, 340–347.
75. Monago-Maraña, O.; Eskildsen, C. E.; Afseth, N. K.; Galeano-Díaz, T.; Muñoz de la Peña, A.; Wold, J. P. Non-destructive Raman Spectroscopy as a Tool for Measuring ASTA Color Values and Sudan I Content in Paprika Powder. *Food Chem.* **2019,** *274*, 187–193.
76. Moskowitz, H. R. Food Quality: Conceptual and Sensory Aspects. *Food Qual. Prefer.* **1995,** *6*, 157–162.
77. Natcha, T.; Panmanas, S. Detection of Adulteration of Soy Sauce by Brine using Near Infrared Spectroscopy. *Agric. Agric. Sci. Procedia* **2014,** *2*, 212–217.
78. Nicolaï, B. M.; Beullens, K.; Bobelyn, E.; Peirs, A.; Saeys, W.; Theron, K. I.; Lammertyn, J. Nondestructive Measurement of Fruit and Vegetable Quality by Means of NIR Spectroscopy: A Review. *Postharvest Biol. Technol.* **2007,** *46*, 99–118.
79. Nordlund, E.; Lille, M.; Silventoinen, P.; Nygren, H.; Seppänen-Laakso, T.; Mikkelson, A.; Aura, A. M.; Heiniö, R. L. Plant Cells as Food – A Concept Taking Shape. *Food Res. Int.* **2018,** *107*, 297–305.
80. Nutrient Analysis of Fruit and Vegetables –Report; Department of Health and Social Care, London, UK, 2013; p 92.
81. Oliveira, R. C. S.; Oliveira, L. S.; Franca, A. S.; Augusti, R. Evaluation of the Potential of SPME-GC-MS and Chemometrics to Detect Adulteration of Ground Roasted Coffee with Roasted Barley. *J. Food Compos. Anal.* **2009,** *22*, 257–261.
82. Oroian, M.; Ropciuc, S. Romanian Honey Authentication using Voltametric Electronic Tongue. Correlation of Voltametric Data with Physicochemical Parameters and Phenolic Compounds. *Comput. Electron. Agric.* **2019,** *157*, 371–379.
83. Oroian, M.; Ropciuc, S.; Paduret, S.; Todosi, E. Rheological Analysis of Honeydew Honey Adulterated with Glucose, Fructose, Inverted Sugar, Hydrolysed Inulin Syrup and Malt Wort. *LWT* **2018,** *95*, 1–8.
84. Orrillo, I.; Cruz-Tirado, J. P.; Cardenas, A.; Oruna, M.; Carnero, A.; Barbin, D. F.; Siche, R. Hyperspectral Imaging as a Powerful Tool for Identification of Papaya Seeds in Black Pepper. *Food Control* **2019**, *101*, 45–52.
85. Osborne, B. G. Near-Infrared Spectroscopy in Food Analysis. In *Encyclopedia of Analytical Chemistry*, 2000; vol *84*, pp 1–14.
86. Ouyang, Q.; Zhao, J.; Chen, Q.; Lin, H. Classification of Rice Wine According to Different Marked Ages Using a Novel Artificial Olfactory Technique Based on Colorimetric Sensor Array. *Food Chem.* **2013,** *138*, 1320–1324.
87. Paraschos, S.; Magiatis, P.; Gikas, E.; Smyrnioudis, I.; Skaltsounis, A. L. Quality Profile Determination of Chios Mastic Gum Essential Oil and Detection of Adulteration in Mastic Oil Products with the Application of Chiral and Non-chiral GC–MS Analysis. *Fitoterapia* **2016,** *114*, 12–17.

88. Pennacchia, C.; Ercolini, D.; Villani, F. Spoilage-Related Microbiota Associated with Chilled Beef Stored in Air or Vacuum Pack. *Food Microbiol.* **2011,** *28*, 84–93.
89. Phattanawasin, P.; Sotanaphun, U.; Sukwattanasinit, T.; Akkarawaranthorn, J.; Kitchaiya, S. Quantitative Determination of Sibutramine in Adulterated Herbal Slimming Formulations by TLC-Image Analysis Method. *Forensic Sci. Int.* **2012,** *219*, 96–100.
90. Polshin, E.; Rudnitskaya, A.; Kirsanov, D.; Legin, A.; Saison, D.; Delvaux, F.; Delvaux, F. R.; Nicolaï, B. M. Electronic Tongue as a Screening Tool for Rapid Analysis of Beer. *Talanta* **2010,** *81*, 88–94.
91. Pomeranian, W. Chemical Composition of the Pepper Fruit Extracts of Hot Cultivars *Capsicum annuum* L. *Acta Sci. Pol., Hortorum Cultus* **2011,** *10*, 171–184.
92. Qiu, S.; Wang, J.; Tang, C.; Du, D. Comparison of ELM, RF, and SVM on E-nose and E-tongue to Trace the Quality Status of Mandarin (Citrus unshiu Marc.). *J. Food Eng.* **2015,** *166*, 193–203.
93. Quelal-Vásconez, M. A.; Pérez-Esteve, É.; Arnau-Bonachera, A.; Barat, J. M.; Talens, P. Rapid Fraud Detection of Cocoa Powder with Carob Flour using Near Infrared Spectroscopy. *Food Control* **2018,** *92*, 183–189.
94. Reis, N.; Franca, A. S.; Oliveira, L. S. Discrimination Between Roasted Coffee, Roasted Corn and Coffee Husks by Diffuse Reflectance Infrared Fourier Transform Spectroscopy. *LWT Food Sci. Technol.* **2013,** *50*, 715–722.
95. Ruiz-Matute, A. I.; Rodríguez-Sánchez, S.; Sanz, M. L.; Martínez-Castro, I. Detection of Adulterations of Honey with High Fructose Syrups from Inulin by GC Analysis. *J. Food Composit. Anal.* **2010,** *23*, 273–276.
96. Sans, S.; Ferré, J.; Boqué, R.; Sabaté, J.; Casals, J.; Simó, J. Determination of Chemical Properties in *Allium cepa* L. by Near Infrared Spectroscopy and Multivariate Calibration. *Food Chem.* **2018,** *262*, 178–183.
97. Schaarschmidt, S. Public and Private Standards for Dried Culinary Herbs and Spices, Part I: Standards Defining the Physical and Chemical Product Quality and Safety. *Food Control* **2016,** *70*, 339–349.
98. Sereshti, H.; Samadi, S. A Rapid and Simple Determination of Caffeine in Teas, Coffees and Eight Beverages. *Food Chem.* **2014,** *158*, 8–13.
99. Shi, J.; Yuan, D.; Hao, S.; Wang, H.; Luo, N.; Liu, J.; Zhang, Y.; Zhang, W. Stimulated Brillouin Scattering in Combination with Visible Absorption Spectroscopy for Authentication of Vegetable Oils and Detection of Olive Oil Adulteration. *Spectrochem. Acta A Mol. Biomol. Spectrosc.* **2019,** *206*, 320–327.
100. Shikuku, K. M.; Okello, J. J.; Wambugu, S.; Sindi, K.; Low, J. W.; McEwan, M. Nutrition and Food Security Impacts of Quality Seeds of Biofortified Orange-Fleshed Sweet Potato: Quasi-Experimental Evidence from Tanzania. *World Dev.* **2019,** *124*, E-article ID: 104646.
101. Simões Da Costa, A. M.; Delgadillo, I.; Rudnitskaya, A. Detection of Copper, Lead, Cadmium and Iron in Wine using Electronic Tongue Sensor System. *Talanta* **2014,** *129*, 63–71.
102. Sinija, V. R.; Mishra, H. N. FT-NIR Spectroscopy for Caffeine Estimation in Instant Green Tea Powder and Granules. *LWT Food Sci. Technol.* **2009,** *42*, 998–1002.
103. Sobrino-Gregorio, L.; Bataller, R.; Soto, J.; Escriche, I. Monitoring Honey Adulteration with Sugar Syrups using an Automatic Pulse Voltametric Electronic Tongue. *Food Control* **2018,** *91*, 254–260.

104. Sobrino-Gregorio, L.; Vilanova, S.; Prohens, J.; Escriche, I. Detection of Honey Adulteration by Conventional and Real-Time PCR. *Food Control* **2019,** *95*, 57–62.
105. Spink, J. Safety of Food and Beverages: Risks of Food Adulteration. In *Encyclopedia of Food Safety 3*; Elsevier Ltd.: New York; pp 413–416; [Online]. DOI: 10.1016/B978-0-12-378612-8.00300-0
106. Sumaira, F.; Sumaira, F. Food Authority Arrests Three for Mixing Bran in Spices, 2022 [Online]. https://www.thenews.com.pk/print/446280-1-000kg-dyed-bran-seized-from-spice-factory (accessed March 16, 2022).
107. Sun, M.; Zhang, D.; Liu, L.; Wang, Z. How to Predict the Sugariness and Hardness of Melons: A Near-Infrared Hyperspectral Imaging Method. *Food Chem.* **2017,** *218*, 413–421.
108. Székács, A.; Wilkinson, M. G.; Mader, A.; Appel, B. Environmental and Food Safety of Spices and Herbs along Global Food Chains. *Food Control* **2018,** *83*, 1–6.
109. Tähkäpää, S.; Maijala, R.; Korkeala, H.; Nevas, M. Patterns of Food Frauds and Adulterations Reported in the EU Rapid Alarm System for Food and Feed and in Finland. *Food Control* **2015,** *47*, 175–184.
110. Tarassenko, L.; Denham, M. In *Sensory Processing, Cognitive Systems - Information Processing Meets Brain Science*, Ed. 4; Elsevier Ltd.: Netherlands, 2006; pp 179–230.
111. Timsorn, K.; Thoopboochagorn, T. Evaluation of Bacterial Population on Chicken Meats Using a Briefcase Electronic Nose. *Biosyst. Eng.* **2016,** *151*, 116–125.
112. UN. UNDP [Online]. https://www.undp.org/content/undp/en/home/sustainable-development-goals.html (accessed March 20, 2022).
113. Vaskova, H.; Buckova, M. Thermal Degradation of Vegetable Oils: Spectroscopic Measurement and Analysis. *Procedia Eng.* **2015,** *100*, 630–635.
114. Villa, C.; Costa, J.; Oliveira, M. B. P. P.; Mafra, I. Novel Quantitative Real-Time PCR Approach to Determine Safflower (*Carthamus tinctorius*) Adulteration in Saffron (*Crocus sativus*). *Food Chem.* **2017,** *229*, 680–687.
115. Villani, T. S.; Reichert, W.; Ferruzzi, M. G.; Pasinetti, G. M.; Simon, J. E.; Wu, Q. Chemical Investigation of Commercial Grape Seed Derived Products to Assess Quality and Detect Adulteration. *Food Chem.* **2015,** *170*, 271–280.
116. Wang, X. B.; Zheng, J.; Li, J. J.; Yu, H. Y.; Li, Q. Y.; Xu, L. H.; Liu, M. J.; Xian, R. Q. Simultaneous Analysis of 23 Illegal Adulterated Aphrodisiac Chemical Ingredients in Health Foods and Chinese Traditional Patent Medicines by Ultrahigh Performance Liquid Chromatography Coupled with Quadrupole Time-of-Flight Mass Spectrometry. *J. Food Drug Anal.* **2018**, *26*, 1138–1153.
117. Wei, Z.; Wang, J. Classification of Monofloral Honeys by Voltametric Electronic Tongue with Chemometrics Method. *Electrochem. Acta* **2011**, *56*, 4907–4915.
118. WHO Global Strategy for Food Safety: Safer Food for Better Health; WHO, Switzerland [Online]. https://apps.who.int/iris/handle/10665/42559 (accessed March 20, 2022).
119. Wilde, A. S.; Haughey, S. A.; Galvin-King, P.; Elliott, C. T. The Feasibility of Applying NIR and FT-IR Fingerprinting to Detect Adulteration in Black Pepper. *Food Control* **2019,** *100*, 1–7.
120. Wine, Olive Oil Has a Fraud Problem—Can the FDA Fix It? [Online] 2019. https://www.msn.com/en-us/foodanddrink/foodnews/olive-oil-has-a-fraud-problem—can-the-fda-fix-it/vi-BBWqwCd (accessed March 20, 2022).

121. Winquist, F. Voltametric Electronic Tongues – Basic Principles and Applications. *Microchim. Acta* **2008,** *163*, 3–10.
122. World Health Organization (WHO)/ FAO Assuring Food Safety and Quality [Online]. https://www.fao.org/documents/card/en/c/92f82d38-5557-4ca1-b361-be14cd129db6/ (accessed March 16, 2022).
123. Xu, B.; Li, P.; Ma, F.; Wang, X.; Matthäus, B.; Chen, R.; Yang, Q.; Zhang, W.; et al. Detection of Virgin Coconut Oil Adulteration with Animal Fats Using Quantitative Cholesterol by GC × GC-TOF/MS Analysis. *Food Chem.* **2015,** *178*, 128–135.
124. Xu, L.; Yan, S. M.; Cai, C. B.; Yu, X. P. Untargeted Detection of Illegal Adulterations in Chinese Glutinous Rice Flour (GRF) by NIR Spectroscopy and Chemometrics: Specificity of Detection Improved by Reducing Unnecessary Variations. *Food Anal. Meth.* **2013,** *65*, 19–25.
125. Yang, Z.; Dong, F.; Shimizu, K.; Kinoshita, T.; Kanamori, M.; Morita, A.; Watanabe, N. Identification of Coumarin-Enriched Japanese Green Teas and their Particular Flavor Using Electronic Nose. *J. Food Eng.* **2009,** ***92***, 312–316.
126. Yılmaz, R.; Bayraç, C.; Başman, A.; Köksel, H. Development of SYBR Green-Based Real Time PCR Assays for Detection and Quantification of Adulteration in Wheat-Based Composite Breads and Their in-House Validation. *J. Cereal Sci.* **2019,** *85*, 91–97.
127. Yu, H.; Wang, J.; Zhang, H.; Yu, Y.; Yao, C. Identification of Green Tea Grade Using Different Feature of Response Signal from E-nose Sensors. *Sens. Actuators B Chem.* **2008**, *128*, 455–461.
128. Yung An, C.; Hossain, M. M.; Alam, F.; Islam, M. A.; Khalil, M. I.; Alam, N.; Gan, S. H. Efficiency of Polyphenol Extraction from Artificial Honey Using C 18 Cartridges and Amberlite® XAD-2 Resin: A Comparative Study. *J. Chem.* **2016,** *201*, 1–6.
129. Zábrodská, B.; Vorlová, L. Adulteration of Honey and Available Methods for Detection – A Review. *Acta Veterinaria Brno* **2014,** *83*, S85–S102.
130. Zaukuu, J. L. Z.; Bazar, G.; Gillay, Z.; Kovacs, Z. Emerging Trends of Advanced Sensor - Based Instruments for Meat, Poultry and Fish Quality – A Review. *Crit. Rev. Food Sci. Nutr.* **2020,** *60*, 3443–3460.
131. Zaukuu, J. L. Z.; Soos, J.; Bodor, Z.; Felföldi, J.; Magyar, I.; Kovacs, Z. Authentication of Tokaj Wine (*Hungaricum*) with the Electronic Tongue and Near Infrared Spectroscopy. *J. Food Sci.* **2019,** *84*, 3437–3444.

CHAPTER 11

ANALYTICAL METHODS FOR NUTRITIONAL COMPOSITION OF HONEY

ZSANETT BODOR, JOHN-LEWIS ZINIA ZAUKUU, BALKIS AOUADI, ZOLTÁN KOVÁCS, and CSILLA BENEDEK

ABSTRACT

Honey is a complex food matrix mainly composed of sugar, water, and different nutritionally active components. Emanating from various origins, honey contains more than 20 different types of mono-, di-, and trisaccharides, and usually less than 20% water content. In addition to sugars, the main nutritive components in honey are vitamins, minerals, organic acids, phenolic compounds, proteins, and amino acids. Due to the wide variety of its bioactive compounds, honey has been used since ancient times as a medical product. The botanical source of honey has a high influence on its composition; however, the effect of geographical origin is also not negligible. Storage and processing (temperature, material of the container) conditions also have a role in the evolution of honey composition. The processing plays a decisive role in both the nutritive and medicinal properties of the product. Determination of honey composition often depends on the target compounds, and the applied techniques used can generally range from classical analytical methods to spectroscopy, chromatography, and nuclear magnetic resonance-based techniques. This chapter focuses on the overview of the composition of honey, with special emphasis on its sugar composition and nutritionally active components, and also provides information on the effect of botanical origin on the amount of these compounds.

Sustainable and Functional Foods from Plants: Health Impact, Bioactive Compounds, and Production Technologies. Megh R. Goyal, Arijit Nath, and Zoltán Kovács (Eds.)
© 2024 Apple Academic Press, Inc. Co-published with CRC Press (Taylor & Francis)

11.1 INTRODUCTION

According to the legislation, honey is a sweet and natural food product made by honeybees (*Apis mellifera*) from the nectar of plants, secretions of plants, or excretions of insects (living on the plants).[26,39] Besides the *Apis mellifera,* other bee species are also known to produce honey such as *Apis dorsata, Apis florea, Apis cerana, Apis adreniformis, Apis binghami, Apis breviligula, Apis koschevnikovi, Apis labriosa, Apis nigrocincta,* and *Apis nuluensis.*[19] Bees collect the nectar, juices, and honeydew and take it to the hive. During the ripening and maturation process, including the deposit, dehydration, and storage, honey is formed.[26,39] The composition of honey consists mainly of water and sugars; however, honey contains other compounds, such as minerals, enzymes, polyphenols, organic acids, vitamins, and other solid particles deriving from the honey gathering.

The physicochemical composition, consistency, color, and aromatic profile highly depend on the botanical source, but the geographical origin also has an influence on the composition. The origin is also influencing the quantity of nutritionally active components of honey.[98] There are two main types of honey according to the legislations: blossom and honeydew honeys.[26,39] In Europe, more than 100 different plants are known to serve as nectar sources for honeybees, thus producing different unifloral honeys.[92] Honeydew can be produced by different sucking insects such as *Metcalfa Pruinosa.*[84]

There are a lot of differences in the composition between honeys produced from nectar (blossom honeys) and honeydew honeys,[88] although honeys from different floral origins can also have significantly different compositions.[6] In Europe and over the world, the most common honey types are the acacia or black locust (*Robinia pseudoacacia*), linden or lime (*Tilia* spp.), rape or canola (*Brassica napus*), honeydew, sunflower (*Helianthus annuus*), thyme (*Thymus* spp.), dandelion (*Taraxacum officinale*), chestnut (*Castanea sativa*), rosemary (*Rosmarinus officinalis*), rhododendron (*Rhododendron* spp.), lavender (*Lavandula* spp.), eucalyptus (*Eucalyptus* spp.), orange blossom or citrus (*Citrus* spp.), and heather or ling (*Calluna vulgaris*).[92] There are several other types of honey from different sources, and each type has its own physicochemical composition and nutritional value.

This chapter provides a description regarding the nutritionally useful components (such as sugars, vitamins, minerals, antioxidants, organic acids, amino acids, and proteins) of different honey types; and provides information about the analytical techniques used for the determination of these biocompounds.

11.2 PHYSICOCHEMICAL COMPOSITION OF HONEY

This section provides information about most of the nutritive compounds of honey, such as sugars, minerals, amino acids, vitamins, and organic acids. Examples are provided for the sources (botanical) of the individual components and the methods used for the identification of such materials.

11.2.1 REGULATIONS AND LEGISLATIONS FOR THE COMPOSITION

Legislations regarding honey are based on the rules of the *Codex Alimentarius* (CA) and on the regulations of the European Council (EC) in the European Union and there are also rules at the national level in several countries.[26,39,112] The CA and EC legislations contain the description of honey and some general compositional limits. At national levels, there are rules also for the unifloral honeys in some countries (Table 11.1). These regulations also provide information about the labeling of the honeys and in Codex Alimentarius, there are methods and sampling of the analysis. The 2014/63/EU contains the amendment and corrections of the 2001/110/EC.[38]

National regulations are used in countries worldwide. Members of the European Union should form their legislations to meet the requirements of the EU legislation; therefore, these are usually the same or stricter than the EU rules. In several countries worldwide, there are not only general rules regarding the composition of the honey but for some honey types there are also compositional criteria such as

- America: honey is divided into grades based on its properties[113].
- Germany: for chestnut, pine, *Erica*, citrus, sunflower, heather, rape, acacia, fir.
- Greece: pine, fir, chestnut, thyme, citrus.
- Hungary: for acacia, linden, and multiflora honey[71].
- Turkey: for acacia, thyme, honeydew, chestnut, rape, linden, citrus, lavender, eucalyptus, rosemary.[112]

TABLE 11.1 Regulation of Honeys by European Council and *Codex Alimentarius* Commission.[26,39]

Parameter	European Council Directives (2001/110/EC)		*Codex Alimentarius* (Revised 2001- CODEX STAN 12-1981)	
	General	Exceptions	General	Exceptions
Moisture %	≤20%	Heather and bakers honey ≤ 23% Bakers honey from heather ≤25%	≤20%	Heather honey ≤20%
Fructose + glucose g/100g	≥60 g/100 g	Honeydew and blossom-honeydew blend honeys ≥45 g/100 g	≥60 g/100 g	Honeydew and blosson-honeydew blend honeys ≥45 g/100 g
Sucrose g/100 g	≤5 g/100g	Lavender ≤15 g/100 g Acacia, alfalfa, citrus, French honeysuckle, leatherwood, Menzies Banskia, red gum ≤10 g/100 g	≤5 g/100 g	Lavender ≤15 g/100 g Acacia, alfalfa, citrus, French honeysuckle, leatherwood, Menzies Banskia, red gum ≤10 g/100 g
Water insoluble solids g/100g	≤0.1 g/100 g	Pressed honey ≤0.5 g/100 g	≤0.1 g/100 g	Pressed honey ≤0.5 g/100 g
Electrical conductivity ms/cm	≤800 μs/cm (except*)	Honeydew, chestnut, and blends of these Except with types listed below ≥0.800 μs/cm *bell heather, eucalyptus, jelly bush, linden, ling heather, manuka, strawberry tree, tea tree	≤800 μs/cm (except*)	Honeydew, chestnut, and blends of these Except with types listed below ≥0.800 μs/cm *bell heather, eucalyptus, jelly bush, linden, ling heather, manuka, strawberry tree, tea tree
Free acidity	≤50 milli-equivalents acid/1000 g	≤80 milliequivalents acid/1000 g	≤50 milli-equivalents acid/1000 g	No exception
Diastase activity (Schade scale)	≥8	Except baker's honey Naturally lower enzyme containing honeys and HMF ≤15 mg/kg - ≥3	≥8	Naturally lower enzyme containing honeys and HMF ≤15 mg/kg - ≥3
HMF mg/kg	≤40 mg/kg	Except baker's honey Honeys from tropical region, or blends of these honeys ≤80 mg/kg	≤40 mg/kg	Honeys from tropical region, or blends of these honeys ≤80 mg/kg

11.2.2 MOISTURE CONTENT AND WATER ACTIVITY OF HONEY

Moisture is one of the main constituents of honey, usually ranging between 13% and 23%. The amount of water in honey is one of the key factors that determine the stability against spoilage and fermentation.[116] Despite, the moisture content of honey being an important indicator of the quality, the water activity determines the power against microbiological contamination. Temperature, pH, and availability of nutrients for the microbes are other factors determining possible microbiological contamination.[5,24,25]

The water activity (a_w) of honey is reportedly below 0.6, and the limiting water activity for yeast is around 0.62–0.61.[24] That implies that, there is a low chance of yeast thriving in honey.

Generally, the most applied and accurate method used to determine the moisture content is the refractometric method, based on a special conversion table.[116] The International Honey Commission mentions this conversion table in its method collection.[16] Similar protocols were also reported by the "AOAC 969.38B – Determination of moisture content.[26]"

11.2.3 SUGAR PROFILE OF HONEY

In addition to water, sugars are the main components of honey, contributing in total up to 95% of the dry weight. About 26 different types of sugar molecules have been detected in honey, including mono-, di-, trisaccharides, and oligosaccharides (in smaller amounts).[17,33,100,116] The sugars that have been detected in honey can be found in Table 11.2.

The oligosaccharides in honey originate from the enzymatic activities in honey. The formation of the oligosaccharides can occur through trans-D-glucosylation and trans-D-fructosylation. During the trans-D-glucosylation, the D-glucopyranosyl molecules are transferred from a D-glucopyranosyl donor to an acceptor that can be mono-, di- or oligosaccharide too. Trans-D-fructosylation means the transfer of D-fructofuranosyl groups to other sugars by means of D-fructofuranosyl-transferring enzyme, to produce oligosaccharides that contain D-fructose.[100]

The two main monosaccharides in honey are the D-glucose, and D-fructose. The ratios of these in flower honeys are equal, but unifloral honey usually has a higher fructose content. The glucose/fructose ratios (F/G ratio) are listed in Table 11.3.

TABLE 11.2 Sugars that have been Detected in Honey-based on the Reviews by Ref. [17,33,100,116].

Sugar	Scientific name	Common name
	Monosaccharides	
Glucose	D-glucopyranose	Grape sugar
Fructose	D-fructopyranose	Fruit sugar
Arabinose [105]	D-arabinofuranose	
	Disaccharides	
Maltose	O-alpha-D-glucopyranosyl-(1→4)-beta-D-glucopyranose	
Isomaltose	O-alpha-D-glucopyranosyl-(1→6)-D-glucopyranose	
Maltulose	O-alpha-D-glucopyranose-(1→4)-D-fructose	
Sucrose	alpha-D-glucopyranosy-l-2-beta-D-fructofranoside	Cane sugar
Nigerose	O-alpha-D-glucopyranosyl-(1→3)-D-glucopyranose	Sakebiose
Turanose	O-alpha-D-glucopyranosyl-(1→3)-beta-D-fructose	
Kojibiose	O-alpha-D-glucopyranosyl-(1→2)-D-glucopyranose	
Laminaribose	O-beta-D-glucopyranosyl-(1→3)-D-glucopyranose	
α,β-Trehalose	alpha-D-glucopyranosyl beta-D-glucopyranoside	
Melibiose	O-alpha-D-galactopyranosyl-(1→6)-D-glucopyranose	
Gentiobiose	O-beta-D-glucopyranosyl-(1→6)-D-glucopyranose	
Leucrose	O-alpha-D-glucopyranosyl-(1→5)-D-fructopyranose	
Isomaltulose [21]	6-O-alpha-D-Glucopyranosyl-D-fructofuranose	Palatinose
	Trisaccharides	
Melezitose	O-alpha-D-glucopyranosyl-(1→3)-beta-D-fructofuranosyl-alpha-D-glucopyranoside	
3- α-Isomaltosylglucose	O-alpha-D-glucopyranosyl-(1→6)-O-alpha-D-glucopyranosyl-(1→3)-D-glucopyranose	
Maltotriose	O-alpha-D-glucopyranosyl-(1→4)-O-alpha-D-glucopyranosyl-(1→4)-D-glucopyranose	
1-ketose	O-alpha-D-glucopyranosyl-(1→2)-beta-D-fructofuranosyl-(1→2)-beta-D-fructofuranoside	
Panose	O-alpha-D-glucopyranosyl-(1→6)-O-alpha-D-glucopyranosyl-(1→4)-D-glucopyranose	
Theandrose	O-alpha-D-glucopyranosyl-(1→6)-alpha-D-glucopyranosyl-beta-D-fructofuranoside	
Centose	O-alpha-D-glucopyranosyl-(1→4)-O-alpha-D-glucopyranosyl-(1→2)-D-glucopyranose	
Isopanose	O-alpha-D-glucopyranosyl-(1→4)-O-alpha-D-glucopyranosyl-(1→6)-D-glucopyranose	
Raffinose	O-alpha-D-glucopyranosyl-(1→6)-O-alpha-D-glucopyranosyl-(1→2)-beta-D-fructofuranoside	
Erlose	O-alpha-D-glucopyranosyl-(1→4)-alpha-D-glucopyranosyl-beta-D-fructofuranoside	

TABLE 11.2 *(Continued)*

Sugar	Scientific name	Common name
	Oligosaccharides	
Isomaltotetraose	O-alpha-D-glucopyranosyl-(1→6)-O-alpha-D-glucopyranosyl-(1→6)-O-alpha-D-glucopyranosyl-(1→6)-D-glucopyranose	
Isomaltopentaose	O-alpha-D-glucopyranosyl-(1→6)-O-alpha-D-glucopyranosyl-(1→6)-O-alpha-D-glucopyranosyl-(1→6)-O-alpha-D-glucopyranosyl-(1→6)-D-glucopyranose	

TABLE 11.3 Fructose/Glucose Ratios for Selected Unifloral Honey Types

Honey type	F/G ratioa	Reference
Acacia (*Robinia pseudoacacia*)	1.12–1.89	[21,43,44,53,82,84,97]
Apiaceae spp.	1.32	[86]
Apple (*Malus domestica*)	1.25	[77]
Canola (*Brassica napus*)	0.32–1.24	[21,43,84,97]
Capparis (*Capparis* spp.)	1.25	[86]
Carob (*Ceratonia siliqua*)	1.49	[2]
Cherry (*Prunus avium*)	1.07	[77]
Chestnut (*Castanea sativa*)	1.36–1.86	[21,84,91,97]
Clover (*Trifolium* spp.)	1.10–1.26	[1,86,91,96]
Coriander (*Coriandirum sativum*)	0.61	[59]
Cotton (*Gossypium*)	1.49	[59]
Dandelion (*Taraxacum officinale*)	0.85–1.15	[84,97]
Egyptian clover (*Trifolium alexandrinum*)	0.88	[59]
Ericaceae	1.13	[86]
Eucalyptus (*Eucalyptus* spp.)	1.13–1.54	[2,10,59,70,84,86,96]
Fir (*Abies*)	1.30	[21]
Heather (*Calluna vulgaris*)	1.17–1.57	[2,43,70,91,97]
Honeydew	1.07–1.77	[70,84,91,97]
Horehound (*Marrubium vulgare*)	1.04	[10]
Jatropha curcas	1.50	[41]
Juazeiro (*Ziziphus joazeiro*)	1.50	[105]
Jujube (*Ziziphus mauritiana*)	1.13	[59]
Jurema branca (*Mimosa arenosa*)	1.1–1.2	[105]
Lavender (*Lavandula* spp.)	1.06–1.32	[21,70,84,91]
Linden (*Tilia* spp.)	1.04–1.41	[21,44,53,84,97]
Litchi (*Litchi chinensis*)	1.00	[59]

TABLE 11.3 *(Continued)*

Honey type	F/G ratioa	Reference
Lotus spp.	1.17–1.36	[86,96]
Malícia (*Mimosa quadrivalvis*)	1.1–1.3	[105]
Melilot (*Melilotus* spp.)	1.11	[86]
Mint (*Mentha piperita*)	1.17	[10]
Mustard (*Brassica campestris*)	0.46	[59]
Myrtaceae spp.	1.21	[86]
Oak (*Quercus* spp.)	1.03–1.31	[97]
Orange and Citrus (*Citrus* spp.)	1.09–1.30	[2,10,70,84,86]
Phacelia (*Phacelia tanacetifolia*)	1.10	[84]
Pine (*Pinus* spp.)	1.13	[82]
Plectranthus rugosus	1.24	[77]
Rhododendron (*Rhododendron* spp.)	1.10–1.39	[84,97]
Rosemary (*Rosmarinus officinalis*)	1.03–1.54	[2,10,70,84]
Rosewood tree (*Delbergia sisoo*)	0.95	[59]
Saffron (*Crocus sativus*)	1.23	[77]
Strawberry tree (*Arbutus unedo*)	1.15–1.67	[2,84]
Sunflower (*Helianthus annuus*)	1.00–1.34	[2,21,43,44,53,59,70,84]
Sweetvetch (*Hedysarum borelae*)	1.20–1.22	[84,86]
Thyme (*Tymus* spp.)	1.11–1.59	[2,10,43,84]
Tree heath (*Erica arborea*)	1.11	[84]
Velame branco (*Croton heliotropiifolius*)	1.20–1.30	[105]

aValues are given based on the cited literatures (minimum and maximum obtained F/G ratios)

11.2.3.1 DISACCHARIDES IN HONEY

Numerous disaccharides have been found in honey. Table 11.2 provides information on the disaccharides found in honey, while Table 11.4 presents some examples of the honey types where these were detected. Spanish researchers studied sunflower, heather, honeydew, eucalyptus, lavender, orange, and rosemary honeys and determined the amount of sucrose, maltose, maltulose, kojibiose, and isomaltose. Orange honey had the highest sucrose content (12%), honeydew honey contained the most maltose, maltulose, and kojibiose.[70]

Oddo et al.[84] created a databank that contained the data of 6719 honey samples from 21 countries. In this, the main physicochemical parameters of honey were evaluated, including the sucrose content. According to this, the highest sucrose content was found in lavender honey, and the lowest in honeydew.

Cotte et al.[21] analyzed seven types of honey with GC-FID and LC-PAD techniques. Seven types of honey, such as acacia, chestnut, linden, lavender, canola, sunflower, and fir were analyzed for their disaccharide content. Among the analyzed disaccharides, they determined the quantity of sucrose, maltose, maltulose, trehalose, isomaltulose, laminaribose, melibiose, isomaltose, and gentiobiose content of the honeys. The highest sucrose content was found in lavender honey. Melibiose was not detected in canola and sunflower honey. Higher maltose content was found in acacia and lavender honey (2.6%). Chestnut honey contained the highest amount of maltulose (2.6%). Acacia honeys were the richest in turanose, fir honey in trehalose (4.5%), and isomaltose (1.8%); however, chestnut also had 1.8% of isomaltose. Laminaribose and palatinose were less than 1% concentration in the examined honey samples except for acacia and fir and linden honeys, where the laminaribose content was 1.2%, 1.2%, and 1.3%, respectively. Summarizing, we can conclude that all types of honey have different disaccharide profiles.

Ruoff[97] evaluated the chemical composition of acacia, rhododendron, chestnut, dandelion, heather linden, and honeydew honey. According to the results, the highest sucrose content was found in acacia, the highest turanose content in rhododendron, and whereas the lowest was detected in heather. Chestnut honey showed the highest nigerose content, while canola honeys showed the lowest. The lowest average maltose and isomaltose content was also found in the canola honey, while *Metcalfa* honeydew had the highest content. Trehalose was found to be in the highest amount in honeydew honey (fir), while the lowest was in canola honey. Generally, each honey type contained the aforementioned disaccharides, but in each type, there were honeys that did not contain some type of sugar. To sum up, sucrose was detected in amounts lower than 0.5%, trehalose, isomaltose between 0.5% and 1%, and maltose, nigerose, turanose were between 1% and 2.1%.

Spanish researchers also found maltose, sucrose, trehalose, turanose, and isomaltose in blossom and honeydew honey. In this study, blossom honey had significantly higher sucrose and turanose content but the content of trehalose and isomaltose was significantly lower.[12]

Escuredo et al.[35] determined the fructose, maltose, sucrose, glucose, and melezitose content of nine different types of honey using HPAEC-PAD method. Maltose and sucrose were found in blackberry, chestnut, eucalyptus, heather, acacia, linden, canola, sunflower, and honeydew honey.

Turkish researchers analyzed blossom and pine honeys and detected sucrose and maltose using HPLC technique.[83] Sucrose was also detected in a Romanian study by HPLC, in acacia, canola, and honeydew honeys, where canola honey had the highest sucrose content. They also detected maltose in acacia, canola, multiflora, linden, and honeydew honey, where however, no significant difference was found between the groups.[43]

Pascual-Maté et al.[91] analyzed the sugar composition of six different honey types, such as chestnut, clover, heather, lavender, honeydew, and multiflora honey using gas chromatography-flame ionization detector (GC-FID). They found that the trehalose content of the samples was below 0.2%, maltose content ranged between 3.34 and 5.14%, whereas heather honey had the lowest and clover honey the highest. Isomaltose and gentiobiose were also under 0.2% but were the highest identified disaccharides in chestnut and honeydew honey.

Saffron, cherry, apple, *Plectranthus rugosus* honeys were examined in an Indian study using HPLC system with a refractive index detector. The study showed that the total disaccharide content of honeys was above 3%, where saffron honey had the highest and cherry honey, the lowest.[77]

TABLE 11.4 Disaccharides in Honey and the Types of Honey: Self-Developed with Concepts from Ref. [17,33,100,116].

Disaccharide	Honeys
Gentiobiose	*Trifolium* spp. [91] Acacia (*Robinia pseudoacacia*) [21] Canola (*Brassica napus*) [21] Chestnut (*Castanea sativa*) [21,91] Fir (*Abies*) [21] Heather (*Calluna vulgaris*) [91] Honeydew [91] Lavender (*Lavandula* spp.) [21,91] Linden (*Tilia* spp.) [21] Multiflora [91] Sunflower (*Helianthus annuus*) [21]
Isomaltose	Acacia (*Robinia pseudoacacia*) [21,97] *Apiaceae* spp. [86] Apple (*Malus domestica*) [77] Canola (*Brassica napus*) [21] Cherry (*Prunus avium*) [77] Chestnut (*Castanea sativa*) [21,91,97] Citrus (*Citrus spp.*) [70]

TABLE 11.4 *(Continued)*

Disaccharide	Honeys
	Dandelion (*Taraxacum officinale*) [97] Eucalyptus (*Eucalyptus* spp.) [70,86] Fir (*Abies*) [21] Heather (*Calluna vulgaris*) [70,91,97] Honeydew [12,70,91,97] Lavender (*Lavandula* spp.) [21,70,91] Linden (*Tilia* spp.) [21,97] *Lotus* spp. [86] *Myrtaceae* spp. [86] *Plectranthus rugosus* [77] Multiflora [86,91] Rhododendron (*Rhododendron* spp.) [97] Rosemary (*Rosmarinus officinalis*) [70] Saffron (*Crocus sativus*) [77] Sunflower (*Helianthus annuus*) [21,70] Sweetvetch (*Hedysarum borelae*) [86] *Trifolium* spp. [86,91]
Isomaltulose	Acacia (*Robinia pseudoacacia*) [21] Canola (*Brassica napus*) [21] Chestnut (*Castanea sativa*) [21] Fir (*Abies*) [21] Linden (*Tilia* spp.) [21] Lavender (*Lavandula* spp.) [21] Sunflower (*Helianthus annuus*) [21]
Kojibiose	Citrus (*Citrus spp.*) [70] Eucalyptus (*Eucalyptus* spp.) [70] Honeydew [70] Heather (*Calluna vulgaris*) [70] Lavender (*Lavandula* spp.) [70] Sunflower (*Helianthus annuus*) [70] Rosemary (*Rosmarinus officinalis*) [70]
Laminaribose	Acacia (*Robinia pseudoacacia*) [21] Canola (*Brassica napus*) [21] Chestnut (*Castanea sativa*) [21] Fir (*Abies*) [21] Lavender (*Lavandula* spp.) [21] Linden (*Tilia* spp.) [21] Sunflower (*Helianthus annuus*) [21]
Maltose	Acacia (*Robinia pseudoacacia*) [21,35,43,82,97] *Apiaceae* spp. [86] Apple (*Malus domestica*) [77] Blackberry (*Rubus* spp.) [35] Canola (*Brassica napus*) [21,35,43] Capparis (*Capparis* spp.) [86]

TABLE 11.4 *(Continued)*

Disaccharide	Honeys
	Carob (*Ceratonia siliqua*) [2] Cherry (*Prunus avium*) [77] Chestnut (*Castanea sativa*) [21,35,91,97] Citrus (*Citrus* spp.) [10,70,86] Dandelion (*Taraxacum officinale*) [97] Eucalyptus (*Eucalyptus* spp.) [2,10,35,70,86] Fir (*Abies*) [21] Heather (*Calluna vulgaris*) [2,35,70,91,97] Honeydew [12,35,43,70,85,91,97] Horehound (*Marrubium vulgare*) [10] *Jatropha curcas* [41] Lavender (*Lavandula* spp.) [21,70,91] Linden (*Tilia* spp.) [21,35,43,97] *Lotus* spp. [86] Melilot (*Melilotus* spp.) [86] Mint (*Mentha piperita*) [10] *Myrtaceae* spp. [86] Pine (*Pinus* spp.) [82] *Plectranthus rugosus* [77] Polyloral [43,91] Rhododendron (*Rhododendron* spp.) [97] Rosemary (*Rosmarinus officinalis*) [2,70] Saffron (*Crocus sativus*) [77] Strawberry tree (*Arbutus unedo*) [2] Sunflower (*Helianthus annuus*) [21,35,70] Sweetvetch (*Hedysarum borelae*) [86] Thyme (*Thymus spp.*) [2,10] *Trifolium* spp. [1,86,91]
Maltulose	Acacia (*Robinia pseudoacacia*) [21] Canola (*Brassica napus*) [21] Chestnut (Castanea sativa) [21] Citrus (*Citrus spp.*) [70] Eucalyptus (*Eucalyptus* spp.) [70] Fir (*Abies*) [21] Heather (*Calluna vulgaris*) [70] Honeydew [70] Lavender (*Lavandula* spp.) [21,70] Linden (*Tilia* spp.) [21] Rosemary (*Rosmarinus officinalis*) [70] Sunflower (*Helianthus annuus*) [21,70]
Melibiose	Acacia (*Robinia pseudoacacia*) [21] Citrus (*Citrus spp.*) [86] Chestnut (*Castanea sativa*) [21]

TABLE 11.4 *(Continued)*

Disaccharide	Honeys
	Ericeae spp. [86] Eucalyptus (*Eucalyptus* spp.) [86] Fir (*Abies*) [21] *Jatropha curcas* [41] Linden (*Tilia* spp.) [21] Lavender (*Lavandula* spp.) [21] Melilot (*Melilotus* spp.) [86] Multiflora [86] *Trifolium* spp. [86]
Nigerose	Acacia (*Robinia pseudoacacia*) [97] Dandelion (*Taraxacum officinale*) [97] Chestnut (*Castanea sativa*) [97] Heather (*Calluna vulgaris*) [97] Honeydew [97] Linden (*Tilia* spp.) [97] Rhododendron (*Rhododendron* spp.) [97]
Sucrose	Acacia (*Robinia pseudoacacia*) [21,35,43,53,82,84,97] *Apiaceae* spp. [86] Apple (*Malus domestica*) [77] Blackberry (*Rubus* spp.) [35] Canola (*Brassica napus*) [21,35,43,46,84] Capparis (*Capparis* spp.) [86] Carob (*Ceratonia siliqua*) [2] Cherry (*Prunus avium*) [77] Chestnut (*Castanea sativa*) [21,35,84,91,97] *Citrus spp*. [10,70,84,86] Coriander (*Coriandirum sativum*) [59] Cotton (*Gossypium*) [59] Dandelion (*Taraxacum officinale*) [84,97] Egyptian clover (*Trifolium alexandrinum*) [59] Eucalyptus (*Eucalyptus* spp.) [2,10,35,59,70,84,86,96] Fir (*Abies*) [21] Heather (*Calluna vulgaris*) [2,35,70,84,91,97] Honeydew [12,35,43,70,84,85,91,97] Horehound (*Marrubium vulgare*) [10] *Jatropha curcas* [41] Juazeiro (*Ziziphus joazeiro*) [105] Jujube (*Ziziphus jujube*) [46] Jujube (*Ziziphus mauritiana*) [59] Jurema branca (*Mimosa arenosa*) [105] Lavender (*Lavandula* spp.) [21,70,84,91] Linden (*Tilia* spp.) [21,35,53,84,97] Litchi (*Litchi chinensis*) [59] *Lotus* spp. [86,96]

TABLE 11.4 *(Continued)*

Disaccharide	Honeys
	Malícia (*Mimosa quadrivalvis*) [105] Melilot (*Melilotus* spp.) [86] Mint (*Mentha piperita*) [10] Multiflora [86,91] Mustard (*Brassica campestris*) [59] *Myrtaceae* spp. [86] Pine (*Pinus* spp.) [82] *Plectranthus rugosus* [77] Rhododendron (*Rhododendron* spp.) [84,97] Rosemary (*Rosmarinus officinalis*) [2,70,84] Rosewood tree (*Delbergia sisoo*) [59] Saffron (*Crocus sativus*) [77] Strawberry tree (*Arbutus unedo*) [2] Sunflower (*Helianthus annuus*) [21,35,53,59,70,84] Sweetvetch (*Hedysarum borelae*) [86] Thyme (*Thymus spp.*) [2,10,84] *Trifolium* spp. [86,91,96] Velame branco (*Croton heliotropiifolius*) [105]
Turanose	Acacia (*Robinia pseudoacacia*) [21,97] Carob (*Ceratonia siliqua*) [2] Chestnut (*Castanea sativa*) [21,97] Dandelion (*Taraxacum officinale*) [97] Fir (*Abies*) [21] Heather (*Calluna vulgaris*) [2,97] Honeydew [12,97] *Jatropha curcas* [41] Lavender (*Lavandula* spp.) [21] Linden (*Tilia* spp.) [21,97] Rhododendron (*Rhododendron* spp.) [97] *Apiaceae* spp. [86] Capparis (*Capparis* spp.) [86] Citrus (*Citrus* spp.) [2,86] Eucalyptus (*Eucalyptus* spp.) [2,86] *Myrtaceae* spp. [86] Sweetvetch (*Hedysarum borelae*) [86] Multiflora [2,86] Melilot (*Melilotus* spp.) [86] *Trifolium* spp. [86] Sunflower (*Helianthus annuus*) [2,21] Rosemary (*Rosmarinus officinalis*) [2] Thyme (*Thymus vulgaris*) [2] Canola (*Brassica napus*) [21]

TABLE 11.4 *(Continued)*

Disaccharide	Honeys
α,β-Trehalose	Acacia (*Robinia pseudoacacia*) [21,97] Apple (*Malus domestica*) [77] Canola (*Brassica napus*) [21] Cherry (*Prunus avium*) [77] Chestnut (*Castanea sativa*) [21,91,97] Dandelion (*Taraxacum officinale*) [97] Eucalyptus (*Eucalyptus* spp.) [86] Fir (*Abies*) [21] *Jatropha curcas* [41] Heather (*Calluna vulgaris*) [91,97] Honeydew [12,91,97] Lavender (*Lavandula* spp.) [21,91] Linden (*Tilia* spp.) [21,97] *Plectranthus rugosus* [77] Multiflora [86,91] Rhododendron (*Rhododendron* spp.) [97] Saffron (*Crocus sativus*) [77] Sunflower (*Helianthus annuus*) [21] *Trifolium* spp. [91]

11.2.3.2 TRISACCHARIDES IN HONEY

The main trisaccharides found in honey are listed in Table 11.2, and examples of the sources of the individual honeys can be found in Table 11.5 Spanish researchers studied sunflower, honeydew, heather, eucalyptus, lavender, orange, and rosemary honeys and found raffinose, melezitose, and erlose in honeys using the gas chromatography technique. In sunflower honey, no melezitose was detected and raffinose was found only in one sample in 0.1%. All of the trisaccharides were found in amounts lower than 0.6% on average. The highest raffinose and melezitose content was detected in honeydew honey, while the highest amount of erlose was found in rosemary honey.[70]

Cotte et al.[21] determined the raffinose, 1-ketose, erlose, melezitose, maltotriose, and panose content of seven types of honey. In canola and sunflower honey, none of 1-ketose, erlose, and melezitose were detected; moreover, none of maltotriose was found in canola honey, and none of 1-ketose in fir honey. Fir honey contained the highest quantity of melezitose (5.7%), while other honeys contained <1%. Higher amount of erlose was 1.9% and 1.4% in acacia and lavender, respectively.

The principal component analysis (PCA) model was built and grouped by the botanical origin of honey using the glucose, raffinose, fructose, trehalose, fructose/glucose ratio, maltose/trehalose ratio, and erlose/maltulose ratio. PCA is a multivariate data analysis approach that can be used to visualize patterns in data in a multidimensional space (principal components). Through the first principal component (45.5%), fir honey was completely discriminated from the other types, the highest contributing factor was the F/G ratio and erlose/maltulose ratio. Separation tendency was found between other types of honey through the second principal component (26.6%), where acacia honey was separated from the lavender, linden, rape, and sunflower honey, and this can be explained by the fact that chestnut and acacia honey are usually present in liquid form and the other ones are likely to crystallize.

Ruoff[97] found the lowest erlose content in heather and canola honey, while rhododendron contained the highest amount of it. Melezitose was found in all of the honey types in lesser than 1 g/kg, and honeydew (fir) had the highest melezitose content. Raffinose and maltotriose were found in almost all the honey types, except heather honey, the highest maltotriose was found in *Metcalfa* honeydew, while fir honeydew had the highest raffinose content. Trisaccharides were detected in quantities lower than 1% when averaging all the types.

Spanish researchers found melezitose in honeydew and blossom honey, where blossom honey had significantly higher melezitose content.[12] In another Spanish study, melezitose was detected in chestnut, blackberry, and honeydew honeys.[35]

Pascual-Maté et al.[91] detected raffinose, melezitose, erlose, maltotriose, panose, and isomaltotriose in six types of honey. Lavender and honeydew honey had the highest erlose content of 0.81% and 0.84%, respectively. Maltotriose and isomaltotriose were found in a really low amount, the highest maltotriose content was 0.11% in the case of the lavender honey, while the lowest value was obtained for the chestnut honey (0.05%). Panose content of heather honey was the lowest (0.125%) and honeydew honey contained the highest amount of 0.25%.

Trisaccharides, such as melezitose, maltotriose, and raffinose were found in cherry, apple, saffron, and *Plectranthus rugosus* honeys in the range of 1.22–2.19%.[77]

TABLE 11.5 Trisaccharides and their Botanical Sources in Honey: Self-developed with Information from Ref. [17,33,100,116].

Trisaccharide	Honey types
1-ketose	Chestnut (*Castanea sativa*) [21] Acacia (*Robinia pseudoacacia*) [21] Lavender (*Lavandula* spp.) [21] Linden (*Tilia* spp.) [21]
Erlose	Acacia (*Robinia pseudoacacia*) [21,97] *Apiaceae* spp. [86] *Capparis* (*Capparis* spp.) [86] Chestnut (*Castanea sativa*) [21,91,97] Citrus (*Citrus* spp.) [70,86] Dandelion (*Taraxacum officinale*) [97] Eucalyptus (*Eucalyptus* spp.) [70,86] Fir (*Abies*) [21] Heather (*Calluna vulgaris*) [70,91,97] Honeydew [70,91,97] Lavender (*Lavandula* spp.) [21,70,91] Linden (*Tilia* spp.) [21,97] *Lotus* spp. [86] Melilot (*Melilotus* spp.) [86] *Myrtaceae* spp. [86] *Multiflora* [86,91] Rhododendron (*Rhododendron* spp.) [97] Rosemary (*Rosmarinus officinalis*) [70] Sunflower (*Helianthus annuus*) [21] Sweetvetch (*Hedysarum borelae*) [86]
Maltotriose	Acacia (*Robinia pseudoacacia*) [21,97] Apple (*Malus domestica*) [77] Cherry (*Prunus avium*) [77] Chestnut (*Castanea sativa*) [21,91,97] Dandelion (*Taraxacum officinale*) [97] Fir (*Abies*) [21] Heather (*Calluna vulgaris*) [91] Honeydew [91,97] *Jatropha curcas* [41] Linden (*Tilia* spp.) [21,97] Lavender (*Lavandula* spp.) [21,91] *Plectranthus rugosus* [77] Multifloral [91] Rhododendron (*Rhododendron* spp.) [97] Saffron (*Crocus sativus*) [77] *Trifolium* spp. [91] Sunflower (*Helianthus annuus*) [21]

TABLE 11.5 *(Continued)*

Trisaccharide	Honey types
Melezitose	Acacia (*Robinia pseudoacacia*) [21,97] *Apiaceae* spp. [86] Apple (*Malus domestica*) [77] Blackberry (*Rubus* spp.) [35] Carob (*Ceratonia siliqua*) [2] cherry (*Prunus avium*) [77] Chestnut (*Castanea sativa*) [21,35,91,97] *Citrus* spp. [2,70] Dandelion (*Taraxacum officinale*) [97] Fir (*Abies*) [21] Eucalyptus (*Eucalyptus* spp.) [2,70,86] Heather (*Calluna vulgaris*) [2,70,91,97] Honeydew [12,35,70,91,97] *Jatropha curcas* [41] Lavender (*Lavandula* spp.) [21,70,91] Linden (*Tilia* spp.) [21,97] *Plectranthus rugosus* [77] Multiflora [2,86,91] Rhododendron (*Rhododendron* spp.) [97] Rosemary (*Rosmarinus officinalis*) [2,70] Saffron (*Crocus sativus*) [77] Strawberry tree (*Arbutus unedo*) [2] Thyme (*Thymus spp.*) [2] *Trifolium* spp. [91]
Panose	*Trifolium* spp. [91] Lavender (*Lavandula* spp.) [21,91] Multifloral [91] Chestnut (*Castanea sativa*) [21,91] Heather (*Calluna vulgaris*) [91] Honeydew [91] Sunflower (*Helianthus annuus*) [21] Linden (*Tilia* spp.) [21] Fir (*Abies*) [21] Canola (*Brassica napus*) [21] Sunflower (*Helianthus annuus*) [21]
Raffinose	Acacia (*Robinia pseudoacacia*) [97] *Apiaceae* spp. [86] Apple (*Malus domestica*) [77] *Capparis* (*Capparis* spp.) [86] Cherry (*Prunus avium*) [77] Chestnut (*Castanea sativa*) [91,97] Citrus (*Citrus* spp.) [70] Dandelion (*Taraxacum officinale*) [97] Eucalyptus (*Eucalyptus* spp.) [70,86]

TABLE 11.5 *(Continued)*

Trisaccharide	Honey types
	Fir (*Abies*) [21]
	Heather (*Calluna vulgaris*) [70,91]
	Honeydew [70,91,97]
	Jatropha curcas [41]
	Linden (*Tilia* spp.) [97]
	Lavender (*Lavandula* spp.) [70,91]
	Plectranthus rugosus [77]
	Multiflora [86] [91]
	Rosemary (*Rosmarinus officinalis*) [70]
	Rhododendron (*Rhododendron* spp.) [97]
	Saffron (*Crocus sativus*) [77]
	Sunflower (*Helianthus annuus*) [70]
	Trifolium spp. [91]

11.2.3.3 OLIGOSACCHARIDES WITH MORE THAN THREE MONOMERS

Sugars with more than three monomers were found in honey. Maltotetraose and maltopentaose, which are composed of glucose monomers are among these oligosaccharides.

Pascual-Maté et al.[91] detected maltotetraose in lavender, multiflora, chestnut, heather, and honeydew honey in the range of 0.0–0.09%. Maltotetraose was also found in saffron, apple, cherry, and *Plectranthus rugosus* honey. The lowest maltotetraose was found in the lowest quantity of 0.49 and 0.63 g/kg in *Plectranthus rugosus* and apple honey, respectively. Cherry and saffron honey had significantly higher amounts.[77]

Morales et al.[76] determined the maltooligosaccharide content of commercial honeys from beekeepers and also tracked adulterating agents, such as sugar syrups using HPAEC-PAD technique. In the study, they found maltose (2 glucose monomers), maltotriose, maltotetraose, maltopentaose, and maltohexose and maltoheptaose in the honeys in low quantity. The average amount of maltotriose was 0.02 g/100 g, those of maltotetraose and maltohexose was 0.01 g/100 g while the amount of maltopentaose was 0.03 g/100 g.

Egyptian researchers studied the sugar syrup adulteration of clover honey and determined its total polysaccharide and trisaccharide content using the HPLC technique. They found 1.5% of trisaccharides and 0.1% of other polysaccharides in the clover honey.[1]

TABLE 11.6 Oligosaccharides in Honey and their Botanical Origin.

Oligosaccharide	Honey types
Maltotetraose	Apple (*Malus domestica*) [77]
	Cherry (*Prunus avium*) [77]
	Chestnut (*Castanea sativa*) [91]
	Heather (*Calluna vulgaris*) [91]
	Honeydew [91]
	Lavender (*Lavandula* spp.) [91]
	Multiflora [91]
	Plectranthus rugosus [77]
	Saffron (*Crocus sativus*) [77]

11.2.3.4 DETERMINATION OF SUGAR IN HONEY

For the sugar determination in honey, various techniques are used. In the beginning, the analysis of sugars was performed using paper chromatography; however, they did not completely reflect the real composition of the samples; therefore, there was a demand for new methods. Selective adsorption enabled a more accurate identification of fructose and glucose,[17,33,100,116] followed by more developed and advanced techniques.

Some of these techniques have been described in the International Honey Commission Book, for the determination of sugars.[16] Reducing sugars can be determined by the reduction of Fehling's reagent; this method can also be found in the Codex Alimentarius protocol handbook. Sugars can also be quantified and detected by high-performance liquid chromatography (HPLC) using a refractive index detector. HPLC can also be used in combination with pulsed amperometric detection as sugars can act as very weak acids at pH 12–14 (pKa 12–13), and thus can be partially or completely ionized and analyzed by an ion exchange mechanism, that is, HPAEC-PAD.[35,76] Gas chromatography technique can also be used after derivatization for the determination of sugars in honey, mainly the predominant ones, such as fructose, glucose, sucrose, turanose, and maltose.[16]

11.2.3.5 CRYSTALLIZATION OF HONEY AND ITS CONNECTION WITH SUGAR AND MOISTURE CONTENT

Crystallization (granulation) is a natural process of honey that is dependent on numerous factors, such as sugar content, moisture content, viscosity, temperature, and nucleation seeds (pollens, beeswax, insoluble particles, and glucose).[34] One of the most decisive properties of crystallization is the fructose/glucose (F/G) ratio of the honey. Honeys with higher fructose/glucose ratio (F/G > 1.31), crystallize slowly, while honeys with a lower F/G than 1.11 crystallize quickly. Another important factor is the glucose/water ratio (G/W), if it is higher than 2.0, honey crystallizes very rapidly.[35]

Generally, honeys that contain more than 280 g/kg of glucose crystallize faster, but other sugars also have an influencing role in this process, such as in the case of the honeydew honeys. Melezitose content of honey, which is a trisaccharide formed from glucose(α1-3)-fructose(β2-1α)-glucose- is another important factor: if the melezitose content of honey is >100 g/kg, honey can crystallize very quickly. This sugar is very common in honeydew honey. The speed of the granulation determines the structure of the honey: the honey that crystallizes quickly generates fine and small crystals, the result being creamy, while slow crystallization can lead to rough and bigger crystals.[18]

11.2.4 AMINO ACIDS AND PROTEINS IN HONEY

Proteins and amino acids have been mentioned as useful indicators of the geographical origin of honey. Proline is the main amino acid of honey and can be applied as an indicator of the authenticity and freshness of honey. Generally, in honey, mostly all the amino acids can be found, but proline is the most important one, and its amount must be over 200 mg/kg. Table 11.7 presents the list of the amino acids and their sources (papers) where they have been investigated. Proline content lower than 180 mg/kg could be suspected as sugar syrups adulterated honey.[17,89]

Enzymes (such as diastase and invertase, glucose oxidase, and catalase) are also present in honey. Legislations state that the diastase number reflecting the diastase content of the honey should be over 8. The enzymes, especially invertase, are damaged by the storage and heat treatment, therefore, they can be used as indicators of freshness and authenticity.[17]

Cometto et al.[23] determined the amino acid composition of honeys from different regions of Argentina using RP-HPLC method, and found significant differences in the amount of the individual amino acids regarding the geographical origin.

Turkish researchers determined the amino acid profile of tualang, gelam, and acacia honeys and found that all of the honeys were rich in phenylalanine, but valine, proline, tyrosine also dominated. In acacia, lysine also was higher than in the two other types.[3] Egyptian researchers found 76.8 mg of proline per 100 g of clover honey.[1]

TABLE 11.7 Amino Acids of Honey and their Sources Self-Developed Components Adapted from da Silva et al.[98]

Amino acids of honey	Source
Alanine	[23,41,49,98,108,116]
Arginine	[3,23,49,108,116]
Aspartic acid	[3,23,41,49,108,116]
Cysteine	[108]
Cystine	[3,116]
Glucosaminic acid	[3,116]
Glutamic acid	[41,49,108,116]
Glutamine	[41]
Glycine	[23,41,49,108,116]
Histidine	[3,41,49,108,116]
Isoleucine	[23,41,49,108,116]
Leucine	[3,23,41,49,108,116]
Lysine	[3,23,41,49,108,116]
Methionine	[3,23,41,49,108,116]
Methionine sulfoxide	[116]
Ornithine	[23,116]
Phenylalanine	[3,23,41,108,116]
Proline	[3,30,41,90,108,116]
Serine	[3,23,41,49,108]
ß-alanine	[116]
Threonine	[3,23,41,49,108]
Tryptophan	[23,41,108,116]
Tyrosine	[3,23,41,49,108,116]
Valine	[3,23,41,49,108,116]
γ-amino butyric acid	[116]

German researchers determined the amino acid profile of *Jatropha curcas* honey using HPLC-DAD and FLD detectors. They did not detect asparagine, arginine, and cysteine. Proline was found in the highest amount (313.2 mg/kg), followed by threonine with 28.1 g/kg.[41]

Popek et al.[90] found that honeydew and nectar honeydew honeys had the highest proline content compared with rape, heather, buckwheat, multiflora, linden, and acacia honey.

Sun et al.[108] determined 18 amino acids in canola, jujube, acacia, lungan, and chaste tree honeys. Chaste tree honeys had the highest total amino acid content and phenylalanine was the dominating amino acid in this type of honey. Chaste tree honeys were followed by jujube honeys, these show higher tyrosine content than other groups. The lowest total amino acid content was obtained for the canola honey, in which serine was the second most abundant amino acid.

Can et al.[30] analyzed the proline content of honeys from different origins and found that chestnut, astragalus, and heather honey contained the highest amount of proline >700 mg/kg; however, other samples also contained more than 200 mg/kg, meeting the aforementioned requirements.

11.2.4.1 DETERMINATION OF PROTEINS AND AMINO ACIDS IN HONEY

Proteins of honey can be analyzed by ion exchange chromatography or immunoassays. Gel electrophoresis is also an available technique and has been popular in the past.[89] Determination of proline, the main amino acid, can be done by colorimetric method by UV/VIS spectrophotometry as described in the AOAC and IHC methods. Amino acids in honey can be determined by ion exchange chromatography or gas chromatography; however, HPLC with fluorescence detection is also available.[89]

11.2.5 ORGANIC ACIDS IN HONEY

It is well known that honeys are in the acidic range according to their pH value, which is the consequence of the presence of organic acids in about 0.57%. These acids can be derived from the enzymes and sugars during the transformation process from nectar to honey. The organic acid composition of the honey depends on its botanical source and a correlation can be found between the flavor and color of the honeys, too. The acidity of the honey also depends on the storage time and elaboration of honey: during long storage time the acidity of the honeys usually increases as a consequence of the fermentation in honey, when the alcohols and sugars in honey are transformed into acids due to the yeast activity.[98]

The most dominant acid in honey is gluconic acid, which is produced during the ripening of the honey by the action of the so-called glucose oxidase enzyme. Another dominating acid is citric acid, the amount of these two has a significant role in the differentiation between blossom and honeydew honey.[98,116]

Other acids (such as formic, fumaric, butyric, glutaric, aspartic, acetic, galacturonic, glutamic, glyoxylic, malic, lactic, malonic, ketoglutaric, α-hydroxyglutaric, α-ketoglutaric, isocitric, methylmalonic, pyruvic, 2-oxopentonic, shikimic, quinic, 2-hydroxybutyric, succinic, oxalic, tartaric, and propionic) can also be found in honey.[98]

Stinson et al.[106] found numerous organic acids in honey (such as acetic, formic, succinic, lactic, malic, citric, butyric, gluconic, and pyroglutamic) and other substances using paper chromatography and infrared spectroscopy.

Suarez-Luque et al.[102] found that organic acids are present in honey in magnitudes of mg/kg. They also extracted fumaric, citric, succinic, citric, malic, and maleic acids from honey.

Kawashima et al.[58] measured organic acids in honey by LC/IRMS method. The honeys were from different botanical and geographical origins and slight differences were found in the $\delta^{13}C$ value of the organic acids among the botanical or geographical groups. Tan et al.[109] also found different organic acids in honey using ion chromatography with electro membrane extraction. In this study, Chinese honey was used, and formic, malic, succinic, maleic oxalic, and fumaric acids were found in it. Generally, formic and malic acids presented the highest peak areas, other types of organic acid were not mentioned. In other studies, amounts and types of organic acids were identified and are presented in Table 11.8.

TABLE 11.8 Main Organic Acids in Honeys and their Botanical Sources.

Organic acid	Botanical sources of honey
	Simple mono- and polycarboxylic acids
Acetic	Honeydew [8] Orange (*Citrus* spp.) [115] Chestnut (*Castanea sativa*) [75,115] Eucalyptus (*Eucalyptus* spp.) [75] Multiflora [75]
Formic	Chestnut (*Castanea sativa*) [75] *Ericaceae* [78,80] Eucalyptus (*Eucalyptus* spp.) [75] Heather (*Calluna vulgaris*) [78,80] Honeydew [8,75] *Lavender* (*Lavandula* spp.) [80] Multiflora [75,80] Oak (*Quercus* spp.) [80] *Onobrychis sativa* [80] Rosemary (*Rosmarinus officinalis*) [80] Thyme (*Thymus* spp.) [78,80]

TABLE 11.8 *(Continued)*

Organic acid	Botanical sources of honey
Fumaric (2-Butenedioic acid)	Asphodel (*Asphodelus microcarpus*) [29] Chestnut (*Castanea sativa*) [48,103] Clover (*Trifolium* spp.) [103] Cotton (*Gossypium*) [48] *Ericaceae* [80] Eucalyptus (*Eucalyptus* spp.) [103] Heather (*Calluna vulgaris*) [80] Hondeydew [8] *Lavender* (*Lavandula* spp.) [80] Multiflora [29,48,80,81,103] Oak (*Quercus* spp.) [48,80] *Onobrychis sativa* [80] Pine (*Pinus* spp.) [48] Red gum (*Eucalyptus camaldulensis*) [29] Rosemary (*Rosmarinus officinalis*) [80] Strawberry tree (*Arbutus unedo*) [29] Sunlflower (*Helianthus annuus*) [48] Thymus (*Thymus spp.*) [80]
Glutaric (Pentanedioic acid)	*Ericaceae* [78,80] Heather (*Calluna vulgaris*) [78] Honeydew [8,78] *Lavender* (*Lavandula* spp.) [78] Multiflora [78] Rosemary (*Rosmarinus officinalis*) [78] Thyme (*Thymus* spp.) [78]
Maleic (cis-butenedioic acid)	Chestnut (*Castanea sativa*) [103] Eucalyptus (*Eucalyptus* spp.) [103] Clover (*Trifolium* spp.) [103] Multiflora [81,103] Honeydew [8]
Malonic (propanedioic acid)	Honeydew [8]
Oxalic (ethanedioic acid)	Multiflora [75,81] Chestnut (*Castanea sativa*) [75] Eucalyptus (*Eucalyptus* spp.) [75] Heather (*Calluna vulgaris*) [78,79] Thyme (*Thymus vulgaris*) [78,79] Multiflora [75] Lavender (*Lavandula* spp.) [78,79] French Lavender (*Lavandula* spp. stoechas) [78,79] Rosemary (*Rosmarinus officinalis*) [78,79,81] Honeydew [78,79] *Ericaceae* [78] Oak (*Quercus* spp.) [78]

TABLE 11.8 *(Continued)*

Organic acid	Botanical sources of honey
Propionic	*Ericaceae* [78,80] Heather (*Calluna vulgaris*) [78,80] Honeydew [8,78] *Lavender* (*Lavandula* spp.) [78,80] Multiflora [78] Oak (*Quercus* spp.) [78,80] Rosemary (*Rosmarinus officinalis*) [78,80] Thymus (*Thymus* spp.) [78,80]
Succinic (butanedioic acid)	Asphodel (*Asphodelus microcarpus*) [29] Acacia (*Robinia pseudoacacia*) [50] Canola (*Brassica napus*) [50,75] Chestnut (*Castanea sativa*) [48,103] Clover (*Trifolium* spp.) [50] *Ericaceae* [80] Eucalyptus (*Eucalyptus* spp.) [75,103] Heather (*Calluna vulgaris*) [80] Honeydew [8] Lavender (*Lavandula* spp.) [50,80] Linden (*Tilia* spp.) [50] Multiflora [29,50,75,80,103] Oak (*Quercus* spp.) [80] Orange (*Citrus* spp.) [50] *Onobrychis sativa* [80] Pine (*Pinus* spp.) [48,50] Red gum (*Eucalyptus camaldulensis*) [29] Rewarewa (*Knightia excelsa*) [117] Rosemary (*Rosmarinus officinalis*) [80] Strawberry tree (*Arbutus unedo*) [29] Thymus (*Thymus* spp.) [80] Wildtobacco (*Nicotina rustica*) [50] Cotton (*Gossypium*) [48]
Alpha-hydroxy acids	
Citric (2-Hydroxypropane-1,2,3-tricarboxylic acid)	Chestnut (*Castanea sativa*) [48,73,75,103,111] Eucalyptus (*Eucalyptus* spp.) [73,75,103] Clover (*Trifolium spp.*) [50,103] Multiflora [29,48,50,73,75,78,80,103] Honeydew [8,78,111] Strawberry tree (*Arbutus unedo*) [29] Asphodel (*Asphodelus microcarpus*) [29] Red gum (*Eucalyptus camaldulensis*) [29] Acacia (*Robinia pseudoacacia*) [50,111] Sunflower (*Helianthus annuus*) [48,50]

TABLE 11.8 *(Continued)*

Organic acid	Botanical sources of honey
	Canola (*Brassica napus*) [50] Linden (*Tilia* spp.) [50] Lavender (*Lavandula* spp.) [50,78,80] Orange (*Citrus* spp.) [50] Pine (*Pinus* spp.) [50] Rubus spp. [73] *Onobrychis sativa* [80] Rosemary (*Rosmarinus officinalis*) [78,80] Thyme (*Thymus* spp.) [80] Oak (*Quercus* spp.) [48,78,80] Heather (*Calluna vulgaris*) [78,80] *Ericaceae* [78,80] Cotton (*Gossypium*) [48]
Citramalic acid (2-hydroxy-2-methylbutanedioic acid)	*Ericaceae* [80] Heather (*Calluna vulgaris*) [80] *Lavender* (*Lavandula* spp.) [80] Oak (*Quercus* spp.) [80] *Onobrychis sativa* [80] Rosemary (*Rosmarinus officinalis*) [80] Thymus (*Thymus* spp.) [80]
Glycolic (2-Hydroxyacetic acid)	Honeydew [8]
Lactic (2-hydroxypropanoic acid)	Honeydew [8] Chestnut (*Castanea sativa*) [75] Eucalyptus (*Eucalyptus* spp.) [75] Multiflora [75]
Malic (2-Hydroxybutanedioic acid)	Asphodel (*Asphodelus microcarpus*) [29] Acacia (*Robinia pseudoacacia*) [50,111] Canola (*Brassica napus*) [50] Chestnut (*Castanea sativa*) [48,75,103,111] Clover (*Trifolium* spp.) [50] Eucalyptus (*Eucalyptus* spp.) [103] Honeydew [8,111] Lavender (*Lavandula* spp.) [50] Linden (*Tilia* spp.) [50] Orange (*Citrus* spp.) [50] Multiflora [29,48,50,75,80,81,103] Pine (*Pinus* spp.) [48,50] Red gum (*Eucalyptus camaldulensis*) [29] Strawberry tree (*Arbutus unedo*) Sunflower (*Helianthus annuus*) [48,50] Wildtobacco (*Nicotina rustica*) [50] Rewarewa (*Knightia excelsa*) [117]

TABLE 11.8 *(Continued)*

Organic acid	Botanical sources of honey
	Thymus (*Thymus* spp.) [80] Oak (*Quercus* spp.) [48,80] Heather (*Calluna vulgaris*) [80] Cotton (*Gossypium*) [48]
Mandelic acid (2-hydroxy-2-phenylacetic acid)	Dandelion (*Taraxacum officinale*) [107] Canola (*Brassica napus*) [107] Buckwheat (*Fagopyrum esculentum*) [107] Honeydew [107] Heather (*Calluna vulgaris*) [107] Pine (*Pinus* spp.) [107]
Tartaric (2,3-Dihydroxysuccinic acid)	Honeydew [8,48] Pine (Pinus spp.) [48] Chestnut (*Castanea sativa*) [48]
Sugar-derived acids	
Shikimic (3,4,5-Trihydroxy-1-cyclohexenecarboxylic acid)	Acacia (*Robinia pseudoacacia*) [50] Multiflora [50] Sunflower (*Helianthus annuus*) [50] Wildtobacco (*Nicotina rustica*) [50] Linden (*Tilia* spp.) [50] Lavender (*Lavandula* spp.) [50] Pine (*Pinus* spp.) [50]
Quinic (1,3,4,5-tetrahydroxycyclo-hexanecarboxylic acid)	Acacia (*Robinia pseudoacacia*) [50] Multiflora [50,78,80] Sunflower (*Helianthus annuus*) [50] Wildtobacco (*Nicotina rustica*) [50] Canola (*Brassica napus*) [50] Linden (*Tilia* spp.) [50] Lavender (*Lavandula* spp.) [50,80] Orange (*Citrus spp.*) [50] Pine (*Pinus* spp.) [50] Rosemary (*Rosmarinus officinalis*) [80] Thymus (*Thymus* spp.) [80] Oak (*Quercus* spp.) [80] Heather (*Calluna vulgaris*) [78,80] *Ericaceae* [80]
Galacturonic (2,3,4,5-tetrahydroxy-6-oxo-hexanoic acid)	Multiflora [78,80] Heather (*Calluna vulgaris*) [78] *Ericaceae* [78]
Gluconic (2,3,4,5,6-Pentahydroxycaproic acid)	Honeydew [8,78] Multiflora [29,74,75,78,80] Strawberry tree (*Arbutus unedo*) [29] Asphodel (*Asphodelus microcarpus*) [29]

TABLE 11.8 *(Continued)*

Organic acid	Botanical sources of honey
	Chestnut (*Castanea sativa*) [74,75] *Ericaceae* [78,80] Eucalyptus (*Eucalyptus* spp.) [74,75] Heather (*Calluna vulgaris*) [78,80] *Lavender* (*Lavandula* spp.) [78,80] Oak (*Quercus* spp.) [78,80] *Onobrychis sativa* [80] Red gum (*Eucalyptus camaldulensis*) [29] Rosemary (*Rosmarinus officinalis*) [78,80] Rubus spp. [74] Thyme (*Thymus* spp.) [78,80]
	Ketoacids
Pyruvic (2-Oxopropanoic acid)	Asphodel (*Asphodelus microcarpus*) [29] *Ericaceae* [80] Eucalyptus (*Eucalyptus* spp.) [75] Heather (*Calluna vulgaris*) [80] *Lavender* (*Lavandula* spp.) [80] Multiflora [29,80] Oak (*Quercus* spp.) [80] *Onobrychis sativa* [80] Red gum (*Eucalyptus camaldulensis*) [29] Rosemary (*Rosmarinus officinalis*) [80] Strawberry tree (*Arbutus unedo*) [29] Thymus (*Thymus* spp.) [80]
	Fatty acids
Stearic (octadecanoic acid)	Acacia (*Robinia pseudoacacia*) [50] Multiflora [50] Sunflower (*Helianthus annuus*) [50] Wildtobacco (*Nicotina rustica*) [50] Canola (*Brassica napus*) [50] Linden (*Tilia* spp.) [50] Clover (*Trifolium* spp.) [50] Orange (*Citrus spp.*) [50] Pine (*Pinus* spp.) [50]
Margaric (heptadecanoic acid)	Acacia (*Robinia pseudoacacia*) [50] Multiflora [50] Linden (*Tilia* spp.) [50] Clover (*Trifolium* spp.) [50] Orange (*Citrus* spp.) [50] Pine (*Pinus* spp.) [50]

The amount of the organic acid varies between the botanical groups, and within botanical origin according to geographical origin. Haroun et al.[48]

used HPLC-DAD method for the detection of organic acids and found that pine honey had higher succinic acid content and citric acid content than the chestnut, cotton, multiflora, and sunflower honeys, while the highest malic acid content was found in chestnut honeys.

Navarrete et al.[74,81] used electrophoretic method and found various maleic (5.10–30.53 mg/kg), malic (207.20–1257.13 mg/kg), oxalic acid (14.1–278.27 mg/kg) content among multiflora honeys. D-gluconic acid content measured by enzymatic method ranged between 3910 and 11710 mg/kg for multiflora honeys, and 5340–8640 mg/kg for eucalyptus honey. Enzymatic method was also used to quantify the citric acid content. The latter was highest in chestnut honey with 827 mg/kg, the results of multiflora honey ranged between 44.2 mg/kg and 428 mg/kg, while eucalyptus honeys had a citric acid content of 48.3–47.9 mg/kg.[73]

In a Spanish study, researchers used capillary zone electrophoresis with direct ultraviolet detection. Chestnut honey was found to have the highest oxalic acid, formic acid, malic, succinic, acetic acid, and citric, gluconic, and total acid content (10.1–16.3 g/kg), while eucalyptus honey had the lowest total acid (3.9–8.9 g/kg), gluconic, citric, and lactic acid content.[75]

Del Nozal et al.[80] used HPLC for the quantification of organic acids and found the highest citramalic acid content in the case of the oak (244 mg/kg) and multiflora (236 mg/kg) honeys, *Erica* spp. honeys had the highest formic (236 mg/kg), quinic (3513 mg/kg), and gluconic (4933 mg/kg) acid content. Oak honey had also the higher malic acid content (873 mg/kg), while thyme honey was the richest in citric acid (650 mg/kg). Honeydew honeys also had higher malic acid content (730–740 g/kg),[111] while acacia honey had lower values (14 mg/kg), being quantified by UV/VIS detector. Spanish researchers also used HPLC method to quantify the organic acids and found that heather and ling honey had the highest oxalic acid content (0.1042–0.336 mg/kg and 0.0855–0.1684 mg/kg, respectively), also these honeys had the highest citric and gluconic acid content. Multiflora honeys were the richest in propionic acid (0.627–2.141 mg/kg).[78]

In honey, not only organic acids, but also inorganic phosphoric acids were detected in the case of acacia, multiflora, sunflower, wild tobacco, canola, linden, lavender, clover, orange, and pine honey.[50]

11.2.5.1 DETERMINATION OF THE ORGANIC ACID CONTENT

In the recent years, different methods have been used to detect or quantify the organic acid content of the honey, such as capillary zone electrophoresis,

HPLC with ultraviolet visible detector or diode array detector, nuclear magnetic resonance (NMR), and gas chromatography–mass spectrometry. In these studies, the limit of detection and limit of quantification were also determined. Brugnerotto et al. summarized the previously used techniques and instruments.[8]

11.2.6 MINERAL CONTENT OF HONEYS

Numerous minerals and trace minerals have been detected in honey (Table 11.9). The quantity of the mineral in honey is affected by the botanical and geographical source and it is reflected in the mineral content. Generally, lighter floral (blossom) honeys contain lower amount of minerals and trace elements, while darker honeys, for example, honeydew and chestnut contain higher amounts. The mineral content, electrical conductivity, and ash content are strongly correlated.[9,15] Mineral content of honey ranges between 0.02% and 1.03% according to White (1978). In honey, potassium was found in the highest amount, counting for about 1/3 of the total mineral content. The environmental conditions and geographical origin (soil) have an important role in the quantity of minerals in honey, due to the heavy metals in honey, such as cadmium, lead, chromium, and nickel originating from anthropogenic source; therefore, the amount and presence of these four elements can serve as an indicator of environmental pollution.[14]

TABLE 11.9 Minerals and Trace Elements in Honey based on Ref. [15,116].

Main minerals	Trace elements
Calcium – Ca	Aluminum – Al
Chlorine – Cl	Arsenic – As
Chromium – Cr	Bromine – Br
Copper – Cu	Barium – Ba
Iron – Fe	Cobalt – Co
Magnesium – Mg	Cadmium – Cd
Manganese – Mn	Fluoride – F
Phosphorus – P	Iodide – I
Potassium – K	Lead – Pb
Selenium – Se	Lithium – Li
Silica – (SiO_2)	Molybdenum – Mo
Sodium – Na	Nickel – Ni
Sulfur – S	Rubidium – Rb
Zinc – Zn	Strontium – Sr
	Vanadium – V
	Zirconium - Zr

11.2.6.1 MINERAL CONTENT OF HONEY BASED ON THE REPORTS OF SOME REPRESENTATIVE STUDIES

Rodriguez-Otero et al.[95] determined the K, Mg, Cu, Na Fe, Mn, P, Ca, Cl, Si, S content of 91 raw honeys from North-West Spain. They found that the honeys contained the highest amount of K followed by Cl, Na, P, Mg, Ca, which were higher than 100 mg/kg. Sulfur content was 68 mg/kg, followed by Si, Fe, Mn, and Cu, which were lower than 10 mg/kg.

Egyptian researchers determined the K, Na, and Ca content of clover honey, obtaining 269 mg/kg Na, 59.5 mg/kg K, and 52.6 mg/kg Ca for the pure clover honey using flame analyzer.[1]

Terrab et al.[110] used the ICP-OES technique to quantify the element content of 24 thyme honeys from various regions of Spain. K, Ca, Si, Mg, Na, P, and S were found to be the most abundant elements. Al, Li, Fe, Cu, Zn, and Mn were found in much lower amounts (> 1 mg/kg, but < 20 mg/kg). Sr, Ba, Cr, As, Pb, Ni, Se, Co, V, Cd, and Mo were trace elements, found in amounts lower than 1 mg/kg. Some of the elements, such as vanadium, molybdenum, cobalt, and cadmium were not detected in all of the honey types. Honeys were classified correctly according to their geographical origin (coastal and Sierra region).

Fernández-Torres et al.[40] applied ICP-AES for the determination of the mineral composition of orange blossom, heather, eucalyptus, and rosemary honeys from different regions of Spain. Eucalyptus and heather honeys had higher mineral content and they were also darker than orange and rosemary honeys. LDA models were built using the results of the B, Zn, Cu, Ca, Mn, Sr, P, Ba, K, and Na for the classification of the four groups. The average correct recognition and prediction abilities were 100% and 97%, respectively.

Hernández et al.[47] used atomic absorption spectrometry (AAS) for the determination of Fe, Zn, K, Na, Mg, Ca, Sr, Li, and Rb of unifloral and multiflora honeys. K was found in the highest amounts, followed by Rb, Ca, Na, Mg, Li, Fe, Zn, Sr, and Cu. The honeys contained different amounts of minerals according to their geographical origin.

Bogdanov et al.[14] studied the minerals content of unifloral honeys from diverse origins (urban, rural, and mountain) using ICP-MS. Pb, Mn, Cd, Fe, Ni, Cr, Zn, and Cu were analyzed in acacia, dandelion, linden, chestnut, canola, rhododendron, fir, mountain blossom, and multiflora honeys. They found significantly higher mineral content in honeydew honey than in blossom honey, except for Ni content. The highest Cd was found in fir honey, Pb, and Cr in rhododendron honey, Mn in chestnut and fir honey, Fe, Ni, Cu, and Zn in fir honey. They also found significant differences in the mineral content according to their geographical origin. Linear discriminant analysis (LDA) model was built for the classification of the botanical sources and they found that acacia, chestnut, and fir honey could have been classified correctly (100%), and the total correct classification was 76%.

Honeys from New Zealand were analyzed by Vanhanen et al.[114] using ICP-OES for the quantification of Al, B, As, Ca, Cr, Fr, Cd, K, Mg, Mn, Cu Mo, Ni, P, Na, Pb, S, and Zn content of local honey types. Honeydew honey contained the highest elemental content (4060 mg/kg) and the lowest was obtained for Viper's bugloss (126 mg/kg). The abundant minerals were K, P, and Ca.

De Alda-Garcilope et al.[4] used AAS to quantify the K, Na, Ca, Mg, Fe, Cu, and Zn content of chestnut, rosemary, thyme, avocado, and orange honey from mountain zone (containing chestnut pollen) and multiflora honeys. No Cu was detected in the honeys, and Fe was found only in chestnut and thyme honey, whereas more Fe was found in chestnut honey. Potassium was found in the largest amount in the honeys followed by Mg, Ca, Na, and Zn. Avocado honey was the richest in K and Zn, while thyme honey contained the most Ca and Na. Chestnut honey was the richest in Mg.

Alves et al.[7] determined the K, Ca, Na, Mg, and Fe content of rosemary, orange, thyme, locust podshrub, strawberry tree, eucalyptus, and heather honeys with AAS. K was found in the highest amount in honeys, and heather honeys were the richest in potassium. Highest calcium and sodium content were obtained in eucalypt honey. Locust podshrub honey was the richest in iron and heather honey in magnesium.

Czipa et al.[31] analyzed the essential and toxic elements in Hungarian honeys from different botanical origins using ICP-MS and ICP-OES techniques. Potassium (K) content was the highest in the samples; acacia honeys (light honey) showed the lowest K content and forest honey had the highest. Calcium was found in the second largest amount after K, showing the highest amount in sunflowers and the lowest in acacia. Other elements, such as Cr, Mo, Se, Cu, Fe, Mg, Mn, P, S, Zn, Al, As, Cd, and Pb were also evaluated. LDA model was built for the classification of acacia, sunflower, and canola honeys, showing complete separation from each other.

Escuredo et al.[36] studied the K, Ca, P, and Mg content of chestnut, eucalyptus, heather, blackberry, and muliflora honeys using AAS. Chestnut honey had the highest K, P, and Mg content, while heather honey had the highest Ca content. Blackberry and eucalyptus honey had significantly lower K content compared with multiflora honey and blackberry honey contained significantly lower Mg content compared with chestnut honey.

Kaygusuz et al.[64] studied the Ca, Fe, K, Cu, and Mn content of heather, chestnut, pine, oak, acacia, astragalus, and lavender honey using ICP-OES method. Each honey type contained different amounts of minerals; moreover, significant differences were found within botanical group too. K was found in the highest amount followed by Ca, Fe, Cu, and Mn. The oak honey was the richest in Mn and Cu. Chestnut honey also contained high amounts of K, Ca, compared with the other types.

Kadri et al.[65] analyzed the mineral content of coffee honey from Brazil using AAS. It contained the most K (944.3 mg/kg) followed by Ca (338.7 mg/kg), Mg (82.21 mg/kg), Na (31.21 mg/kg). Fe, Cu, Li, and Zn were found below 1.02 mg/kg.

Karabagias et al.[63] used ICP-OES to quantify the Al, Ba, Ca, B, Cu, Li, Mg, Fe, Mn, Mo, P, Sb, Se, Ni, Si, Sr, Sn, Ti, and Zn content of citrus honeys of different geographical origins from countries of the Mediterranean region. Significant differences were found between the different geographical sources except for Zn, Mg, Mn, Cu, Ba, and Al content. Greek honeys had the highest total mineral content, followed by Egyptian, Moroccan, and Spanish honeys. Based on the minerals, honeys were classified according to their country of origin with a correct classification rate of 83.8%.

Biluca et al.[13] determined the Na, K, Ca, Mg, Mn, Al, Fe, Co, Cu, Zn, Li, Ni, and Ba content of Brazilian honey types using ICP-MS method. They found that Mg, Cu, Zn, Ba, Mn showed the highest bioaccessibility but contained the most K (262–4980 mg/kg), Ca (88.6–138 mg/kg), Na (12.7–261 mg/kg), and Mg (25.9–231 mg/kg). Fe, Mn, Zn, Cu, and Co were found at trace levels, lesser than 6 mg/kg.

Louppis et al.[67] determined the elemental composition of Greek fir, pine, orange blossom, and thyme honeys using ICP-OES technique. They found significant differences in the Al, As, Ba, Cu, Fe, Mg, Ag, Ca, Mn, Sb, Ni, Se, and Si content, while significant differences were not found between the four botanical groups in the Zn, Tl, Ti, Pb, Mo, Hg, Cr, and B content. Beryllium was found only in fir honeys, Ag only in pine and fir honeys, titanium in thyme honey, and no thallium and chromium were found in orange blossom honey. Cd, Co, and V were not found in the honeys at all. The classification model was built using the data of the mineral evaluation to discriminate the four botanical types resulting in 93.5% correct classification for the whole sample set, fir and pine honeys showed correct classification.

Bergamo et al.[20] used a capillary electrophoresis system equipped with a diode array detector to determine the K, Mg, Na, Ca, and Mn content of bracatinga honeydew and blossom honeys and their mixture. K was in the highest amount in both blossom and honeydew honeys. The total mineral content was higher in the honeydew honey (5532.06 mg/kg) than blossom honeys (2076.16 mg/kg). Honeydew honeys contained more K and Mg, while blossom honeys were richer in Ca, Mn, and Na.

Nayik et al.[82] determined the K, Na, P, and Ca content of pine, acacia, and Multiflora honeys using air acetylene flame atomic absorption spectrometer (FAAS). Pine honey contained the highest amount of the four minerals, significantly higher than the multiflora and acacia honey contained significantly lower amounts compared to the other two groups.

Lanjwani et al.[66] analyzed the Na, Co, Mg, Ca, K, Fe, Zn, Cu, Pb, Cr, Mn, Ni, Cd, quantity of different honey samples from diverse botanical and geographical region of Pakistan using FAAS. Potassium was found in the highest amount, ranging between 225 and 439 mg/kg, followed by sodium (77.5–200 mg/kg), Ca (46.1–98.1 mg/kg), and Mg (31.3–71.8 mg/kg). Fe was also found to be relatively higher than the other microelements, ranging between 2.98 and 16.2 mg/kg. Zn also was above 1 mg/kg, while Co, Cu, Mn, Cr, Ni, Pb, and Cd were <1 mg/kg.

El-Haskoury et al.[37] determined the Na, K, Ca, Zn, Mg, Fe, Cu, Zn, Ni, Pb, Cd, and Co content of carob honey. Cd, Pb, and Co were not detected in the honeys. Honeys were richest in K and Na, followed by Ca, Mg, Fe, Zn, Cu, and Ni.

Rodríguez-Flores et al.[94] analyzed the K, Zn, Fe, Mg, Na, Ca, P, and Cu content of honeydew and chestnut honeys from Spain using atomic emission and atomic absorption spectrometer. Researchers found that chestnut honey had significantly higher Na (48 mg/kg) than honeydew honey, while honeydew honey had significantly higher Mg content (195 mg/kg). Significant differences were not obtained for the rest of the analyzed elements.

Sakač et al.[101] determined the K, Na, Ca, Mg, Fe, Cu, and Zn content of acacia, sunflower, and meadow honeys using AAS. Acacia honeys contained significantly lower K, Mg, Zn, and Ca than the meadow and sunflower honey. Sunflower honey had the highest average K, Mg, and Ca content, while meadow honey was the richest in Zn, and acacia honey in Fe. Cu was found <1 mg/kg and Na <50 mg/kg.

Sajtos et al.[99] investigated the elemental content of Hungarian honeys from various botanical and geographical origins. Acacia, honeydew, forest, phacelia, solidago, milkweed, chestnut, linden, sunflower, canola, and multiflora honeys were analyzed according to their Al, B, Zn, Ca, Mn, K, Na, Pb, and Mg, content using ICP-OES and MP-AES techniques. Researchers found K, Na, Ca, and Mg, in the highest amount. Honeys were also ordered according to their total element concentration in the following ascending order: phacelia, milkweed, acacia, canola, multiflora, solidago, sunflower, linden, forest, honeydew, and chestnut honeys. They also found differences according to the geographical origin of the honeys: Western Transdanubian and Central Hungarian honeys differed from the rest of the parts of the county.

11.2.6.2 TECHNIQUES FOR THE ANALYSIS OF MINERAL COMPOSITION OF HONEY

Elemental analysis of honey can be performed with various methods. The total mineral content can be determined using gravimetric analysis of the ash content, as described in the International Honey Commission Method Book and by the AOAC (method 920.181). Determination of electrical conductivity is faster, and it can be used as an alternative to ash determination. Sancho et al. determined linear correlation between total ash content and electrical conductivity as the following: % total ash = 0.083 × electrical conductivity – 0.092 (r = 0.964).[104]

Moreover, gravimetric, titrimetric, and colorimetric methods were also applied to determine the chloride and phosphate content of honey samples. Specific anions, such as molybdate, barium sulfate, and vanadate were analyzed with titrimetric and colorimetric methods.[89]

Currently, spectroscopic methods are the most commonly applied techniques for the quantification of minerals and trace elements of honey. Atomic absorption spectroscopy (AAS) is often used for the analysis of honey. Two types of this technique are available: the flame atomic absorption and the electrothermal absorption spectroscopy. Atomic emission spectroscopy (AES) is also often applied and relies on measuring the intensity of the emitted light, which is proportional to the number of the ions or atoms. Inductively coupled plasma mass spectrometry (ICP-MS) enables the accurate determination of minerals at very low concentrations. These three methods are the most applied techniques[87] although chromatographic methods with electrochemical or conductivity detectors are also available. Capillary zone electrophoresis provides a quick, accurate, and sensitive approach for the reliable determination of the anions and cations in honey.[89]

11.2.7 VITAMINS IN HONEY

Honeys are not rich in vitamins; however in the recent 80 years, some studies have been performed to check their vitamin content. Based on the articles,

B1 (thiamine), B2 (riboflavin), vitamin C (ascorbic acid), B5 (pantothenic acid), B3 (niacin), B6 (pyridoxine) folic acid, and biotin were detected in honey in different amounts. Vitamin C can be found in most of the honey types, but it is sensitive, its amount can decrease through the oxidation caused by the hydrogen peroxide (produced by the glucose oxidase), the filtration, storage, and process method. Niacin and vitamin C are usually present in honey at milligram levels, while others are at microgram order of magnitude.[11,27,45,51,57,60,61,68,69,72,98]

Results of the studies were collected, and averages were calculated for the different groups of honey showing a high variability among the botanical groups; however, the standard deviation was only high within one botanical group. This can be because of the variability of the geographical origin, storage conditions, and also the determining method which has a high effect on the results; therefore, a clear conclusion cannot be stated regarding the vitamin content of the honey. Nonetheless, honeys contain mainly vitamins from the B group and vitamin C, but only few studies have been performed on this topic.[11,27,45,51,57,60,61,68,69,72,98]

11.2.7.1 DETERMINATION OF VITAMIN CONTENT OF HONEY

The first vitamin-related studies regarding honey were done in the years of 1930–1940.[45] During that time, mostly biological assays were used for the determination, but in the decade of 2010, a new rapid method was found to measure the vitamin content of the honey with high-performance liquid chromatography-reverse phase (HPLC-RP) assays.[98]

11.2.8 ANTIOXIDANTS IN HONEY

Honey is rich in antioxidants mainly containing polyphenols. In our previous chapter, we elaborated on the different types of antioxidant compounds – phenolic acids and flavonoids – occurring in honey. In general, lighter honeys contain less polyphenols than darker honey. The phenolic components can be divided into two principal groups, the non-flavonoid and flavonoids compounds.

Numerous types of flavonoids have been identified in honeys, among which quercetin, chrysin, pinobanskin, kaempferol, pinocembrin, luteolin, galangin, apigenin, and myricetin are the most common. Naringenin, hesperetin, acacetin, catechin, epicatechin, tricetin, genistein, isorhamnetin or ellagic acid were also detected only in some types of honey.

The non-flavonoid components are usually phenolic acids and can be classified into groups based on which compound they derived from. The main benzoic acid derivatives are benzoic acid, syringic acid, methyl syringate, 4-hydroxybenzoic acid, 3-hydroxybenzoic acid, methyl-4-hydroxybenzoate, protocatechuic acid, gentisic acid, gallic acid, vanillic acid, eusdemic acid, and o-anisic-acid. The latter two have been found only in Manuka (*Leptoseprum scoparium*) honey. The most important hydroxyphenylacetic acid derivatives include 4-hydroxyphenylacetic acid homogentisic acid and homoanisic acid. Trans-cinnamic acid, m-coumaric acid, caffeic acid, ferulic acid, isoferulic acid, rosmarinic acid, and chlorogenic acid, p-coumaric acid are belonging to the group of cinnamic acid derivates. There are also metabolites of phenolic acids, such as, phenylacetic acid, 3-phenillactic acid, phenylpropanoic acid, and mandelic acid.[6,22,28,32,42,52,54,55,56,62,93]

11.3 SUMMARY

Honeys are composed of nutritionally active components, such as sugars, amino acids, minerals, organic acids, vitamins, and antioxidants. The amount of which is dependent on the botanical and geographical origin of the honeys. In this chapter, an overview of the most important constituents was given and relevant literature referring to the occurrences of these substances in honey was also summarized.

The water content of the honey was generally lower than 20%, honey having a relatively low water activity $a_w < 0.6$. Besides moisture, sugar is the main compound of honey with more than 20 different types of sugars belonging to the group of mono-, di-, tri,- and tetra- or pentasaccharides. Monosaccharides were present in the highest proportion (glucose and fructose), and their ratio strongly influences the crystallization of honey. Other sugars (e.g., disaccharides, sucrose, maltose, isomaltose, turanose) have been found in numerous types of honey in a range that varies between and within the botanical groups and was also dependent on the geographical source. With regards to trisaccharides, melezitose, raffinose, and erlose were the most abundant ones. Tetra-and pentasaccharides are present only in few amounts.

Amino acids and proteins are other components of honey that could be found only in low amounts, depending on the type of the amino acid and also on the origin of the honey. Almost all the physiologically important amino acids were available in honey, but only in small amounts. Proline is the most important amongst amino acids in honey. It can be stated that the amino acid composition of honeys is not very well studied.

Organic acids were also available in honey in about 0.57%. The most important organic acids reported in several types of honey were fumaric, succinic, citric, malic, quinic, gluconic, and oxalic acid. Besides these, maleic, glutaric, formic, acetic, shikimic, and tartaric acids were also detectable in honeys.

With regards to the mineral content, honeys contain minerals approximately in the range of 0.01–1.03%. Generally, potassium could be found in the highest amount, but magnesium, sodium, and calcium could also be present in significant amounts. However, the quantity and the order of these latter three depend on the botanical and geographical source of honey.

Among the vitamins, mainly vitamins of the B group were present in honey: B1 (thiamine), B2 (riboflavin), vitamin C (ascorbic acid), B5 (pantothenic acid), B3 (niacin), B6 (pyridoxine) folic acid, and biotin were detected in honey in different amounts. Vitamin C (ascorbic acid) could also be found in honeys. Niacin and vitamin C were generally present at mg/kg magnitude, while the others were at μg/kg.

Beyond all of these, polyphenols, including flavonoids and non-flavonoids were also important phytochemicals in honey.

It can be concluded that honey is a complex matrix, rich in nutritionally active components like amino acids, organic acids, vitamins, and minerals. Botanical and geographical source has a decisive role in the amount of the individual compounds.

ACKNOWLEDGMENTS

Supported by the ÚNKP-19-3-I-SZIE-71 (Zsanett Bodor) and the ÚNKP-19-4 (Zoltan Kovacs) New National Excellence Program of the Ministry for Innovation and Technology. This chapter was supported by the Doctoral School of Food Science, Szent István University. (Zsanett Bodor) and by the Janos Bolyai Scholarship of the Hungarian Academy of Sciences. The Project is supported by the European Union and co-financed by the European Social Fund (grant agreement no. EFOP-3.6.3-VEKOP-16-2017-00005. This research was supported by the Ministry for Innovation and Technology within the framework of the Higher Education Institutional Excellence Program (NKFIH-1159-6/2019) in the scope of plant breeding and plant protection researches of Szent István University.

KEYWORDS

- **botanical origin**
- **composition**
- **geographical origin**
- **honey**
- **identification**
- **physicochemical**
- **nutrients**

REFERENCES

1. Abdel-Aal, E. S. M. S. M. M.; Ziena, H. M.; Youssef, M. M. Adulteration of Honey with High-fructose Corn Syrup: Detection by Different Methods. *Food Chem.* **1993,** *48* (2), 209–212.
2. Abramovič, H.; Jamnik, M.; Burkan, L.; Kač, M. Water Activity and Water Content in Slovenian Honeys. *Food Control* **2008,** *19* (11), 1086–1090.
3. Adnan, N. A.; Suan Chua, L.; Sarmidi, M. R. Thermal Treatment Effect on Free Amino Acids in Honey Samples. *J. Teknol.* **2014,** *69* (4), 29–33.
4. Alves, A.; Ramos, A.; Gonçalves, M. M.; Bernardo, M.; Mendes, B. Antioxidant Activity, Quality Parameters and Mineral Content of Portuguese Monofloral Honeys. *J. Food Compost. Anal.* **2013,** *30* (2), 130–138.
5. Anjos, O.; Campos, M. G.; Ruiz, P. C.; Antunes, P. Application of FTIR-ATR Spectroscopy to the Quantification of Sugar in Honey. *Food Chem,* **2015,** *169*, 218–223.
6. Anklam, E. A Review of the Analytical Methods to Determine the Geographical and Botanical Origin of Honey. *Food Chem.* **1998,** *63* (4), 549–562.
7. Becker, E.; Kardos, R. F. *Über den Vitamin C-Gehalt von Honig* (About the vitamin C content of honey). *Zeitschrift für Untersuchung der Lebensmittel* **1939,** *78* (4), 305–308.
8. Bentabol - Manzanares, A.; García, Z. H.; Galdón, B. R.; Rodríguez, E. R.; Romero, C. D. Differentiation of Blossom and Honeydew Honeys using Multivariate Analysis on the Physicochemical Parameters and Sugar Composition. *Food Chem.* **2011,** *126* (2), 664–672.
9. Bergamo, G.; Seraglio, S. K. T.; Gonzaga, L. V.; Fett, R.; Costa, A. C. O.; Tischer Seraglio, S. K.; Gonzaga, L. V.; Fett, R. Mineral Profile as a Potential Parameter for Verifying the Authenticity of Bracatinga Honeydew Honeys. *LWT Food Sci. Technol.* **2018,** *97* (3), 390–395.
10. Biluca, F. C.; de Gois, J. S.; Schulz, M.; Braghini, F.; Gonzaga, L. V.; Maltez, H. F.; Rodrigues, E.; Vitali, L. Phenolic Compounds, Antioxidant Capacity and Bioaccessibility of Minerals of Stingless Bee Honey (Meliponinae). *J. Food Composit. Anal.* **2017,** *63* (2), 89–97.

11. Bodor, Z.; Benedek, C.; Kaszab, T.; Zaukuu, J. L. Z.; Kertész, I.; Kovacs, Z. Classical and Correlative Analytical Methods for Origin Identification of Hungarian Honeys. *Acta Aliment.* **2019,** *48* (4), 477–487.
12. Bogdanov, S.; Haldimann, M.; Luginbühl, W.; Gallmann, P. Minerals in Honey: Environmental, Geographical and Botanical Aspects. *J. Apic. Res.* **2007,** *46* (4), 269–275.
13. Bogdanov, S.; Jurendic, T.; Sieber, R.; Gallmann, P. Honey for Nutrition and Health: A Review. *J. Am. Coll. Nutr.* **2008,** *27* (6), 677–689.
14. Bogdanov, S. Harmonized Methods of the International Honey Commission. *Int. Honey Commission* **2009**, *2009* (5), 1–62.
15. Bogdanov, S. Honey Composition. In *Book of Honey*; Bogdanov, S., Ed.; Bee Hexagon Knowledge Network: Switzerland, 2014; pp 27–36.
16. Bogdanov, S. Liquefaction of Honey. *Apiacta* **1993,** *28* (1), 4–10.
17. Boussaid, A.; Chouaibi, M.; Rezig, L.; Hellal, R.; Donsì, F.; Ferrari, G.; Hamdi, S. Physicochemical and Bioactive Properties of Six Honey Samples from Various Floral Origins from Tunisia. *Arabian J. Chem.* **2018,** *11* (2), 265–274.
18. Bradbear, N. Bee Species Description, In *Bees and Their Role in Forest Livelihoods*; Bradbear, N., Ed.; Food and Agriculture Organization of the United Nations (FAO): Rome, 2009; pp 5–16.
19. Brugnerotto, P.; Della Betta, F.; Gonzaga, L.V.; Fett, R.; Oliveira Costa, A. C. Capillary Electrophoresis Method to Determine Aliphatic Organic Acids in Bracatinga Honeydew Honey and Floral Honey. *J. Food Composit. Anal.* **2019,** *82* (4), 103243.
20. Can, Z.; Yildiz, O.; Sahin, H.; Akyuz Turumtay, E.; Silici, S.; Kolayli, S. An investigation of Turkish Honeys: Their Physicochemical Properties, Antioxidant Capacities and Phenolic Profiles. *Food Chem.* **2015,** *180*, 133–141.
21. Chan, C. W.; Deadman, B. J.; Manley-Harris, M.; Wilkins, A. L.; Alber, D. G.; Harry, E. Analysis of the Flavonoid Component of Bioactive New Zealand Mānuka (*Leptospermum scoparium*) Honey and the Isolation, Characterisation and Synthesis of an Unusual Pyrrole. *Food Chem.* **2013,** *141* (3), 1772–1781.
22. Chen, C. Relationship Between Water Activity and Moisture Content in Floral Honey. *Foods* **2019,** *8* (1), 30–35.
23. Cherchi, A.; Spanedda, L.; Tuberoso, C.; Cabras, P. Solid-phase Extraction and High-Performance Liquid Chromatographic Determination of Organic Acids in Honey. *J. Chromatogr. A* **1994,** *669* (1–2), 59–64.
24. Chirife, J.; Zamora, M. C.; Motto, A. The Correlation Between Water Activity and % Moisture in Honey: Fundamental Aspects and Application to Argentine Honeys. *J. Food Eng.* **2006,** *72* (3), 287–292.
25. Ciulu, M.; Solinas, S.; Floris, I.; Panzanelli, A.; Pilo, M. I.; Piu, P. C.; Spano, N.; Sanna, G. RP-HPLC Determination of Water-Soluble Vitamins in Honey. *Talanta* **2011,** *83* (3), 924–929.
26. Ciulu, M.; Spano, N.; Pilo, M. I.; Sanna, G. Recent Advances in the Analysis of Phenolic Compounds in Unifloral Honeys. *Molecules* **2016,** *21* (4), 1–32.
27. *Codex Alimentarius* Commission Codex Standard for Honey. *Codex Alimentarius*, [Online] 2001. http://teca.fao.org/sites/default/files/resources/Annex A Codex Alimentarius Honey Standard.pdf (accessed August 02, 2022).

28. Cometto, P. M.; Faye, P. F.; Di Paola Naranjo, R. D.; Rubio, M. A.; Aldao, M. A. J. Comparison of Free Amino Acids Profile in Honey from Three Argentinian Regions. *J. Agric. Food Chem.* **2003,** *51* (17), 5079–5087.
29. Cotte, J. F.; Casabianca, H.; Chardon, S.; Lheritier, J.; Grenier-Loustalot, M. F. Chromatographic Analysis of Sugars Applied to the Characterisation of Monofloral Honey. *Anal. Bioanal. Chem.* **2004,** *380* (4), 698–705.
30. Czipa, N.; Andrási, D.; Kovács, B. Determination of Essential and Toxic Elements in Hungarian Honeys. *Food Chem.* **2015,** *175* (12), 536–542.
31. Da Silva, P. M.; Gauche, C.; Gonzaga, L. V.; Costa, A. C. O.; Fett, R. Honey: Chemical Composition, Stability and Authenticity. *Food Chem.* **2016,** *196*, 309–323.
32. De Alda-Garcilope, C.; Gallego-Picó, A.; Bravo-Yagüe, J. C.; Garcinuño-Martínez, R. M.; Fernández-Hernando, P. Characterization of Spanish Honeys with Protected Designation of Origin "miel de Granada" According to their Mineral Content. *Food Chem.* **2012,** *135* (3), 1785–1788.
33. de Sousa, J. M. B.; de Souza, E. L.; Marques, G.; Benassi, M. de T.; Gullón, B.; Pintado, M. M.; Magnani, M. Sugar Profile, Physicochemical and Sensory Aspects of Monofloral Honeys Produced by Different Stingless Bee Species in Brazilian Semi-arid Region. *LWT Food Sci. Technol.* **2016,** *65*, 645–651.
34. Del Nozal, M. J.; Bernal, J. L.; Marinero, P.; Diego, J. C.; Frechilla, J. I.; Higes, M.; Llorente, J. High Performance Liquid Chromatographic Determination of Organic Acids in Honeys from Different Botanical Origin. *J. Liq. Chromatogr. Relat. Technol.* **1998,** *21* (20), 3197–3214.
35. Dettori, A.; Tappi, S.; Piana, L.; Dalla Rosa, M.; Rocculi, P. Kinetic of Induced Honey Crystallization and Related Evolution of Structural and Physical Properties. *LWT Food Sci. Technol.* **2018,** *95*, 333–338.
36. Dimitrova, B.; Gevrenova, R.; Anklam, E. Analysis of Phenolic Acids in Honeys of Different Floral Origin by Solid-Phase Extraction and High-Performance Liquid Chromatography. *Phytochem. Anal.* **2007,** *18* (1), 24–32.
37. Doner, L. W. The Sugars of Honey - A Review. *J. Sci. Food Agric.* **1977,** *28* (5), 443–456.
38. El-Haskoury, R.; Kriaa, W.; Lyoussi, B.; Makni, M. Ceratonia Siliqua Honeys from Morocco: Physicochemical Properties, Mineral Contents, and Antioxidant Activities. *J. Food Drug Anal.* **2018,** *26* (1), 67–73.
39. Escuredo, O.; Dobre, I.; Fernández-González, M.; Seijo, M. C. Contribution of Botanical Origin and Sugar Composition of Honeys on the Crystallization Phenomenon. *Food Chem.* **2014,** *149*, 84–90.
40. Escuredo, O.; González-Martín, M. I.; Rodríguez-Flores, M. S.; Seijo, M. C. Near Infrared Spectroscopy Applied to the Rapid Prediction of the Floral Origin and Mineral Content of Honeys. *Food Chem.* **2015,** *170*, 47–54.
41. European Council (EU). European Parliament Directive 2014/63/EU of The European Parliament and of the Council of 15 May 2014 Amending Council Directive 2001/110/EC Relating to Honey, *Official Journal of the European Communities*, 2014; vol *L164*, pp 1–5.
42. Fernández-Torres, R.; Pérez-Bernal, J. L.; Bello-López, M. Á.; Callejón-Mochón, M.; Jiménez-Sánchez, J. C.; Guiraúm-Pérez, A. Mineral Content and Botanical Origin of Spanish Honeys. *Talanta* **2005,** *65* (3), 686–691.

43. Fröschle, M.; Horn, H.; Spring, O. Characterization of Jatropha Curcas Honeys Originating from the Southern Highlands of Madagascar. *LWT Food Sci. Technol.* **2018,** *93* (2), 525–533.
44. Gašić, U. M.; Milojković-Opsenica, D. M.; Tešić, Ž. L. Polyphenols as Possible Markers of Botanical Origin of Honey. *J. AOAC Int.* **2017,** *100* (4), 852–861.
45. Geană, E. I.; Ciucure, C. T.; Costinel, D.; Ionete, R. E. Evaluation of Honey in Terms of Quality and Authenticity Based on the General Physicochemical Pattern, Major Sugar Composition and δ13C Signature. *Food Control* **2020,** *109* (7), 106919.
46. Griebel, C.; Heß, G. Vitamin C Enthaltende Honige. *Zeitschrift für Untersuchung der Lebensmittel* (*J. Food Stud.*) **1939,** *78* (4), 308–314.
47. Guelpa, A.; Marini, F.; du Plessis, A.; Slabbert, R.; Manley, M. Verification of Authenticity and Fraud Detection in South African Honey using NIR Spectroscopy. *Food Control* **2017,** *73*, 1388–1396.
48. Guo, W.; Zhu, X.; Liu, Y.; Zhuang, H. Sugar and Water Contents of Honey with Dielectric Property Sensing. *J. Food Eng.* **2010,** *97* (2), 275–281.
49. Habib, H. M.; Al Meqbali, F. T.; Kamal, H.; Souka, U. D.; Ibrahim, W. H. Physicochemical and Biochemical Properties of Honeys from Arid Regions. *Food Chem.* **2014,** *153*, 35–43.
50. Haroun, M. I.; Konar, N.; Poyrazoglu, E. S.; Hospolat, I.; Art, N.; Honey, A. In *Organic Acid Profiles and Contents of Turkish Honeydew and Floral Honeys*, International Conference on Environment, Agriculture and Food Sciences (ICEAFS'2012) Aug 11–12, 2012 Phuket (Thailand) Organic, 2012; pp 55–58.
51. Haydak, M. H.; Palmer, L. S.; Tanquary, M. C.; Vivino, A. E. Vitamin Content of Honeys. *J. Nutr.* **1942,** *23* (6), 581–588.
52. Hernández, O. M.; Fraga, J. M. G. G.; Jiménez, A. I.; Jiménez, F.; Arias, J. J. Characterization of Honey from the Canary Islands: Determination of the Mineral Content by Atomic Absorption Spectrophotometry. *Food Chem.* **2005**, *93* (3), 449–458.
53. Horváth, K.; Molnár-Perl, I. Simultaneous GC-MS Quantitation of o-phosphoric, Aliphatic and Aromatic Carboxylic Acids, Proline, Hydroxymethylfurfurol and Sugars as their TMS Derivatives in Honeys. *Chromatographia* **1998,** *48* (1–2), 120–126.
54. Huidobro, F.; Suarez-luque, S.; Sancho, T.; Suárez-Luque, S.; Mato, I.; Huidobro, J. F.; Simal-Lozano, J.; Sancho, M. T. Rapid Determination of Minority Organic Acids in Honey by High-performance Liquid Chromatography. *J. Chromatogr. A* **2002,** *955* (2), 207–214.
55. Istasse, T.; Jacquet, N.; Berchem, T.; Haubruge, E.; Nguyen, B. K.; Richel, A. Extraction of Honey Polyphenols: Method Development and Evidence of Cis isomerization. *Anal. Chem. Insights* **2016,** *2016* (11), 49–57.
56. Jandrić, Z.; Frew, R. D.; Fernandez-Cedi, L. N.; Cannavan, A. An Investigative Study on Discrimination of Honey of Various Floral and Geographical Origins using UPLC-QToF MS and Multivariate Data Analysis. *Food Control* **2017,** *72*, 189–197.
57. Jasicka-Misiak, I.; Poliwoda, A.; Dereń, M.; Kafarski, P. Phenolic Compounds and Abscisic Acid as Potential Markers for the Floral Origin of Two Polish Unifloral Honeys. *Food Chem.* **2012,** *131* (4), 1149–1156.
58. Joerg, E.; Sontag, G. Multichannel Coulometric Detection Coupled with Liquid Chromatography for Determination of Phenolic Esters in Honey. *J. Chromatogr. A* **1993,** *635* (1), 137–142.

59. Juan-Borrás, M.; Domenech, E.; Hellebrandova, M.; Escriche, I. Effect of Country Origin on Physicochemical, Sugar and Volatile Composition of Acacia, Sunflower and Tilia Honeys. *Food Res. Int.* **2014,** *60*, 86–94.
60. Kadri, S. M.; Zaluski, R.; Pereira Lima, G. P.; Mazzafera, P.; De Oliveira Orsi, R. Characterization of *Coffea arabica* Monofloral Honey from Espírito Santo, Brazil. *Food Chem.* **2016,** *203*, 252–257.
61. Kalimi, M. Y.; Kalama, S. Studies in Mahabaleshwar Honey, Part III: Vitamin Contents (Ascorbic Acid, Thiamine, Riboflavin and Niacin) and Effect of Storage on these Vitamins. *Indian J. Nutr. Diet.* **1964,** *2* (1), 9–11.
62. Karabagias, I. K.; Louppis, A. P.; Karabournioti, S.; Kontakos, S.; Papastephanou, C.; Kontominas, M. G. Characterization and Geographical Discrimination of Commercial Citrus spp. Honeys Produced in Different Mediterranean Countries based on Minerals, Volatile Compounds and Physicochemical Parameters, Using Chemometrics. *Food Chem.* **2017,** *217*, 445–455.
63. Kawashima, H.; Suto, M.; Suto, N. Stable Carbon Isotope Ratios for Organic Acids in Commercial Honey Samples. *Food Chem.* **2019,** *289* (8), 49–55.
64. Kaygusuz, H.; Tezcan, F.; Bedia Erim, F.; Yildiz, O.; Sahin, H.; Can, Z.; Kolayli, S. Characterization of Anatolian Honeys Based on Minerals, Bioactive Components and Principal Component Analysis. *LWT Food Sci. Technol.* **2016,** *68*, 273–279.
65. Kim, L.; Brudzynski, K. Identification of Menaquinones (Vitamin K2 Homologues) as Novel Constituents of Honey. *Food Chem.* **2018,** *249* (10), 184–192.
66. Kiztes, G.; Schuette, H. A.; Elvehjem, G. A. The B Vitamins in Honey. *J. Nutr.* **1943,** *26* (3), 241–250.
67. Kıvrak, Ş.; Kıvrak, İ. Assessment of Phenolic Profile of Turkish Honeys. *Int. J. Food Prop.* **2017,** *20* (4), 864–876.
68. Kumar, A.; Gill, J. P. S.; Bedi, J. S.; Manav, M.; Ansari, M. J.; Walia, G. S. Sensorial and Physicochemical Analysis of Indian Honeys for Assessment of Quality and Floral Origins. *Food Res. Int.* **2018,** *108* (4), 571–583.
69. Lanjwani, M. F.; Channa, F. A. Minerals Content in Different Types of Local and Branded Honey in Sindh, Pakistan. *Heliyon* **2019,** *5* (7), e02042.
70. León-Ruiz, V.; Vera, S.; González-Porto, A. V.; Andrés, M. P. S. Vitamin C and Sugar Levels as Simple Markers for Discriminating Spanish Honey Sources. *J. Food Sci.* **2011,** *76* (3).
71. Louppis, A. P.; Karabagias, I. K.; Kontakos, S.; Kontominas, M. G.; Papastephanou, C. Botanical Discrimination of Greek Unifloral Honeys Based on Mineral Content in combination with Physicochemical Parameter Analysis, Using a Validated Chemometric Approach. *Microchem. J.* **2017,** *135*, 180–189.
72. Lüttge, U. Über den Vitamin C-Gehalt von Bienenhonig. *Zeitschrift für Lebensmittel-Untersuchung und Forschung* **1962,** *117* (4), 289–296.
73. Magyar Élelmiszerkönyv Bizottság. *2-100 számú irányelv Megkülönböztető jelöléssel ellátott mézfélék* (Guideline 2-100 Honey with a distinctive mark). Hungary: Magyar Élelmiszerkönyv Bizottság, 2009; pp 1–4.
74. Matei, N.; Birghila, S.; Dobrinas, S.; Capota, P. Determination of C Vitamin and Some Essential Trace Elements (Ni, Mn, Fe, Cr) in Bee Products. *Acta Chem. Slov.* **2004,** *51*, 169–175.
75. Mateo, R.; Bosch-Reig, F. Sugar Profiles of Spanish Unifloral Honeys. *Food Chem.* **1997,** *60* (1), 33–41.

76. Mato, I.; Huidobro, J. F.; Cendón, V.; Muniategui, S.; Fernández-Muiño, M. A.; Sancho, M. T. Enzymatic Determination of Citric Acid in Honey by Using Polyvinyl Polypyrrolidone Clarification. *J. Agric. Food Chem.* **1998,** *46* (1), 141–144.
77. Mato, I.; Huidobro, J. F.; Sánchez, M. P.; Muniategui, S.; Fernández-Muiño, M. A.; Sancho, M. T. Enzymatic Determination of Total D-gluconic Acid in Honey. *J Agric. Food Chem.* **1997,** *45* (9), 3550–3553.
78. Mato, I.; Huidobro, J. F.; Simal-Lozano, J.; Sancho, M. T. Rapid Determination of Nonaromatic Organic Acids in Honey by Capillary Zone Electrophoresis with Direct Ultraviolet Detection. *J. Agric. Food Chem.* **2006,** *54* (5), 1541–1550.
79. Morales, V.; Corzo, N.; Sanz, M. L. L. HPAEC-PAD Oligosaccharide Analysis to Detect Adulterations of Honey with Sugar Syrups. *Food Chem.*, **2008,** *107* (2), 922–928.
80. Navarrete, M.; Casado, S.; Minelli, M.; Segura, A.; Bonetti, A.; Dinelli, G. Direct Determination of Aliphatic Acids in Honey by Coelectroosmotic Capillary Zone Electrophoresis Direct Determination of Aliphatic Acids in Honey by Coelectroosmotic Capillary Zone Electrophoresis. *J. Apic. Res.* **2005,** *44* (2), 65–70.
81. Nayik, G. A.; Dar, B. N.; Nanda, V. Physico-Chemical, Rheological and Sugar Profile of Different Unifloral Honeys from Kashmir Valley of India. *Arabian J. Chem.* **2019,** *12* (8), 3151–3162.
82. Nayik, G. A.; Suhag, Y.; Majid, I.; Nanda, V. Discrimination of High-Altitude Indian Honey by Chemometric Approach According to their Antioxidant Properties and Macro Minerals. *J. Saudi Soc. Agric. Sci.* **2018,** *17* (2), 200–207.
83. Nozal, M. J.; Bernal, J. L.; Diego, J. C.; Gómez, L. A.; Higes, M. HPLC Determination of Low Molecular Weight Organic Acids in Honey with Series-Coupled Ion-Exclusion Columns. *J. Liq. Chromatogr. Relat. Technol.* **2003,** *26* (8), 1231–1253.
84. Nozal, M. J.; Bernal, J. L.; Gómez, L. A.; Higes, M.; Meana, A. Determination of Oxalic Acid and Other Organic Acids in Honey and in Some Anatomic Structures of Bees. *Apidologie* **2003,** *34*, 181–188.
85. Oddo, L. P.; Piro, R. Main European Unifloral Honeys: Descriptive Sheets. *Apidologie* **2004,** *35*, S38–S81.
86. Oroian, M.; Ropciuc, S.; Paduret, S.; Todosi, E. Rheological Analysis of Honeydew Honey Adulterated with Glucose, Fructose, Inverted Sugar, Hydrolysed Inulin Syrup and Malt Wort. *LWT Food Sci. Technol.* **2018,** *95* (4), 1–8.
87. Ouchemoukh, S.; Schweitzer, P.; Bachir Bey, M.; Djoudad-Kadji, H.; Louaileche, H.; Bey, M. B.; Djoudad-Kadji, H.; Louaileche, H. HPLC Sugar Profiles of Algerian Honeys. *Food Chem.* **2010,** *121* (2), 561–568.
88. Özbalci, B.; Boyaci, I. H.; Topcu, A.; Kadilar, C.; Tamer, U. Rapid Analysis of Sugars in Honey by Processing Raman Spectrum using Chemometric Methods and Artificial Neural Networks. *Food Chem.* **2013,** *136* (3–4), 1444–1452.
89. Pascual-Maté, A.; Osés, S. M.; Marcazzan, G. L.; Gardini, S.; Fernández Muiño, M. A.; Sancho, T. M. Sugar Composition and Sugar-Related Parameters of Honeys from the Northern Iberian Plateau. *J. Food Composit. Anal.* **2018,** *74* (8), 34–43.
90. Pascual-Maté, A.; Osés, S. M.; Fernández-Muiño, M. A.; Sancho, M. T. Methods of Analysis of Honey. *J. Apic. Res.* **2018,** *57* (1), 38–74.
91. Persano Oddo, L.; Piana, L.; Bogdanov, S. Botanical Species Giving Unifloral Honey in Europe. *Apidologie* **2006**, *37*, 452–461.
92. Pita-Calvo, C.; Vázquez, M. Differences between Honeydew and Blossom Honeys: A Review. *Trends Food Sci. Technol.* **2017,** *59*, 79–87.

93. Pohl, P.; Bielawska-Pohl, A.; Dzimitrowicz, A.; Jamroz, P.; Welna, M.; Lesniewicz, A.; Szymczycha-Madeja, A. Recent Achievements in Element Analysis of Bee Honeys by Atomic and Mass Spectrometry Methods. *Trends Anal. Chem.* **2017,** *93*, 67–77.
94. Popek, S.; Halagarda, M.; Kursa, K. A New Model to Identify Botanical Origin of Polish Honeys Based on the Physicochemical Parameters and Chemometric Analysis. *LWT Food Sci. Technol.* **2017,** *77*, 482–487.
95. Pyrzynska, K.; Biesaga, M. Analysis of Phenolic Acids and Flavonoids in Honey. *Trends Anal. Chem.* **2009,** *28* (7), 893–902.
96. Ramos, O. Y.; Salomón, V.; Libonatti, C.; Cepeda, R.; Maldonado, L.; Basualdo, M. Effect of Botanical and Physicochemical Composition of Argentinean Honeys on the Inhibitory Action Against Food Pathogens. *LWT Food Sci. Technol.* **2018,** *87*, 457–463.
97. Rodríguez-Flores, M. S.; Escuredo, O.; Míguez, M.; Seijo, M. C.; Míguez, M.; Seijo, M. C. Differentiation of Oak Honeydew and Chestnut Honeys from the Same Geographical Origin Using Chemometric Methods. *Food Chem.* **2019,** *297* (6), 124979.
98. Rodriguez-Otero, J. L.; Paseiro, P.; Simal, J.; Cepeda, A. Mineral Content of the Honeys Produced in Galicia (North-west Spain). *Food Chem.* **1994,** *49* (2), 169–171.
99. Ruoff, K. In *Authentication of the Botanical Origin of Honey*; Universitx of Helsinki, 2006; p 86.
100. Sajtos, Z.; Herman, P.; Harangi, S.; Baranyai, E. Elemental Analysis of Hungarian Honey Samples and Bee Products by MP-AES Method. *Microchem. J.* **2019,** *149* (2), 103968.
101. Sakač, M. B.; Jovanov, P. T.; Marić, A. Z.; Pezo, L. L.; Kevrešan, Ž. S.; Novaković, A. R.; Nedeljković, N. M. Physicochemical Properties and Mineral Content of Honey Samples from Vojvodina (Republic of Serbia). *Food Chem.* **2019,** *276* (9), 15–21.
102. Sancho, M. T.; Muniategui, S.; Sánchez, M. P.; Huidobro, J. F.; Simal, J. Relationship Between Electrical Conductivity and Total and Sulphated Ash Contents in Basque Honeys. *Apidologie* **1991,** *22*, 67–76.
103. Siddiqui, I. R. The Sugars of Honey. *Adv. Carbohydr. Chem. Biochem.* **1970,** *25*, 285–309.
104. Steeg, E.; Montag, A. Originalarbeiten Quantitative Bestimmung Aromatischer Carbonsiiuren in Honig (Original Work Quantitative Determination of Aromatic Carboxylic Acids in Honey). *Zeitschrift für Lebensmittel-Untersuchung und Forschung* (*J. Food Invest. Res.*) **2000,** *187*, 115–120.
105. Stinson, E. E.; Subers, M. H.; Petty, J.; White, J. W. The Composition of Honey, Part V: Separation and Identification of the Organic Acids. *Arch. Biochem. Biophys.* **1960,** *89* (1), 6–12.
106. Suárez-Luque, S.; Mato, I.; Huidobro, J. F.; Simal-Lozano, J.; Huidobro, F.; Suarez-Luque, S. Solid-Phase Extraction Procedure to Remove Organic Acids from Honey. *J. Chromatogr. B Anal. Technol. Biomed. Life Sci.* **2002,** *770* (1–2), 77–82.
107. Sun, Z.; Zhao, L.; Cheng, N.; Xue, X.; Wu, L.; Zheng, J.; Cao, W. Identification of Botanical Origin of Chinese Unifloral Honeys by Free Amino Acid Profiles and Chemometric Methods. *J. Pharm. Anal.* **2017,** *7* (5), 317–323.
108. Tan, T. Y.; Basheer, C.; Low, K. M.; Lee, H. K. Electro Membrane Extraction of Organic Acids in Undiluted Honey with Ion Chromatographic Analysis. *Microchem. J.* **2018,** *143* (6), 234–242.

109. Terrab, A.; Hernanz, D.; Heredia, F. J. Inductively Coupled Plasma Optical Emission Spectrometric Determination of Minerals in Thyme Honeys and their Determination of Geographical Discrimination. *J. Agric. Food Chem.* **2004,** *52* (11), 3441–3445.
110. The European Council (EU). Council Directive 2001/110/EC of 20 December 2001 relating to honey; Official Journal of the European Union, 2001; vol *L10*, pp 47–52.
111. Thrasyvoulou, A.; Tananaki, C.; Goras, G.; Karazafiris, E.; Dimou, M.; Liolios, V.; Kanelis, D.; Gounari, S. Legislation of Honey Criteria and Standards. *J. Apic. Res.* **2018,** *57* (1), 88–96.
112. Tourn, M. L.; Lombard, A.; Belliardo, F.; Buffa, M. Quantitative Analysis of Carbohydrates and Organic Acids in Honeydew, Honey and Royal Jelly by Enzymic Methods. *J. Apic. Res.* **1980,** *19* (2), 144–146.
113. USDA. United States Standards for Grades of Extracted Honey. *Federal Register*, 1985; vol *50FR15861*, pp 1–12.
114. Vanhanen, L. P.; Emmertz, A.; Savage, G. P. Mineral Analysis of Mono-floral New Zealand Honey. *Food Chem.* **2011,** *128* (1), 236–240.
115. Verzera, A.; Campisi, S.; Zappalà, M.; Bonaccorsi, I. SPME-GC-MS Analysis of Honey Volatile Components for the Characterization of Different Floral Origin. *Am. Lab.* **2001,** *33*, 18–21.
116. White, J. W. Honey. In *Advances in Food Research*; Chichester, C. O., Ed.; Elsievier: New York, 1978; vol *24*, pp 287–374.
117. Wilkins, A. L.; Lu, Y.; Tan, S. T. Extractives from New Zealand Honeys, Part 5: Aliphatic Dicarboxylic Acids in New Zealand Rewarewa (*Knightea excelsa*) Honey. *J. Agric. Food Chem.* **1995,** *43* (12), 3021–3025.

CHAPTER 12

POTENTIAL OF AQUAPHOTOMICS FOR DETERMINATION OF QUALITY OF NATURAL FOODS

JELENA MUNCAN, BALKIS AOUADI, and ROUMIANA TSENKOVA

ABSTRACT

Water, a major component of numerous biosystems and food matrices, was proven to have a crucial role in conferring different functionalities as a result of different molecular structures. This dynamic behavior was long considered a limiting factor until a novel scientific discipline—aquaphotomics—showed that it can be utilized as a unique tool in assessing the quality of studied matrices. Aquaphotomics is a spectroscopy-based field, dedicated to studying the water–light interaction phenomena using spectroscopic techniques to reach a better understanding of the water molecular structure in aqueous and biological systems in relation to their state and functionality. The water molecular structure, being highly sensitive to any changes within the aqueous-bio system or its surroundings, in aquaphotomics serves as a sensor, its spectral pattern a marker of the system state, directly related to its functionality. This innovative approach has proven effective in the areas of biodiagnosis of abiotic and biotic stress in plants, elucidation of mechanisms underlying stress response to temperature changes and extreme desiccation, and diverse aspects of food quality evaluation. These findings hold great promise for future applications, reinforcement, and development of other targeted functionalities. This chapter introduces this scientific discipline, presents the basics of aquaphotomics analytical methodology, and provides a comprehensive review of existing aquaphotomics works, done in the past decade and related to plants and plant-based food quality determination.

Sustainable and Functional Foods from Plants: Health Impact, Bioactive Compounds, and Production Technologies. Megh R. Goyal, Arijit Nath, and Zoltán Kovács (Eds.)
© 2024 Apple Academic Press, Inc. Co-published with CRC Press (Taylor & Francis)

12.1 AQUAPHOTOMICS: A NOVEL SCIENTIFIC DISCIPLINE

Water is in constant interaction with its surroundings. The stability of this well-structured molecular network is defined by the strength of hydrogen bonds under ever-changing conditions. The study of this dynamic behavior has, however, been a major challenge. Being a vital component of biological systems, extracting valuable information contained in the water molecular network is imperative. In fact, over the course of the years, an array of different techniques was used to examine water structure and state in biosystems. Near-infrared spectroscopy has been particularly used for this purpose. The reason has been the low absorbance of water molecules in this range, very little need or no need for sample preparation at all, as well as the non-invasive nature of the method which enables the analysis of a wider range of not only aqueous samples but biological systems at all levels of organization from cell to the whole organisms.

The advent of aquaphotomics added a new dimension to the research in this direction. The surge of aquaphotomics came as a result of the need for complementary knowledge to the conventional "omics" disciplines which, for the most part, overlooked the role of water in conferring distinctive attributes to the aqueous system.[99]

Aquaphotomics is based on determining the spectral pattern of water as an integrative indicator of the state of the respective studied system. The correlation between water structure and its resultant functionality provides important insights into the dynamics of the system.[142]

Water structure is, however, easily subjected to changes that are mainly due to the high reactivity of its two basic elements; hydrogen and oxygen. The interactions with neighboring molecules or changes in the system surroundings have a traceable effect on existing water molecular species, which is detected through the monitoring of the spectra.

To describe the state or dynamics of any aqueous system from the aspect of its water molecular structure, first, the correspondence between the water absorbance bands (WABs) and the relative water species needs to be established. Consequently, assigning the observed behavioral change to the previously determined spectral pattern will serve as a descriptor for the functionality of the respective water molecular conformations. Reproducing these steps under various environmental and operational conditions will provide a thorough database for both current and future endeavors.

Relying solely on water spectra for the description of system dynamics can be a huge step toward more cost-effective integrative approaches. In this respect, extending the applicability of aquaphotomics from the basic elucidation of biological phenomena to more complex applications can resolve many rising issues in the 21st century.

This paradigm shift, regarding the role of water molecules has already contributed to cross-disciplinary applications. Ranging from the diagnosis of mammary gland inflammation, where alterations were surprisingly visible in water structure but not in the concentration of the main biomolecules in milk, to designing novel measurement instruments based on the interaction of water with light; aquaphotomics has shown immense potential.[141]

On the contrary, water was long considered a hindrance to the determination of properties of other targeted constituents in food matrices as it entails additional sample preparation steps and added costs. Alternatively, having the possibility to indirectly quantify substances by the measurement of water absorbance changes, offers the advantage of overcoming these limitations as well as the poor quantification limits of conventional analytical methods and the detection of components that do not absorb the incident light. Adopting an aquaphotomics approach and relying on an indirect mode of measurement using WABs, instead of absorbance bands of analytes, can substantially improve the accuracy of the built model.[42,111,142]

Additionally, looking beyond the measurements of single molecules, aquaphotomics with its integrative approach offers a new way for the characterization of systems functionalities. Aquaphotomics aims to replace the current state of art approach of measuring multiple parameters (physical, chemical, biological) describing the state or quality by only one—integrative marker—Water Spectral Pattern (WASP) that characterizes the state of water molecular matrix as a dominant component of the samples influenced by all known and unknown factors (Fig. 12.1).

This novel concept offers huge possibilities in quality control, monitoring, biodiagnostics, and biomonitoring. The technology developed based on the innovative ideas that aquaphotomics provided has already been proven effective in these areas as this chapter will illustrate.

The main objectives of this chapter are to explain the basics of the phenomena of water–light interaction and how they resulted in the development of aquaphotomics, to briefly introduce the principles of measurement and aquaphotomics analytical methodology, and to provide a comprehensive review of existing aquaphotomics works, done in the past decade and related to plants and plant-based food quality determination. These findings hold great promise for future applications, reinforcement, and development of other targeted functionalities and would hopefully inspire the readers to envisage further expansion.

12.2 AQUAPHOTOMICS ANALYTICAL METHODOLOGY

Behind the concept of aquaphotomics is a well-thought-out methodology that enables the extraction of valuable information from studied systems as much as possible. For this purpose, a thorough guide has been developed for researchers working in this ever-growing field.[140]

An initial step would involve setting up an experimental design that responds to the aim of the conducted study. Defining the perturbations to be considered and screening the available literature for expected outcomes in terms of identifying WABs and their corresponding structures serves as a good starting point. The spectral database containing an estimated number of more than 500 calculated or experimentally discovered WABs in the near-infrared spectral region (750–2500 nm) and the corresponding water structures have already been established.[95,140] Spectral regions where these absorbance bands are most probable constitute water matrix coordinates (WAMACS).[142] Originally, 12 WAMACS proved to be highly important for the screening of various biosystems (Table 12.1). However, with continuous data mining advancements and the study of novel matrices, new absorbance bands are constantly discovered.[76]

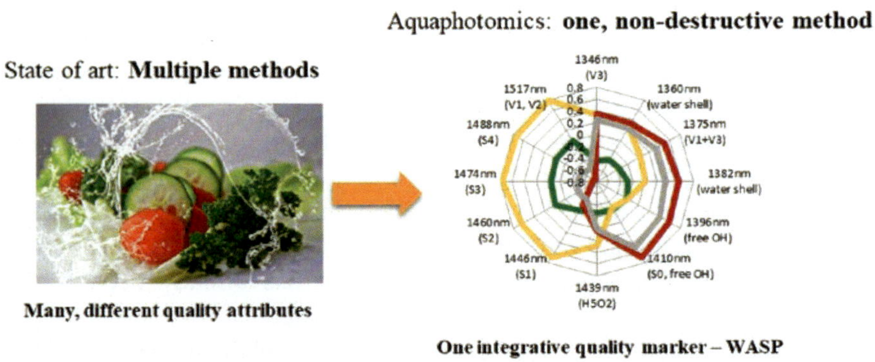

FIGURE 12.1 Aquaphotomics: an integrative, systemic approach for the exploration of aqueous and biosystems through the study of the state of water molecular matrix under perturbing factors.

Visualizing the spectral information is the next step toward identifying any pattern obtained under the studied perturbations. This WASP is the fingerprint that links the structure of water to its respective function within the aqueous or biological system. The graphical representation of the WASP

TABLE 12.1 The WAMACS—Characteristic Water Absorbance Bands in the Region of First Overtone of OH Stretching Vibrations (~1300 to 1600 nm). Source: Self-developed with Concepts from Ref. [19,51,74,76,94,142].

WAMACS	Wavelength range	Assignment
C1	1336–1348 nm	v_3: H_2O Asymmetric stretching vibration
C2	1360–1366 nm	$OH^-(H_2O)_n$, where n = 1, 2, 4: Hydroxylated water clusters; Water solvation shell
C3	1370–1379 nm	$v_1 + v_3$: H_2O symmetric stretching vibration H_2O asymmetric stretching vibration Protonated water clusters
C4	1380–1388 nm	$OH^-(H_2O)_n$, where n = 1, 4: Water solvation shell $O_2^-(H_2O)_n$, where n = 4: Hydrated superoxide clusters
C5	1396–1412 nm	Trapped water (TW): Water confined in the local field of ions (1396–1403 nm) S_0: Free water molecules
C6	1421–1430 nm	H–OH bond O–H: Water molecules involved with hydration
C7	1432–1444 nm	S_1: Water molecules with one hydrogen bond
C8	1448–1458 nm	$OH^-(H_2O)_n$ where n = 4, 5: Water solvation shell
C9	1460–1464 nm	S_2: Water molecules with two hydrogen bonds
C10	1472–1482 nm	S_3: Water molecules with three hydrogen bonds
C11	1492–1494 nm	S_4: Water molecules with four hydrogen bonds
C12	1506–1516 nm	v_1, v_2: Strongly bound water

is termed "aquagram." Aquagrams exist in different forms[140] and show the normalized absorbance at WABs of interest. The normalized absorbance values are obtained according to the following eq 12.1:

$$A' = \frac{(A\lambda - \mu\lambda)}{\sigma\lambda} \qquad (12.1)$$

where λ represents specific wavelengths when WABs are activated by respective perturbation and have a noticeable effect on the results of the analysis[99]; A_λ is the scatter corrected absorbance; μ_λ is the mean of all obtained spectra; and σ_λ stands for the standard deviation of all obtained spectra.

The aquagrams serve as comparative graphical tools that can easily pinpoint the differences in water absorbance at particular bands generated by the applied perturbations. As opposed to other scientific disciplines, where these perturbations are usually identified and eventually eliminated or compensated for, in aquaphotomics approach, these are sometimes even solicited to identify how they affect the water molecular system being a mirror of the rest of the system and to provide a contribution to the enrichment of the previously mentioned database of WABs.

This perception shift about perturbation comes as a result of research done by Tsenkova,[139] who proved that even minor perturbations such as consecutive illuminations of the same pure water sample reveal different water species resulting in different spectra. In another example, the application of light as a perturbing factor contributed to the determination of the functions of different protein isoforms.

These studies evidence that, when performed, aquaphotomics analysis is able to trace even subtle changes and distinctive functionalities. These findings also reflect the potential of research done on more complex matrices for the discovery of new phenomena.

12.3 PLANT AQUAPHOTOMICS

NIR spectroscopy coupled with aquaphotomics has proven to be highly effective as a novel method for various applications and provided a better understanding of natural phenomena in a wide range of disciplines (biodiagnosis, biomonitoring, etc.).[99] Expanding its applications to unexplored or less-studied fields can help to elucidate basic phenomena and potentially lead to new breakthroughs.

Today, with the recurrent environmental issues and the urgency for better crop management, understanding the dynamic behavior of water molecular networks in plants and plant-based foods is of utmost importance. Studies on this matter remain scarce but plant aquaphotomics holds the promise of exploring a novel set of applications to alleviate these concerns.

The interest in undertaking such research stems from the fact that plants play a fundamental role in the food chain; in supplying basic elements such as oxygen, nitrogen, etc., as well as their contribution to bioremediation.[118] Although some plant species exhibit distinctive coping strategies when faced with different stressors,[54] most of the mechanisms governing these processes are not fully deciphered. Another reason why plants can be considered suitable matrices for aquaphotomics analysis is mainly their significant water content which in green (non-woody) plants can be as high as 80–90% of the biomass.[40]

Water is also a central molecule in most physiological and metabolic processes and an essential transporter of metabolites and nutrients.[59] A better knowledge of the water structures responsible for these attributes will not only allow the biomonitoring of plants and the control of the biological processes but also offer the opportunity for new plant-based product development with enhanced functionalities.

The estimation of plant physiological status has importance to decide on measures to be taken, that is, it should provide feedback and an opportunity to react in good time. Water stress is among the prevalent limitations in food or crop production; hence, the non-destructive measurements especially on the canopy level would be of great interest.[114] However, the nondestructive measurements and the results obtained by aquaphotomics analysis which, will be described here, have a great potential in precision agriculture to move far beyond just the estimation of plant water status and for example, the need for irrigation. They can also be used now to move toward necessary actions like certain treatments, for example, the application of the right therapy to prevent spreading of the crop diseases, precise amounts of fertilizers to stimulate growth at only certain locations, and other similar treatments.

12.3.1 BIOTIC STRESS

Comprising of conditions caused by various pathogenic microorganisms such as fungi, bacteria, viruses, nematodes, and herbivores, biotic stress was proven to affect water behavior on a cellular level and in a systematic way. To combat this form of stress, plants have developed innate adaptation mechanisms which were investigated by several methods. For the most part, they were based on univariate analysis approaches relying mainly on macroscale indicators rather than defining multidimensional markers.

For instance, a multitude of techniques have been used to detect viral infections affecting crops, ranging from enzymatic assays, and polymerase chain reactions to other sophisticated techniques. Unfortunately, some of these are costly, time consuming, require the use of chemicals, and may lack the required accuracy.[63]

Alternatively, the observation of the WASP was proven to be quite revelatory for the detection of mosaic virus infection in soybeans before any disease symptoms were observed.[63] Specific absorbance bands situated at 910, 937, and 970 nm were specifically sensitive to the disease, had typical absorbance values depending on the health status of the soybean cultivars, and in combination, helped to discriminate infected plants from healthy ones with 91.6% sensitivity and 95.8% specificity using multivariate soft independent modeling of class analogy (SIMCA). Such outcomes were obtained after applying several mathematical transformations such as mean centering, second derivatives, and baseline corrections to the original spectra. The spectra were also averaged in an attempt to account for any differences due to anatomical attributes. With these spectral transformations and their consequent performances, one can attest to the efficiency of detecting plant diseases and tracking their real-time development while monitoring the absorbance of specific water species.

Of the three most important bands for discrimination, the first one 910 nm can be ascribed to the third overtone of CH stretching vibrations, but the other two 937 nm and especially 970 nm are WABs.[49] For example, other studies have reported that a wavelength range of 950–970 nm[114] is strongly related to the water status of the plants, while the shift of the band located at 970 nm is reported as a consequence of water stress in plants,[113] which is in agreement with the observation of the peak shift in the spectra of the infected plants in the work of Jinendra et al.[63]

The absorbance band at 970 nm (second overtone, corresponding to 1455 nm in the first overtone region—bulk water[53]) appears as a significant variable also for the diagnosis of Grapevine leafroll disease, a viral disease, in two wine grape cultivars.[101] Especially in the model for discriminating between the infected leaves in the non-symptomatic stage and healthy leaves, the importance of this water band remains vital. The interesting finding similar to the work on the detection of mosaic virus infection in soybean, this WAB was important for the timely detection of viral disease, while the plants are still in the asymptomatic stage.

Another interesting study explored the potential of near-infrared spectroscopy in the diagnosis of potato virus-Y (PVY) infected plants.[93] The authors acquired the reflectance spectra of healthy and virus-infected leaves of potato plants from two cultivars. The spectral analysis indicated the regions of 900–1100 nm to be sensitive to the PVY infection and a good choice to look further for the spectral signature of infection. Employing principal component analysis (PCA) several absorbance bands were identified as sensitive (1020, 1090, 960, 985, and 996 nm), and then those were selected to be used in the linear discriminant analysis (LDA) of infection status. The use of different absorbance bands for discrimination yielded different accuracy—the highest accuracy of 100% was achieved when wavelengths 1020 and 1090 nm were used. These bands could be related to the absorption of lipids and/or proteins[49,50,100]; but since both are in the area of large water absorbance (second overtone and first overtone) in this region, there is a possibility that they also can be attributed to water.[91] Slightly less accuracy was achieved when 960, 985, and 996 nm were used for discrimination, that is, 78.57, 92.86, and 85.71%, respectively.

The bands could be thought of as the overtones of bands 1440 nm (C7: 1432–1444 nm, water dimer), 1447.5 nm (C8: 1448–1454 nm, first overtone DDA symmetric stretch [OH-$(H_2O)_4$]), and 1494 nm (C11: 1472–1482 nm, water molecules making 4 hydrogen bonds), corresponding to the well-defined WAMACS ranges in Table 12.1.[142] These results indicate that water molecular rearrangement is closely followed by the changes in other molecular components which are changed due to the presence of the virus. The authors of the study concluded this too, relating the reflectance variation in this region to changes in cell structure and water activity.

The importance of WABs for disease detection was confirmed in the studies of other, not only virus-induced diseases. For example, Bauriegel et al.[7] utilized near-infrared hyperspectral imaging over the 400–1000-nm region for head blight disease detection in wheat. This disease is caused by the fungi *Fusarium* ssp. and is a common problem worldwide in wheat, rye, oat, and barley production. The study revealed that healthy and infected plants exhibited spectral patterns that were considerably different at the regions of the spectra corresponding to the absorption of carotenoids, chlorophylls, and water (range of 927–931 nm). This latter spectral range clearly revealed differences in the tissue water content between the healthy and diseased plants.[7]

Another study on the early detection of fungal infection was performed by Li et al.[79] who employed hyperspectral imaging in the 325–1100-nm range for the detection of early symptoms of decay in citrus fruits. The PCA performed on the averaged spectra of decayed and sound fruit tissues showed separation of the two classes along the first principal component, whose loading vector indicated the importance of four wavelengths for separation—575, 69, 81, and 969 nm. Again, the WAB (969 nm) indicated that just like the other components of the tissue, such as pigments, which are changing with the decay, so does the water.

Further, Saranwong et al.[134] examined the possibility of applying near-infrared spectroscopy to detect a fly infestation in intact mango fruits. Using a portable spectroscopic FQA-NIR gun instrument (Shizuoka Shibuya Seiki, Hamamatsu, Japan), they acquired the spectra of unripe and fully mature mangos at several time points after artificially induced infestation.

Examination of the second derivative of acquired spectra showed that for both unripe and ripe fruits, absorbance bands of water (negative peaks observed at 740, 829, and 951 nm) show changes after the infestation. The observed bands were assumed to be shifted from typically observed WABs at 760, 844, and 970 nm[60,64] due to the specific sample matrix as well as the instrument geometry. The shifts of the bands as a consequence of differences in a sample matrix are a very common occurrence.[153] With the infestation, the absorbance bands of water showed different behavior in ripe and unripe fruits. For green mangos, the peak at 947 nm shifted to 949 nm 2 days after the infestation, shifting more, even to 951 nm after the noticeable signs of infestation. In yellow mangos, the shift from the 949-nm band to 951 nm happened within 12 h after the infestation. The most distinctive feature of the spectra of infested fruits (both green and ripe) compared to non-infested was the change in the magnitude of the 951-nm band. This band can be thought of as the overtone of the 1426 nm band (C6 WAMACS: 1421–1430 nm, related to hydration, that is, absorption of water molecules hydrating other molecules[142]). The authors observed the change at this band (951 nm) 24 h after the infestation, when they concluded the absorption at this band cannot be due to the fruit fly eggs but because of already developed larvae whose moisture content (MC) is lower than the fruit tissue and has different composition, mainly due to the protein and fat content. Compared to carbohydrates which are the main components in the fruit flesh, this would have caused the diffusion of water molecules from the fruit tissue structures toward the larvae bodies and forming of the hydrogen bonds with proteins and fat, that is, hydration of these molecules. This is in agreement with observations by Tsenkova[142] that lower moisture materials exhibit absorbance at this particular band owing to the engagement of water molecules in hydration.

While most of the studies on biotic stress employed short wavelength ranges which, is more convenient due to the better light penetration into the plant tissues, there are reports of the utilization of longer wavelengths near-infrared ranges in the studies of insect infestation of grain. For example, the water band at 1450 nm was one of the three most significant bands for the detection of insect-infested mung bean kernels[69]; and also 1447.6 nm is one of the most important three wavelengths for the detection of wheat kernels infested by different types of insects in red flour and rusty grain beetle, rice weevil, and lesser grain borer.[129]

Wang et al.[150] were investigating sour skin disease in onions, a bacterial disease caused by the species *Burkholderia cepacia*. The authors were developing a hyperspectral imaging-based method for disease diagnosis, where they employed near-infrared reflectance imaging in the region 950–1650 nm. The images of onion samples were acquired before and consecutively each day for 7 days after the inoculation with bacteria. Subsequently, PCA was performed on selected regions of interest in images and the loading vectors of principal components were examined. The results showed that two spectral regions provide a good separation between the normal and infected onion samples as follows: The first one was 1370–1420 nm (a feature of the first principal component) and the second one was 1050–1100 nm (a feature of the second principal component). The main bands at 1400 and 1070 nm were chosen to construct the ratio images for detection of skin disease which was further used for discrimination of healthy and sour-skinned onions resulting in accuracy of detection of 95% 3–4 days after inoculation. In the region which was informative about the sour skin disease 1370–1420 nm, there are several WABs (C3-C5 WAMACS[142]). The absorbance at 1400 nm falls within the range of 1396–1403 nm which can be allocated to the absorbance of confined molecules of water[76] and is indicative of water expulsion from the cells due to the rupture of cell walls.[142] The authors of the study also noticed a decreasing trend in the intensity of reflectance for the onion samples in the late stages of the disease and they credited this change to the elevated water content of the scales due to bacteria–exactly the information provided by the WAB at 1400 nm.

What all these studies showed is that following the induction of disease (or damage by biotic factors—such as by insects in kernels), reorganization of the water matrix in the plant tissues is quite an early and important event. Furthering the studies of water-focused spectroscopy, within other plant matrices, would help elucidate the mechanism of biotic stress and hopefully lead to novel ideas in incorporating protective measures for better crop breeding and yield optimizations.

12.3.2 ABIOTIC STRESS

Adverse effects caused by extreme conditions of radiation, salinity, drought, and temperature besides other non-living factors are defined as abiotic stresses.[52] As documented by several plant biology studies, these stress conditions alter the plants systematically, resulting in biochemical, molecular, morphological, and physiological changes.[98]

In this context, with a constant increase in climate change and an aggravation of the intensity of abiotic stresses on a global scale, there is an urgency to develop crop varieties that are more resilient to ensure both food security and safety. Since mechanisms of resistance against some of these abiotic extremes are not yet completely understood, having a more in-depth knowledge of the water molecular structures responsible for such responses is important.

Until recently, the extraction of valuable information regarding the dynamics of the water matrix has been challenging due to limitations such as light scattering, existing leaf pigments, and variable plant leaves anatomy. However, the emergence of near-infrared spectroscopy's recent approach enabled the overcoming of such factors by offering complementary information contained in WASPs under different perturbations.

One of the pioneering studies in the detection of drought stress was performed by Penuelas et al.[119] They found that reflectance was informative in the near-infrared range 950–970 nm, about the water content in plants and canopies; therefore, can be useful for the detection of water stress. The study developed a way to measure water content, by using water index (WI) defined as the ratio of reflectance at the reference wavelength of 900 nm (where there is no absorbance by water) and 970 nm (WAB) for ground-based estimation of plant water concentration which was tested in the field during 1 year cycle and additionally in seeds grown in pots (nine different plant species) submitted to desiccation. They also found that with decreased plant water concentration the maximum water absorption at 970–980 nm shifts to a lower wavelength range of 930–950 nm.[119]

The earlier study of Penuelas et al.[114] also found an interesting correlation between the changes in the 950–970 nm WI and changes in the temperature of the plants' leaves (infrared thermal measurements). This finding is important because it indicates that not only water content in the leaves is changed in response to stress, but there is a change in the structural organization of the tissue matrix of which water is the main element.

Examples of applying the aquaphotomics approach for the detection and investigation of abiotic stress are well documented. For instance, soybean cultivars undergoing mild cold stress conditions, caused by a decrease of only 5°C, were successfully distinguished from non-stressed ones using multivariate analysis tools such as difference spectra, PCA as well as soft independent modeling by class analogy (SIMCA).[61] These applied methods enabled the identification of 815, 9, and 944 nm wavelengths as specific for the stressed plants. When compared to genetically modified cultivars exhibiting cold tolerance, the studied aquagrams showcased the importance of water species in a non-hydrogen bonded state for the development of cold tolerance. Interestingly, at the same temperatures, pure water does not exhibit the same structure as the water matrix in soybean leaves. Such findings accentuate the significance of water interactions with its surrounding molecules in attributing different functionalities.[61]

An equally interesting example of survival is manifested in the case of plants that withstand long periods of extreme desiccation. For instance, Kuroki et al.[75] monitored dehydration and consequent rehydration processes of *Haberlea rhodopensis* and *Deinostigma eberhardtii* plants (both from the same botanical family, but only the first one can survive extreme desiccation) and concluded that they do not react in similar manner to this particular kind of abiotic stress. The resurrection-like behavior was exclusively manifested by *H. rhodopensis* plant. This plant drastically reduced free water, accumulated dimers, and water molecules with four hydrogen bonds and kept other water species at a nearly constant level throughout the desiccation process (Fig. 12.2). When faced with abiotic stress conditions, the reduction of free water content does not seem to be the only effect.[75] The effects are very diverse, affect different organizational levels in the plant system and can also be different depending on the plant species, their developmental stage, duration of stress, and its intensity.[62]

Other plants develop tolerance to water deficits through morphological changes. For example, cacti plants growing in dry regions have adopted inherent management strategies to cope with water deficit, consisting of accumulation, absorption, and storage by developing trichomes, spines, and mucilage, respectively. These adaptive processes minimize the loss of absorbed water and confer resilient behavior to these plants.[71] Furthering the research on the molecular structures of absorbed water can presumably, help reproduce such behavior in other crops undergoing similar stress conditions but lacking this kind of adaptive mechanism.

Other researchers demonstrated that oxidative stress is almost always a part of the plant response to abiotic stresses such as temperature, salinity, or water stress and is manifested by generating several reactive oxygen species (ROS) (peroxides, superoxides, hydroxyl radicals, singlet oxygen, etc.).[59] These findings demonstrate that the deprotonation of water into these reactive species can be indicative of undergoing abiotic stress conditions.

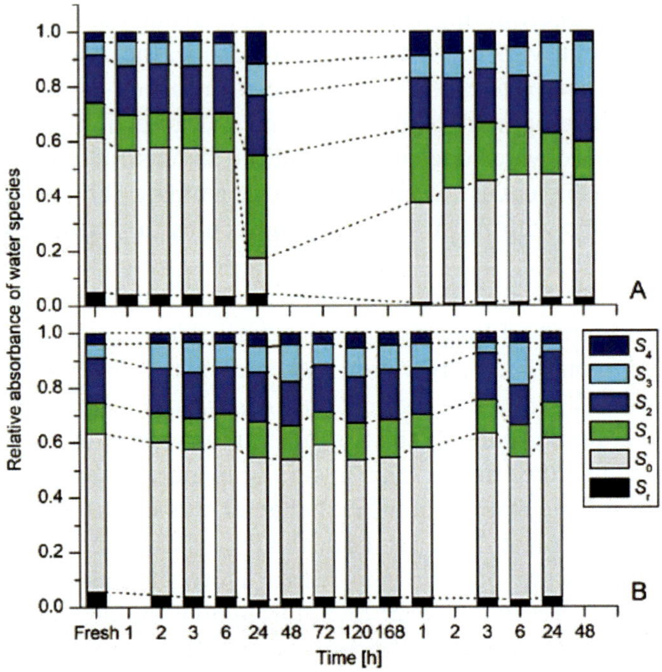

FIGURE 12.2 Dynamic behavior of the different water species of *H. rhodopensis* (top: A) and *D. eberhardtii* (bottom: B), in the course of dehydration followed by rehydration. **Sr**, protonated clusters of water; **S0**, free water molecules; **S1**, water dimers; **S2**, **S3,** and **S4**, water molecules containing, respectively, 2, 3, and 4 hydrogen bonds.

In the case of drought, Basu et al.[9] classified plant coping strategies into a combination of stress avoidance and stress tolerance, where the first one refers to the plant's ability to minimize adverse effects, while the second one refers to the capacity to endure unfavorable conditions. The latter is particularly interesting from aquaphotomics point of view since they involve regulation of metabolic processes, which were proven to be controlled by water.

In fact, in a study of plants with drought tolerance, the dehydration process was linked to changes in water molecular interactions[80] while other studies confirmed that in cells with the desiccation tolerance trait, molecular dynamics are closely controlled in such a way that loss of water is counter-acted by reversible interactions with other surrounding molecules. In plant organisms, among bryophytes, and terrestrial or marine intertidal algae, desiccation tolerance is very common,[124] but it is also well-known in some angiosperms (known as resurrection plants).[28] Although the resistance of these plants has already been demonstrated, the mechanisms by which these responses are ensured are yet to be thoroughly studied on a water molecular level in a way similar to the one developed for *H. rhodopensis*.[95]

12.3.3 DETECTION OF DAMAGE, DEFECTS, AND ADULTERATION OF FOODS

Harvest, transport, and handling of the food can cause damage and result in diminished quality and market value. In the field of food control and detection of defects and damage, hyperspectral imaging systems have been extensively used because of the advantage of being capable of rapid online applications.[34,81,108]

Hyperspectral imaging combines conventional imaging and spectroscopy, offering one more—spatial dimension which is of interest for obtaining information about the distribution of defects or damage in food and food products—large amounts of not only spectral but also spatial information can be obtained simultaneously from the studied objects.[58,156] It is rapidly advancing technology because it provides a non-destructive, real-time quality, and safety assessment.

One of the pioneering studies on the detection of damage in food which combined hyperspectral imaging and utilized aquaphotomics principles was performed by Gowen et al.[38] In their study, the damage in white mushrooms was induced by mechanical vibration for different time periods (five damage (perturbation) levels) to simulate the damage that can occur during transport and handling. The mushrooms were scanned using a hyperspectral imaging device operating in the 900–1700-nm spectral region and then examined using difference spectra (determined through subtraction of the average control mushrooms [not damaged] spectrum) and then by means of partial least squares regression (PLSR) models, the time period of exposure to vibration was predicted. The difference spectra revealed that major changes in the spectra of damaged mushrooms compared to the control ones happen around the absorbance bands located at 950, 1100, and 1170 nm and in the spectral region 1390–1450 nm. The first and the last indicated region correspond to the water absorbance regions.[109,112] The closer inspection revealed the particularly increased absorbance at 1398 nm for higher levels of vibration damage—the absorbance band associated with the dehydration, that is, removal of water from the cell.[142] It was later discovered that this band correspond to quazi-free water molecules, that is, subpopulation of free water molecules.[76]

The PLSR model developed for prediction of duration of exposure to vibration revealed more information. Examination of PLSR vector coefficients confirmed the increase of absorbance at 1398 nm while the opposing trend was observed at the wavelengths corresponding to absorbance bands of more strongly bonded water—1433, 145, 149, and 1510 nm (C7, C8, C9, and C12 WAMACS). These findings together indicated that mechanical vibration leads to disruption of cell walls, which resulted in decrease of more strongly hydrogen bonded water, followed by increase of weakly hydrogen-bonded water, that is, increased water motility and visible appearance of water on the surface of the mushroom caps as a result of damage. One of the very important contributions of this work is finding the direct connection between the water species absorbing at 1398 nm and the functionality, that is, damage.

The same authors furthered the study of physical damage to the aging of mushrooms.[31] In this case, they used a reflectance mode of measurement in the 400–2498-nm range, one perturbation level to induce the damage, and afterward, the mushrooms were stored in the ambient conditions. After storage, the acquisition of spectra was performed for 4 h, every 1 h. The mass of mushrooms was also measured to build a PLSR model for the prediction of weight loss during storage. The analysis was focused only on the range of 1300–1600 nm (the first overtone of water). Consistency with the previous study was confirmed by finding the differences in WASP of damaged and undamaged mushrooms at the following WABs: 1408–1410 nm (C5, molecules of free water), 1434 nm (C7, intermediate strongly-bound water; water dimers), 1462–1464 nm (C9, water trimers, that is, water molecules with 2 hydrogen bonds), and 1520 nm (C12, strongly bound water).

The PLSR models for weight loss were built separately for damaged and undamaged mushrooms and showed differences. The PLSR model for weight loss in damaged mushrooms needed only 4 latent variables and had lower accuracy (R^2 = 0.84, root-mean-standard error for cross-validation [$RMSE_{CV}$] = 1.15%); while the PLSR model for undamaged mushrooms required 6 latent variables and had better accuracy of prediction (R^2 = 0.92, $RMSE_{CV}$ = 0.65%). A comparison of relevant absorbance bands found by inspection of PLSR vector coefficients revealed that the band located at 1420 nm had an opposing trend for the damaged and undamaged mushrooms and that the band at 1520 nm was important for undamaged mushrooms but not for damaged mushrooms. It is an interesting finding that weight loss is related to different WABs for damaged and undamaged mushrooms, which illustrates well that different systems, like this case of damaged or undamaged mushrooms are described by different WASPs that could be used as integrative biomarkers with practical importance for applications in food quality evaluation and/or biodiagnosis.[142]

Another study employed visible near-infrared (Vis/NIR) hyperspectral imaging utilizing the region 400–1000 nm for the monitoring of sprouting in wheat kernels.[159] The study was performed by acquiring images of single kernels of Canada Western Red Spring (CWRS) wheat. The PCA of the spectra of wheat kernels could distinguish the different quality parameters while, the loadings plot helped to identify four wavelengths (475, 525, 670, and 875 nm) with the highest influence on successful discrimination. Of the four wavelengths, the first three were attributed to pigments and their residual variations, but the last one 875 nm was related to the variations in water content[18,159] which happens when kernels begin sprouting. Furthering the analysis, the authors demonstrated that the ratio of mean reflectance at wavelengths 878 and 728 nm could be used as a highly efficient indicator enabling successful separation of sprouted and not-sprouted, sound kernels. The entire sound kernels were successfully identified using only this indicator, which shows its suitability for practical applications in kernel sorting systems.

Hyperspectral imaging was also applied to study coffee beans roast defects.[14] The spectra of five different types of coffee bean defects (medium roasted [control], underdeveloped, overroasted, interior underdeveloped, and interior scorched) were acquired with a hyperspectral apparatus in the 1000–1700-nm region, and partial least squares discrimination analysis (PLS-DA) enabled discrimination of roasting defects with the accuracy of 86.2%. Examination of variables in the spectra which mostly contributed to successful discrimination revealed that the absorbance band at 1420 nm was the most important. This band was assigned to the O–H bond of free water,[14,43] which is in agreement with previous findings.[1]

Another problem in food quality control, which is rapidly gaining attention in the latest years, is the intentional disruption of food quality by adulteration. The latter is a phenomenon by which the nature or the quality of food is altered by the intentional addition of usually inferior food substances (adulterants) or by the removal of vital food components. It is a serious concern worldwide because it may lead to unwanted health effects and diminishes the quality of the product while it is sold at the price of the original (non-adulterated) one. Regular use of methods for screening food adulteration is a good strategy to prevent such occurrences; vibrational spectroscopy is a great candidate for this role because it is non-destructive and can provide very rapid detection of food authenticity and adulteration.

Honey is a naturally sweet food substance, consumed alone for its beneficial health effects or as an ingredient in multiple foodstuffs. Due to the high market demand for honey and the high price of this commodity, honey is often subjected to adulteration which may not necessarily pose a health risk (however, there are honey adulterants that may be very harmful), but is fraudulent and impacts honey quality negatively. NIR spectroscopy has been employed for various purposes in honey quality evaluations—for example; in the investigation of floral origin,[22,32] detection of adulteration with glucose and fructose,[30] beet invert syrup and high fructose corn syrup (HFCS).[23,70] It was also proved reliable for the rapid prediction of mineral content (potassium, calcium, magnesium, and phosphorus) of honey.[32] It is interesting that the mineral content in the mentioned study was successfully predicted with very high values of ratio performance deviation (RPD) between 4.0 and 5.2. The spectral range used was 1100–2498 nm, and while the study did not focus on the reasons behind the successful prediction capacity, it is possible that mineral content was determined indirectly by means of the impact that the minerals leave on the water structure of the honey samples which is, reflected in this spectral region.

The information from the water molecular system of honey enabled tracking of HFCS adulteration in *R. pseudoacacia* (acacia or black locust) honey collected from four geographical regions in Hungary.[10] Using a fiber-optic immersion probe, the acquisition of spectra of the honey samples was performed in the region of 1300–1800 nm where both absorption bands of carbohydrates and water are located. Different levels of adulteration with corn syrup (ranging from 0 to 40%) were detected with high accuracy ($RMSE_{CV} = 1.48$, $R^2_{CV} = 0.98$). The aquaphotomics evaluation of the water molecular system in honey samples (Fig. 10.3) revealed that, compared to sugar syrup, samples of non-adulterated honey contained highly organized water probably due to the high chemical content (proteins, vitamins, minerals, etc.) of the honey (Fig. 12.3a).

Water molecular structure in HFCS showed high absorbance at WAB related to trapped water (1398–1402 nm,[76]), hydration band (1426 nm–1428 nm), and water band which can be ascribed to absorbance of water dimers (1434–1436 nm, S_1[142]) or weakly hydrogen-bonded water in the first hydration shell.[76] This latter band of water is found often in sugar-water systems related to the phase transition.[77,86] Adulteration led to a decrease of several water molecular species (Fig. 12.3b), but water trimers were particularly

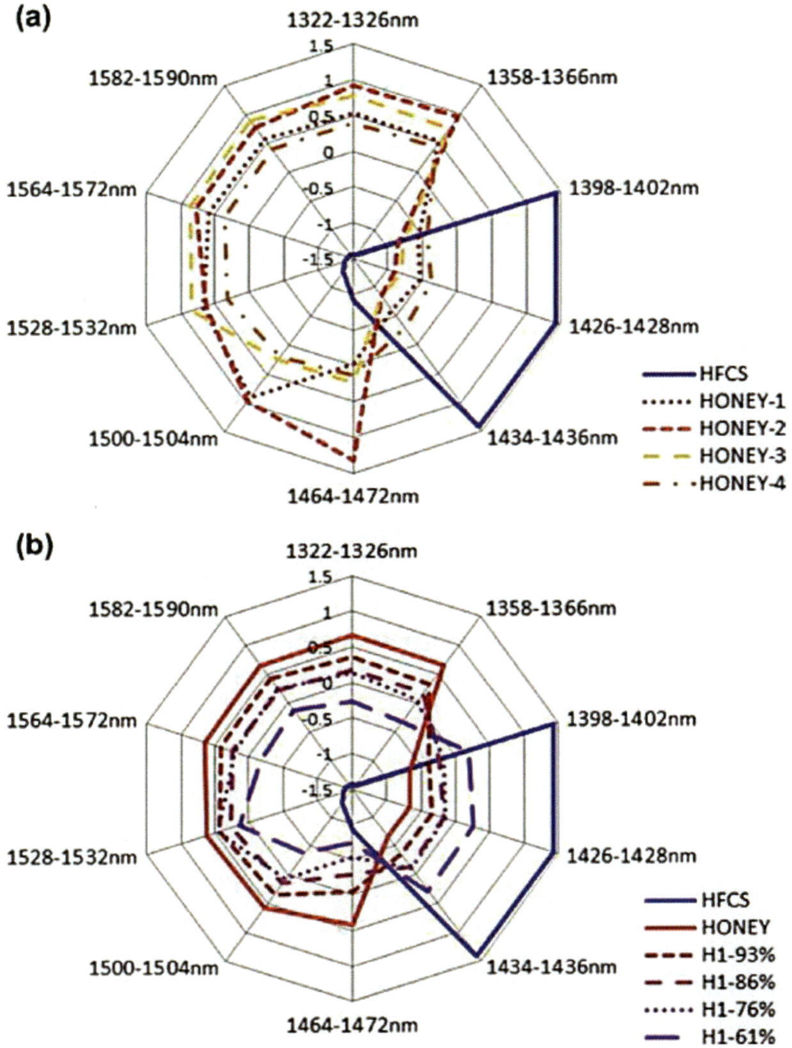

FIGURE 12.3 (Top: a) Aquagram of the four pure *Robinia pseudoacacia* honey—of different geographical origin and HFCS—the water matrices of pure honey are characterized by strongly bound water compared to the syrup. (Bottom: b) Aquagram of the tampered honey mixtures (H1: % of purity) along with the aquagram of the HFCS used as an adulterant and the mean aquagram of pure honey.

affected (C9: 1458–1472 nm, S_2[142]). Figure 12.3b shows the mean aquagram of all pure honey samples (labeled as HONEY), aquagrams of one honey type adulterated with syrup in different percentages (61–93% purity) compared with aquagram of adulterant (HCFS). It can be observed from the graph that with the increasing amount of adulterant in honey the absorbance of hydrogen-bonded water (left part of the aquagram) decreases and the absorbance at absorbance bands which are characteristic of HFCS increases (1434–1472, 142, 1398–1402 nm)—this is particularly evident in the aquagram of the adulterated honey with the highest percentage of adulterant (H1-61%). The decrease in hydrogen-bonded water corresponds well even with the observation of the increasingly liquid appearance of adulterant and adulterated honey.[10]

Another work on honey adulteration investigated how adulteration with water affects the honey from the genus *Apis* and stingless bees by utilizing the Vis-NIR (350–1050 nm) absorbance spectroscopy measurements.[107] Examination of spectra revealed that with water dilution, two main spectral features change—the visible part, due to the fading color of honey when it was diluted with water, and the other region located around 974 nm, due to the strong water absorption. Adulteration of honey by dilution led to the shifting of this peak; the absorbance peak of pure honey was located at higher wavelengths, and with dilution shifted to lower wavelengths. This observation was in agreement with the observed shift in water-sugar samples' spectra with decreasing sugar concentration observed in other works as well.[102]

The absorbance band at 975 nm can be assigned to the absorption of water trimer (S2, water molecules making 2 hydrogen bonds[137]). The corresponding absorbance band in the first overtone region would be 1461 nm, that is, exactly the same C9 WAMACS, water trimers band, which Bazar et al.[10] also found to be particularly affected by adulteration.

12.3.4 FRUIT QUALITY EVALUATION

Near-infrared spectroscopy has a long history of application in the determination of fruit quality due to the possibility for non-destructive measurements of many direct quality indices, the most important ones being soluble solids content (SSC) and sugar content (in degree Brix), pectin content and acidity, but also for damage and bruises detection in intact fruit, while lately, attributes like firmness, ripeness, texture, which do not seem to be directly correlated with the near-infrared spectra are also being measured.[82] Many of the mentioned quality parameters of fruit are related to the structure of the water in the fruit matrix.

Fruits are very high water-containing systems (usually above 80% of water) and their NIR spectra are strongly influenced by water absorption presenting main absorption peaks at around 760, 970, 1190, 1450, and 1940 nm.[5,12,103]

Acidity and SSC are one of the most common quality parameters of the fruit related to the maturity of the product, and information about these parameters can also be of help in the determination of the most appropriate harvesting time. One of the aquaphotomics studies investigated the potential of the aquaphotomics approach in the determination of acids and pH of aqueous acid solutions to translate the findings to measurements in intact fruits.[103] The authors of the study aimed to achieve the assignment of the wavelengths which best contribute to the pH measurements of most common fruit acids—citric ($C_6H_8O_7$), oxalic ($C_2H_2O_4$), tartaric ($C_4H_6O_6$), and malic acid ($C_4H_6O_5$). The spectra of acidic solutions were acquired in the range 700–1000 nm. The analysis was performed utilizing the aquaphotomics approach by focusing on the range where the contribution of water in the NIR signal is the highest, which yielded prediction results of all examined acids with R^2 higher than 0.9. The important wavelengths for successful calibration were acid dependent, however, the 975 nm—one of the peak absorptions of water was important for all examined acids. This wavelength can be attributed to the absorption of water trimer (S2, water molecules making 2 hydrogen bonds,[137] recalculation of the corresponding absorption in the first overtone region produces wavelength 1462 nm, C9 WAMACS: 1458–1468 nm, S2 water trimer[142]). Besides, also regions 918–925 and 990–996 nm significantly contributed to the measurements of pH.

Similar fundamental studies of aqueous solutions were performed with the purpose of measurements of the most common sugars in fruit—sucrose, fructose, and glucose[8,41,102,104] and contributed significantly to a better understanding of how NIR water absorbance pattern is influenced by these compounds which can help future non-destructive measurements in intact fruits. The sugar content is one of the fundamental quality characteristics of fruit and for practical purposes, it is usually estimated through the measurements of SSC (expressed in °Brix).[16] However, °Brix is the representation of all the soluble content of the fruit including different sugars—sucrose, glucose, and fructose of which sucrose is usually the most dominant, and also acids, vitamins, minerals, proteins, and other soluble materials. The sweetness of the fruit, which is one of the main attributes contributing to consumer satisfaction, depends on the total sugar content, but also to the specific sugar profile, because sugars like fructose or glucose have very small sweetening power compared to sucrose.[16]

Quantification of individual sugars, as well as total sugar content and SSC are important tasks for applications of near-infrared spectroscopy in fruit quality evaluation, and the early studies on simple one- and two-component water-sugar solutions indicated that region of first overtone of water significantly contributes to the quantification.[6]

One of the aquaphotomics studies aimed at the quantification of individual sugars in mixed sugar solutions.[8] The study employed near-infrared spectra of four mixed sugar solutions (glucose, fructose, sucrose, and lactose) in millimolar concentrations for identification of individual sugars and their quantification (individual content and total content), and the results showed good separation as well as high accuracy of predicted sugar content. The spectral regions responsible for good separation of individual sugars were 1400–1425 and 1445–1470 nm (findings based on examination of difference spectra, PCA loading vectors, and PLS regression vectors). Examination of PLSR vector coefficients from PLSR models built for the prediction of individual sugars in sugar mixture solutions showed a repeatable pattern of 24 absorbance bands out of which only 4 were CH bands, the rest were all related to specific water structures affected differently by each sugar. The results clearly showed that even very small compositional changes in mono- or disaccharide solutes affect the water matrix of the solutions to a large degree which allowed their indirect measurements by utilization of WABs in addition to absorbance bands related to sugars.

A similar fundamental study of aqueous sugar solutions (sucrose, glucose, and fructose) was performed by Omar et al.[102]; but they performed measurements in the different NIR ranges (360–1100 nm). The purpose of the work was to identify the absorbance bands in this region which contribute most to the successful prediction of sugar concentration (in °Brix). While NIR prediction of total sugars was proved to be good, the prediction of individual sugars usually produces poorer sensitivity and accuracy.[102,122] Wavelength region between 940 and 985 nm was found to provide meaningful results and detailed quantification analysis was performed within that range. Absorbance bands at 959, 960, and 961 nm generated the best coefficients of determination and the lowest standard error for the prediction of sucrose, glucose, and fructose concentrations, respectively, when using linear regression. The linear relationship was lost for wavelengths longer than 960 nm.

Further, multiple linear regression (MLR) analysis was performed and the most efficient calibration algorithm has been identified and included slightly different combinations of four wavelengths for the following each individual sugar quantification: For sucrose 730, 830, 915, and 960 nm; for glucose 830, 909, 960, and 965 nm; and for fructose 730, 912, 960, and 965 nm, meaning that combination of C–H bands (730, 830, and 909–915 nm) with water bands (960 and 965 nm) provided enough information to predict each individual sugar concentration. The absorption bands in the 909–915-nm region are actually quite complex and can also be attributed to different water species.[51,94,123,155]

Furthermore, increased sugar concentration (for the range of Brix values between approximately 1 and 35°Brix) resulted in the highly noticeable shifting of the WAB peak which was different for different sugars. Initially located at 975.5 nm, the peak of sucrose shifted to 985.5 nm, for fructose and glucose, the shift occurred, respectively, from 976.8 nm to 987.2 nm and from 974.5 nm to 984.5 nm. The shifting of the peak was for each sugar approximately 10 nm toward longer wavelengths. This phenomenon found in the short wavelength region was in agreement with the observations of the behavior of the water peak in the longer range as noticed by Giangiacomo.[39]

The same influence of increasing sugar concentration on strengthening the hydrogen bonds network in water was observed in both studies—by Giangacomo[39] and by Omar[102] and was used as an additional source of information about the concentration of individual sugars in aqueous solutions. However, for practical applications in non-destructive individual sugar measurements in food quality evaluation the shorter wavelength range due to the better light penetration depth is of more importance.

Using the transmittance spectra of the apple juice samples in the first overtone of the water region, Kaur et al.[65,67] reached the conclusion that increased Brix level leads to an increase of water molecular species (S_2, S_3, and S_4: water molecules with two, three, and four hydrogen bonds) characterized by strong hydrogen bonds. The same authors reported similar conclusions when they performed prediction of SSC in kiwifruit juice and sucrose solutions, where the better prediction was obtained using longer cell path length and short wavelength region (870–1100 nm).[66]

Delwiche et al.[26] also utilized the disruption of the water molecular network by sugars and soluble solids in intact mango fruits for the prediction of soluble solutes content—SSC and individual amounts of sucrose, fructose, and glucose. The spectra of intact fruit were acquired during a period of 11 days (from unripe stage to a few days after optimal ripeness) using NIR interactance mode in the range 750–1088 nm and developed PLSR models for prediction of SSC, total sugar, and individual sugar contents. The analysis yielded the best results for prediction of SSC (R^2 = 0.70–0.88, standard error of cross-validation [SE_{CV}] = 0.93–1.10), and sucrose (R^2 = 0.75, SE_{CV} = 15.6), glucose (R^2 = 0.67, SE_{CV} = 1.4), fructose (R^2 = 0.70, SE_{CV} = 6.9), and total sugars (R^2 = 0.82, SE_{CV} = 16.8). Examination of first three loadings of the PLSR analysis and the second derivative spectra (11-points, Savitzky–Golay second-order polynomial) of the pure sugars in a sodium–sulfate aqueous matrix pointed out how each individual sugar produces different effects in the spectra. In the first loading, the major difference appeared in the 900–1000-nm region ascribed to water but the overall response of all the sugars occurred at the same 966-nm peak. The second loading showed variation among the sugars where the most prominent feature was a downward peak within 950–960 nm whose position varied between the sugars. Also, a noticeable pattern was present in the third loading in the region 948–956 nm. This examination testified to the unique influence of each of the sugars on the disruption of the hydrogen-bond-based network in the water matrix of the mango fruit. What the authors concluded based on their results is that prediction of SSC content compared with individual or total sugar content could be satisfactorily predicted using the applied technology and has value for assessing the ripeness and quality of mango.

Another study about the prediction of SSC in intact fruit was performed using reflectance and interactance near-infrared spectra of carambola fruit.[106] The study also used a novel spectral linearization technique[102,105] for quantification of SSC from the spectra. The spectral linearization technique is based on the observed phenomenon that the steepness of the spectral profile is directly linked to the water content. This steepness is equally related to the linearity of the absorbance curve; hence the linearity of the spectrum can be used for quantification. The results obtained by this technique were also compared to one of the commonly used quantification methods, namely, MLR. The interactance spectra produced better prediction results of SSC in intact fruit when spectral linearization was used, compared to MLR and for the wavelength range between 940 and 1025 nm. The reason for better prediction using the interactance technique was explained by the much higher contribution of water signal to the spectra acquired in the interactance mode, as opposed to the reflectance mode.

Xia et al.[158] also investigated the possibility of predicting SSC in intact apples, using diffuse reflectance spectra in 550–950-nm region. SSC in apples varied within the limits 8.00–13.6°Brix. Several variable selection algorithms were employed to find the most effective wavelengths for modeling of SSC of apples. The Competitive Adaptive Reweighted

Sampling—Successive Projection Algorithm (CARS-SPA) method was deemed the best in this regard giving the best prediction accuracy based on the analysis of 37 chosen effective wavelengths. Among those, three were related to known water overtones—748.35 nm (third overtone); 944.25 and 945.48 nm (second overtone). Similarly, an important contribution of WABs for the prediction of the SSC content of apples was reported in another work by the same authors.[157]

Omar and MatJafri[105] used the Vis-NIR reflectance spectroscopy between 400 and 1000 nm to monitor decay in apples and pears. The objective of the study was to find a single wavelength that would yield the highest coefficient of determination, R^2 of the relationship between the absorbance and the dehydration period (in hours) of fruits. For four different fruit types examined (red and green apples, and yellow and green pear) the results showed that the wavelength range from 960 nm to 980 nm produces the highest R^2, with the wavelength of 975 nm being the most important for estimation of dehydration in hours,[105] very similar to the findings by Penuelas et al.[119]

Giangiacomo et al.[39] investigated the process of osmotic dehydration of apples in concentrated sugar solutions. The first major observation was that the max absorbance peak in the first overtone region of water shifts around 7 nm to higher wavelengths as the concentration of sugars increases. This was explained as an increase in the number of hydrogen bonds due to the binding of sugars progressively during water removal. More detailed analysis revealed that during osmotic dehydration changes occur at particular WABs: 1396 nm (confined water[76]), 1459 nm (C9 WAMACS, water trimer[142]), 1936 nm (free water[142]), and 1518 nm (C12 WAMACS, strongly bound water). Using two-dimensional correlation spectroscopy, the authors were able to establish the time pattern of changes in the following water matrix of the apples with dehydration: First, there is a change in free water molecules (moisture) (1396 nm band), then, this change is followed by the presence of confined water (1396 nm) and strongly bound water (1518 nm). The process looks similar to the changes, that is, dehydration of mushrooms due to the mechanical vibration.[38]

The texture is another good fruit quality indicator, a very complex, multidimensional trait related to the composition of fruits, as well as their mechanical properties and geometry.[149] The usual textural attributes used for description include juiciness, firmness, hardness, crispness, mealiness, chewiness, stringiness, melting, and fibrosity.[149]

The objective of the study performed by Vanoli et al.[147] was to use aquaphotomics methods to relate the WASP of apples with the sensory texture.

Similar to kernel hardness,[56] texture is not something usually perceived as a parameter related to water molecular structure. Fruit freshness can be dictated by the textural properties of the fruit, and have a direct influence on the consumer's choice—crunchiness and juiciness are considered positive attributes while mealiness is perceived as negative.[115]

Vanoli et al.[148] analyzed sensory texture attributes of three different apple cultivars (Braeburn, Gala, and Kanzi), and using cluster analysis on sensory textures four different texture profiles were found for each cultivar. On average, the water absorbance pattern related to mealiness in apples was characterized by absorbance values that were the highest at 1364, 1372, and 1382 nm water bands and the lowest in the 1438–1492-nm region indicating dominance of weakly hydrogen-bonded water species. The opposite was observed in juicy and crispy apples which were characterized by the dominance of medium to strong hydrogen-bonded water species. The juicy, firm, crispy apples had the smallest intercellular spaces, resulting in higher firmness, while the flesh of mealy apples had larger intercellular spaces, that is, they were less firm/softer. The process of softening actually is a process of increased hydration of cell wall polymers due to the extensive structural modifications of pectin.[46] All these changes, highly dependent on the strength of hydrogen bonding, affect the water molecular organization in fruit tissues, consequently resulting in changes in the WASP of the fruit.

12.3.5 MOISTURE CONTENT AND WATER ACTIVITY

When it comes to dried food, MC together with a water activity (a_w) are the most used and most important parameters related to the quality of stored products and their stability.[89] These two parameters are in a complex nonlinear relationship, and while MC refers to the content of water in the food product (and indirectly to the yield, texture, and other properties), the water activity is a measure of water available to participate in chemical reactions and therefore to microbial safety and spoilage. Traditionally, MC and a_w are measured using destructive and time-consuming techniques which are not applicable in online processing, while NIR has become a widely used method, apt for MC quantification due to the important absorption of water in this region.[1]

Since, in the majority of cases, food products are unlikely homogenous, hyperspectral imaging as an emerging platform technology that provides spatial information in addition to spectral information is an excellent tool for examination of MC and a_w.

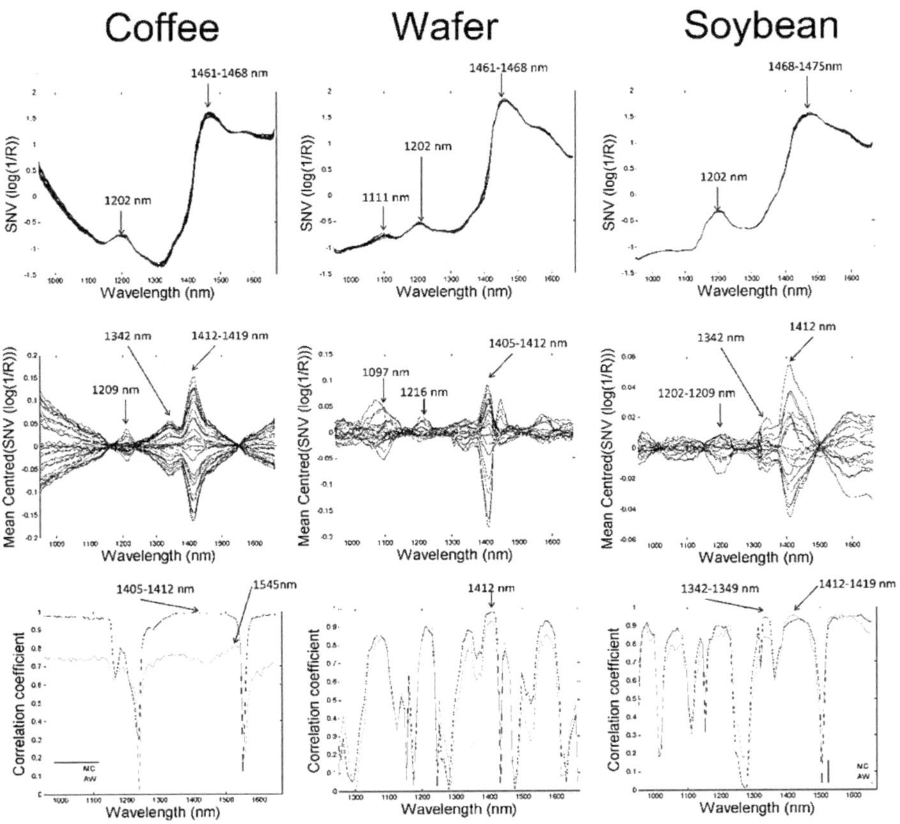

FIGURE 12.4 Pre-treated mean spectra (standard normal variate [SNV]) (top row); Mean centered SNV spectra (subtraction of the mean SNV pretreated spectrum from each group) of coffee, wafer, and soybean (middle row). Calibration is set in blue and independent prediction is set in red. Correlation coefficients between SNV (log[1/R]) at the studied wavelengths and water activity (green) or MC (black) (bottom row).

In a study by Achata et al.[1] the relationship between MC and a_w was investigated in three types of food products—wafer, coffee, and soybeans using hyperspectral imaging (Fig. 12.4). It was obvious that differences in the spectra due to different food matrix composition exist, some absorption peaks were common in all the three products in relation to the changes in hydration: 1202–1216, 1342, and 1402–1419 nm which were interpreted to be due to the first overtone of asymmetric and symmetric stretch vibration of water, and in particular, the latter one as the band of free water molecules (C5 WAMACS: 1398–1418 nm). This band was found to be increased as the samples were hydrated and are related to free water that is able to partake in chemical reactions such as those related to spoilage, microbial growth, and enzymatic browning. In other words, this absorbance band carried information about the quality and shelf life of the food products. For each product, a little bit of specific correlation was found between the absorbance bands and the MC and a_w. For example, in coffee, 1545 nm was most correlated to the a_w while the spectral region 1405–1412 nm was essentially correlated to MC. For wafer, both MC and a_w were found to be highly associated with the wavelength 1412 nm, while for soybean, for MC, the highest correlation was found with 1342–1349 nm, and for a_w with the 1412–1419 nm band.

However, in other regions, a relatively high correlation was found with both MC and a_w, too, indicating that hydration affects many different WABs, that is, different water species. Prediction of MC for all three products was highly accurate for all three products, while for a_w the best model was developed for coffee samples. Prediction plots also allowed visualization of spatial differences in the distribution of water activity and MC, which was mainly prominent in soybean.

In a similar study, Gowen[48] applied aquaphotomics in combination with hyperspectral imaging to investigate the role of water in food quality. This type of investigation could be of importance for controlling the drying process where MC affects the product shelf life and sensory properties. Gowen investigated the dehydration of different foodstuff (bread and banana) as well as the hydration of chickpeas. In samples exposed to drying, the major spectral feature appeared at water band 1398–1400 nm—the sign of free water decreasing. Similarly, during the hydration of chickpeas, at the same band changes were observable, only with the progression of hydration, the absorbance was increasing. So, e study also confirmed is the importance of this water band for dehydration–rehydration processes.

Estimation of water content in different fruit types usually shows activation of slightly different water bands in response to dehydration/rehydration which is simply the consequence of different sample matrices.[153] For example, out of eight wavelengths (440, 525, 633, 672, 709, 760, 925, and 984 nm) needed to predict MC in bananas at various maturity stages, Rajkumar[126] identified 760, 925, and 984 nm water bands.

In determining the MC of peach fruits, Williams and Norris[154] confirmed the prominent role of water absorption bands at 834, 938, 958, 978, 986, and 994 nm, while Lu and Peng[85] found that for the same fruits, the absorption band of water located at 950 nm was important for moisture determination.

A recent study on MC in tapioca starch also contributed to the identification of water bands related to water content.[116] In tapioca starch production,

MC has been used as a sign of the quality of the finished product and Phetpan et al.[116] were interested in the development of a sensor for in-line measurements which can be used to signal the completion of the drying process. Examination of the average second derivative transformed spectra of tapioca showed there were 4 prominent regions: around 985, 1200, 1430, and 1580 nm. The authors ascribed the absorbance in these regions to the second overtone attributed to the OH stretching of starch (990 nm), bands of starch (1202 nm), the first overtone ascribed to OH stretching of starch and water (1450 nm), and the first overtone attributed to the OH stretching of starch (1580 nm), respectively. However, as their PLS prediction models showed, the most important absorbance band with the highest coefficient of determination was found at 1410 nm, the band of free water molecules (C5).[142] Additionally, important bands were also found at 1385, 1405, 1430–1450 nm, all of which fall within the ranges of C4, C5, and C6-C7 WAMACs.1[42]

12.3.6 QUALITY EVALUATION OF SEEDS, GRAIN, AND STARCH

The main components of agricultural products (e.g., starch, protein, cellulose, lipids, and minerals) are all profoundly influenced by the medium, in which they are brought together: **(a) water**: it essentially contributes to their behavior and their spectra in near-infrared region, even in the case of the low-moisture systems (typically up to 15%) where water it is not a dominant constituent.[153] Due to the small size of water molecules which are forming hydrogen bonds with main constituents, water becomes a huge influential factor in shaping the chemical and even the physical properties of agricultural materials.[153] The spectra of such materials will, regardless of the small water content, still exhibit main WABs located around 1410 and 1940 nm, but the exact position will show changes dependent on the material.[153]

A particularly interesting study whose significance seems overlooked was performed by Hong et al.[56] In this study, wheat with different kernel hardness was studied using near-infrared reflectance spectroscopy to investigate how it affects milling behavior, the size of the produced particles and the resultant water absorption, and the properties of produced dough. In earlier studies, Simmonds showed that the hardness of wheat depends on water-soluble components.[128]

Further biochemical basis of grain hardness was also investigated by Mattern and Sandstet,[87] who showed that the main determining factor in the mixing requirement of wheat flour was the water-soluble portion of wheat. Pentosans were identified as a major component of this water-soluble part. This is what attracted the attention of Hong et al.[56] to perform an NIR investigation using wheat water extracts to estimate pentosans content and to identify which chemical component is responsible for the correlation found with whole grain hardness. Their study found three wavelengths—1366, 1436, and 2108 nm which contributed to significantly high correlation coefficients between plain water extract and grain hardness ($r = 0.86$). In a series of experiments designed to identify the active component in water extracts which correlates with hardness, it was proven that it was not the protein, lipid, pentosan (hemicellulose) or any of the fermentable sugars.

Their study also showed one more thing: the extraction of water significantly improved the accuracy of pentosan prediction (enzyme-extractable, water-soluble, and total pentosan content) in comparison to predictions based on the spectra of dry wheat flour. This work significantly contributes to the aquaphotomics field. First, the better quantification of pentosans in water extracts compared to dry samples shows that obviously, the interaction of pentosans-water contributes to the information potential of near-infrared spectra. The water molecules involved with the hydration of pentosans help their more accurate estimation. Second, the bands identified as important with high correlation to grain hardness and pentosan levels were exactly WABs: 1366 nm is an absorbance band of water hydration shells (WAMACS C2: 1360–1366 nm, $WS_{1,2,4}$[142]) and 1436 nm is also WAB (WAMACS C7: 1432–1444 nm, S1[142]). This band was also found important for phase transition from amorphous to crystalline state (for lactose[77] and sucrose[86]), hence we can consider that this particular water band is related to the degree of crystallinity which is known to have a direct influence on the hardness of materials.

The third absorbance band was found to be related to hardness was 2108 nm and it is usually assigned to cellulose[127] or hemicellulose,[24] but since the hemicellulose was disregarded, it is likely that this is also a WAB, that corresponds to a combination of OH deformation and stretching vibrations.[146] While it is generally well recognized that water is an effective plasticizer in food matrices affecting the food hardness/softness,[117] the fact that specific water molecular structures are directly correlated to this mechanical property is not really considered.

Seed viability or seed vitality is a very significant aspect of seed quality, which presents the potential for germination, directly affecting sprouting and growth. It is often the case that seed vitality changes throughout the storage period due to the lack of proper conditions, for instance, adequate moisture, temperature, or atmosphere. Rapid and effective methods for the estimation of seed vitality and germination potential prior to cultivation are something the current agricultural practice needs.

There are multiple reports of the application of either spectroscopy or hyperspectral imaging in the near-infrared region for the determination of seed viability.[160] In one such study, NIR hyperspectral imaging was applied to identify the vitality of rice seeds.[55] Using the least squares support vector machine analysis on raw spectral data, the authors achieved 94.38% accuracy of classification between viable and non-viable seeds using selected wavelengths shown to have the highest contribution (968, 988, 1204, 1301, 1409, 1463, 1629, 1646, and 1659 nm).[55] Out of the 9 selected wavelengths, at least 4, namely, 96, 988, 1409 , and 1463 nm are most likely WABs; the first one is close to the 967 nm band reported as water solvation in the concentrated environment,[53] the second one can be related to large protonated clusters,[152] while the last two fall within the ranges of known WAMACS of free water molecules (C5) and water trimers (C9), respectively.[142]

Rittiton et al.[125] developed a non-destructive NIR system for the quality evaluation of individual rice kernels using transmittance mode with the purpose of improvement in breeding programs. The spectra of rice kernels of different cultivars, in different moisture ranges were examined in the 1100–1800 nm spectral region, which was selected to cover the most important bands of water (1450 nm) and protein (1735 nm). The authors developed simple MLR models which relied on only a few wavelengths for the prediction of moisture (1414 nm, 1474 nm, 1370 nm, 1336 nm) and protein content (1738 nm, 1696 nm, 1652 nm, 1686 nm). These wavelengths were selected from the variables with the highest coefficients in the regression vector of PLSR analysis. Using just MLR models (to speed up the process of sorting) they achieved sufficiently accurate results.

Soltani et al.[130] employed near-infrared reflectance spectroscopy in the 780–2500-nm region combined with multivariate data analysis for investigating the viability of oriental beechnuts, *Fagus orientalis*. They first noticed that raw absorbance spectra were consisting of major water bands near 1450 nm, and also 1940 nm. The spectra of non-viable seeds showed slightly higher values of absorbance in the 750–1350-nm region, while the viable seeds had higher absorbance in the 1350–2200-nm region. Examination of the difference spectrum between viable seeds and non-viable showcased differences in absorbance at 1440, 1898, and 1980 nm for viable seeds, whereas non-viable at the entire 750–950-nm range with the highest peak at 896 nm. Partial least squares (PLS) discrimination models were further developed for discrimination between the two classes of seeds, using raw and pretreated data (multiplicative scatter correction, orthogonal signal correction). All the models had 100% discrimination accuracy. Analysis of the PLS weights showed consistency with what was observed in the difference spectrum, the spectral profile looked very similar and indicated the most important regions for separation at 750–950, 1400–1500, and 1880–2000 nm. The first region corresponded to third and second overtone regions of water (although authors reported the peak at 890 nm in this absorption region to be characterized by C–H stretch third overtone, yet this was also related to water), the second to the first overtone region of water and the third indicated was a combination band of water. The authors concluded that the information about the viability of seeds is related to moisture.[130] The difference in water content can be explained by the structural damage in non-viable seeds, which leads to the loss of water. It is possible that due to this loss, the spectra of non-viable seeds contained more information coming from the other components in the 750–950-nm region.[130]

In a similar study, Tigabu and Oden[138] attempted to distinguish viable seeds of *Pinus patula* "Schiede" and "Deppe" from non-viable ones using near-infrared reflectance and transmittance modes. Employing PLS discriminant analysis they also had 100% accuracy of discrimination between viable and non-viable seeds when the reflectance spectra were used. Examination of difference spectrum and PLS loadings displayed large differences in the spectra between classes of seeds due to the protein and fatty acids content, components that empty seeds did not have. The authors also examined the 1300–1600-nm region and noticed weak bumps around 1394 and 1510 nm which they ascribed to the C–H combination as well as the first overtone of N–H stretching and although mentioning that the absorption band of water may highly perturb other bands in this region, they assumed that protein hydration could be the basis of variation in this part of the spectra. However, the two bands, 1394 and 1510 nm, can be attributed to the absorption of the confined water molecules,[76] (band 1396 nm–1403 nm, which may show a little shift due to the seed sample matrix) and strongly bound water (C12 WAMACS, [142]). Together, these two bands indicate damage and dehydration, as previously observed in several cases.[38,39] The authors also observed the changes in the region 1850–2050 nm, with one band situated near 1926 nm, and concluded that most probably this band correlates with changes in water because the empty seeds are more dry than viable. Kandpal et al.[72] applied NIR hyperspectral imaging coupled with multivariate analysis to estimate the viability and seed vigor (judged by the duration of the germination period) of muskmelon seeds. The second derivative transformed mean spectra of non-viable seeds, and seeds after 5- and 3-day germination periods showed major differences at 1362, 1398, and 1430 nm and between 1910 and 1935 nm. Although the authors assigned these bands to C–H and N–H bonds, it is more probable that they represented differences in water structure of the seeds—1362 nm (C2 WAMACS: 1360–1366 nm[142]), 1398 nm is trapped water[76]—related to dehydration, while 1430 nm (C6 WAMACS: 1421–1430 nm[142]) is related to protein hydration.19 The variation at 1910 and 1935 nm is most probably, like in the previous two studies related to moisture, however, the authors stated an interesting observation—that the change in this spectral region in connection with the loss of viability and vigor is not due to the simple change in water content but rather, change in the water environment, that is, water restructuring. That this was indeed the case and was supported by the observation of the modifications occurring at the absorbance bands of water in the first overtone of the water region (1362, 1398, and 1430 nm). The PLS discrimination model was built for the classification of seeds depending on the vigor and viability and the classification accuracy achieved was 97.9% for the calibration set and 93.4% for the validation set. The highest values in regression coefficients provided information about which wavelengths in the spectra were important for successful discrimination. They found those located at 1398, 1430, and 1949 nm, equally observed in the second derivative transformed spectra, where the last absorbance band is also related to water.[12] The study also reported that in the shorter wavelength range, the 970-nm band, related again to water (important for water content and shown to be an indicator of stress) provides information about the germination ability of seeds.

A similar study by McGoverin et al.[92] about wheat, whole barley, and sorghum grain viability using hyperspectral imaging in the NIR region reported that successful results in viable/non-viable differentiating were based on the spectral differences in 1920–1940-nm region—OH stretching and deformation combination band. This band was directly related to the loss of kernel viability, and not just in terms of loss of moisture, but due to the changed interactions between water and other biomolecular components.

Cattaneo et al.[13] performed one more interesting study where they applied aquaphotomics to inquire about the differences between rice varieties with different tocopherol content. Aquaphotomics approach was successful to detect changes in the water matrix of rice seeds with different micro-lipo compounds. The WASP presented in aquagrams showed a regular decrease of absorbance at all water absorbance bands (C1–C12 WAMACS) as the tocopherol content was increased in rice. The explanation for decreased water absorbance was connected to the very low hydrophilicity of tocopherols.

12.3.7 FOOD PRESERVATION

Chemical, enzymatic, and microbial activities can naturally affect food products causing spoilage (deterioration) and changes in appearance such as color, texture, and nutritional values, all of which could be described as loss of quality.[3,131] To retain the quality of food products, different conventional and in recent years advanced and innovative techniques have been adopted. In addition to preserving food by drying, chilling, freezing, or pasteurization, recent advancements include sophisticated but highly advanced techniques for food preservation like packaging, coating, and food irradiation.[3,25.]

Food coatings and packaging have a crucial role in the protection and preservation of food and new requirements have brought novel packaging solutions which utilize so-called smart materials, active materials, and/or eco-friendly materials.[4] The active packaging should not only prevent interaction with the environment which could lead to microbial spoilage but also, enable significant extension of shelf life and positively affect the product quality through the use of active molecules.

Edible coatings are the latest technologies introduced to preserve the quality of food, as excellent alternatives to chemical treatments. In addition, they are environmentally friendly.[27]

In edible films or coatings, a multitude of active ingredients with different functionalities can be introduced in the polymer matrix to preserve quality, ensure safety, and even improve nutritional and/or sensory features.[27]

In one particular study, Barzhagi et al.[4] used aquaphotomics to explore how the presence of bioactive material, propolis, in different types of packaging materials with different thicknesses, with and without polyethylene layers produced in both laboratory and industrial conditions affects the properties of the material. First, there was higher absorbance of water molecules with two (S2, WAMACS C9) and three (S3, WAMACS C10) hydrogen bonds whether the propolis was spread on a packaging material surface or used directly in the paper pulp. Second the spectra of packaging materials with and without polyethylene sheets were divided into two sets and compared using aquagrams to detect the presence of propolis compounds. For papers without the polyethylene sheet, the papers with incorporated propolis differed from papers without propolis or spread on the surface at water bands corresponding to the absorbance of free water molecules (1410 nm, C5) and strongly bound water (1518 nm, C12). Compared to normal paper, adding propolis to the paper pulp resulted in an increase in free water and a decrease in strongly bound water. The decrease in strongly bound water was observed also for papers with propolis spread on the surface. For the materials with polyethylene sheet, the results showed different spectral patterns—if the propolis was spread on the surface, it led to differences at more hydrogen-bonded water (region 1474 nm–1518 nm) and at 1438 and 1444 nm (S1, water dimers, C7). In summary, it could be concluded that not only the presence of propolis could be detected, but also the way it was used in packaging paper. It was shown that propolis influences the distribution of water molecular species in paper and mostly affects strongly bound water and free water molecules. The future focus of the study could be to further examine the role of these two water species with respect to the influence on the shelf life of products packed in the papers with and without propolis. A similar interplay of free water molecular species and strongly bound water was observed in hydrogels for example, and it was related to porosity of the material and ability for oxygen transport.[96,132]

An aquaphotomics study on innovative coatings by Cattaneo et al.[21] investigated how two types of organic polymer coatings with different permeability to oxygen influence the preservation and ripening of winter melons. Coating C_2 was a water-soluble mixture constituted of cellulose polymer with lower oxygen permeability of 0.02 $cm^3m^{-2}d^{-1}bar^{-1}$, while coating C_3 was a polyvinyl alcohols mixture with somewhat higher oxygen permeability of 0.44 $cm^3m^{-2}d^{-1}bar^{-1}$. Using aquagrams to inspect WASP of winter melons, it was determined that C_3-coating does not result in much differences compared to uncoated melons. The spectral pattern of uncoated and C_3-coated melons was characterized by weakly hydrogen-bonded water (WAMACS: C1, C2, and C3) and medium-strong hydrogen-bonded water (WAMACS: C7 and C10). In contrast, the spectral pattern of C_2-coated winter melons showed lowest absorbance of weakly hydrogen-bonded water species (WAMACS: C1–C3) and a large increase at all other WAMACS coordinates. The spectral patterns of C_2-coated melons over time (during storage period of 2 weeks), were further compared with C_3-coated and uncoated melon patterns averaged together (we will refer to this further as control). At the beginning of storage, spectral patterns of C_2-coated and control melons were similar; characterized by high absorbance of weakly hydrogen-bonded water (C1–C3 WAMACs). During storage, control samples did not show any marked changes in absorbance pattern until the nineth day when redistribution of water molecular species was observed, that is, marked increase of all water species absorbing at C4–C12 WAMACS. At the end of storage time, the pattern reversed and it was characterized again by the increase of absorbance at C1–C3 WAMACS, but still with high absorbance at C7 and C8 WAMACs.

The spectral pattern of C_2-coated winter melons showed marked changes already after 6 days, and the changes were similar to what the control samples showed after slightly longer period (9 days), that is, decrease of absorbance at C1–C3 WAMACs and increase at C4–C12 WAMACs. After 9 days, the spectral pattern of C_2-coated melons showed increased absorbance at C1–C3 WAMACs, decrease at C4–C12 but this spectral pattern could be described as a balanced ratio of all water species which was kept until the end of storage time. Reference analysis showed that there were differences in winter melons depending on the coating: C_2-coating melons showed higher weight loss, despite the lower oxygen permeability of the coating material; the internal oxygen values dropped much more than in control samples and triggered fermentation processes and after 6 days of storage strong increase in fermentative metabolites was observed.[145] These changes confirmed by other methods are consistent with the observed changes in the spectral pattern of C_2-coated melons, that is, the water matrix reflected well the changes in structure and processes occurring during storage.

Recent years have seen an increased trend in the application of ultraviolet (UV) light for the treatment of fresh produce (fruits and vegetables) as an alternative to washing raw products with water rich in disinfectants or hot water to prevent pathogen-borne diseases.[144] UV light treatment is proposed as a minimal processing technique for fruit and vegetable surfaces, which keeps the integrity and freshness while effectively reducing microbial loads.[144] A number of publications show improved and extended postharvest life, for example, in strawberry,[121] longan fruit,[136] tomato,[110] carambola fruit,[90] and peach[47] without adversely affecting quality parameters. Similarly, fruit juices have been exposed to UV irradiation aiming to prolong the shelf life of juices and liquid foods by inactivation of microorganisms[35] with usually minimal changes in juice properties (e.g., in lime juice[97]).

However, it was previously shown that UV light irradiation of the tissues of plants produces abiotic stress and in particular, UVB light irradiation (UV light with wavelengths in the 280–315-nm range) increases the production of enzymes responsible for the biosynthesis of flavonoids, compounds that prevent damage caused by UV radiation.[15] Several works have shown altered metabolism in different plants (apples,[29,73] parsley,[33] maize,[135] cherries[2,68]) in response to UV irradiation as means of preventing UV-induced damage. Also, plants lacking phenolic compounds which can act as sunscreens have shown oxidative damage and injury upon irradiation.[78,84]

The use of ultraviolet-C (UVC) light in the range 100–280 nm) treatments is envisioned as a "promising non-thermal process in postharvest technology,[17]" a very good "alternative to thermal treatment.[35]" There are reports about UV treatment of water that is increasingly common due to the effective action against microorganisms. In this regard, it is said that "overdose is not possible, chemical residues or by-products are avoided, and water quality is unaffected.[20]" UV irradiation treatment for food and fruits or vegetables is usually considered as a non-thermal treatment and a great course of action against microorganisms which does not leave chemical residuals and does not affect quality negatively nor add to it.

While certainly more studies are needed, one aquaphotomics study could serve as an advisory note on the influence of UVC irradiation. In a study about the influence of UVC irradiation of DNA solutions[36] the results showed that UVC light changes the water structure of solutions and that even the dose of irradiation can be predicted with excellent accuracy (Fig. 12.5). If the regression vector of the PLSR model built for the estimation of the dose of irradiation is examined (Fig. 12.5b) it can be seen that irradiation influences the water structure similar to the influence of temperature (for comparison see loading vectors of PCA performed on spectra of temperature-perturbed water in works of Gowen et al.[44] Martens,[88] or Wentz[151]). The irradiation disrupts the entire hydrogen bond network of water, similarly to the influence of decreasing temperature, increasing the existing hydrogen bonded species of water (labeled as C7, C8, C9, C10, and C11 in Fig. 12.5b) at the expense of free water molecules (C5) (Table 12.2).

TABLE 12.2 Representation of the WAMACS along with the Corresponding UVC Irradiation Effect on Each: Self-developed with Concepts from Goto et al.[37]

WAMACS	Wavelength range [nm]	Assignment	Effect of UVC irradiation
C5	1396–1418	Trapped water, TW—Water confined in the local field of ions (1396–1403 nm) S_0: Free water molecules	↓ Decrease
C7	1432–1444	S_1: Water molecules with 1 hydrogen bond	↑ Increase
C8	1448–1454	$OH\text{–}(H_2O)_n$ ($n = 4, 5$) Water solvation shell	↑ Increase
C9	1458–1468	S_2: Water molecules with two hydrogen bonds	↑ Increase
C10	1472–1482	S_3: Water molecules with three hydrogen bonds	↑ Increase
C11	1482–1495	S_4: Water molecules with four hydrogen bonds	↑ Increase

Similar effects could be expected to happen with the structure of water in irradiated food. However, at the moment there are no investigations performed with the aim to establish what is happening to the molecular structure of water in irradiated food, nor what could be the possible health effects. However, what this study showed is that water is affected by irradiation, in a similar way to the decrease in temperature, and we cannot say that the state of water was not affected by irradiation, and consequently, one more factor may come into play when we describe the quality of water or food. The results of aquaphotomics study on the influence of irradiation bring a new level of awareness to the practice which is becoming widespread despite not being understood completely.

12.4 SUMMARY AND FUTURE PERSPECTIVES

Near-infrared spectroscopy (NIR) is an acknowledged, widely used measurement technique in food quality assessment. However, it has not reached its full potential yet, especially as a tool for biomonitoring and food quality evaluation in agriculture.

It is generally thought that the NIR spectra of most biological samples and agro–food commodities, being highly aqueous systems, are dominated by water absorption which hides the absorbance bands of analytes of interest, and therefore require sample preparation techniques aimed to reduce the water content.[83] However, there are many new discoveries that showed that there is no need for water removal, and the development of aquaphotomics together with the realization that water is not an obstacle as usually thought,[83] but an important contributor and source of information that could help further advancements and applications of NIRS in the quality control and precision agriculture. While it is true that many compounds have low sensitivity in the NIR region, like minerals or salts[11] aquaphotomics showed that through the effect on the hydrogen bonds network of the water matrix in the samples, their quantification can be achieved. An additional and, not previously considered possibility is that the WASP of the samples can serve as an integrative quality indicator or an integrative biomarker.

This possibility circumvented one of the major limitations of vibrational spectroscopy in some cases—the need to rely on reference method analysis, that is, secondary analytical methods for chemical or physical analysis that provides data on which calibrations are performed. If the WASP is used as an integrative marker, there will be no need for reference measurements of single compounds in the sample. Some attributes of interest of the sample can easily be correlated with the state of the sample water matrix.

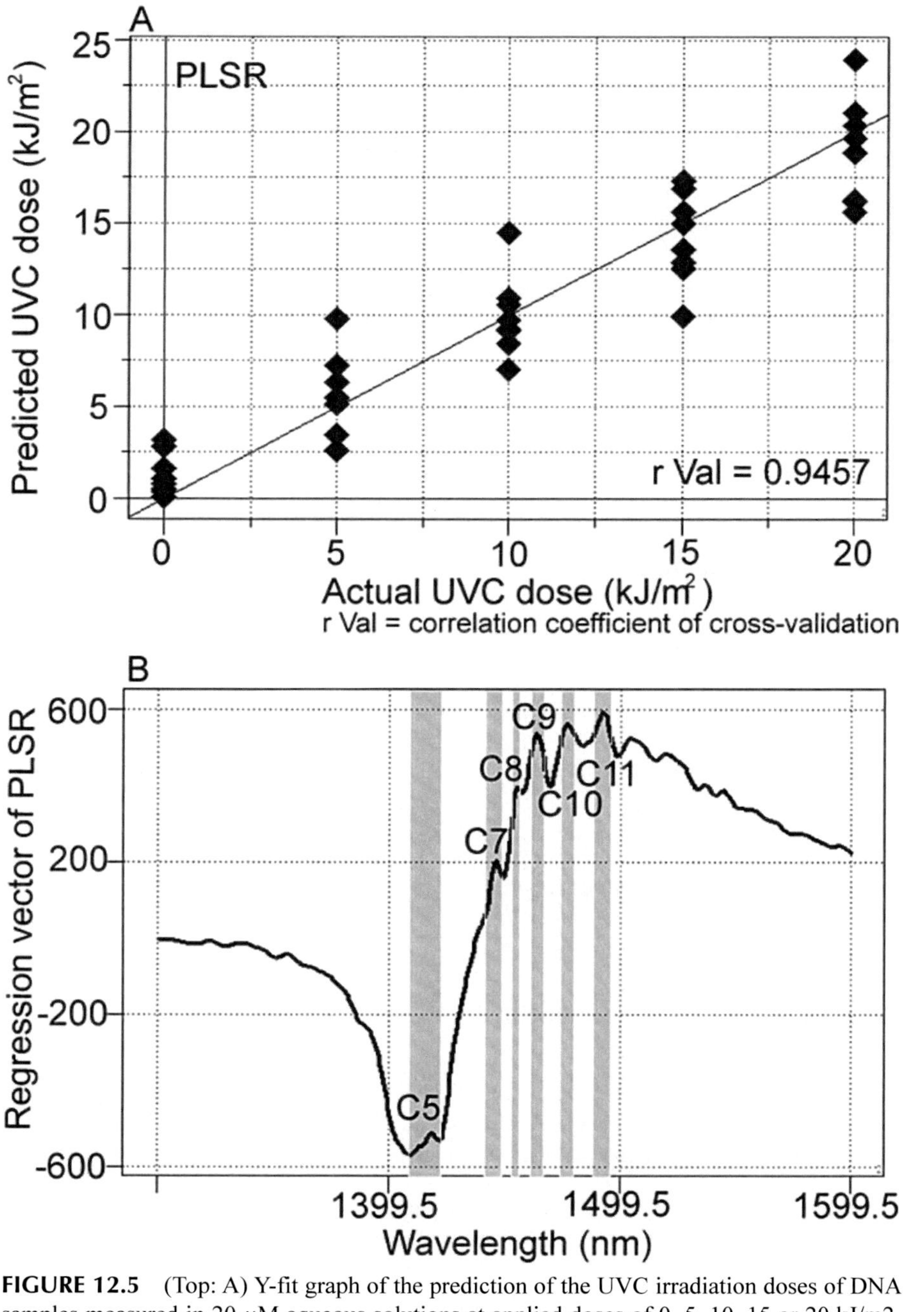

FIGURE 12.5 (Top: A) Y-fit graph of the prediction of the UVC irradiation doses of DNA samples measured in 20-μM aqueous solutions at applied doses of 0, 5, 10, 15 or 20 kJ/m2. (Bottom: B) Loading vectors showcasing the most influential WAMACS on the prediction accuracy.[36]

Technological developments in NIRS such as miniaturization,[57,120] hyperspectral imaging, and fiber-spectrometers[45] continuously increase the potential of NIRS technology. In the last decade, there have been an increasing number of new mobile applications and miniaturized instruments for this purpose.[120] The availability of quality analytical methods at all times and ease of use are the present-day public requirements for sustainable technology.[120] The construction of miniature, cost-effective instruments suitable for integration even into cell phones is in development and expected to be in the market.[133] With these technological advancements and novel knowledge, with which aquaphotomics contributed, major steps forward in quality analysis and biodiagnosis could be expected.

KEYWORDS

- **aquaphotomics**
- **biodiagnosis**
- **biomonitoring**
- **food quality**
- **integrated marker**
- **near-infrared spectroscopy**
- **non-destructive**
- **plant stress**
- **water molecular structure**

REFERENCES

1. Achata, E.; Esquerre, C.; O'Donnell, C.; Gowen, A.; Achata, E.; Esquerre, C.; O'Donnell, C.; Gowen, A. Study on the Application of Near Infrared Hyperspectral Chemical Imaging for Monitoring Moisture Content and Water Activity in Low Moisture Systems. *Molecules* **2015,** *20* (2), 2611–2621.
2. Amit, S. K.; Uddin, M. M.; Rahman, R.; Islam, S. M. R.; Khan, M. S. Review on Mechanisms and Commercial Aspects of Food Preservation and Processing. *Agric. Food Secur.* **2017,** *6* (1), 51–73.
3. Arakawa, O. Effect of Ultraviolet Light on Anthocyanin Synthesis in Light-Colored Sweet Cherry. *J. Japanese Soc. Hortic. Sci.* **1993,** *62* (3), 543–546.

4. Bakier, S. Application of NIR Spectroscopy for the Analysis of Water-Carbohydrate Interactions in Water Solutions. *Acta Agrophys.* **2008,** *11* (1), 7–21.
5. Barzaghi, S.; Cremonesi, K.; Cattaneo, T. M. P. Influence of the Presence of Bioactive Compounds in Smart Packaging Materials on Water Absorption using NIR Spectroscopy and Aquaphotomics. *NIR News* **2017,** *28* (2), 21–24.
6. Basu, S.; Ramegowda, V.; Kumar, A.; Pereira, A. Plant Adaptation to Drought Stress. *F1000 Res.* **2016,** *5*, 1–10.
7. Bauriegel, E.; Giebel, A.; Geyer, M.; Schmidt, U.; Herppich, W. B. Early Detection of Fusarium Infection in Wheat using Hyper-Spectral Imaging. *Comput. Electron. Agric.* **2011,** *75* (2), 304–312.
8. Bázár, G.; Kovacs, Z.; Tanaka, M.; Furukawa, A.; Nagai, A.; Osawa, M.; Itakura, Y.; Sugiyama, H. Water Revealed as Molecular Mirror when Measuring Low Concentrations of Sugar with Near Infrared Light. *Anal. Chem. Acta*, **2015,** *896*, 52–62.
9. Bázár, G.; Romvári, R.; Szabó, A.; Somogyi, T.; Éles, V.; Tsenkova, R. NIR Detection of Honey Adulteration Reveals Differences in Water Spectral Pattern. *Food Chem.* **2016,** *194*, 873–880.
10. Berentsen, S.; Stolz, T.; Molt, K. Analysis of Aqueous Solutions by Near-infrared Spectrometry (NIRS), Part IV: One- and Two-Component Systems of Organic Compounds in Water. *J. Mol. Struct.* **1997,** *410–411* (96), 581–585.
11. Büning-Pfaue, H. Analysis of Water in Food by Near Infrared Spectroscopy. *Food Chem.* **2003,** *82* (1), 107–115.
12. Burns, D. A.; Ciurczak, E. Eds.; In *Handbook of Near-Infrared Analysis*; CRC Press: Boca Raton, 2008; pp 8–9.
13. Cantos, E.; García-Viguera, C.; Pascual-Teresa, S.; Tomás-Barberán, F. A. Effect of Postharvest Ultraviolet Irradiation on Resveratrol and Other Phenolics of cv. Napoleon Table Grapes. *J. Agric. Food Chem.* **2000,** *48* (10), 4606–4612.
14. Castillo, M.; Payne, F. A.; López, M. B.; Ferrandini, E.; Laencina, J. Optical Sensor Technology for Measuring Whey Fat Concentration in Cheese Making. *J. Food Eng.* **2005,** *71* (4), 354–360.
15. Cattaneo, T. M. P.; Cortellino, G.; Barzaghi, S. In *Relationship Between Rice Tocopherols Content and Water Absorption in Near InfraRed Region*, Near Infrared Spectroscopy: Proceedings of the International Conference, 2015; vol *1*, pp 83–86.
16. Cattaneo, T. M. P.; Vanoli, M.; Grassi, M.; Rizzolo, A.; Barzaghi, S. The Aquaphotomics Approach as a Tool for Studying the Influence of Food Coating Materials on Cheese and Winter Melon Samples. *J. Near Infrared Spec.* **2016,** *24* (4), 381–390.
17. Chatani, E.; Tsuchisaka, Y.; Masuda, Y.; Tsenkova, R. Water Molecular System Dynamics Associated with Amyloidogenic Nucleation as Revealed by Real Time Near Infrared Spectroscopy and Aquaphotomics. *PLoS One* **2014,** *9* (7), 1–10.
18. Chen, L.; Wang, J.; Ye, Z.; Zhao, J.; Xue, X.; Heyden, Y. Vander; Sun, Q. Classification of Chinese Honeys According to their Floral Origin by Near Infrared Spectroscopy. *Food Chem.* **2012,** *135* (2), 338–342.
19. Chen, L.; Xue, X.; Ye, Z.; Zhou, J.; Chen, F.; Zhao, J. Determination of Chinese Honey Adulterated with High Fructose Corn Syrup by Near Infrared Spectroscopy. *Food Chem.* **2011,** *128* (4), 1110–1114.
20. Cho, J. S.; Bae, H. J.; Cho, B. K.; Moon, K. D. Qualitative Properties of Roasting Defect Beans and Development of its Classification Methods by Hyperspectral Imaging Technology. *Food Chem.* **2017,** *220*, 505–509.

21. Cirilli, M.; Bassi, D.; Ciacciulli, A. Sugars in Peach Fruit: A Breeding Perspective. *Hortic. Res.* **2016,** *3*, 15067.
22. Civello, P. M.; Vicente, A.; Martinez, G. A. UV-C Technology to Control Postharvest Diseases of Fruits and Vegetables. In *Recent Advances in. Alternative Postharvest Technologies to Control Fungal Diseases in Fruits & Vegetables*, 2007; vol *2007*, pp 71–102.
23. Cutler, T. D.; Zimmerman, J. J. Ultraviolet Irradiation and the Mechanisms Underlying its Inactivation of Infectious Agents. *Anim. Health Res. Rev* **2011,** *12* (1), 15–23,
24. Czuchajowska, Z.; Pomeranz, Y. Changes in Maturing Wheat as Determined by Near-infrared Reflectance Spectroscopy. *Cereal Chem.* **1989,** *66* (5), 432–435.
25. Dehghani, S.; Hosseini, S. V.; Regenstein, J. M. Edible Films and Coatings in Seafood Preservation: A Review. *Food Chem.* **2018,** *240*, 505–513.
26. Delwiche, S. R.; Mekwatanakarn, W.; Wang, C. Y. Soluble Solids and Simple Sugars Measurement in Intact Mango Using Near Infrared Spectroscopy. *HortTechnology* **2008,** *18* (3), 410–416.
27. Dhall, R. K. Advances in Edible Coatings for Fresh Fruits and Vegetables: A Review. *Crit. Rev. Food Sci. Nutr.* **2013,** *53* (5), 435–450.
28. Dinakar, C.; Bartels, D. Desiccation Tolerance in Resurrection Plants: New Insights from Transcriptome, Proteome, and Metabolome Analysis. *Front. Plant Sci.* **2013,** *4*, 1–14.
29. Dong, Y. hu; Mitra, D.; Kootstra, A.; Lister, C.; Lancaster, J. Postharvest Stimulation of Skin Color in Royal Gala Apple. *J. Am. Soc. Hortic. Sci.* **1995,** *120* (1), 95–100.
30. Downey, G.; Fouratier, V.; Kelly, J. D. Detection of Honey Adulteration by Addition of Fructose and Glucose Using near Infrared Transflectance Spectroscopy. *J. Near Infrared Spec.* **2003,** *11* (6), 447–456.
31. Eckey-Kaltenbach, H.; Heller, W.; Sonnenbichler, J.; Zetl, I.; Schäfer, W.; Ernst, D.; Sandermann, H. Oxidative Stress and Plant Secondary Metabolism: 6-O-malonylapiin in Parsley. *Phytochemistry* **1993,** *34* (3), 687–691.
32. Escuredo, O.; González-Martín, M. I.; Rodríguez-Flores, M. S.; Seijo, M. C. Near Infrared Spectroscopy Applied to the Rapid Prediction of the Floral Origin and Mineral Content of Honeys. *Food Chem.* **2015,** *170*, 47–54.
33. Esquerre, C.; Gowen, A.; O'Donnell, C. P.; Downey, G. Water Absorbance Pattern of Physical Damage in Mushrooms. *Biosyst. Eng. Res. Rev.* **2009,** *14*, 68–71.
34. Falguera, V.; Pagán, J.; Garza, S.; Garvín, A.; Ibarz, A. Ultraviolet Processing of Liquid Food: A Review. Part 2: Effects on Microorganisms and on Food Components and Properties. *Food Res. Int.* **2011,** *44* (6), 1580–1588.
35. Feng, Y. Z.; Sun, D. W. Application of Hyperspectral Imaging in Food Safety Inspection and Control: A Review. *Crit. Rev. Food Sci. Nutr.* **2012,** *52* (11), 1039–1058.
36. Giangiacomo, R. Study of Water-Sugar Interactions at Increasing Sugar Concentration by NIR Spectroscopy. *Food Chem.* **2006,** *96* (3), 371–379.
37. Giangiacomo, R.; Pani, P.; Barzaghi, S. Sugars as a Perturbation of the Water Matrix. *J. Near Infrared Spec.* **2009,** *17* (6), 329–335.
38. Gimenez, C.; Gallardo, M.; Thompson, R. B. Plant–Water Relations. In *Encyclopedia of Soils in the Environment*, 2005; vol *2005*, pp 231–238.
39. Goda, K.; Jalali, B. Dispersive Fourier Transformation for Fast Continuous Single-Shot Measurements. *Nat. Photonics* **2013,** *7* (2), 102–112.

40. Golic, M.; Walsh, K.; Lawson, P. Short-Wavelength Near-Infrared Spectra of Sucrose, Glucose, and Fructose with Respect to Sugar Concentration and Temperature. *Appl. Spec.* **2003,** *57* (2), 139–145.
41. Gonzalez-Aguilar, G.; Wang, C. Y.; Buta, G. J. UV-C Irradiation Reduces Breakdown and Chilling Injury of Peaches During Cold Storage. *J. Sci. Food Agric.* **2004,** *84* (5), 415–422.
42. Goto, N.; Bazar, G.; Kovacs, Z.; Kunisada, M.; Morita, H.; Kizaki, S.; Sugiyama, H.; Tsenkova, R. Detection of UV-Induced Cyclobutane Pyrimidine Dimers by Near-Infrared Spectroscopy and Aquaphotomics. *Sci. Rep.* **2015,** *5* (1), 11808.
43. Goulao, L. F.; Oliveira, C. M. Cell Wall Modifications During Fruit Ripening: When a Fruit is not the Fruit. *Trends Food Sci. Technol.* **2008,** *19* (1), 4–25.
44. Goulao, L. F.; Santos, J.; de Sousa, I.; Oliveira, C. M. Patterns of Enzymatic Activity of Cell Wall-Modifying Enzymes During Growth and Ripening of Apples. *Postharvest Biol. Technol.* **2007,** *43*, 307–318.
45. Gowen, A. A. Water and Food Quality. *Contemp. Mater.* **2012,** *1* (3), 31–37.
46. Gowen, A. A.; Amigo, J. M.; Tsenkova, R. Characterization of Hydrogen Bond Perturbations in Aqueous Systems Using Aquaphotomics and Multivariate Curve Resolution-Alternating Least Squares. *Anal. Chim. Acta* **2013,** *759*, 8–20.
47. Gowen, A. A.; Esquerre, C.; O'Donnell, C. P.; Downey, G.; Tsenkova, R. Use of Near Infrared Hyperspectral Imaging to Identify Water Matrix Co-ordinates in Mushrooms (*Agaricus bisporus*) Subjected to Mechanical Vibration. *J. Near Infrared Spec.* **2009,** *17* (6), 363–371.
48. Gowen, A. A.; Marini, F.; Tsuchisaka, Y.; De Luca, S.; Bevilacqua, M.; O'Donnell, C.; Downey, G.; Tsenkova, R. On the Feasibility of Near Infrared Spectroscopy to Detect Contaminants in Water using Single Salt Solutions as Model Systems. *Talanta* **2015,** *131*, 609–618.
49. Gowen, A. A.; O'Donnell, C. P.; Cullen, P. J.; Bell, S. E. Recent Applications of Chemical Imaging to Pharmaceutical Process Monitoring and Quality Control. *Eur. J. Pharm. Biopharm.* **2008,** *69* (1), 10–22.
50. Hall, J. W.; Quaresima, V.; Ferrari, M. In *Can We Get More Tissue Biochemistry Information from In Vivo Near infrared Spectra ?*, Proceedings of SPIE 2387, Advances in Laser and Light Spectroscopy to Diagnose Cancer and Other Diseases II, 1995; vol *2387*, pp 225–231.
51. He, M.; He, C. Q.; Ding, N. Z. Abiotic Stresses: General Defenses of Land Plants and Chances for Engineering Multistress Tolerance. *Front. Plant Sci.* **2018,** *871*, 1–18.
52. He, X.; Feng, X.; Sun, D.; Liu, F.; Bao, Y.; He, Y.; He, X.; Feng, X.; et al. Rapid and Nondestructive Measurement of Rice Seed Vitality of Different Years Using Near-infrared Hyperspectral Imaging. *Molecules* **2019,** *24* (12), 2227–2230.
53. Headrick, J. M.; Diken, E. G.; Walters, R. S.; Hammer, N. I.; Christie, R. A.; Cui, J.; Myshakin, E. M.; Duncan, M. A. Spectral Signatures of Hydrated Proton Vibrations in Water Clusters. *Science* **2005,** *308* (5729), 1765–1769.
54. Heiman, A.; Licht, S. Fundamental Baseline Variations in Aqueous Near-Infrared Analysis. *Anal. Chim. Acta* **1999,** *394* (2–3), 135–147.
55. Heribert, H.; Shinozaki, K. In *Plant Responses to Abiotic Stress. Topics in Current Genetics*, 2003; vol *4*, pp 1–8.

56. Hong, B. H.; Rubenthaler, G. L.; Allan, R. E. Wheat Pentosans, Part II: Estimating Kernel Hardness and Pentosans in Water Extracts by Near-Infrared Reflectance. *Cereal Chem.* **1989,** *66* (5), 374–377.
57. Huang, H.; Liu, L.; Ngadi, M. Recent Developments in Hyperspectral Imaging for Assessment of Food Quality and Safety. *Sensors* **2014,** *14* (4), 7248–7276.
58. Huck, C. W. Theoretical and Technical Advancements of Near-infrared Spectroscopy and its Operational Impact in Industry. *NIR News* **2017,** *28* (8), 17–21.
59. Ismail, M. M.; Hiroshi, H. Water Stress. *InTech* **2012**, *2012*, 314-320.
60. Iwamoto, M.; Kawano, S.; Abe, H. Analysis of Hydrogen Bonding Related to Water in Foods. *NIR News* **1995,** *6* (3), 10–12.
61. Jaleel, C. A.; Manivannan, P.; Wahid, A.; Farooq, M.; Al-Juburi, H. J.; Somasundaram, R.; Panneerselvam, R. Drought Stress in Plants: A Review on Morphological Characteristics and Pigments Composition. *Int. J. Agric. Biol.* **2009,** *11* (1), 100–105.
62. Jinendra, B. *Near Infrared Spectroscopy and Aquaphotomics: Novel Tool for Biotic and Abiotic Stress Diagnosis of Soybean*; Unpublished Report for Kobe University, 2011; p 21.
63. Jinendra, B.; Tamaki, K.; Kuroki, S.; Vassileva, M.; Yoshida, S.; Tsenkova, R. Near Infrared Spectroscopy and Aquaphotomics: Novel Approach for Rapid *in vivo* Diagnosis of Virus Infected Soybean. *Biochem. Biophys. Res. Commun.* **2010,** *397* (4), 685–690.
64. Kaliramesh, S.; Chelladurai, V.; Jayas, D. S.; Alagusundaram, K.; White, N. D. G.; Fields, P. G. Detection of Infestation by *Callosobruchus maculatus* in Mung Bean Using Near-infrared Hyperspectral Imaging. *J. Stored Prod. Res.* **2013,** *52*, 107–111.
65. Kandpal, L. M.; Lohumi, S.; Kim, M. S.; Kang, J. S.; Cho, B. K. Near-infrared Hyperspectral Imaging System Coupled with Multivariate Methods to Predict Viability and Vigor in Muskmelon Seeds. *Sens. Actuators B Chem.* **2016,** *229*, 534–544.
66. Kataoka, I.; Beppu, K.; Sugiyama, A.; Taira, S. Enhancement of Coloration of "Satohnishiki" Sweet Cherry Fruit by Postharvest Irradiation with Ultraviolet Rays. *Environ. Control Biol.* **1996,** *34* (4), 313–319.
67. Kaur, H.; Kunnemeyer, R.; McGlone, A. In *Assessment of Apple Juice by Aquaphotomics in a Temperature Dependent Environment*, The 3rd Australia New Zealand Conference on Optics and Photonics (ANZCOP), 2017; vol *2017*, pp 2–3.
68. Kaur, H.; Kunnemeyer, R.; McGlone, A. In *Investigating Aquaphotomics for Fruit Quality Assessment*, The Third International Aquaphotomics Symposium, Awajishima, Japan, 2018; vol *2018*, pp 19–25.
69. Kaur, H.; Künnemeyer, R.; McGlone, A. Investigating Aquaphotomics for Temperature-Independent Prediction of Soluble Solids Content of Pure Apple Juice. *J. Near Infrared Spec.* **2020,** *28* (2), 103–112.
70. Kawano, S.; Abe, H.; Iwamoto, M. Development of a Calibration Equation with Temperature Compensation for Determining the Brix Value in Intact Peaches. *J. Near Infrared Spec.* **1995,** *3* (4), 211–218.
71. Kelly, J. D.; Petisco, C.; Downey, G. Potential of Near Infrared Transflectance Spectroscopy to Detect Adulteration of Irish Honey by Beet Invert Syrup and High Fructose Corn Syrup. *J. Near Infrared Spec.* **2006,** *14* (2), 139–146.
72. Kim, K.; Kim, H.; Park, S. H.; Lee, S. J. Hydraulic Strategy of *Cactus trichome* for Absorption and Storage of Water Under Arid Environment. *Front. Plant Sci.* **2017,** *8*, 1–8.

73. Kojić, D.; Tsenkova, R.; Tomobe, K.; Yasuoka, K.; Yasui, M. Water Confined in the Local Field of Ions. *Chem. Phys. Chem.* **2014,** *15* (18), 4077–4086.
74. Kootstra, A. Protection from UV-B-Induced DNA Damage by Flavonoids. *Plant Mol. Biol.* **1994,** *26* (2), 771–774.
75. Kraats, E. B. Van De; Munćan, J. S.; Tsenkova, R. N. Aquaphotomics Origin, Concept, Applications and Future Perspectives. *Substantia* **2019,** *3*, 13–28.
76. Kuroki, S.; Tsenkova, R.; Moyankova, D.; Muncan, J.; Morita, H.; Atanassova, S.; Djilianov, D. Water Molecular Structure Underpins Extreme Desiccation Tolerance of the Resurrection Plant *Haberlea rhodopensis*. *Sci. Rep.* **2019,** *3049*, 1–12.
77. Landry, L. G.; Chapple, C. C. S.; Last, R. L. Arabidopsis Mutants Lacking Phenolic Sunscreens Exhibit Enhanced Ultraviolet-B Injury and Oxidative Damage. *Plant Physiol.* **1995,** *109* (4), 1159–1166.
78. Lane, R. A.; Buckton, G. The Novel Combination of Dynamic Vapour Sorption Gravimetric Analysis and Near Infrared Spectroscopy as a Hyphenated Technique. *Int. J. Pharm.* **2000,** *207* (1–2), 49–56.
79. Li, J.; Huang, W.; Tian, X.; Wang, C.; Fan, S.; Zhao, C. Fast Detection and Visualization of Early Decay in Citrus Using Vis-NIR Hyperspectral Imaging. *Comput. Electron. Agric.* **2016,** *127*, 582–592.
80. Lidon, Z. Z.; Cebola, F. An Overview on Drought Induced Changes in Plant Growth, Water Relations and Photosynthesis. *Emir. J. Food Agric.* **2012,** *24* (1), 57–72.
81. Lin, H.; Ying, Y. Theory and Application of Near Infrared Spectroscopy in Assessment of Fruit Quality: A Review. In *Sensing and Instrumentation for Food Quality and Safety*, 2009; vol *3* (2), pp 130–141.
82. Liu, Y.; Pu, H.; Sun, D. W. Hyperspectral Imaging Technique for Evaluating Food Quality and Safety During Various Processes: A Review of Recent Applications. *Trends Food Sci. Technol.* **2017,** *69*, 25–35.
83. Lohumi, S.; Lee, S.; Lee, H.; Cho, B. K. A Review of Vibrational Spectroscopic Techniques for the Detection of Food Authenticity and Adulteration. *Trends Food Sci. Technol.* **2015,** *46* (1), 85–98.
84. Lois, R.; Buchanan, B. B. Severe Sensitivity to Ultraviolet Radiation in an Arabidopsis Mutant Deficient in Flavonoid Accumulation - II. Mechanisms of UV-Resistance in Arabidopsis. *Planta* **1994,** *194* (4), 504–509.
85. Lu, R.; Peng, Y. Hyperspectral Scattering for Assessing Peach Fruit Firmness. *Biosyst. Eng.* **2006,** *93* (2), 161–171.
86. Luner, P. E.; Seyer, J. J. Assessment of Crystallinity in Processed Sucrose by Near-Infrared Spectroscopy and Application to Lyophiles. *J. Pharm. Sci.* **2014,** *103* (9), 2884–2895.
87. Martens, H.; Stark, E. Extended Multiplicative Signal Correction and Spectral Interference Subtraction: New Preprocessing Methods for Near Infrared Spectroscopy. *J. Pharm. Biomed. Anal.* **1991,** *9* (8), 625–635.
88. Mathlouthi, M. Water Content, Water Activity, Water Structure and the Stability of Foodstuffs. *Food Control* **2001,** *12* (7), 409–417.
89. Mattern, P. J.; Sandstedt, R. M. The Influence of the Water-Soluble Constituents of Wheat Flour on its Mixing and Baking Characteristics. *Cereal Chem.* **1957,** *34* (4), 252–267.

90. McGlone, V. A.; Kawano, S. Firmness, Dry-matter and Soluble-Solids Assessment of Postharvest Kiwifruit by NIR Spectroscopy. *Postharvest Biol. Technol.* **1998,** *13* (2), 131–141.
91. McGoverin, C. M.; Engelbrecht, P.; Geladi, P.; Manley, M. Characterisation ofNon-viable Whole Barley, Wheat and Sorghum Grains using Near-Infrared Hyperspectral Data and Chemometrics. *Anal. Bioanal. Chem.* **2011,** *401* (7), 2283–2289.
92. Mizuse, K.; Fujii, A. Tuning of the Internal Energy and Isomer Distribution in Small Protonated Water Clusters H + (H_2O) 4–8: An Application of the Inert Gas Messenger Technique. *J. Phys. Chem. A* **2012,** *116* (20), 4868–4877.
93. Mohd-Hanif, H.; Shamsudin, R.; Adzahan, N. M. UVC Dosage Effects on the Physicochemical Properties of Lime (*Citrus aurantifolia*) Juice. *Food Sci. Biotechnol.* **2016,** *25*, 63–67.
94. Moreno, C.; Andrade-Cuvi, M. J.; Zaro, M. J.; Darre, M.; Vicente, A. R.; Concellón, A. Short UV-C Treatment Prevents Browning and Extends the Shelf-life of Fresh-Cut Carambola. *J. Food Qual.* **2017,** *2017*, 1–9.
95. Moslemkhani, C.; Hassani, F.; Azar, E. N.; Khelgatibana, F. Potential of Spectroscopy for Differentiation between PVY Infected and Healthy Potato Plants. *J. Crop Prot.* **2019,** *8* (2), 143–151.
96. Muncan, J.; Kuroki, S.; Moyankova, D.; Morita, H.; Atanassova, S.; Djilianov, D.; Tsenkova, R. Protocol for Aquaphotomics Monitoring of Water Molecular Structure in Leaves of Resurrection Plants During Desiccation and Recovery. *Protoc. Exch.* **2019,** *1*, 1–17.
97. Muncan, J.; Mileusnić, I.; Šakota Rosić, J.; Vasić-Milovanović, A.; Matija, L. Water Properties of Soft Contact Lenses: A Comparative Near-infrared Study of Two Hydrogel Materials. *Int. J. Polym. Sci.* **2016**, *2016*, 1–8.
98. Muncan, J.; Tsenkova, R. Aquaphotomics: From Innovative Knowledge to Integrative Platform in Science and Technology. *Molecules* **2019,** *24* (15), 2742–2750.
99. Murtaza, G.; Rasool, F.; Habib, R.; Javed, T.; Sardar, K.; Mohsin, M.; Ayub, M. A.; Rasool, A. A Review of Morphological, Physiological and Biochemical Responses of Plants Under Drought Stress Conditions. *Imp. J. Interdiscip. Res.* **2016,** *12*, 1600–1606.
100. Naidu, R. A.; Perry, E. M.; Pierce, F. J.; Mekuria, T. The Potential of Spectral Reflectance Technique for the Detection of Grapevine Leafroll-Associated Virus-3 in Two Red-Berried Wine Grape Cultivars. *Comput. Electron. Agric.* **2009,** *66* (1), 38–45.
101. Nortvedt, R.; Torrissen, O. J.; Tuene, S. Application of Near-Infrared Transmittance Spectroscopy in the Determination of Fat, Protein and Dry Matter in Atlantic Halibut Fillet. *Chemom. Intell. Lab. Syst.* **1998,** *42* (1–2), 199–207.
102. Omar, A.; MatJafri, M. Z. In *Monitoring of Spectrum Responsivity on Apples and Pears Decay through VIS-NIR (Visible and Near Infra Red) Reflectance Spectroscopy*, Proceedings of the International Society for Southeast Asian Agriculture Sciences, 2008; vol *2008*, pp 63–69.
103. Omar, A. F. Quantifying Water-Sucrose Solutions Through NIR Spectral Absorbance Linearisation and Gradient Shift. *J. Opt.* **2013,** *42* (3), 189–193.
104. Omar, A. F.; Atan, H.; MatJafri, M. Z. NIR Spectroscopic Properties of Aqueous Acids Solutions. *Molecules* **2012,** *17* (6), 7440–7450.
105. Omar, A. F.; Atan, H.; Matjafri, M. Z. Peak Response Identification Through Near-Infrared Spectroscopy Analysis on Aqueous Sucrose, Glucose, and Fructose Solution. *Spectrosc. Lett.* **2012,** *45* (3), 190–201.

106. Omar, A. F.; Mardziah Yahaya, O. K.; Tan, K. C.; Mail, M. H.; Seeni, A. In *The Influence of Additional Water Content Towards the Spectroscopy and Physicochemical Properties of Genus Apis and Stingless Bee Honey*, Optical Sensing and Detection IV, 2016; vol *9899*, 98990Y.
107. Omar, A. F.; Matjafri, M. Z. Near Infrared Spectral Linearization in Quantifying Soluble Solids Content of Intact Carambola. *Sensors* **2013,** *13* (4), 4876–4883.
108. Opara, U. L.; Pathare, P. B. Bruise Damage Measurement and Analysis of Fresh Horticultural Produce - A Review. *Postharvest Biol. Technol.* **2014,** *91*, 9–24.
109. Osborne, B.; Fearn, T.; Hindle, P. In *Practical NIR Spectroscopy with Applications in Food and Beverage Analysis*; UCL Discovery, 1993; vol *1993*, pp 2–10.
110. Palmer, K. F.; Williams, D. Optical Properties of Water in the Near Infrared. *J. Opt. Soc. Am.* **1974,** *64* (8), 1107–1110.
111. Pan, J.; Vicente, A. R.; Martínez, G. A.; Chaves, A. R.; Civello, P. M. Combined Use of UV-C Irradiation and Heat Treatment to Improve Postharvest Life of Strawberry Fruit. *J. Sci. Food Agric.* **2004,** *84* (14), 1831–1838.
112. Pasquini, C. Near Infrared Spectroscopy: A Mature Analytical Technique with New Perspectives. *Anal. Chim. Acta* **2018,** *1026*, 8–36.
113. Péneau, S.; Hoehn, E.; Roth, H. R.; Escher, F.; Nuessli, J. Importance and Consumer Perception of Freshness of Apples. *Food Qual. Prefer.* **2006,** *17* (1–2), 9–19.
114. Peñuelas, J.; Filella, I. Visible and Near-Infrared Reflectance Techniques for Diagnosing Plant Physiological Status. *Trends Plant Sci.* **1998,** *3* (4), 151–156.
115. Penuelas, J.; Filella, I.; Biel, C.; Serrano, L.; Save, R. The Reflectance at the 950–970 nm Region as an Indicator of Plant Water Status. *Int. J. Remote Sens.* **1993,** *14* (10), 1887–1905.
116. Peñuelas, J.; Piñol, J.; Ogaya, R.; Filella, I. Estimation Plant Water Concentration by the Reflectance Water Index (WI R900 / R970). *Int. J. Remote Sens.* **1997,** *18* (13), 2869–2875.
117. Peralta-Videa, J. R.; Lopez, M. L.; Narayan, M.; Saupe, G.; Gardea-Torresdey, J. The Biochemistry of Environmental Heavy Metal Uptake by Plants: Implications for the Food Chain. *Int. J. Biochem. Cell Biol.* **2009,** *41* (8–9), 1665–1677.
118. Phetpan, K.; Udompetaikul, V.; Sirisomboon, P. In-line Near Infrared Spectroscopy for the Prediction of Moisture Content in the Tapioca Starch Drying Process. *Powder Technol.* **2019,** *345*, 608–615.
119. Pinheiro, J.; Alegria, C.; Abreu, M.; Gonçalves, E. M.; Silva, C. L. M. Use of UV-C Postharvest Treatment for Extending Fresh Whole Tomato (*Solanum lycopersicum*) Shelf-Life. *J. Food Sci. Technol.* **2015,** *52* (8), 5066–5074.
120. Pittia, P.; Sacchetti, G. Antiplasticization Effect of Water in Amorphous Foods: A Review. *Food Chem.* **2008,** 4, 106–110.
121. Pügner, T.; Knobbe, J.; Grüger, H. Near-Infrared Grating Spectrometer for Mobile Phone Applications. *Appl. Spectros.* **2016,** *70* (5), 734–745.
122. Rajkumar, P.; Wang, N.; Elmasry, G.; Raghavan, G. S. V.; Gariepy, Y. Studies on Banana Fruit Quality and Maturity Stages Using Hyperspectral Imaging. *J. Food Eng.* **2012,** *108* (1), 194–200.
123. Rambla, F. J.; Garrigues, S.; De La Guardia, M. PLS-NIR Determination of Total Sugar, Glucose, Fructose and Sucrose in Aqueous Solutions of Fruit Juices. *Anal. Chim. Acta* **1997**, *344* (1–2), 41–53.

124. Rebecchi, L.; Altiero, T.; Guidetti, R. Anhydrobiosis: The Extreme Limit of Desiccation Tolerance. *Invertebr. Surv. J.* **2007,** *4* (2), 65–81.
125. Rittiron, R.; Saranwong, S.; Kawano, S. Useful Tips for Constructing a Near Infrared-Based Quality Sorting System for Single Brown-Rice Kernels. *J. Near Infrared Spectros.* **2004,** *12* (2), 133–139.
126. Robertson, W. H.; Diken, E. G.; Price, E. A.; Shin, J. W.; Johnson, M. A. Spectroscopic Determination of the OH- Solvation Shell in the OH-(H2O)$_n$ Clusters. *Science* **2003,** 299 (5611), 1367–72.
127. Šakota Rosić, J.; Munćan, J.; Mileusnić, I.; Kosić, B.; Matija, L. Detection of Protein Deposits Using NIR Spectroscopy. *Soft Mater.* **2016,** *14* (4), 264–271.
128. Saranwong, S.; Thanapase, W.; Suttiwijitpukdee, N.; Rittiron, R.; Kasemsumran, S.; Kawano, S. Applying Near Infrared Spectroscopy to the Detection of Fruit Fly Eggs and Larvae in Intact Fruit. *J. Near Infrared Spectrosc.* **2010,** 1*8* (4), 271–280.
129. Schwanninger, M.; Rodrigues, J. C.; Fackler, K. Review of Band Assignments in Near Infrared Spectra of Wood and Wood Components. *J. Near Infrared Spectrosc.* **2011,** *19* (5), 287–308.
130. Sharif, Z.; Mustapha, F.; Jai, J.; Mohd Yusof, N.; Zaki, N. Review on Methods for Preservation and Natural Preservatives for Extending the Food Longevity. *Chem. Eng. Res. Bull.* **2017,** *19*, 145–150.
131. Simmonds, D. H. Chemical Basis of Hardness and Vitreosity. In *Bakers Digest*, 1974; vol *48* (5), pp 16–20.
132. Singh, C. B.; Jayas, D. S.; Paliwal, J.; White, N. D. G. Detection of Insect-Damaged Wheat Kernels Using Near-Infrared Hyperspectral Imaging. *J. Stored Prod. Res.* **2009,** *45* (3), 151–158.
133. Soltani, A.; Lestander, T. A.; Tigabu, M.; Odén, P. C. Prediction of Viability of Oriental Beechnuts, *Fagus orientalis*, Using Near Infrared Spectroscopy and Partial Least Squares Regression. *J. Near Infrared Spectros.* **2003,** *11* (5), 357–364.
134. Sørensen, K. M.; Khakimov, B.; Engelsen, S. B. The Use of Rapid Spectroscopic Screening Methods to Detect Adulteration of Food Raw Materials and Ingredients. *Curr. Opin. Food Sci.* **2016,** *10,* 45–51.
135. Stapleton, A. E.; Walbot, V. Flavonoids can Protect Maize DNA from the Induction of Ultraviolet Radiation Damage. *Plant Physiol.* **1994,** *105* (3), 881–889.
136. Suiubon, S.; Supapvanich, S.; Promyou, S. Postharvest Quality Maintenance of Longan Fruit by Ultra Violet-C Incorporated with Salicylic Acid Application. *Emirates J. Food Agric.* **2017,** *29* (3), 179–187.
137. Suzuki, T. Hydrogen-Bond Studies of Thin Film Water Using Near-Infrared Spectroscopy in the 970 nm Spectral Region. *Appl. Surf. Sci.* **2002,** *187* (3–4), 261–265.
138. Tigabu, M.; Odén, P. C. Discrimination of Viable and Empty Seeds of *Pinus patula* Schiede and Deppe with Near-Infrared Spectroscopy. *New Forests* **2003,** *25* (3), 163–176.
139. Tsenkova, R. Aquaphotomics: Dynamic Spectroscopy of Aqueous and Biological Systems Describes Peculiarities of Water. *J. Near Infrared Spectros.* **2009,** *17* (6), 303–313.
140. Tsenkova, R. Aquaphotomics: Water Absorbance Pattern as a Biological Marker for Disease Diagnosis and Disease Understanding. *NIR News* **2007,** *18* (2), 14–16.
141. Tsenkova, R.; Munćan, J.; Pollner, B.; Kovacs, Z. Essentials of Aquaphotomics and Its Chemometrics Approaches. *Front. Chem.* **2018,** *6* , 1–25.

142. Tsenkova, R. N.; Iordanova, I. K.; Toyoda, K.; Brown, D. R. Prion Protein Fate Governed by Metal Binding. *Biochem. Biophys. Res. Commun.* **2004,** *325* (3), 1005–1012.
143. Tsenkova, R.; Kovacs, Z.; Kubota Y. Aquaphotomics: Near Infrared Spectroscopy and Water States in Biological Systems in Membrane Hydration: Role of Water in the Structure and Function of Biological Membranes. **2015,** *71*, 189–193.
144. Turtoi, M. Ultraviolet Light Treatment of Fresh Fruits and Vegetables Surface: A Review. *J. Agroaliment. Processes Technol.* **2013,** *19* (3), 325–337.
145. Valente, M.; Ribeyre, F.; Self, G.; Berthiot, L.; Assemat, S. Instrumental and Sensory Characterization of Mango Fruit Texture. *J. Food Qual.* **2011,** *34* (6), 413–424.
146. Vanoli, M.; Grassi, M.; Buccheri, M.; Rizzolo, A. Influence of Edible Coatings on Postharvest Physiology and Quality of Honeydew Melon (*Cucumis melo*). *Adv. Hortic. Sci.* **2015,** *29*, 34–40.
147. Vanoli, M.; Lovati, F.; Grassi, M.; Buccheri, M.; Zanella, A.; Cattaneo, T. M. P.; Rizzolo, A. Water Spectral Pattern as a Marker for Studying Apple Sensory Texture. *Adv. Hortic. Sci.* **2018,** *32* (3), 343–351.
148. Vanoli, M.; Rizzolo, A.; Grassi, M.; Farina, A.; Pifferi, A.; Spinelli, L.; Torricelli, A. Time-Resolved Reflectance Spectroscopy Nondestructively Reveals Structural Changes in 'Pink Lady®' Apples During Storage. *Procedia Food Sci.* **2011,** *1*, 81–89.
149. Via, B. K.; Jiang, W. Nonlinear Multivariate Modeling of Strand Mechanical Properties with Near-Infrared Spectroscopy. In *The Forestry Chronicle*, 2013; vol *89* (5), pp 621–630.
150. Wang, W.; Li, C.; Gitaitis, R.; Tollner, E. W.; Rains, G.; Yoon, S. C. In *Near-Infrared Hyperspectral Reflectance Imaging for Early Detection of Sour Skin Disease in Vidalia Sweet Onions*, American Society of Agricultural and Biological Engineers Annual International Meeting 2010, 2010; vol *4* (10), pp 3415–3436.
151. Weber, J. M.; Kelley, J. A.; Nielsen, S. B.; Ayotte, P.; Johnson, M. A. Isolating the Spectroscopic Signature of a Hydration Shell with the Use of Clusters: Superoxide Tetrahydrate. *Science* **2000,** *287* (5462), 2461–2463.
152. Wei, D.; Salahub, D. R. Hydrated Proton Clusters: *Ab initio* Molecular Dynamics Simulation and Simulated Annealing. *J. Chem. Phys.* **1998,** *106* (14), 6086.
153. Wenz, J. J. Examining Water in Model Membranes by Near Infrared Spectroscopy and Multivariate Analysis. *Biochim. Biophys. Acta Biomembr.* **2018,** *1860* (3), 673–682.
154. Williams, P. Influence of Water on Prediction of Composition and Quality Factors: The Aquaphotomics of Low Moisture Agricultural Materials. *J. Near Infrared Spectros.* **2009,** *17* (6), 315–328.
155. Williams, P.; Norris, K. In *Near-Infrared Technology in the Agricultural and Food Industries*; American Association of Cereal Chemists, 1987; vol *1987*, pp 2–11.
156. Wu, D.; Sun, D. W. Advanced Applications of Hyperspectral Imaging Technology for Food Quality and Safety Analysis and Assessment, Part I: Fundamentals. *Innov. Food Sci. Emerg. Technol.* **2013,** *19*, 1–14.
157. Xia, Y.; Huang, W.; Fan, S.; Li, J.; Chen, L. Effect of Fruit Moving Speed on Online Prediction of Soluble Solids Content of Apple using Vis/NIR Diffuse Transmission. *J. Food Process Eng.* **2018,** *41* (8), 1–10.
158. Xia, Y.; Huang, W.; Fan, S.; Li, J.; Chen, L. Effect of Spectral Measurement Orientation on Online Prediction of Soluble Solids Content of Apple Using Vis/NIR Diffuse Reflectance. *Infrared Phys. Technol.* **2019,** 97, 467–477.

159. Xia, Y.; Xu, Y.; Li, J.; Zhang, C.; Fan, S. Recent Advances in Emerging Techniques for Non-destructive Detection of Seed Viability: A Review. *Artif. Intell. Agric.* **2019,** *1*, 35–47.
160. Xing, J.; Symons, S.; Shahin, M.; Hatcher, D. Detection of Sprout Damage in Canada Western Red Spring Wheat with Multiple Wavebands Using Visible/Near-Infrared Hyperspectral Imaging. *Biosyst. Eng.* **2010,** *106* (2), 188–194.

INDEX

A

Ackee *(B. sapida),* 104
- chemical composition, 109–110
- commercialization
  - challenges and opportunities for, 121–123
- development and distribution, 105–107
- domestication, 109
- health benefits, 116
- history, 104–105
- methylene cyclopropyl acetic acid (MCPA), 110
- morphology, 105–107
- nutritional composition, 110
  - amino acid profile, 113–114
  - fatty acids, 115
  - mineral composition, 113
  - proximate composition, 111–113
  - sensory characteristics, 115
- physiology of, 108
- phytochemicals, pharmacological effects, 116–117
  - current uses, 118
  - food products, 118–119
  - loaf, 119
  - non-food uses, 120
  - potential uses, 120–121
- safety related issues, 109–110
- varieties and propagation, 107–108

Adulteration
- in natural foods, detection, 293
  - adulterants, 297
  - advantages, 304
  - application, 307
  - codex alimentarius commission (CAC), 300
  - constituents manipulated, 295–296
  - electronic nose, 308–309
  - electronic tongue, 306–308
  - emerging instruments, 304–305
  - Food and Agricultural Organization (FAO), 300
  - food control, 299–300
  - Fourier Transform Near-infrared Spectroscopy (FT-NIR), 310
  - gas chromatography (GC), 301
  - high liquid performance chromatography and mass spectrometry (HPLC-MS), 301
  - high-performance liquid chromatography (HPLC), 294
  - International Electrotechnical Commission (IEC), 300
  - International Organization for Standardization (ISO), 300
  - liquid chromatography (LC), 294
  - monitoring, 309
  - near infrared spectroscopy (NIRS), 310
  - pant-based raw materials, 298
  - plant-based food products, 296–299
  - prominent quality control methods, 302–303
  - quality control methods, 301–304
  - regulations, 300–301
  - spectroscopy, 309–310
  - type, 311
  - ultraviolet visible spectroscopy (UV-VIS), 310
  - United Nations Development Programme (UNDP), 295
  - World Health Organization (WHO), 301

Allyl-isothiocyanate (AITC), 179

Aquaphotomics, 371
- analytical methodology, 374–376
- Canada Western Red Spring (CWRS), 387
- competitive adaptive reweighted sampling successive projection algorithm (CARS-SPA), 394–395
- future perspectives, 408–410
- high fructose corn syrup (HFCS), 388
- linear discriminant analysis (LDA), 378
- moisture content (MC), 380
- multiple linear regression (MLR), 392

near-infrared spectroscopy (NIR), 408
novel scientific discipline, 372–373
partial least squares regression (PLSR), 385
plant, 376–377
abiotic, 381–385
adulteration, 385–390
biotic stress, 377–381
damage, detection, 385–390
defects, 385–390
food preservation, 404–408
fruit quality evaluation, 390–396
grain, 399–403
moisture content, 396–399
seeds, 399–403
starch, 399–403
water activity, 396–399
potato virus-Y (PVY), 378
principal component analysis (PCA), 378
ratio performance deviation (RPD), 388
reactive oxygen species (ROS), 383
representation, 407
self-developed with concepts, 375
soft independent modeling of class analogy (SIMCA), 378
soluble solids content (SSC), 390
ultra violet (UV), 406
water absorbance bands (WABs), 372
water index (WI), 382
water spectral pattern (WASP), 373
Atomic absorption spectrometry (AAS), 354
Atomic emission spectroscopy (AES), 358

B

Beer
bioactive polyphenols in, 253
anticancer agent, 277
bFGF (basic fibroblast growth factor), 270
cancer research, 269–271
cholangiocarcinoma, 273–274
colorectal cancer, 274–275
content, 266
controversial effects, 281–282
flavonoids, 255
HOP essential oil, 265
isohumulones, 255–265
melanoidins, 266
non-flavonoid phenolic compounds, 255
origin, 256–260
pharmacological aspects, 277–280
phytoestrogen, 280–281
prenylated flavonoids, 267–269
9-prenylnaringenin, 280–281
self-developed with concepts, 261–264
vascular endothelial growth factor (VEGF), 270
vitamins, 266
xanthohumol, 271
xanthohumol
cervix cancer, 272
cholangiocarcinoma, 273
gastric cancer, 271–272
hepatic cancer, 273
larynx cancer, 272
mammary carcinoma, 273
pancreas cancer, 272
Berry Sensory Analysis (BSA), 78
BFGF (basic fibroblast growth factor), 270
Brassica nigra (Mustard)
allyl-isothiocyanate (AITC), 179
constituents, 178
essential oils and plant sterols, 179
glucosinolates, 179
phytochemicals, 178
sinigrin (allyl-glucosinolate or 2-propenyl-glucosinolate), 179
spices, 178

C

Canada Western Red Spring (CWRS), 387
Case studies
traditional medicinal herbs
biological activity, 229–232
Centre for Aromatic Plants (CAP), 237
conservation assessment and management plan (CAMP), 226
Herbal Pharmaceutical Development Unit (HPDU), 236
Herbal Research And Development Institute (HRDI), 237
India Himalayan Region (IHR), 226
Indian scenario, herbal medicines, 226–227
local names, 238–241
marketing strategies, 236–242

medicinal and aromatic plants (MAPs), 236
minimum support prices, 237
North-western Region of Himalayas, 223
Sate Medicinal Plants Board (SMPB), 237
state of art, 227–236
Centre for Aromatic Plants (CAP), 237
Codex alimentarius (CA), 325
Codex alimentarius commission (CAC), 300
Competitive adaptive reweighted sampling successive projection algorithm (CARS-SPA), 394–395
Conservation assessment and management plan (CAMP), 226
Coriandrum sativum (Coriander)
antibacterial activity, 173
anti-diabetic activity, 173
dose-dependance antifertility activity, 174
dried coriander, 173
enzyme activity, 173
essential oil, 172
use, 172
vitamins, 172
Cuminum cyminum (Cumin), 180
dietary constituent, 182
drought-tolerant crop, 181
medicinal uses, 181
oil, 182
seeds, 181
structure, 181
studies, 182
Cymbopogon citratus (Lmongrass)
cryoprotective roles, 175
leaves uses, 174
oral application, 175
pain reliever, 174
research, 174–175
studies, 174, 176

D

Descriptive analysis (DA), 77
Destination marketing organizations (DMOs), 85
Detection methods of adulteration in natural foods, 293
adulterants, 297
advantages, 304
application, 307
codex alimentarius commission (CAC), 300
constituents manipulated, 295–296
emerging instruments, 304–305
electronic nose, 308–309
electronic tongue, 306–308
spectroscopy, 309–310
Food and Agricultural Organization (FAO), 300
food control, 299–300
Fourier Transform Near-infrared Spectroscopy (FT-NIR), 310
gas chromatography (GC), 301
high liquid performance chromatography and mass spectrometry (HPLC-MS), 301
high-performance liquid chromatography (HPLC), 294
International Electrotechnical Commission (IEC), 300
International Organization for Standardization (ISO), 300
liquid chromatography (LC), 294
monitoring, 309
near infrared spectroscopy (NIRS), 310
pant-based raw materials, 298
plant-based food products, 296–299
prominent quality control methods, 302–303
quality control methods, 301–304
regulations, 300–301
type, 311
ultraviolet visible spectroscopy (UV-VIS), 310
United Nations Development Programme (UNDP), 295
World Health Organization (WHO), 301
Diospyros lysioides, 163

E

European Council (EC), 325

F

Foeniculum vulgare (Fennel)
anethole, 177
consumption, 176

injection, 177
medical use, 178
studies, 177
United States Department of Agriculture (USDA), 176
usage, 176
Food and Agricultural Organization (FAO), 4, 300
Food and nutrient insecurity, 25
millets
domestication and available varieties, 27–31
food security crop, 31–36
functional and pasting properties, 44–46
nutrient and health benefits, 39–44
production, 36–39
species of, 26
Fourier Transform Near-infrared Spectroscopy (FT-NIR), 310
Fresh grapes, 63

G

Gas chromatography (GC), 301
Gas chromatography-flame ionization detector (GC-FID), 332
Glycemic index (GI), 195
baking industry, composite flours, 213–214
advantages, 214–215
cassava, 205–206
cocoyam, 208–209
food crops, 198
factors, affect, 205
production, 199–204
types, 198–199
food identification criteria, 198
future prospects, 215–216
International Fund for Agricultural Development (IFAD), 217
metric tons (MT), 199
processed state, 212
relevance, 196–198
sweet potato, 209–212
world, commonly consumed, 201
yam, 206–207
Good Manufacturing Practice (GMP), 88
Grape seed proanthocyanidin extract (GSPE), 153
Grapes and wines
demand
cellars and vineyards, 69–71
consumers, motivation and expectations, 64–68
selling opportunities, 68–69
dried grapes (raisins)
methods, used, 62–63
table grapes, 63
environmental factors, 56
European grape, 56
events for, 71–73
factors influencing quality
biotic factors, 80
climatic factors, 80
destination marketing organizations (DMOs), 85
environmental features, 81
labeling, 84–86
physical environment, 79–81
physiographic factors, 80
production methods, 81–83
soil (edaphic) factors, 80
weather conditions, 80
nutrients in, 56
producing regions, 57
production worldwide, 59
China, 60
Italy, 61
Romans, 61
quality assurance, 74
Berry Sensory Analysis (BSA), 78
descriptive analysis (DA), 77
organoleptic analysis, 79
perception of, 75–77
quality turn, 75
Quantified Descriptive Sensory Analysis (QDSA), 77
sensory analyses, 77
total soluble solids (TSS), 75
quality assurance standards
Good Manufacturing Practice (GMP), 88
hazard analysis and critical control points (HACCP), 88, 91–93
ISO system, 89–91
quality and safety, 87
quality-management system (QMS), 88
Total Quality Management (TQM), 87

species for cultivation
Mediterranean climatic conditions, 58
suitable climatic conditions, 57, 58
thematic routes, 73–74
V. vinifera
asian species, 58
Australia, 58
caribbean, 58
central and east China, 58
middle asia and euro-mediterranean, 58
New Zeeland, 58
North Africa, 58
North America, 59
The Green Revolution of the 1970s, 17

H

Hazard analysis and critical control points (HACCP), 88, 91–93
Herbal Pharmaceutical Development Unit (HPDU), 236
Herbal Research And Development Institute (HRDI), 237
High fructose corn syrup (HFCS), 388
High liquid performance chromatography and mass spectrometry (HPLC-MS), 301
High-performance liquid chromatography (HPLC), 294, 342
High-performance liquid chromatography-reverse phase (HPLC-RP), 359
Honey
mineral
representative studies, 354–358
techniques, 358
nutritional composition, 323
amino acids, 345
atomic absorption spectrometry (AAS), 354
atomic emission spectroscopy (AES), 358
botanical sources, 339–341
codex alimentarius (CA), 325
European Council (EC), 325
gas chromatography-flame ionization detector (GC-FID), 332
high-performance liquid chromatography (HPLC), 342
high-performance liquid chromatography-reverse phase (HPLC-RP), 359
inductively coupled plasma mass spectrometry (ICP-MS), 358
linear discriminant analysis (LDA), 355
nuclear magnetic resonance (NMR), 353
organic acids, 346–351, 352–353
origin, 342
physicochemical composition, 325
principal component analysis (PCA), 338
proteins, 345
regulation, 326
self-developed components, 344
sugars, 328–330
types, 332–337
physicochemical composition
amino acids, 343–345
antioxidants, 359–360
legislations, 325–326
mineral, 353
moisture content, 327
organic acids, 345–352
proteins, 343–345
regulations, 325–326
sugar profile, 327–330
vitamins, 358–359
water activity, 327
sugar profile
crystallization, 342–343
determination, 342
disaccharides, 330–337
oligosaccharides with more than three monomers, 341–342
trisaccharides, 337–341
Hungarian context
bioactive compounds, 147
flavonoid phenols, 149
minerals, 150
non-flavonoid phenolic compounds, 148–149
polyphenol compounds, 148
tannins, 149
vitamins, 150–151
by-products of wine processing

grape marc, 152–153
grape seed oil, 154
grape seed proanthocyanidin extract (GSPE), 153
grape seeds, 153–154
grape skin, 154
grape processing technologies, 141
aging of, 144
harvesting and processing, 142
mash treatment, 142
must extraction, 143–144
pressing, 142–143
wine treatment, 144
grapes and wine
Balaton wine region, 137
consumption, 130
Duna wine region, 137
Eger wine region, 137
historical evidences, 131–133
Northern transdanubian wine region, 137
Pannon wine region, 137–138
Sopron wine region, 138
Tokaj wine region, 138
and types, 138–141
viniculture and wine regions, 134–137
Vinum album, 131
viticulture, 138–141
viticulture, ecological factors, 133–134
wine production, 138–141
vinotherapy, 154
massaging body, 155
wine
consumption habits in, 146–147
production, 145–146

I

India Himalayan Region (IHR), 226
Indian cuisine, 163
Inductively coupled plasma mass spectrometry (ICP-MS), 358
International Electrotechnical Commission (IEC), 300
International Fund For Agricultural Development (IFAD), 217
International Organization for Standardization (ISO), 300

L

Lactic acid bacteria (LAB), 11
Laurus nobilis (Bay leaf)
antibacterial activity, 171
antimutagenic, 171
diabetes and migraine, 171
essential oil, 170
ethanol extract, 171
export, 169
health benefits, 170
immune system, 170
powder, 171
Linear discriminant analysis (LDA), 355, 378
Liquid chromatography (LC), 294

M

Medicinal and aromatic plants (MAPs), 236
Methylene cyclopropyl acetic acid (MCPA), 110
Metric tons (MT), 199
Millets
in baby foods
complementary food, 10
extrusion technology, 11
lactic acid bacteria (LAB), 11
Moringa oleifera leaf, 10
protein solubility, 10
challenges and probable solutions, 16
The Green Revolution of the 1970s, 17
consumption figures, 18
economic decline, 17
knowledge, 19
pricing, 18
side-lining millet, 17
time and effort, 19
West African regions, 17
food diversification and processing techniques
Food and Agricultural Organization (FAO), 4
United Nations Development Program (UNDP), 4
nutrient content and acceptability
acceptable probiotic fruit bar, 8
celiac disease, 9
cereals/roots and tubers, 7
cookies, 7

Dhokla, 8
dough meal, 7
finger millet flour, 6
glycemic index (GI), 9
malting technique, 8
millet-based meals, 9
Ogi, 5, 6
RTC (Ready-to-Cook), 9
soybean, 6
nutrients processing methods, effects, 11
amylase inhibitors, 14
antioxidant activities, 16
boiling process, 14
composite blending, 16
cooling processes, 15–16
extrusion, 15
fermentation process, 13, 14
fermentation traditional methods, 12
fortified millet with whey protein, 15
germination, process, 16
germination traditional methods, 12
heat application methods, 14
malting, process, 16
malting traditional methods, 12
mechanical methods, 12
pearl millet, 13
phytates and tannins, removal, 12
preservation treatments, 15
puffing, 14
roasting, 14
soaking, process, 16
soaking traditional methods, 12
thermal processing, 15
Moisture content (MC), 380
Multiple linear regression (MLR), 392
Myristica fragrans (Nutmeg)
antibacterial activity, 180
brain cells, 180
health benefits, 180
Mace, 179
powder use, 180
structure, 180

N

Near infrared spectroscopy (NIRS), 310
Near-infrared spectroscopy (NIR), 408
Nuclear magnetic resonance (NMR), 353

O

Ocimum tenuiflorum (Tulsi), 167
antifungal and antibacterial activities, 166
chemical structure, 165
and gastric problems, 166
health benefits, 165
hepatic disorders, 166
oil, 165
reactive oxygen species (ROS), 166

P

Partial least squares regression (PLSR), 385
Petroselinum crispum (Parsley)
constituents of, 169
juice, 168
oil, 167
phytochemical compounds, 168
Potato virus-Y (PVY), 378
Principal component analysis (PCA), 338, 378

Q

Quality-management system (QMS), 88
Quantified Descriptive Sensory Analysis (QDSA), 77

R

Ratio performance deviation (RPD), 388
Reactive oxygen species (ROS), 166, 383

S

Sate Medicinal Plants Board (SMPB), 237
*S*inigrin (allyl-glucosinolate or 2-propenyl-glucosinolate), 179
Soft independent modeling of class analogy (SIMCA), 378
Soluble solids content (SSC), 390

T

Table grapes, 63. *See also* Fresh grapes
Total Quality Management (TQM), 87
Total soluble solids (TSS), 75
Traditional medicinal herbs
North-western Region of Himalayas, 223
biological activity, 229–232
centre for aromatic plants (CAP), 237

conservation assessment and management plan (CAMP), 226
Herbal Pharmaceutical Development Unit (HPDU), 236
Herbal Research And Development Institute (HRDI), 237
India Himalayan Region (IHR), 226
Indian scenario, herbal medicines, 226–227
local names, 238–241
marketing strategies, 236–242
medicinal and aromatic plants (MAPs), 236
minimum support prices, 237
Sate Medicinal Plants Board (SMPB), 237
state of art, 227–236
Trigonella foenumgraecum (Fenugreek)
availability, 183
beneficial nutrients, 182–183
composition, 183
medicinal uses, 184
phytochemicals, 183
purified methanolic extract, 184
seeds, 183
studies and research, 184

U

Ultra violet (UV), 406
Ultraviolet visible spectroscopy (UV-VIS), 310
United Nations Development Program (UNDP), 4
United Nations Development Programme (UNDP), 295
United States Department of Agriculture (USDA), 176

V

Vascular endothelial growth factor (VEGF), 270

W

Water absorbance bands (WABs), 372
Water index (WI), 382
Water spectral pattern (WASP), 373
World Health Organization (WHO), 301

X

Xanthohumol, 271
cervix cancer, 272
cholangiocarcinoma, 273
gastric cancer, 271–272
hepatic cancer, 273
larynx cancer, 272
mammary carcinoma, 273
pancreas cancer, 272

For Product Safety Concerns and Information please contact our EU representative GPSR@taylorandfrancis.com Taylor & Francis Verlag GmbH, Kaufingerstraße 24, 80331 München, Germany